**FLORAL BIOLOGY
OF TEMPERATE ZONE
FRUIT TREES
AND SMALL FRUITS**

FLORAL BIOLOGY OF TEMPERATE ZONE FRUIT TREES AND SMALL FRUITS

Edited by

József NYÉKI, DSc

PANNON University of Agricultural Sciences,
Georgikon Faculty of Agriculture,
Keszthely, Hungary

and

Miklós SOLTÉSZ, DSc

University of Horticulture and Food Industry
College Faculty of Horticulture,
Kecskemét, Hungary

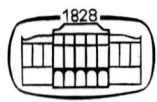

AKADÉMIAI KIADÓ, BUDAPEST

Translated by

Iringó K. KECSKÉS (Chapters 1–4)

and

Gábor OLÁH (Chapters 5–9)

Hungarian text revised by

Pál KOZMA

Member of the Hungarian Academy of Sciences
University of Horticulture and Food Industry, Budapest, Hungary

and

Pál TOMCSÁNYI

Member of the Hungarian Academy of Sciences
Institute for Agricultural Qualification, Budapest, Hungary

English text revised by

Miklós FAUST

Professor of Plant Physiology
Fruit Laboratory, Beltsville,
Maryland Agricultural Research Center, United States Department of Agriculture

The publication of the present volume has been sponsored by the Hungarian Academy of Sciences.

ISBN 963 05 6896 9

© Akadémiai Kiadó, 1996
© English translation—Iringó K. Kecskés and G. Oláh, 1996

All rights reserved. No part of this book may be reproduced by any means, or transmitted or translated into machine language without permission of the publisher.

Published by
Akadémiai Kiadó, H-1117 Budapest, Prielle Kornélia u. 19–35

Printed in Hungary
by Akadémiai Kiadó és Nyomda, Budapest

Contents

List of contributors vii

1. Introduction (*J. Nyéki* and *M. Soltész*) 1

2. Flower development and formation of sexual organs (*T. Bubán*) 3
2.1. Endogenous factors of flower bud development 3
2.2. Site and time of flower bud development 5
2.3. Realization of flower induction (flower bud differentiation) 9
2.4. Development of the sexual organs of the flower 33
2.4.1. Formation of the stamen and pollen 33
2.4.2. Structure of the pistil 38
2.4.3. Development of the ovule and its longevity 44
2.4.4. Organization of the embryo sac 47

3. Morphological characteristics of flowers and fruits (*E. Dibuz*) 55
3.1. General characteristics of flowers of fruit species 55
3.2. Characteristics of the perianth, androecium and gynoecium 72
3.3. Important characteristics of fruits 76

4. Flowering (*M. Soltész*) 80
4.1. Time of bloom 80
4.1.1. Time of bloom and its succession in fruit species 80
4.1.2. Beginning of bloom in various cultivars 84
4.1.3. The relative bloom time of cultivars 85
4.1.4. Full bloom of cultivars 88
4.1.5. Duration of bloom in various cultivars 93
4.2. Groups of blooming time and the extent of joint blooming 97
4.3. Factors influencing the beginning and the course of bloom 104
4.3.1. Length of dormancy and requirement for cold 104
4.3.2. Calculation of the amount of heat necessary to start and sustain blooming .. 106
4.3.3. The role of geographical latitude and height above sea level 111
4.3.4. Cultivar properties influencing bloom 114
4.3.5. The effect of cultivation on bloom time 119
4.4. Methods for observing and evaluating phenology of bloom 125
4.4.1. Determining the time of blooming 125
4.4.2. Blooming phenograms 126
4.4.3. Overlap in blooming and evaluation of joint blooming 128
4.4.4. Classifying cultivars into blooming groups 131

5. Receptivity of sexual organs (*M. Soltész, J. Nyéki* and *Z. Szabó*) 132
5.1. Stigma receptivity 132
5.2. The period of pollen shedding 138
5.3. The overlap of pollen shedding and stigma receptivity 144

6. Pollination and fertilization ... 153
6.1. Means of pollination and factors affecting pollination (*J. Nyéki*) ... 153
6.1.1. Pollination types ... 153
6.1.2. Factors affecting pollination ... 154
6.2. Pollen adhesion on stigmatic surfaces and the outset of pollen tube growth *(T. Bubán)* ... 156
6.3. Pollen tube growth in the style ... 159
6.4. Double fertilization ... 162
6.5. Period between pollination and fertilization ... 169
6.6. The effect of temperature and environmental pollution on fertilization ... 170
6.7. Effective pollination period ... 172
6.8. Embryo development ... 173
6.9. Research methodology ... 182

7. Fertilization conditions (*J. Nyéki*) ... 185
7.1. Self-fertility and self-sterility ... 185
7.2. Cross-pollination ... 215
7.3. Incompatibility and inter-incompatibility ... 229
7.4. Parthenocarpy ... 241
7.5. Research methodology ... 248

8. Requirements for successful fruit set in orchards (*M. Soltész*) ... 257
8.1. How to determine cultivar combinations? ... 257
8.2. Xenia and metaxenia ... 265
8.3. Locating cultivars in an orchard ... 274

9. Insect pollination of fruit crops (*P. Benedek*) ... 287
9.1. The possible role of wind in pollination of insect-pollinated fruit species ... 287
9.2. Insect pollinators of fruit species ... 289
9.3. The relative efficiency of insects in pollination ... 293
9.4. The role of nectar and pollen production in pollination ... 296
9.5. The activity and behaviour of insect pollinators in different fruit species ... 298
9.6. Factors affecting the distribution of honeybees in orchards and the intensity of their flower visitation ... 301
9.7. The effect of insect pollination on fruit set, yield and fruit characteristics ... 307
9.8. Demand for pollination in fruit species and features of their insect pollination ... 313
9.9. The placing and management of bee colonies ... 337
9.10. Concluding remarks ... 340

10. Cited literature ... 341

11. Index ... 370

List of Contributors

Pál Benedek, PhD, DSc

PANNON University of Agricultural Sciences, Faculty of Agricultural Sciences, Vár u. 4, Mosonmagyaróvár, H-9201, Hungary

Tamás Bubán, PhD, DSc

Research Station for Fruit Growing, P. O. Box 38, Újfehértó, H-4244, Hungary

Erzsébet Dibuz, PhD

University of Horticulture and Food Industry, College Faculty of Horticulture, Erdei Ferenc tér 1–3, Kecskemét, H-6000, Hungary

József Nyéki, DSc

PANNON University of Agricultural Sciences, Georgikon Faculty of Agriculture, Festetics u. 7, Keszthely, H-8361, Hungary

Miklós Soltész, DSc

University of Horticulture and Food Industry, College Faculty of Horticulture, Erdei Ferenc tér 1–3, Kecskemét, H-6000, Hungary

Zoltán Szabó, PhD

PANNON University of Agricultural Sciences, Georgikon Faculty of Agriculture, Festetics u. 7, Keszthely, H-8361, Hungary

1. Introduction

The knowledge of blooming and pollination and its use are indispensable in maximizing of cropping potential of fruits in economical fruit production. In attaining maximum yield a greater attention has to be focused on choosing cultivar combinations, and results of experiments on blooming, pollination and fertilization must be applied carefully.

When the idea of writing this book arose we wanted to integrate our own research results of several decades with the extensive literature to demonstrate the generalities and correlations supported by numerous experimental data and observations and present them in a way that it is useful for the practical orchardist.

This is indicated by the general layout of the book and in the arrangement and proportion of the chapters. We deal in detail with the structure and formation of the flower: the development, viability and functional capacities of the sexual organs; and we describe the flowering, pollination and fertilization processes. We placed great emphasis on the fertilization conditions of fruit-bearing plants, as well as on the factors influencing fertilization and fruit set.

We hope that the information gained from this book will be useful for those planning orchards and elaborating new cultural practices as well as those involved in teaching and research. We included chapters for researchers where the observations, investigations and evaluation methods are presented and where the methodology concerning flower biology of fruit species is summarized.

We also present the conditions for applying knowledge of flowering and fertilization in accordance with the biological factors and production technology. This chapter will be useful to those who wish to plant an orchard of any size. Flowering and fruit set are prerequisites for successful fruit production. There is no cultivation manipulation, regardless how well it is carried out, that could supplement fruit set, therefore blooming, pollination and fertilization have to have greater attention when planning and establishing an orchard.

Insect pollination, with the exception of a very small proportion of wind-pollinated fruit species, is indispensable in species requiring cross-fertilization but even in self-fertile species it can increase the extent of fruit set and the amount of yield.

To have efficient bee pollination requires attention at the time of designing an orchard. It requires further attention at the time of bloom of any of the fruit-bearing species. Markets demand new types of fruit which forces constant changes in the cultivar composition of orchards. The blooming, pollinating and fertilization characteristics of cultivars chosen have to be known before an orchard is set up. Apart from the general knowledge of trees considered to be planted, there is a great need to know the flowering, pollinating and fertilization characteristics of each cultivar in detail. This is why in this book the biological and physiological correlations are illustrated with widely used cultivars.

We would like to express our appreciation to the Department of Agronomy, Hungarian Academy of Sciences for enabling us to prepare this work and for supporting us financially and morally. We are most grateful to academicians, Pál Kozma and Pál Tomcsányi and professor Miklós Faust for reading the manuscript and making a thorough review. The Publishing House of the Hungarian Academy of Sciences deserves to be mentioned with gratitude for their co-operation, meticulous editing and beautiful layout as well as for the quick publication.

We dedicate this book to the memory of Pál Maliga (1913–1987). He was the founder of flower biology and floral phenological research in Hungary. He carried out systematic methodological investigations on fertilization over many decades. His experimental results are useful even today. Beyond his own research he created a school of researchers who are devoted to this field of science. Pál Maliga, we thank you for your endeavours in flower biology of fruit species. Without your work we would know much less today.

József Nyéki
Miklós Soltész

2. Flower Development and Formation of Sexual Organs
T. Bubán

2.1. Endogenous Factors of Flower Bud Development

The most important characteristics of flower formation, to this day, have remained unsolved for both theoretical biology and applied horticulture. Therefore, the statement of Wellensiek (1977) still applies that "the principles of flower formation can be defined as anything which happens between flower-forming genes and meiosis." Flower initiation also can be looked upon as a process that stops the inhibition of repressed genes, responsible for flower formation and in this sense induction means the same as deblocking. Luckwill (1974) defined induction as a qualitative alteration, probably applicable also to fruit trees, that is realized through the alteration of hormone balance.

Our knowledge on flower bud development is largely confined to apples and pears. This, however, does not mean that this knowledge can simply be transmitted to other fruits.

The formation of flowers, appearing in fruit trees in spring, actually, begins the previous summer. Flowers are induced in the apices of buds which had been vegetative previously. The apical meristem becomes an embryonic flower (within the bud scales) in the second half of summer. The development of the flower primordia is quite intensive until leaf fall, slows down during winter and at about the end of winter, accelerates again, during early spring. Therefore, it is customary to speak of flower bud development in the case of fruit trees instead of the more general term flower formation. Conditions for flower bud development, however, are not the same in young or in mature trees.

The bearing potential of *young trees* is determined by their growth characteristics. Here we have to consider apical dominance and the function of acrotony. The branching system determining the growth habit and bearing capacity of fruit trees, in itself, cannot be explained by apical dominance. The trees are characterized by a trunk and for the trunk to form it is necessary to preserve the role of the uppermost buds. This acrotony establishes a hierarchy among the branches forming the canopy of the tree.

Despite the fact that acrotony is transmitted and extended by the apical dominance during the vegetation period, it is connected to the dormancy of buds. Unlike apical dominance, acrotony is required as a predisposition of vigorousness prior to any kind of growth. Apical dominance, in contrast, is the active mechanism of inhibition in the shoots exerted by the growing point located in terminal position (Crabbe 1981, 1984; Faust 1989).

The conditions for flower formation in young trees are set by the modifications of these growth correlations. Moderation of both acrotony and apical dominance leads to the development of a larger number of meristematic growing points, and the growth capacity is distributed among these growing points. The benefits of these manifested in favourable strength and periodicity of shoot growth, higher number of shoots with optimal strength for generative development and increased susceptibility to flower induction in the lateral buds of long shoots.

Flower formation in *bearing trees* is fundamentally determined by the presence of hormones. "The failure of flower initiation in trees carrying a heavy crop of fruit which, in the past was attributed directly to the effect of the crop in depleting the carbohydrate and nitrogenous reserves of the tree is now widely recognized to be due to hormonal rather than nutritional causes" (Luckwill 1974). The hormones important here are synthesized in the seedlets of developing fruits.

The inhibitory effect of seedlets in flower bud formation was first observed by Tumanov and Gareev (1951, cit. Luckwill 1977) but it became well known through Chan and Cains' (1967) publication.

The auxin level of the seeds reaches a maximum four–five weeks after flowering (Luckwill 1970) which is followed by a second auxin peak—probably originating from the embryo seven weeks later (Grochowska and Karasewska 1974). The amount of auxin flowing out from the fruit through the fruit pedicle is not determined by the number of seeds only. Beyond the fact, that on an average, in fruits of apple trees bearing annually there are five seedlets and in alternately bearing trees there are eight, the intensity of the outflow of the auxin in the latter was 60% higher. Another important point is that there is no hormone degradation in apple pedicle in contrast with that of the strawberry (Grochowska and Karasewska 1976, 1978*a*). Furthermore, the amount of indoleacetic acid in the buds of apple trees prior to the appearance of flower meristems drops to half the amount which can be measured prior to flower development (Werziloff et al. 1978).

The auxins are able to intensify the effect of the gibberellins. It is particularly interesting to have a look at the investigations on the effect of gibberellins. In seedlets of young fruits Nitsch (1958, cit. Sinska et al. 1973) was the first to find gibberellins and to identify the GA_4 and GA_7 as particular gibberellins (Dennis and Nitsch 1966, cit. Sinska et al. 1973). The gibberellins begin appearing in the seeds four–five weeks after bloom and they reach a peak at the ninth week after bloom. Their concentration, measured in GA_3 units at this time was 15–500 times higher than in the leaves and shoots (Luckwill 1970, 1974). Since, a large number of gibberellins have been identified (Dennis 1976, Hoad et al. 1977) the interconversion of gibberellins has become of focal interest. The physiological effect and role of gibberellins of various polarity are different (Pharis 1977, cit. Grochowska and Karasewska 1978*b*, Looney et al. 1978), and by converting a gibberellin into another its role could change.

The activity of gibberellins in June could only be detected in leaves of trees bearing a crop (Lacey et al., 1976). In the fruits of 'Laxtons Superb' apple cultivar the gibberellin concentration was much higher than in the more regularly bearing 'Cox's Orange Pippin' (Hoad and Donaldson 1977, Hoad 1978). It has been shown by Grochowska (1968) that when five-week-old seeds were substituted by gibberellins the inhibition of flower bud formation was the same as that of intact fruits. Gibberellins inhibited the initial stages of flower formation (Tromp 1976). However, it should be emphasized that the flower bud inhibition cannot be attributed solely to gibberellins, but to the mutual effect of gibberellins and other endogenous hormones.

It has also been shown in pear varieties that the severe inhibition of flower formation originated from the fruit and from this aspect the critical period, according to Griggs et al. (1970), is within 30 days after full bloom. Huet and Lemoine (1972) claim that the critical period is later in time. It is important from the point of view of fruit production (Huet 1974) that flower bud initiation will be inhibited or be insufficient, even in the presence of seedless fruits, if there are less than six leaves or less than 70 cm^2 leaf area on a spur. The possible role of other endogenous hormones in flower formation has been reviewed earlier (Bubán and Faust 1982).

2.2. Site and Time of Flower Bud Development

In apple: The primary site of flower formation is the apical buds of spurs (or brachyblasts); (Horavka 1961, Huet 1974). The spurs are characterized by short internodes and their whole size is less than 10 cm in length. The growth of the spurs ends two to four weeks after flowering with the formation of terminal buds in which, under appropriate conditions, the differentiation of inflorescence primordium begins. On the spur during subsequent years newer spurs appear and this is how several year-old brachyblasts form. The flowers on the younger spurs have a better fertilization capacity. Those flowers appearing on older spurs have more well-developed stamens which produce viable pollen (Milutinovic 1974, Gosh 1970, cit. in Bubán and Faust 1982). Flowers in the apical buds of long shoots are similar to those of spurs (Streitberg 1978, cit. Bubán and Faust 1982). Sometimes, these flowers have the highest functional value (Rudloff and Lucke 1958, cit. in Bubán and Faust, 1982), but the frequency of their occurrence is much less (Bubán and Faust 1982).

In the axillary (lateral) buds of elongated shoots, flower formation may be characteristic of the apple cultivar or it may be insignificant, and its extent declines with advancing age of trees (Gribanovski 1970). On elongated shoots, the flower development is the earliest in the middle zone buds, which is not parallel with the bud forming order (Zeller 1961). In lateral buds the flower development takes place much later than in the terminal buds of the spurs and it can be as late as October or the end of winter (Zeller 1961). However, the morphogenesis of flower primordium is faster in the lateral buds than in the spur buds (Eliseeva 1970).

In apical buds the pistil primordium can already be found at the beginning of winter, while in the lateral buds only the anther primordia can be identified at this time (Pogorelov 1970). This difference in development between the two types of flowers remains during the course of winter (Reichel 1964*b*) thus the flowers of one-year-old shoots open later reducing the risk of frost that may occur at bloom (Zeller 1960*c*). It has long been known, however, that the fruit setting potential of the flowers of one-year-old shoots is much smaller (Reichel 1964*b*). We will return to these causes revealed by Zeller (*see* Subsection 2.4.3).

With respect to other fruit species we will restrict ourselves to a few examples. Of 29 *pear* cultivars described by Carrera (1982) in 17 the flower development was realized primarily in the spurs, in two cultivars in the terminal buds of *elongated* shoots and in 12 cultivars on both spurs and shoots. A detailed analysis has shown (Huet 1974) that the relative growth rate of shoots of pear cultivars, in the month before growth is completed, is inversely related to the flower bud number/shoot value. The terminal bud of shoots may often be flower buds and in the lateral buds flower development is not general. In *apricot* trees the internode elongation is increased by low light intensity. There is a negative correlation between the length of the internodes and flower initiation (Jackson and Sweet 1972). At the same time, the total number of nodes on the tree showed a positive correlation with the number of flower buds formed. Fewer flowers were induced in buds in the axillary position of heavy bearing trees (Fedtsenkova 1970). The ratio of flower buds/leaf buds in *sour cherry* trees decreases with increase in shoot length (Rasmussen et al. 1983). In one-year-old shoots longer than 40 cm the quality of the flowers also declines significantly judged from fruit-setting data. In *plum* trees more flower buds are formed on shoots of the upper part of the canopy, on shoots located horizontally and on short, or slowly developing nodes than shoots in other positions (Hassibb 1966).

The male flowers of the *walnut* constitute the inflorescence called catkin which, depending on the cultivar, consists of 7–105 male flowers, and are formed in the axillary buds of shoots. The female flowers are differentiated in the terminal buds of short shoots. The developed female flowers have a rudimentary calyx and the corolla is missing (Sartorius 1990). The female flowers are initiated at the end of summer three to five months later than the male flowers (Ramina 1969, Blasse 1976, cit. Bubán 1980*a*). More precisely, the first signs of differentiation of male flowers is observable 35–45 days after spring growth begins, whereas the first indication of female flowers is observable after 125–135 days. About ten days before this period (i.e on the 25th–35th and 115th–125th days, resp.) flower induction may be considered irreversible (Ramina 1969). Flower induction is inhibited by the leaves at the beginning of the vegetation period probably because of the change in the ratio of the endogenous growth substances. The fruits only have an influence at the beginning of the vegetation period. Their removal, until the beginning of June, increases the number of flower buds, especially that of catkins (Ramina 1970). The number of nodes and their insertion in *black currant* shoots (Karnatz 1971) definitely determines the probability of flower development in the bud located on the nodes.

The beginning and the temporal course of flower development cannot be determined according to the calendar. The variations in environmental factors and the annual deviations of fruit crop and the physiological condition of the trees all have influence on the development of flower buds. Nevertheless, a certain order in time among the species according to Zeller (1955, 1960*a*) undoubtedly exists. This order, from early to late development, is: cherry, apple, pear, peach and quince. Since the first signs of the beginning flower development can be observed over a rather long period of 4–13 weeks (Zeller 1954) the beginning of flower development may be indicated when flower primordia are observable in 50% of the buds (Carrera 1982, Raseira and Moore 1986).

The start of flower development on the spurs of *apple* trees in Sardinia is at the end of June (Deidda and Pisanu 1968), in Hungary at the beginning and middle of July (Bubán 1967, Gyuró 1959) and in colder regions at the end of July, beginning of August (Ghosh 1970). Flower bud differentiation on shoots takes place 10–20 days later (Pogorelov 1970), one to three weeks later (Reichel 1964*b*, Ghosh 1970) or even two months later (Golikova 1969). The delay according to Zeller (1961) is at least three weeks, but in the axial buds of shoots flower development may commence as late as February or March. In some cases flower initiation may also take place in the autumn, e.g. in apple trees if there is no crop in that year (Tromp 1968) or after an early harvest (Luckwill 1974, Zatykó 1974). In autumn, flower bud development commonly occurs in axillary buds of shoots in apple (Zeller 1955), and even in spurs of quince (Zeller 1960*b*).

Flower formation in *pear* trees in Sardinia and Bulgaria begins in the middle or at the end of June (Deidda and Pisanu 1968, Gornevsky 1976) in Spain, depending on the cultivars and location, between the second half of June and the first half of August (Carrera 1982). There is an apparent correlation between the early flower initiation characteristic of the cultivar and the annual regular flowering (Carrera 1982).

Flower bud development of *sour cherry* and *sweet cherry* begins in Hungary in the first half of July, in Poland in July, in Romania at the end of June and in the former Soviet Union, depending on the region, in the first half of July or at the end of June. Primordia of all flower parts appear 87–100 days after initiation in Hungary, after 68–112 days in Romania and after 90–95 or 115–125 days in Russia and about after 150 days in Poland (Elekné 1974).

Flower development of *peach* in the Czech and Slovak Republics begins at the end of July, beginning of August (Hladik 1972), in Hungary, Italy and California in the second half of July (Bubán and Zeller 1974). Flower bud formation of *apricot* in Romania begins at the end of July, beginning of August, but on secondary shoots one month later. This developmental period in apricot entails 42–85 days (Tarnavschi et al. 1963, Cociu and Bumbac 1973). In Turkey after a similar start the development of the flower primordia ends only in December (Gülcan and Askin 1990) whereas in Hungary it ends by the end of October (Surányi 1977a). Flower development in the *plum* cultivars occurs the earliest at the end of June, beginning of July (Cociu and Bumbac 1973). As for spurs and shoots of quince, flower bud development never starts before the end of September or end of October (Zeller 1960b). In June bearing *strawberry* flower buds develop only from the middle of September (Naumann 1964). In the southern part of Finland the differentiation of flowerlets of the *black currant* begins in the middle of August and by the beginning of October the pistil primordium can also be observed. However, near the north pole region the process starts much later and the primordium formation only reaches the petal stage (Zeller 1968). Gooseberry flower bud formation in Hungary begins at the end of July, beginning of August (Bubán 1980a).

Often the time of flower development is *connected* to other *phenological* and *developmental* processes. In pomaceous species there generally is a correlation between shoot growth and flower induction (Li et al. 1989) and this, with respect to the apple, has been confirmed by several authors (Gyuró 1959, Eliseeva 1970, Abdulkadyrov et al. 1972). Even in spurs the time of flower initiation is based on the differences in the completion of their growth (Zeller 1954). Contradiction to this rule must be considered as an exception (Benko 1967). Shoot growth generally stops on peach and apricot trees before flower initiation (Jackson and Sweet 1972). However, according to Cociu and Bumbac (1973), flower development in peach and apricots starts rather before than after shoot growth ends. The results of detailed investigations carried out on three peach cultivars (Li et al. 1989) showed that the shorter the shoot the earlier the flower initiation, but the period of induction is longer on short shoots and in buds in basal positions on the shoot. Furthermore, flower initiation begins already in the intensive shoot growth period and when they reach 70–80% of their final length, one–two weeks before growth ends, the majority of the buds has gone beyond the physiological differentiation stage. Cociu and Bumbac (1973) determined the beginning of flower development five weeks before shoot growth ended in plum trees.

Another basis for comparison is the *length of time* from flowering to flower bud differentiation. In peach trees it is 116–144 days after bloom (Stadler and Strydom 1967, cit. Bubán and Zeller 1974), in the average of three years it is 92 days after bloom (Hadj-Hassan 1969, cit. in Bubán and Zeller 1974) or 97–110 days after bloom (Bumbac 1970, cit. in Bubán and Zeller 1974). In Hungary 105–115 days are necessary between bloom and the beginning of flower development (Bubán and Zeller 1974). Flower bud formation in the plum begins two months after flowering (Cociu and Bumbac 1973). In apple varieties this kind of correlation could not be established (Reichel 1964b, Bubán 1967).

In apple, the start of flower initiation may coincide with fruit abscission in June (Elekné 1966) when the embryo in the developing seeds is in the globular stage (Zeller 1964b), but more often flower initiation begins two–five weeks after the embryo has developed completely (Zeller 1964a, b). No correlation was found between the flower bud formation and time of fruit ripening in apple (Benko 1967) and peach (Raseira and Moore 1987).

Only a few examples will be mentioned relating to the role of *environmental factors* in initiating flower development. Within a species the differences between cultivars are less in those years when the period of flower initiation is compressed. Flower initiation is also compressed in those growing areas where flower development begins early (Neumann 1962). There is little correlation between the time of flower initiation and the chilling requirements in peaches (Raseira and Moore 1987). However, the internal temperature of buds may have a role in this (Draczynski 1958, cit. by Surányi 1978b). Raseira and Moore (1986) grew three peach varieties on the same latitudes of the northern and southern hemispheres (N 35°, S 32°). The number of days which passed between the starting date of summer (June 22 and December 22) and the starting date of flower development did not show a greater difference than it occurs in a given orchard from year to year.

Grupce (1966) observed earlier flower bud differentiation at lower altitudes in apple trees. In warmer areas the number of inflorescences forming on the apple trees as well as the number of flowers within the inflorescence are less. One of the explanations for this problem (Tromp 1990) may be found in the fact that the night temperatures are high (18–22 °C) which prevent the development of flowers.

The *rootstocks* of sour cherry (Elekné 1974) and that of the apple (Neumann 1962, Reichel 1964b, Bubán 1967, Keremidarska 1968) hardly or not at all influenced the time of flower development with the exception of the dwarf rootstock of the apple, reported by Nesterov et al. (1972), which had an accelerating effect.

Pruning, depending on its extent, delays the flower bud differentiation by one–three weeks (Gyuró 1959). In fruitless trees or in trees with little fruit flower bud formation starts earlier than in trees with heavy crop (Zeller 1955, Rudenko 1958, Gyuró 1959, Keremidarska 1968).

We should also mention how flower initiation is realized under extreme conditions. In Finland flower bud development on apple trees, depending on the cultivar, begins in the middle of July and early September (Zeller 1964b). In tropical Ceylon the periodically occurring monsoon twice a year creates a wet and a dry zone within the island. Flower induction occurs once in the dry zone (in January). However, there are two periods of flower induction within a year in the wet zone: first, in July–August and in January again (Zeller 1973). Following the monsoons the time between the flower initiation and flowering is altogether eight to ten weeks in 'Rome Beauty' apple cultivar and Japanese pear trees. Since cold requirement is unsatisfied the bud break is delayed and the flower primordia may die partly or totally. In this case the period between the differentiation of the archesporium and the meiosis of the mother cells of the pollen are decisive (Zeller 1973). A similar problem arises with the growing of the Asian pears (*Pyrus serotina* Rehd.) in Taiwan (Lin et al. 1987). There, cultivars with high chilling requirements can only be grown at levels higher than 1800 m above sea level. Those with low chilling requirements can be grown below 500 m above sea level. Flower initiation in the high chilling-requiring cultivars starts one month later, at the beginning of July, than the low chilling-requiring cultivars.

The time between flower bud differentiation and flowering in the strawberry grown in Ceylon is less than eight weeks and the flower organization is undisturbed if cultivation takes place at 2000 m above sea level. At lower regions (100–520 m above sea level) the flowers are few and distorted although the daylight hours everywhere are 12 hours (Zeller 1969). In the Arctic zone, *Rubus* species overwinters in Lapland (i.e. in Finland where the vegetation period is very short) with root-buds (*R. arcticus*) and with below ground shoots (*R. chamaemorus*) (Zeller 1964c). Flower bud differentiation begins in the middle of July or early August and all flower primordia develop exceptionally quickly. They are complete by the beginning of September. In Finland the time between *Prunus padus* flower initiation, which occurs at the end of July, and archesporium differentiation in the anther primordia is just six weeks (Zeller 1964b). In the Helsinki region *Ribes alpinum* flower formation is one to two weeks before that of the black currant in the first half of August (Zeller 1968).

2.3. Realization of Flower Induction (Flower Bud Differentiation)

In the first half of the vegetation period all buds in fruit trees are vegetative buds even on the spurs irrespective of the presence of the fruit (Marro and Ricci 1962). After flower induction, which is a still unknown process, with successive steps of physiological, histological and morphological differentiation the apex of the bud becomes organized into a primordium of flower (or inflorescence). This transformation, however, only takes place if the structure of the vegetative bud becomes complete. In apple, for example, primordium differentiation may only begin when nine bud scales, three transitional leaves, six leaflets and three bractea are formed. These of course may vary depending on the cultivar, for example, in the case of 'Cox's Orange Pippin' the number is 20 but only 16 in 'Golden Delicious' (Abbott 1970). Furthermore, the rate of development of the vegetative buds should be properly rapid, that is, the time between the initiation of successive organ primordia in the apex, the plastochron, should not be more than five–seven days long. The length of the plastochron is regulated by the hormonal interaction of the foliage leaves and the leaflets of different age within the developing bud (Fulford 1965). The inhibiting effect of high temperatures on flower bud development is based on this. Tromp (1976) stated that the lengthening of the plastochron is attributed to greater amounts of available gibberellins at a higher temperature and this prevents the development of the flower. The flower-forming inhibitory effect of fruits (*see* Chapter 2.1) is realized through the same kind of correlation (Fulford 1966, 1973a, b, cit. in Bubán and Faust 1982).

Bernier (1970) distinguished three main phases of flower formation. The first is when the stimulus for flower initiation reaches the apex and the essential nucleic acid and protein synthesis begins. The second step is the increase in the mitotic activity when first the cell nuclei being in the G_2 phase divide. Finally, the morphogenetic phase when the flower primordium develops.

A more recent review (Gasser 1991) reports on three phases of flower development:

1. induction and evocation, that is, the re-organization of the meristem, considered the other aspect of induction;
2. organ specification at floral apex;
3. differentiation of tissues within organs.

After transformation of the vegetative meristem to the floral meristem, (first phase) the floral meristem produces organ primordia laterally which become differentiated into various organs of the flower. Which primordium will produce which flower organ is genetically determined. After the fate of the primordium is specified (second phase) the various organs attain their final form through a series of programmed cell divisions and elongations. During the course of this development, the cells become differentiated anatomically and biochemically and in this way it becomes possible for the flower organs to carry out their special functions.

The basis of flower initiation in apple was reviewed by Bubán and Faust (1982). We have followed flower development in *apple* trees for years. After preliminary methodological studies (Hesemann and Bubán 1973) we investigated the apex of terminal buds with cytochemical methods (Bubán and Hesemann 1975, 1979; Bubán 1981). We also described the histological differentiation of the apex (Bubán 1965) and the formation of flower primordium as well as their further development in winter (Bubán et al. 1979). Flower formation in apple trees can be summed up in the following manner.

In the apex of the still vegetative bud (*see* Fig. 2.2*a*) there are two exterior rows of cells, dermatogen and subdermatogen, labelled as Z_1 in Fig. 2.1. Within this Z_1 layer the subdermatogen has the primary function for organogenetic activity. The initiation of primordia of new organs in all cases begins in the lateral part of the subdermatogen with the formation of periclinal cell walls. The deeper layers consisting of one-three rows of cell (Fig. 2.1, Z_2) has a variable role. Below this layer is the central meristem (Z_3) and below that is the pith-rib meristem (Z_4). At flower induction in the apex of the terminal buds of fruitless spurs the nucleic acid levels are higher than in apical buds of fruit-bearing spurs (Fig. 2.1). The terminal buds of fruitless spurs are not inhibited to form flowers, the apical buds of the fruit-bearing spurs are inhibited to form flowers. Thus, there is a correlation between nucleic acid content and flower development in the apices of spur buds. The reverse is true for the level of nucleohistones inhibiting protein synthesis dependent on nucleic acids and mitoses, resp. Beside the increased mitotic activity, indispensable for flower initiation, the nucleic acids have a special organogenic function. Flower initiation can be inhibited with base analogues of nucleic acids (in molecular biological experiments); or moderated in apple trees with bromouracyl (Bubán 1969) but, according to our own experiment, this was not a real selective effect in the morphophysiological sense (Bubán et al. 1974).

After the cytochemical changes there is an increased mitotic activity in the apex, the central meristem is more extended and is now located directly below the subdermatogen (Fig. 2.2*b*). With this rearrangement, but without morphological changes in the apex, a histological transformation takes place and from this point on the flower initiation is irreversible.

In addition to the decisive factors mentioned before, according to the "nutrient diversion hypothesis" (Sachs 1977), the presence of assimilates and their distribution is also essential in activating the central zone.

Although it is generally accepted that the hormones and nucleic acids are responsible for mitotic activity before differentiation Sachs (1977) considers that according to the nutrient diversion hypothesis, "induction causes activation of the central zone, a requisite for floral initiation and early development through greater availability of nutritional factors (assimilates)". "Thus control of flowering by chemical or environmental factors may be an indirect result of influence on assimilate supply and distribution and not upon specific morphogenetic influences in the shoot apical meristems."

At a later stage of development, through morphological changes, the flower meristem forms (Fig. 2.2*c*). From this meristem a column-like formation develops (Fig. 2.3*a*) which is termed the *Pflock-Stadium* in German. The first morphological changes of the apex are also called secondary flower induction in which the introductory step is the reduction in the number of the cell rows and its essential aspect is the morphological differentiation (Tombesi 1965, cit. Martinez-Tellez et al. 1982, *see also* Hilkenbaumer and Buchloh 1954). The *Pflock-Stadium* is important for the development of the full value of the inflorescence. This meristem block is weakly developed in the lateral buds of shoots and the flowers formed here are undeveloped (Zeller 1960*c*).

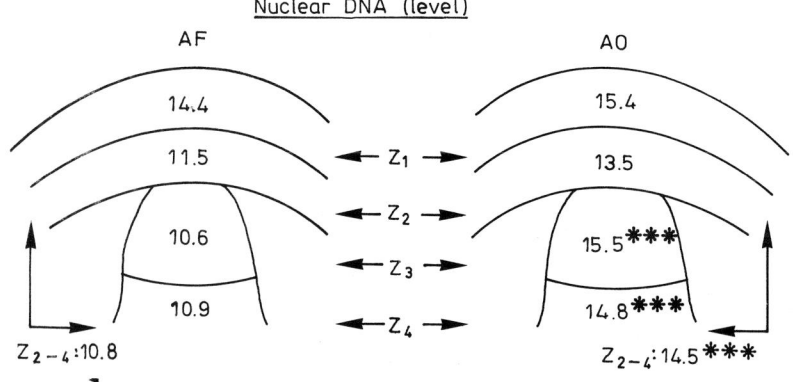
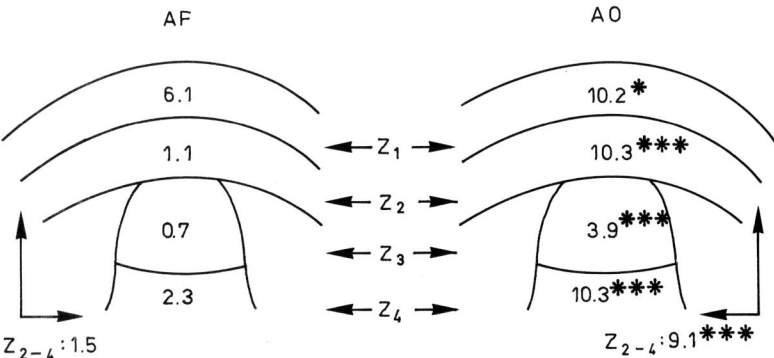
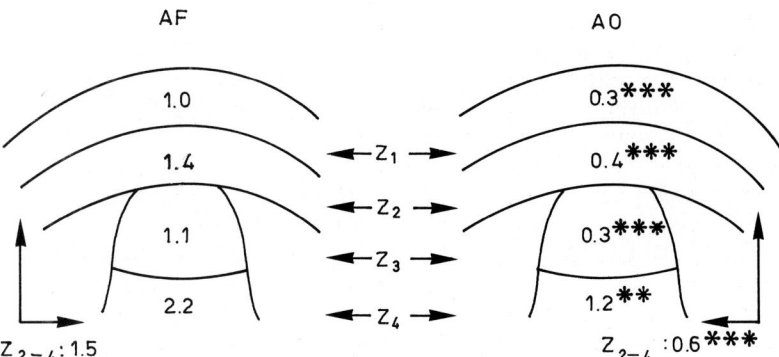

Fig. 2.1. Nucleic acid and nucleohistone level in the terminal buds of bearing and nonbearing spurs of apple trees. Above: DNA level of nuclei; middle: DNA and RNA levels of nuclei; below: nucleohistone level. AF: spur with fruit; AO: spur without fruit. Significance level between similar zones of AF and AO buds; *** 0.1% and ** 1% level, after Bubán and Hesemann (1979)

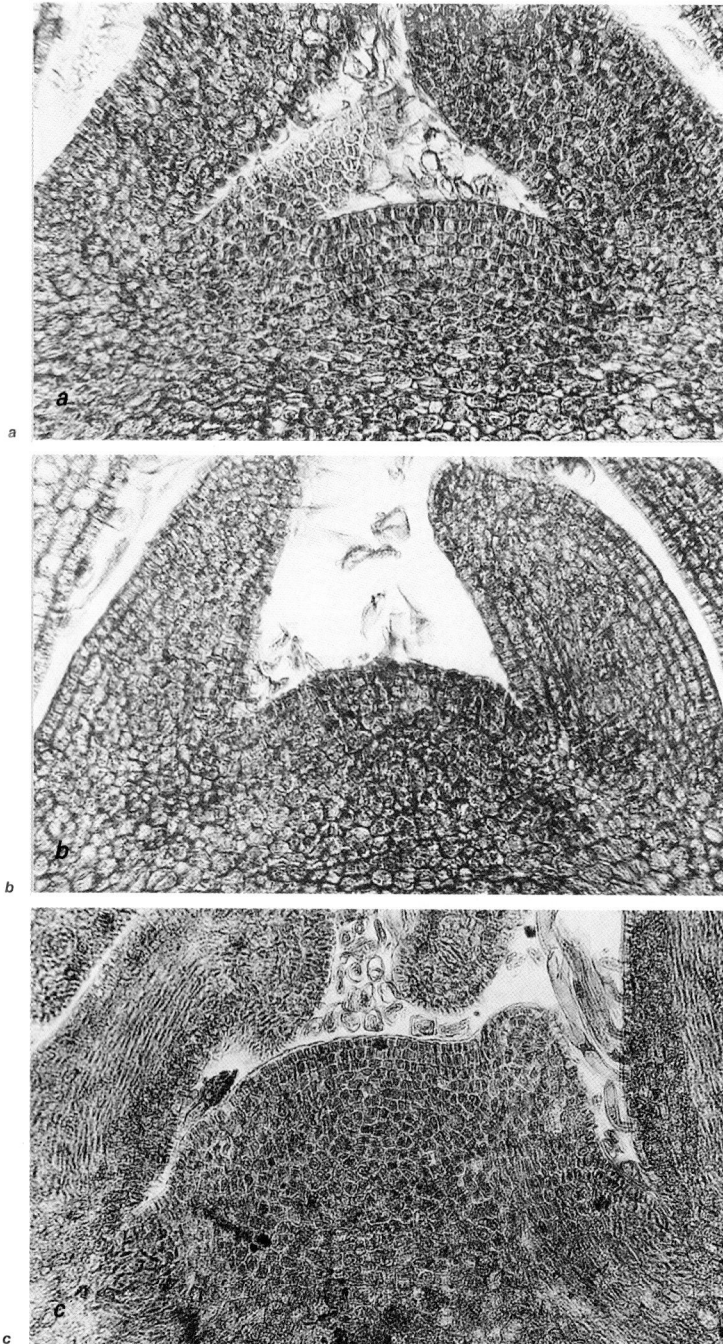

Fig. 2.2. Flower bud differentiation in the terminal buds of spurs in apple trees I. (*a*) Vegetative apex; (*b*) histological differentiation; (*c*) flower meristem: beginning of morphological differentiation (after Bubán 1965)

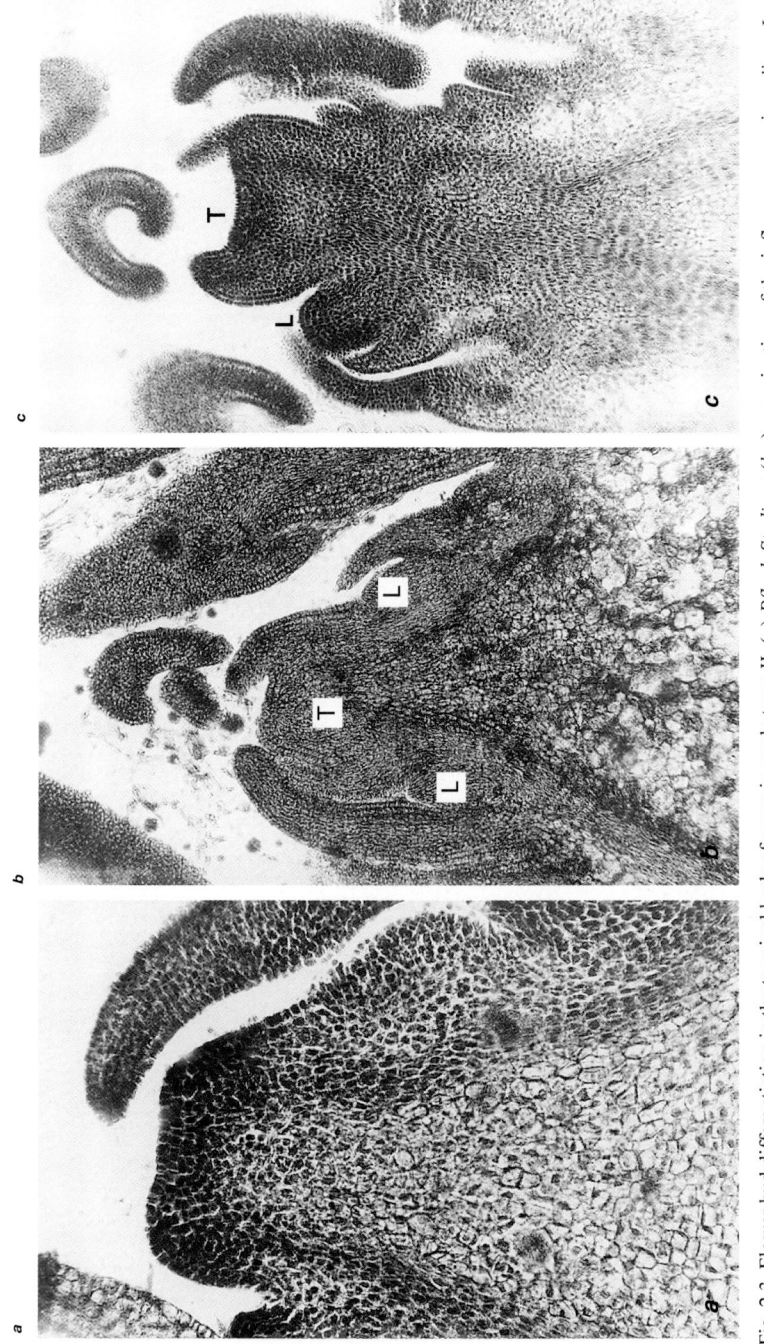

Fig. 2.3. Flower bud differentiation in the terminal buds of spurs in apple trees II. (*a*) *Pflock-Stadium*; (*b,c*) organization of the inflorescence primordium; L = lateral flower primordium; T = terminal flower primordium (after Bubán 1965)

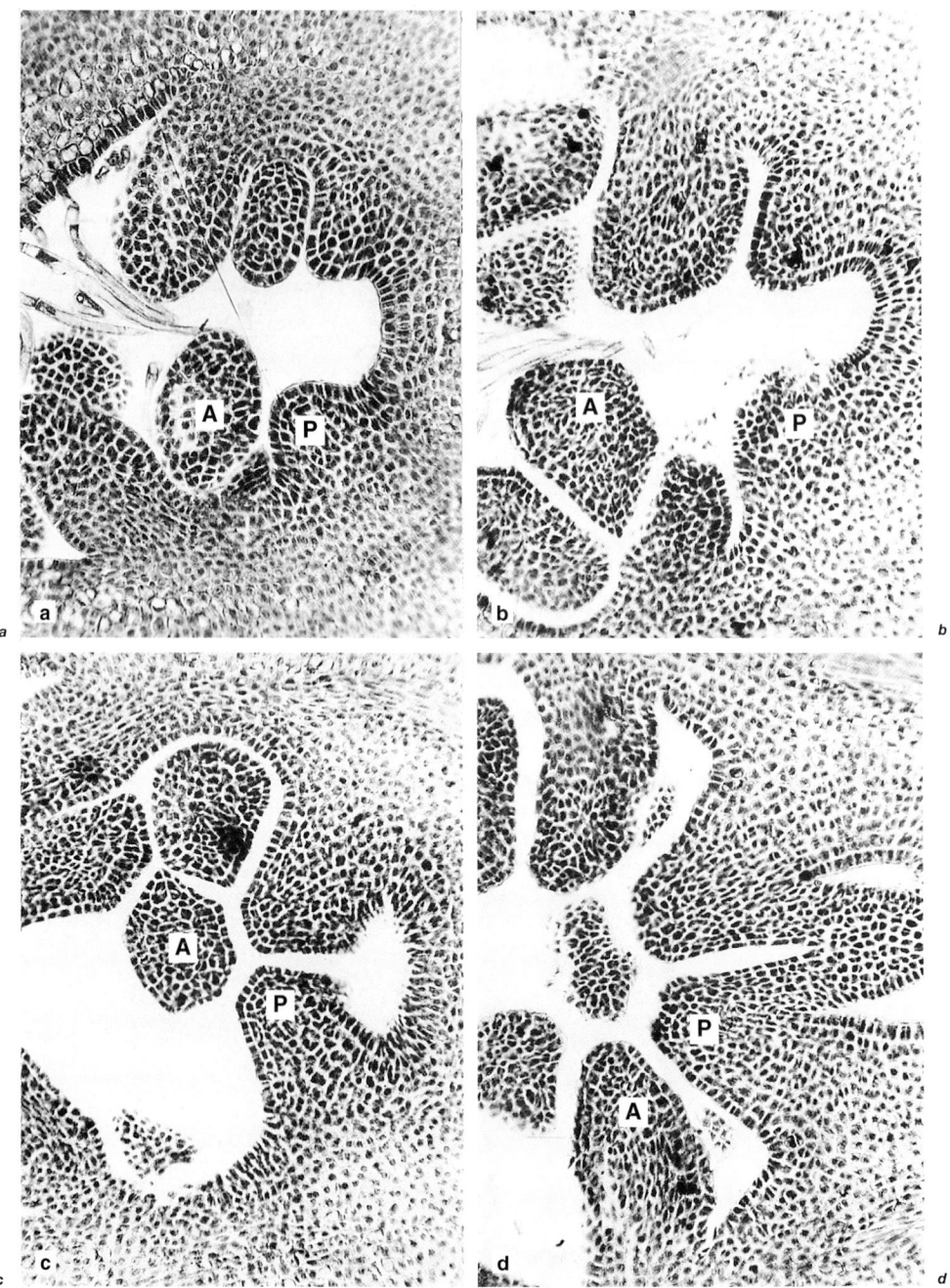

Fig. 2.4. Winter development of flower primordia in the terminal buds of spurs in apple trees. Terminal (*a, c, e*) and lateral (*b, d, f*) flower primordia late autumn (*a, b*) in the middle of February (*c, d*) and at bud burst

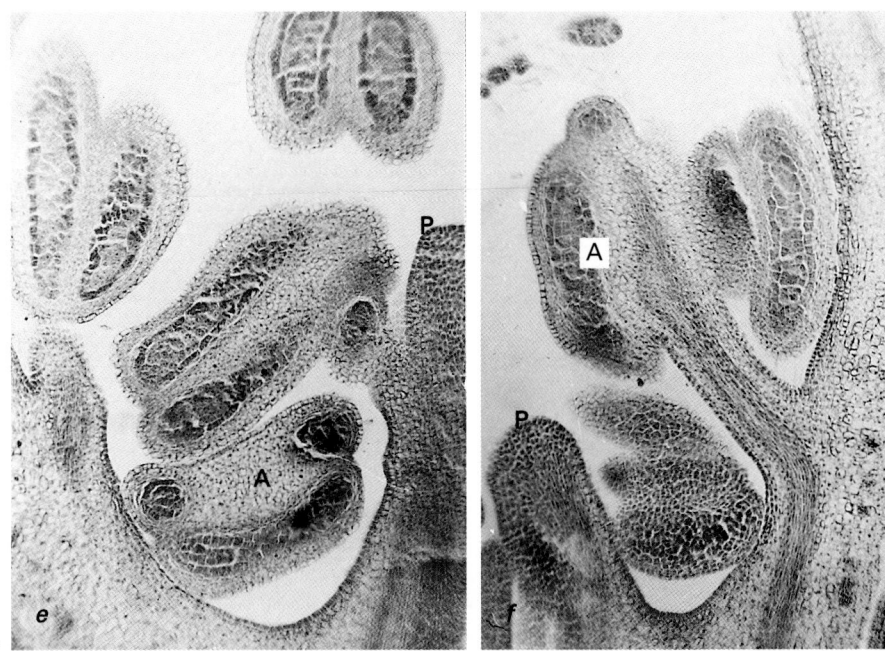

Fig. 2.4. (e, f). A = anther primordia; P = pistil primordium (after Bubán et al. 1979)

It is obvious, from the *Pflock-Stadium* on, that the flower in the apical position of the inflorescence is in a dominant position (Fig. 2.3b, c). The heterochronic nature of flower formation is characteristic of the Pomoideae subfamily but in Prunoideae flower formation is more synchronous (Zeller 1955). The morphological differentiation is a fast developmental process. By the beginning of October the pistil primordia can be recognized (Fig. 2.4a, b). This quick development starting with the *Pflock-Stadium* and the intensity of the RNA synthesis show an obvious parallelism (Schmidt 1978). The development of the inflorescence primordia is slower but continues later on, as well (Fig. 2.4c–f).

The increased mitotic activity in pear trees is an indispensable condition for flower initiation, too (Huet 1974). The flower initiation of the *pear* begins in July (Zeller 1983). By the middle of August there are inflorescence primordia in the buds (Fig. 2.5a). At the end of September the pistil primordia appear in the terminal flowers (Fig. 2.5b) and by winter we can find morphologically differentiated anther primordia (Fig. 2.5c). By the middle of March there are developing ovules in the ovary and pollen grains in the anthers (Fig. 2.5d).

The axillary buds of the *peach* shoots are single buds until the middle of summer (Bubán 1992) which have vegetative apices (Fig. 2.6a). Next to the vegetative apex, an apex initial of the prospective lateral bud, which is a potential flower bud, can be seen seven–eight weeks after full bloom (Fig. 2.6b). Such initial apices develop on both sides of the leaf bud. After a rapid development the apices of the lateral buds are complete by the end of 10th–12th week after full bloom (Fig. 2.6c, d). By the 13–14th week after full bloom the apices of the side buds become identical in size and structure with the apex of the middle leaf bud (Fig. 2.6e). This is followed by increase in size (Fig. 2.7a) and morphological changes resulting in flower meristem development (Fig. 2.7b–d). The sepal,

15

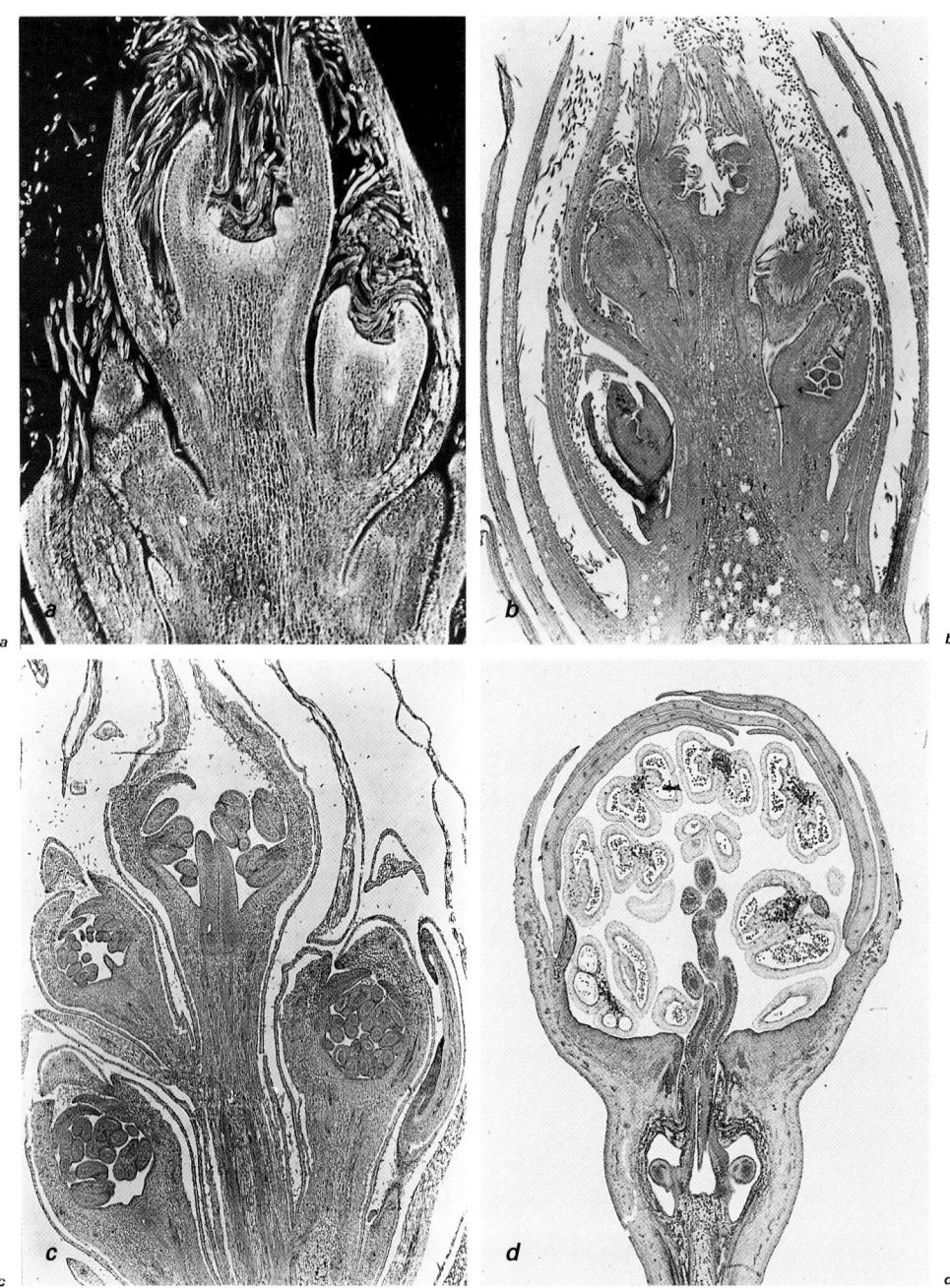

Fig. 2.5. Flower bud differentiation of the pear. Inflorescence primordium (*a*) in the middle of August; (*b*) at the end of September; (*c*) in winter and (*d*) a flower primordium in the middle of March (after Zeller 1983)

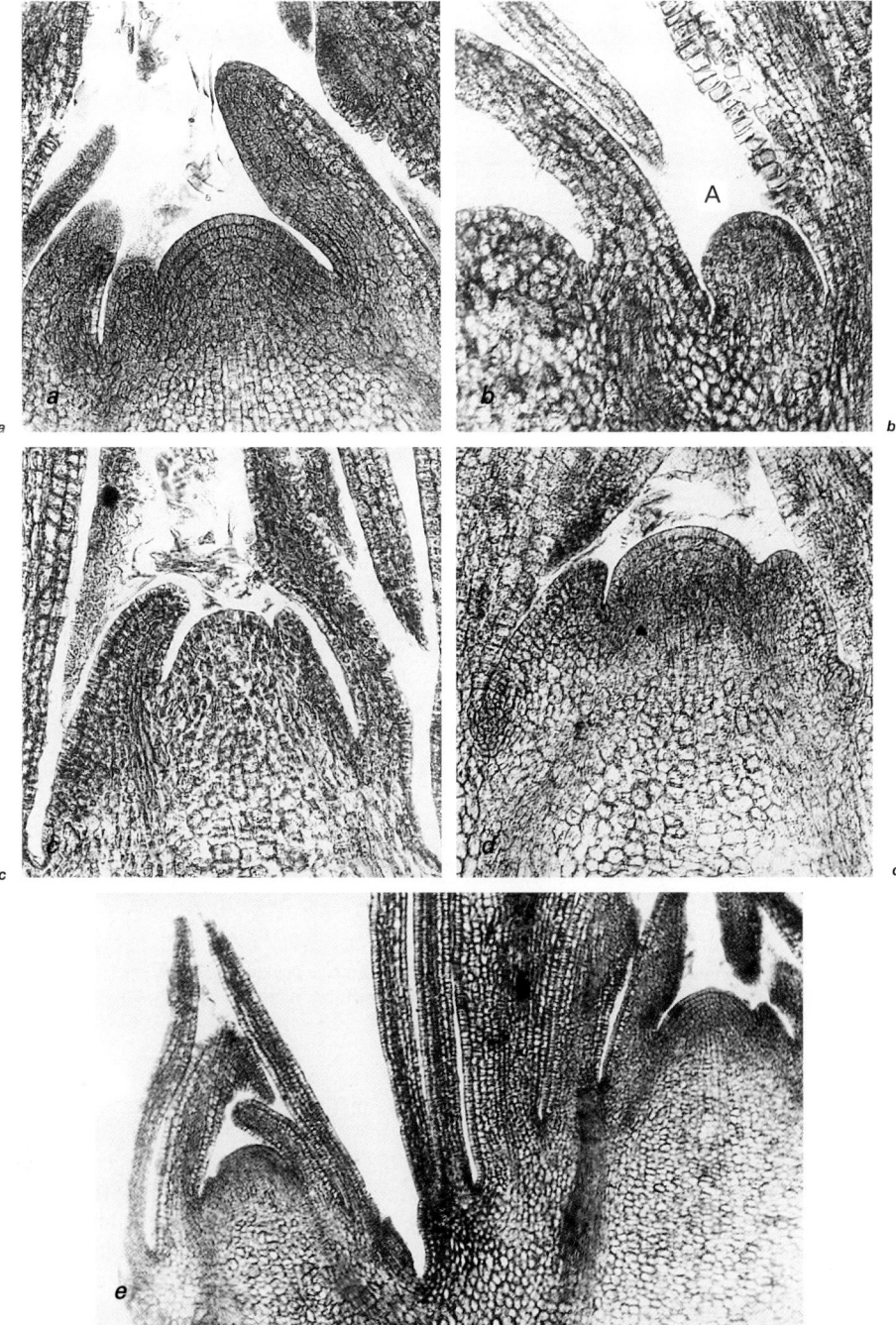

Fig. 2.6. Development of the lateral bud (potential flower bud) next to the axillary leaf bud of peach. (*a*) Vegetative apex of the axillary leaf bud; (*b*) first sign of initiation of the lateral bud apex (marked with A) seven to eight weeks after full bloom; (*c, d, e*) development of the lateral bud apex and its identical structure with that of the middle bud 13–14 weeks after full bloom (after Bubán 1992)

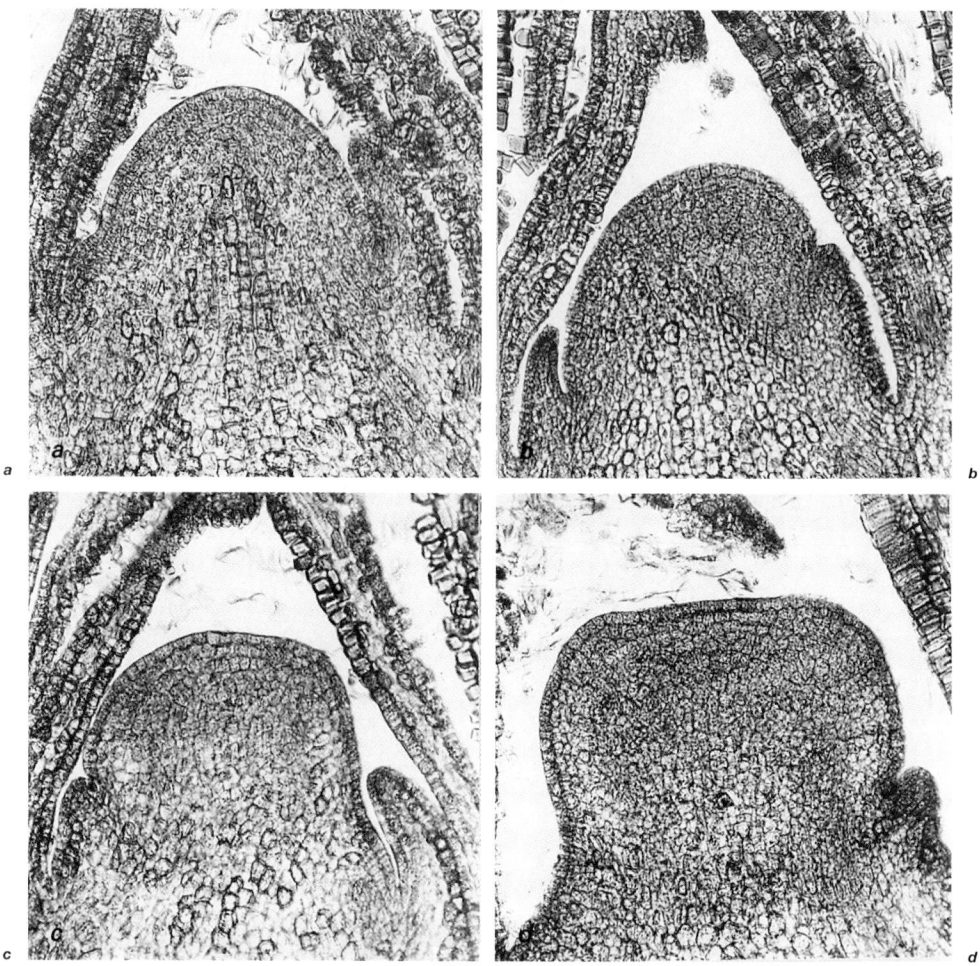

Fig. 2.7. Differentiation of the flower bud of the peach I. Vegetative apex of the lateral bud in the axillary bud group on shoots becomes enlarged (*a*) and then organized into flower meristem (*b-d*) starting 14–15 weeks after full bloom (after Bubán 1992)

petal and anther primordia and in September the pistil primordium appears in an external to internal order (Fig. 2.8*a–d*).

In the buds of *sour cherry* trees (Zeller 1983) the apices are still vegetative in June. However, after their early and rapid differentiation by the middle of July the first flower primordia can be detected (Fig. 2.9*a, b*). The simultaneity in the development of flower primordia is conspicuous and typical for Prunoideae subfamily (Zeller 1955) (Fig. 2.9*c, d*). During winter a further development can be observed in the ovules and the pollen formation is completed by the end of March (Fig. 2.9*e, f*).

The flower bud formation of the June bearing *strawberry* is known from Zeller's (1969) and Sattler's (1973, cit. by Bubán 1980*a*) work. At the end of August, apices are still in a vegetative stage (Fig. 2.10*a*). Two to three weeks after the decrease in daylength

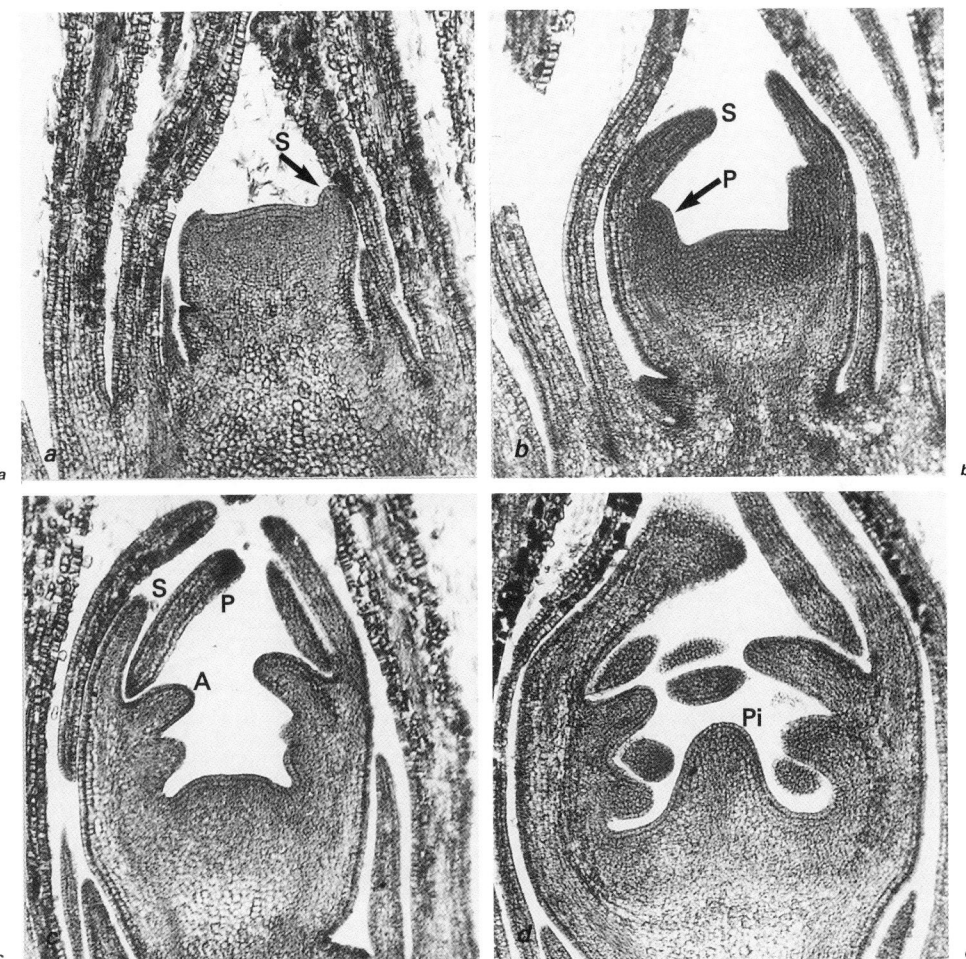

Fig. 2.8. Differentiation of the flower bud of the peach II. Appearance of the sepal (S), petal (P), anther (A) primordia in August and the pistil (Pi) primordium in the middle of September (after Bubán 1992)

the inflorescence primordium is observable (Fig. 2.10b). After the appearance of the anthera primordia the innermost part of the flower elongates and evolves into a conical formation, the receptacle, on which large numbers of pistil primordia develop (Fig. 2.10c). During the winter (Fig. 2.10d) we can find morphologically differentiating pistil primordia, and pollen tetrades in the strawberry. A few days before flowering the flower becomes completely organized (Fig. 2.11a) and after flowering the tiny achenes begin to develop on the surface of the receptacle (Fig. 2.11b). In the achenes already in the second week after anthesis there is an embryo in the globular stage (Fig. 2.11c, d).

In the buds of the *gooseberry* (Bubán 1980a) occasionally there is one flower but mostly two–three flowers in each inflorescence. When a single flower is initiated by the

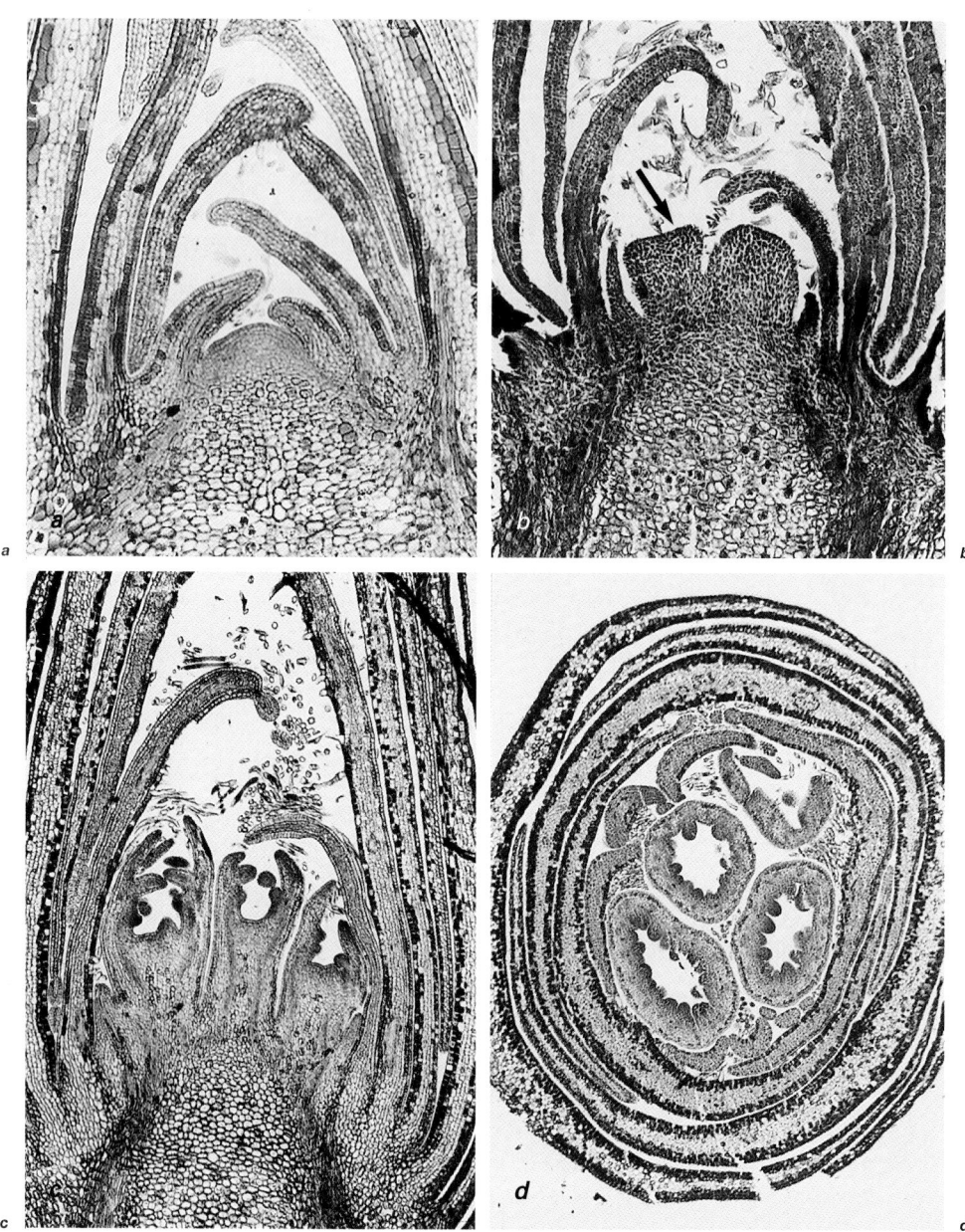

Fig. 2.9. Differentiation of the flower bud of the sour cherry. (*a*) In June the apex is still vegetative; (*b*) first flower primordium appears in mid-July (arrow); (*c*) longitudinal and (*d*) cross-section of flower bud in September

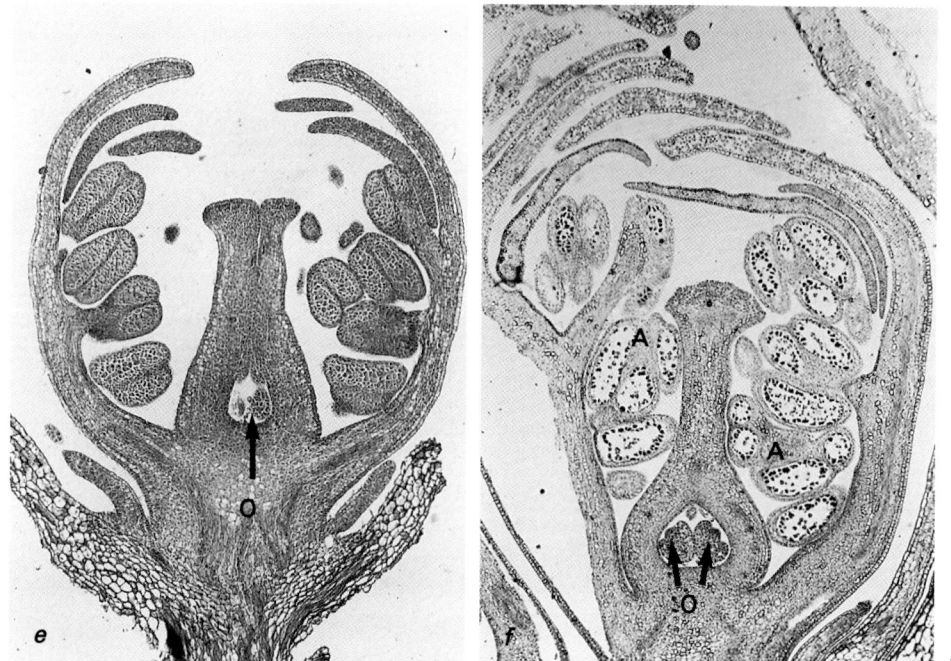

Fig. 2.9. (e) The still rudimentary ovule, which will be morphologically differentiated by March; (f) there are pollen grains in the anther primordia. A = anther primordia, O = ovule (after Zeller 1960a, 1983)

second half of July there is a protrusion of the vegetative apex (Fig. 2.12a) and primary thickening (Fig. 2.12b, c) and then calyx primordium initiation begins (Fig. 2.12d).

Before the formation of a raceme consisting of two–three flowers, the apex becomes asymmetrical (Fig. 2.12e) and in the lateral apex region intensive cell division can be seen. The cell division results in the protuberance like flower meristem (Fig. 2.12f) from which the flower will develop in basal position within the inflorescence. In September, although not in all cultivars, we can find ovule primordia in the ovary (Fig. 2.13a, b). Until March, when bud break occurs, pollen can be found in the anthera, but the development of the ovules is still slow (Fig. 2.13c). The structure of the ovules located in large number on the placenta becomes complete one week before anthesis (Fig. 2.13d, 2.14a, b).

Flower initiation in the axillary buds of the *black currant* shoots in the Helsinki region (Zeller 1968) begins in August and first the flower primordia having basal position appear in the future inflorescence (Fig. 2.15a). By the beginning of November the archesporium develops in the anthers (Fig. 2.15b) and the pollen mother cells form in March. By this time ovule development also begins (Fig. 2.15c). A few days before anthesis the flower parts complete their development (Fig. 2.15d).

We have already mentioned (*see* Chapter 2.2; Zeller 1964c) the flower-forming characteristics of the *Arctic blackberry, Rubus arcticus*. The flower formation seen in the figures (Fig. 2.16a, b, c) is realized within one month from flower initiation (Zeller 1983). Flower initiation in *cow berry, Vaccinium vitis-idae*, which is also considered an Arctic

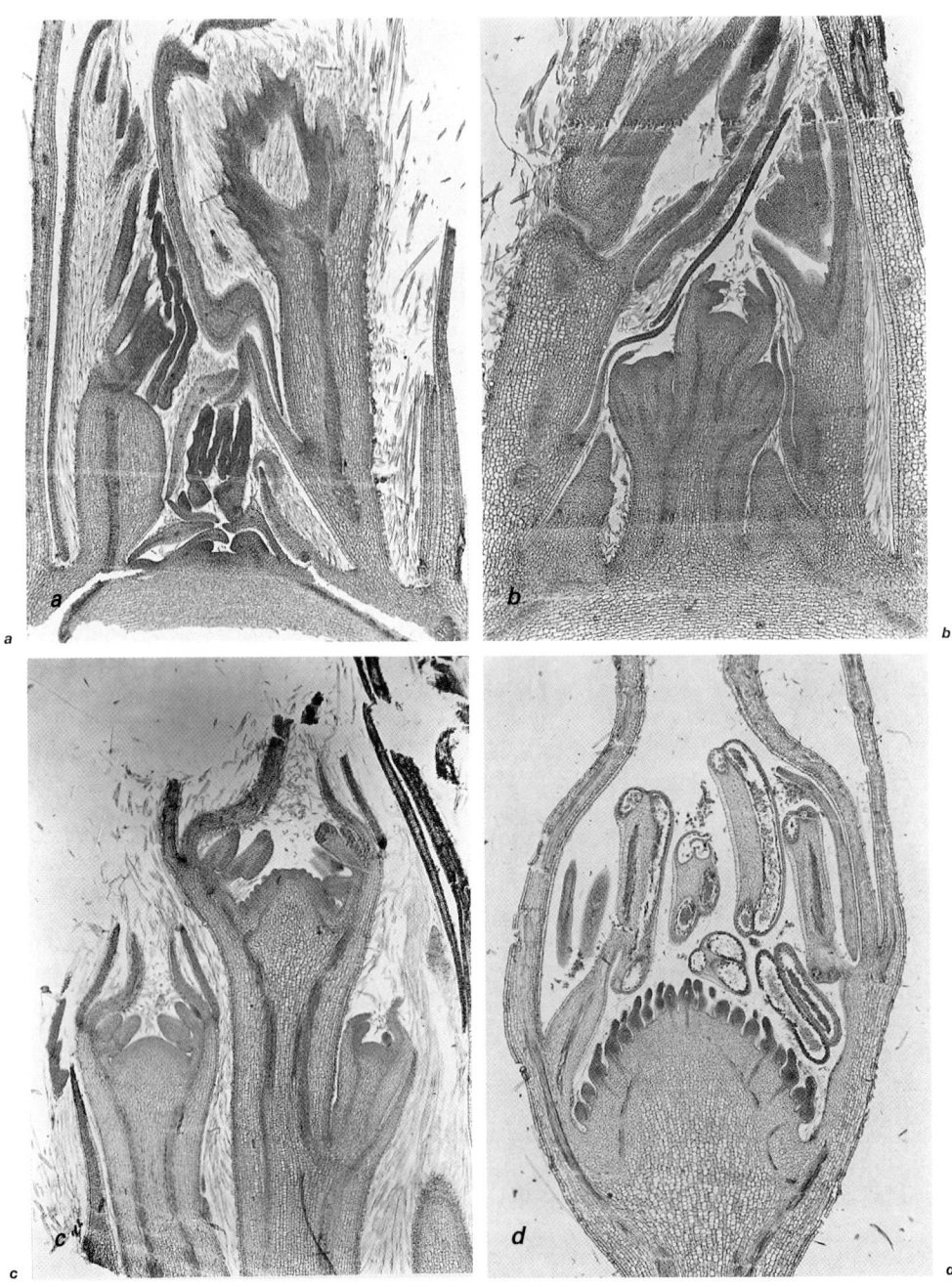

Fig. 2.10. Differentiation of flower buds in strawberry I. (*a*) The vegetative apex at the end of August; inflorescence primordium (*b*) in the middle of September and (*c*) in October; (*d*) during the winter pistil primordia and pollen tetrades in the anthers are present (after Zeller 1969)

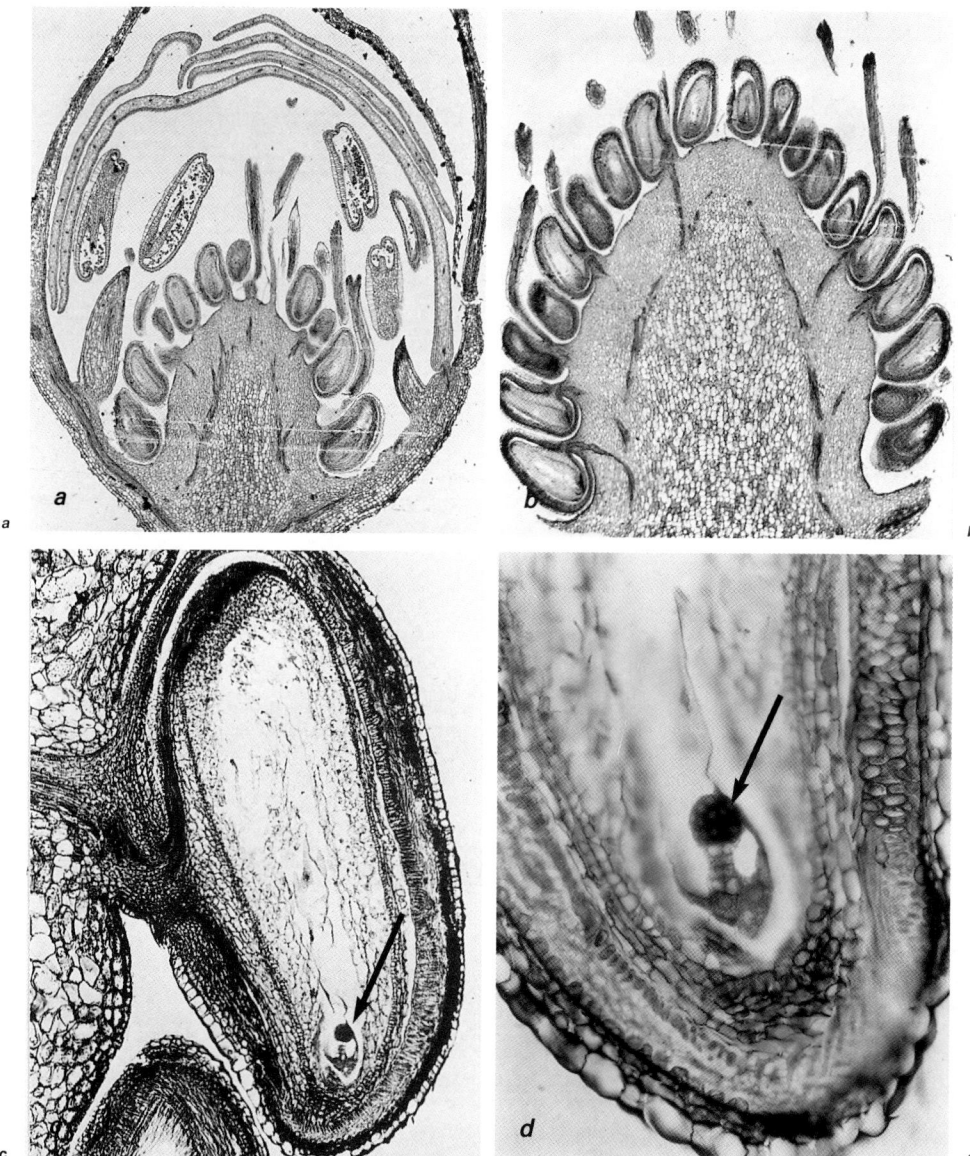

Fig. 2.11. Differentiation of the flower buds in strawberry II. (*a*) Completely developed structure of flower a few days before anthesis; (*b*) in the second week after flowering a large number of quickly developing achenes are present; and (*c, d*) there is a proembryo (arrows) in the achene (after Zeller 1969)

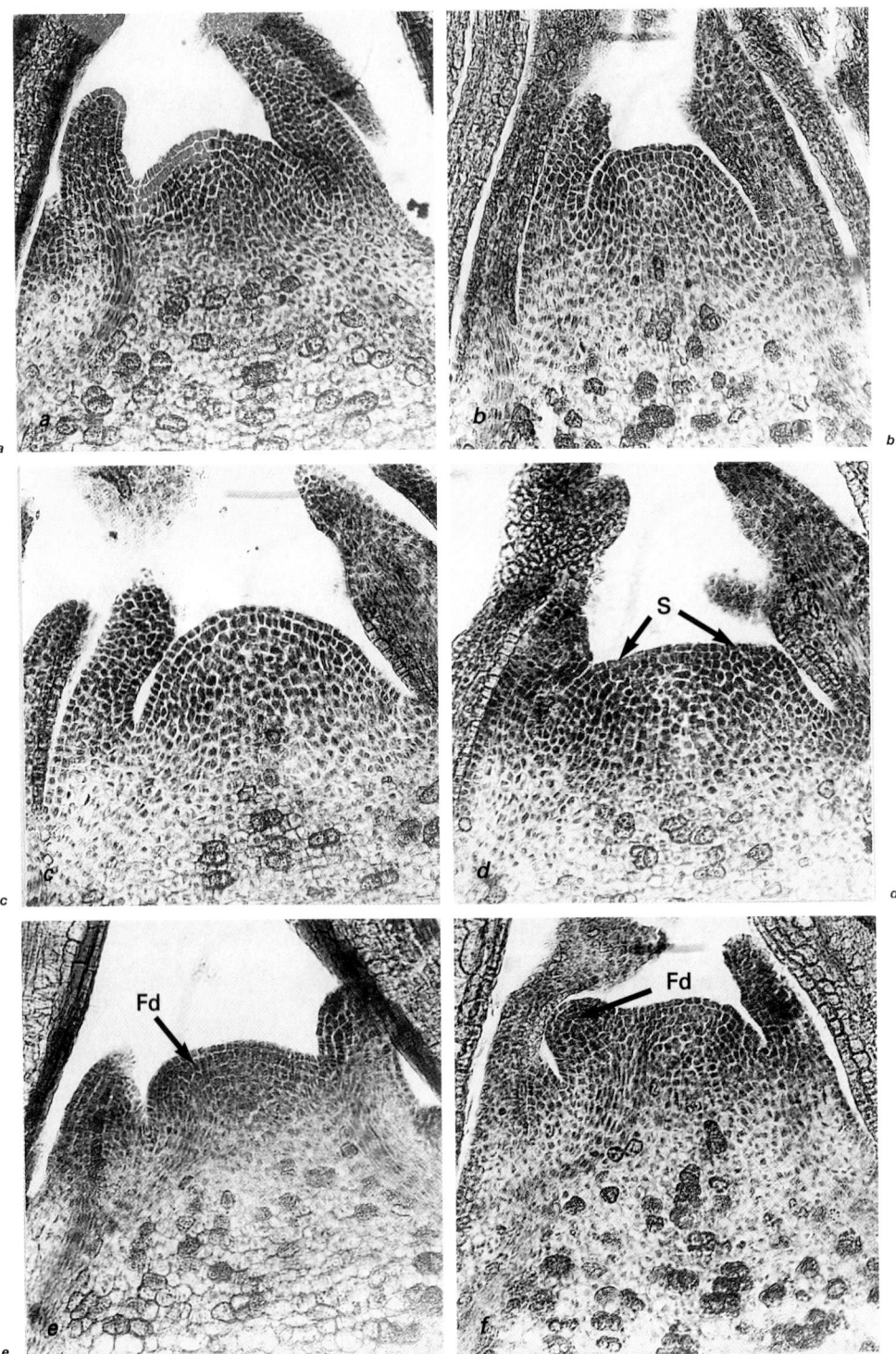

Fig. 2.12. Differentiation of the flower buds in gooseberry I. On developing only one flower: (*a*) the vegetative apex will be transformed into (*b, c*) the flower meristem, then (*d*) initiation of sepal primordia begins at the end of July. When an inflorescence forms in the bud (*e, f*) first the meristem of flower in basal position within the inflorescence appears. Fd = start of the first (basal) flower differentiation; S = sepal primordium initiation (after Bubán 1980*a*)

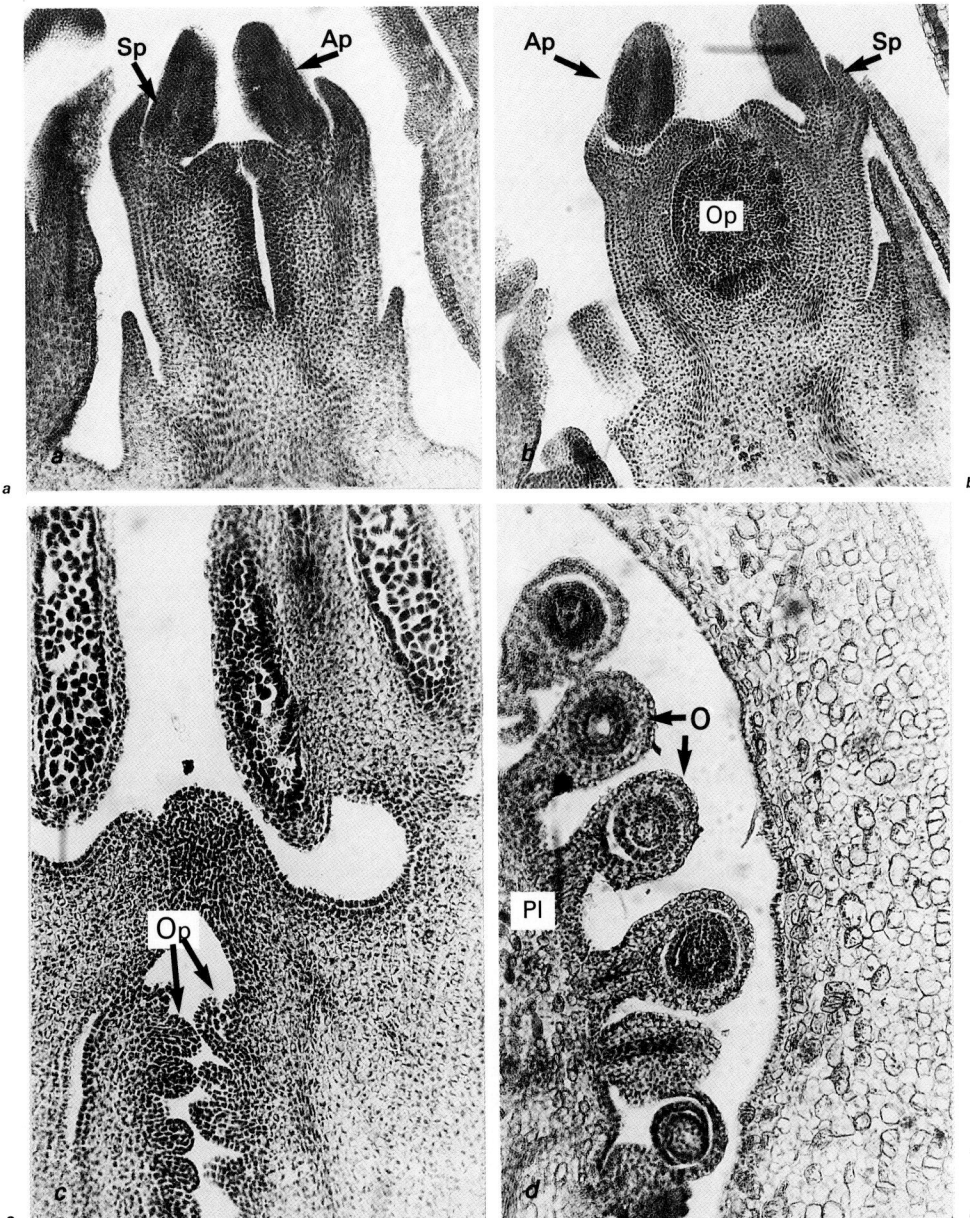

Fig. 2.13. Differentiation of the flower buds in gooseberry II. (*a, b*) Flower primordium at the end of September; (*c*) ovule primordia at bud burst; (*d*) ripe ovules one week before flowering. Ap = anther primordium, O = ovules; Op = ovule primordia; Pl = placenta; Sp = sepal primordium (after Bubán 1980*a*)

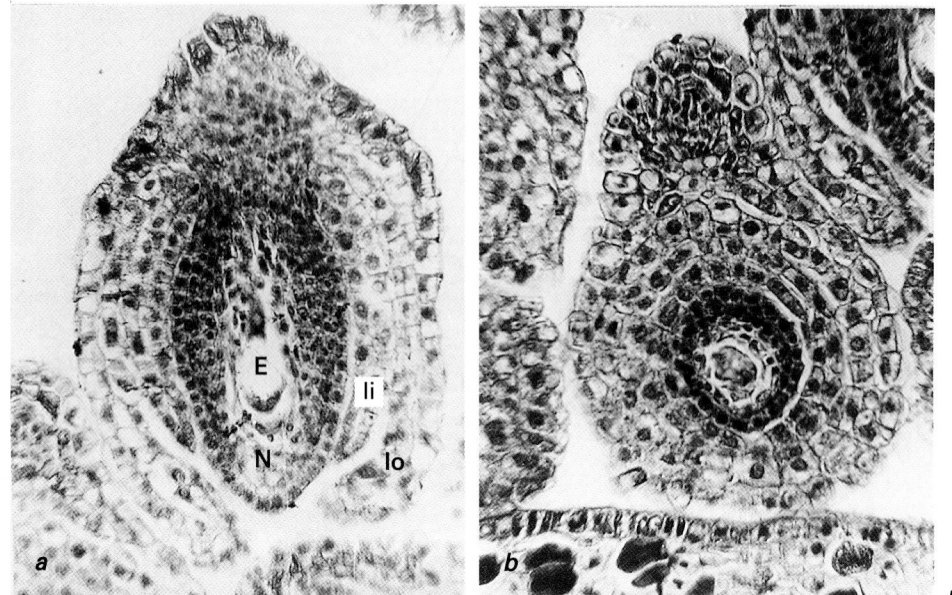

Fig. 2.14. Longitudinal and cross-section of the ovule of the gooseberry. (*a*) Longitudinal; (*b*) cross-section. Io and Ii = external and internal integuments; E = embryo sac; N = nucellus (after Bubán 1980*a*)

fruit species, begins early in May–June, in the terminal buds of young shoots (Zeller 1983). In the overwintering bud there is a well-developed inflorescence (Fig. 2.17*a, b*).

The flower bud formation of *walnut* was mentioned in Chapter 2.2. The first signs of flower differentiation of the *chestnut, Castanea sativa* L., is the typical structural modifications taking place in the apical meristem of the annual shoot resulting in the appearance of the catkin primordia. The initiation of the primordia can be observed in histological sections 30–40 days after the start of spring growth (Bergamini and Ramina 1971). The female inflorescence primordium (glomerule) can first be detected towards the end of July on the basis of the already elongated catkin primordia. At this time, they are undifferentiated protuberation of tissue only, but by end of October they develop into a simple apex-like structure. The formation of the male flowers within the catkin as well as of the female inflorescences takes place in the first period of spring growth before flowering.

In the middle and terminal zones of shoots there are axillary inserted catkins with male inflorescences only and which are mixed inflorescences, i.e. catkins, too, whose bottom part also contains female inflorescences. There are three–ten female flowers in the female inflorescence. It is favourable from the point of view of crop value if there are two–three flowers which are fertilized only. The flowering of the flowers of the two sexes takes place not at the same time. Protandreous flowering is more frequent. The receptivity of the pistil cannot be visually determined because it does not have stigma exudate and the maturity of the pistil is not accompanied by discolouration. It has been found, however, that female flowers become fertilizable in 8–21 days after their appearance but not in the first week (Szentiványi 1976*b*).

The male inflorescence of the *hazelnut, Corylus avellana* L., the catkin, develops in the flower buds and the female inflorescence, known as glomerule, forms in the com-

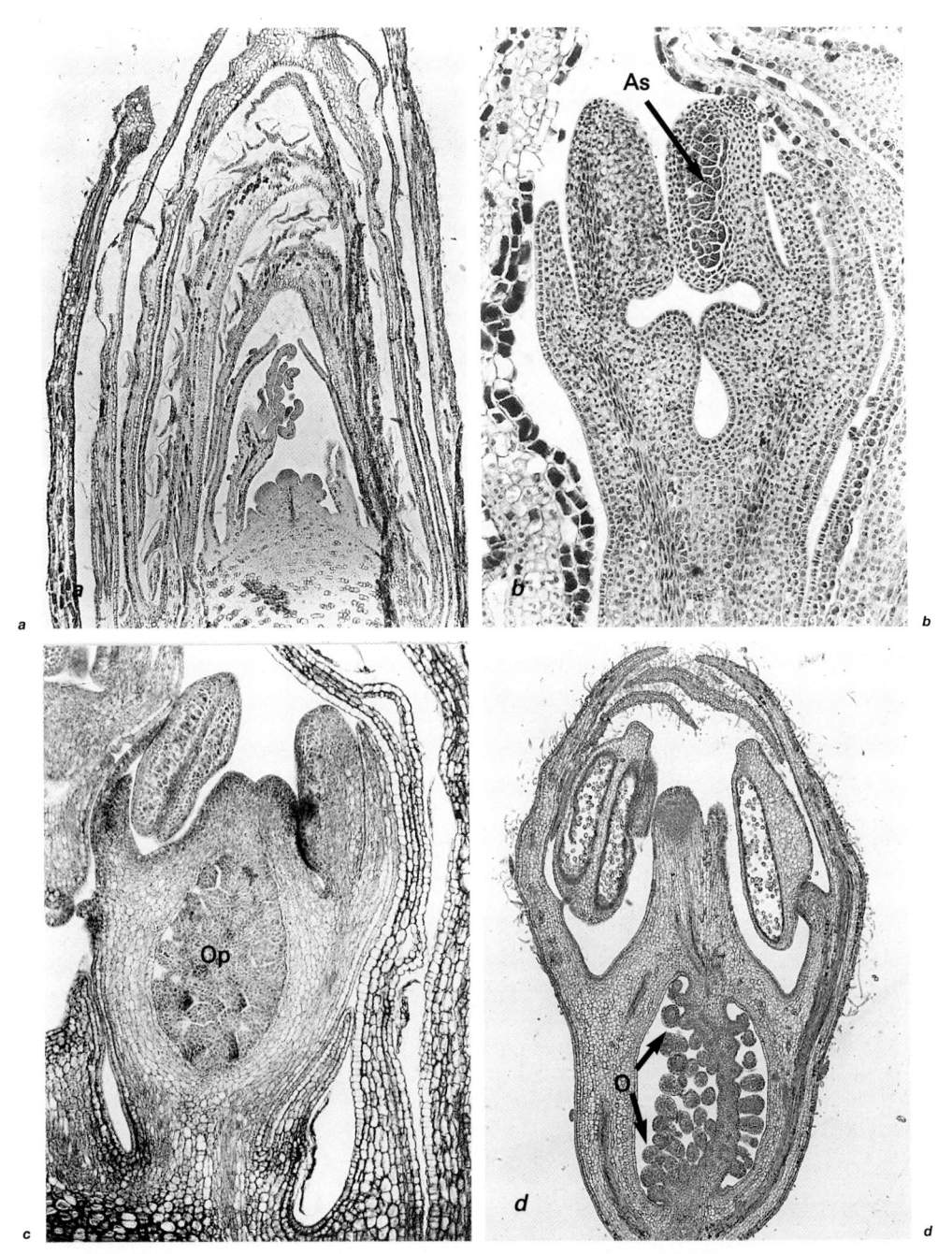

Fig. 2.15. Flower formation of the black currant. (*a*) Beginning of flower development in August; (*b*) flower primordium in November, with archesporium; (*c*) in March there are ovule primordia present; (*d*) completely developed flower just before anthesis, As = archesporium; O = ovules; Op = ovule primordium (after Zeller 1968)

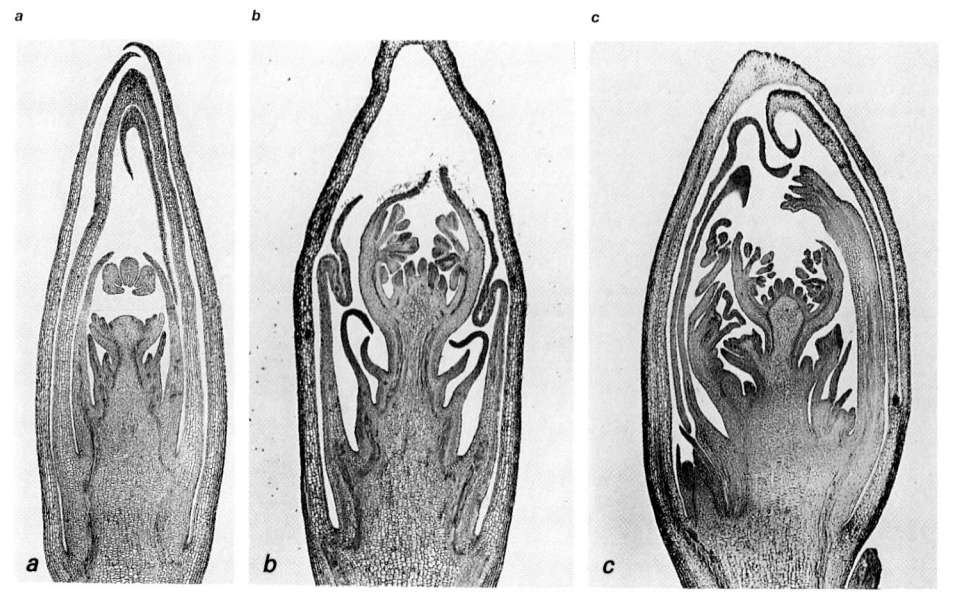

Fig. 2.16. Flower bud differentiation of the Arctic blackberry (*Rubus arcticus*). (*a*) Primordia at the beginning of August; (*b*) end of August; (*c*) beginning of September (after Zeller 1964c, 1983)

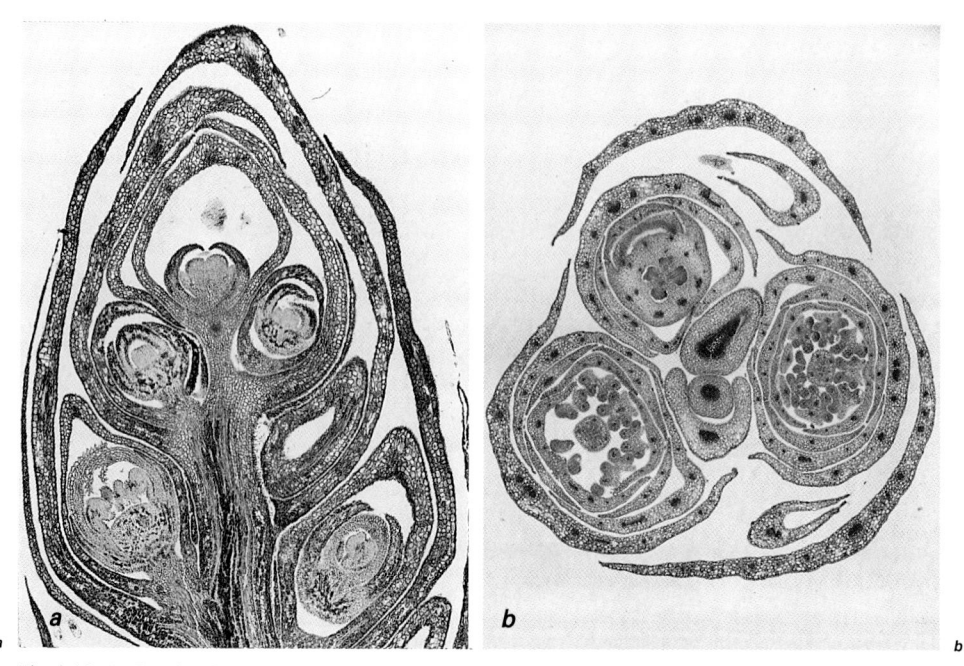

Fig. 2.17. (*a*) Longitudinal and (*b*) cross-section of the inflorescence primordium of the cowberry (*Vaccinium vitis-idae*) (after Zeller 1983)

posite buds. The differentiation of the catkin starts first around the middle of May (Trotter 1951, cit. in Germain 1983, Dimoulas in Geraci 1974, cit. in Germain 1983), while the female inflorescences start to develop at the end of June or beginning of July. According to other observations (Pisani et al. 1970, Romisondo 1977), the order is the reverse. The female inflorescences differentiate from mid-May to mid-June and the male inflorescence differentiates beginning from the middle of June. The results of Mussano et al. (1983*a*) seem to resolve this contradiction. According to them, differentiation of the catkins occurs since May. The female inflorescences in the axillary buds differentiate at the end of May, but only one month later in the buds on the base of peduncles of catkins.

The further development of the catkins and the glomerule is well known from the work of Dimoulas in Geraci 1974 (cit. by Germain 1983, Mussano et al. 1983*a*). The stamen primordia appear at the end of June, early July and the anther follows this two weeks later. The pollen mother cells can be seen by the end of July and meiosis proceeds in the first half of August. During the next four–six months until flowering the further development of the catkins is slow. The outlines of the female inflorescence and the bracteole formation can only be detected in August and the outlines of the style somewhat later. The style begins to intensify its growth in September. By the beginning of October both the male and female inflorescences have formed.

The male inflorescences of hazelnut, the catkins, are located in an axillary position and they form a group containing two–ten catkins. One catkin has 130–260 male flowers (Pisani and Guilivo 1968, cit. in Germain 1983, Barbeau 1972, cit. in Germain 1983). Four stamens in a flower mean altogether eight anthera and in every anther contain 10,000–20,000 pollen grains (Pisani and Guilivo 1968, cit. in Romisondo 1977). Every flower is surrounded by a bract and two bracteoles. The female inflorescence is short and is located terminally in an apical shoot consisting of seven to ten internodes. Next to every four bracts there are two flowers (Germain 1983). The glomerule according to Pisani (1968, cit. in Romisondo 1977) consists of seven to ten flowers but in extreme cases there may be 20 flowers also (Barbeau 1973, cit. in Horn 1976).

The formation of the female inflorescences is determined by several factors. Many authors consider (Germain 1983) that the ratio of the glomerule to the male inflorescences depends on the length of the shoots. The ratio is less on shoots 15 cm and shorter, or 40 cm and longer. The origin of the shoots is also important. The shoot can derive from a vegetative bud or from the glomerule. If the shoot developed form the glomerule the presence of the nuts strictly inhibits the formation of female inflorescences (Barbeau 1977, cit. in Germain 1983). In those shoots which get a lot of sunlight one-and-a-half to three times as many female inflorescences develop than in shoots lacking sunlight in shady parts of the plants (Bergougnoux et al. 1978, cit. in Germain 1983).

The particularly unique flowering and fertilization properties of the hazelnut will be dealt with in Chapter 6.4.

Mention should be made of some considerations which may be of *interest to the practising fruit grower* as well.

The bud population on our trees is the raw material for the next year's harvest (Abbott, cit. in Krause 1973). But even at sites preferable for flower formation, as for example in the spurs of apple trees, the flower primordium does not always develop. Therefore, it is advisable to study the ratio of real flower buds before pruning, then we can regulate the crop by the strength and way of pruning (Bubán 1967, Krause 1973, Streitberg and Handschack 1983).

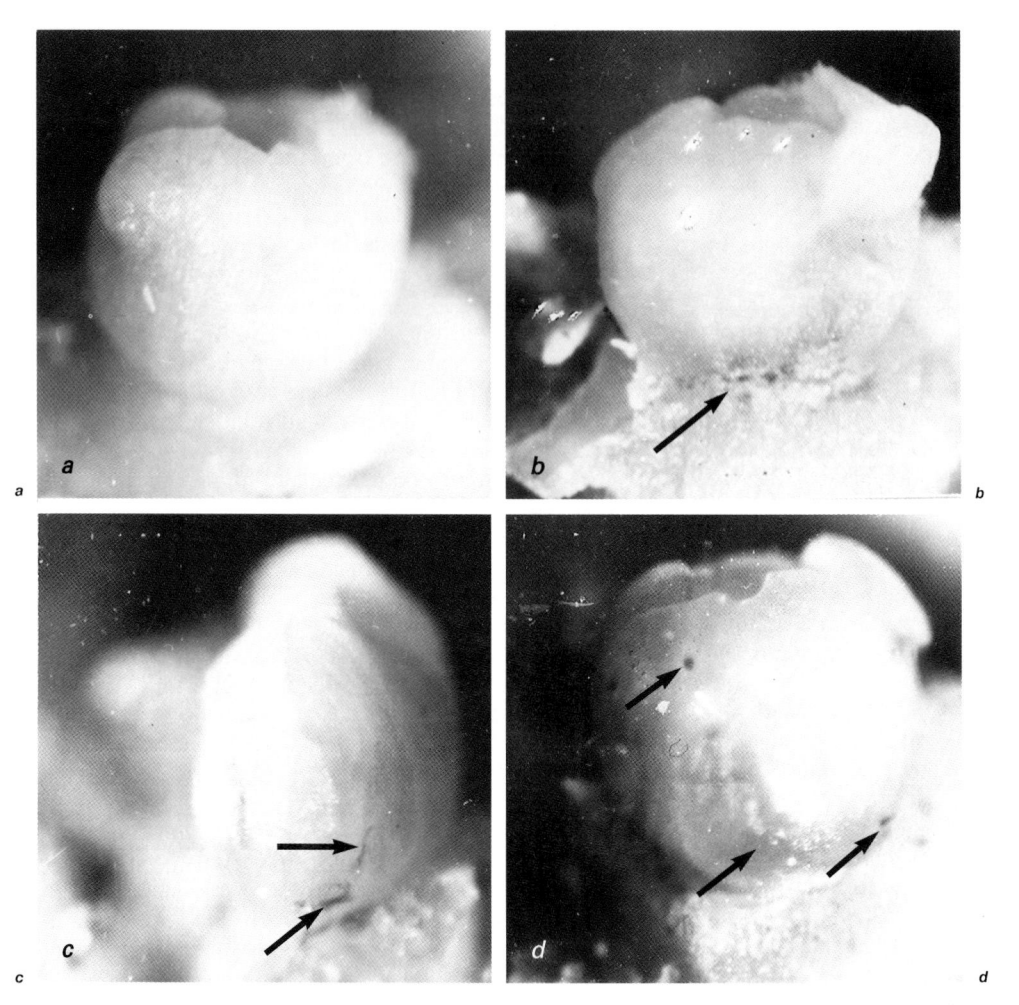

Fig. 2.18. Damage of apricot primordium in autumn I. (*a*) Symptomless primordium; (*b*) first sign of necrotic discolouration (arrow) on the base of the primordium; (*c, d*) later signs of injury on the surface of the sepal primordia (after Bubán et al. 1982)

In pome fruit species (apple, pear, quince) the differentiation of certain flower parts in the buds, called qualitative development, is continued during the winter months (Zeller 1955, 1960*c*, Reichel 1964*b*). Although there are exceptions, e.g. *Malus baccata* (Zeller 1955), the winter development occurs in most cultivated forms. The qualitative development (*see* Fig. 2.4*a–f*) is accompanied by considerable growth in size of buds. The diameter of the apical and lateral flower primordium of the apple inflorescence between October and early December increase by 23–36%. By the middle of February the growth was slower, 16–17% and 6–18%, respectively, but following this from mid-February until the middle of March the flower primordia more than doubled (Bubán et al. 1979). The sudden growth in the middle of February makes it probable that by then the bound water in the tissues of flower buds is already converted to free water (Faust et al. 1991) and the real vascularization developed between the flower primordium and the spurs replacing the hitherto functioning procambium (Ashworth 1984, Ashworth and Rowse 1982).

With respect to the stone fruit species we cannot generalize. There is hardly any further development in the flower buds of the cherry and peach during the winter according to Zeller (1955). Feucht (1955 cit. in Reichel 1964b), however, observed a quantitative growth. Tarnavschi et al. (1963) as well as Cociu and Bumbac (1973) have reported on growth in certain flower parts or in the tissues and archesporium of the apricot, mainly by cell elongation.

The flower buds in winter are endangered not only by frost. Bacterial decay of the flower buds of the pear tree are well known throughout the world. Of the *Pseudomonas syringae* bacteria isolated from these buds 40% were ice-nucleation active (INA) and 70% was pathogenic to pear trees. Of the strains 39% had both properties (Montesinos and Vilardell 1991). The destruction of flower buds in apricot by an unknown cause was already observed in autumn (Bubán et al. 1982). In September and October before the frost, dark necrotic spots appeared on the basal part of the flower primordia (Fig. 2.18b) and later on parts of the sepal primordia (Fig. 2.18c, d). Our histological investigations also found that the symptoms occurred first in the basal region of the bud especially in the basal part of the ovary (Fig. 2.19b), then extended to other flower parts (Fig. 2.19c). We could not isolate pathogenic fungi or bacteria from the flower primordium. Medeira

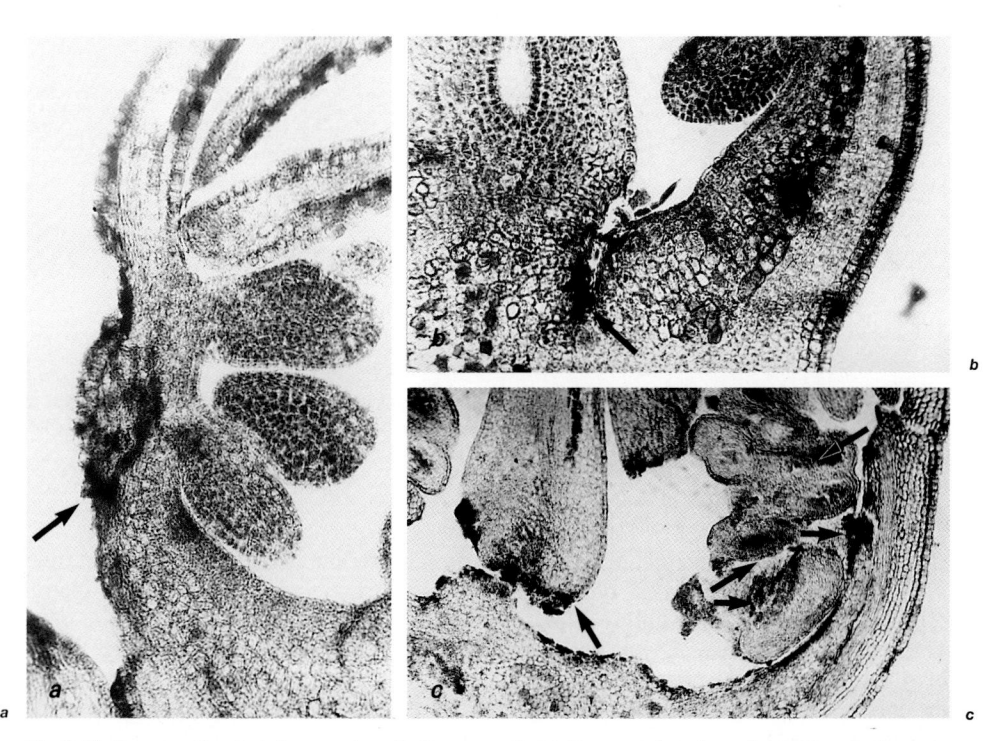

Fig. 2.19. Damage of apricot flower primordia in autumn II. (*a*) Damage of surface origin; (*b*) early tissue destruction at the base of the ovary; (*c*) damage spreading to the other parts of the primordium (arrow) (after Bubán et al., 1982)

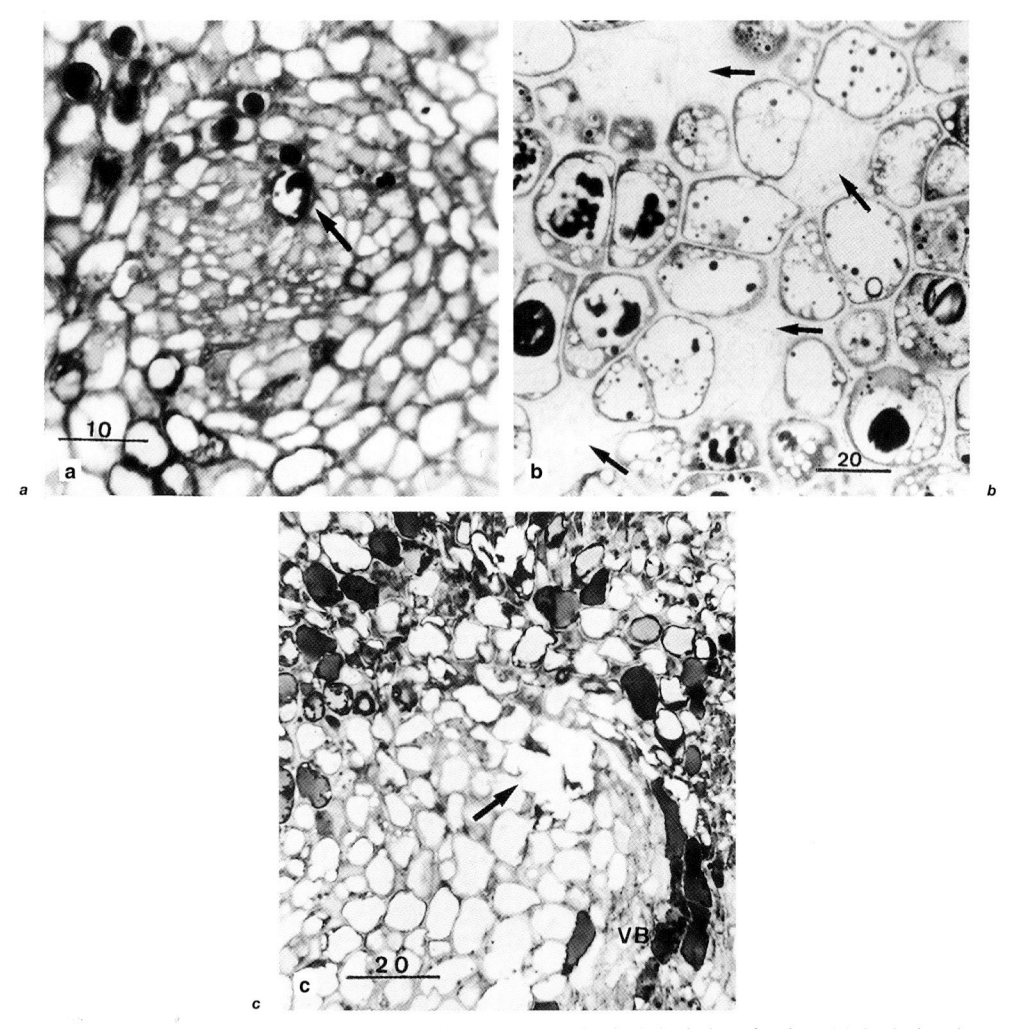

Fig. 2.20. Symptoms (arrows) in November preceding the December bud abscission of apricot. (*a*) Occlusions in the vascular bundles of the receptacle; (*b*) alteration in the pith tissues; (*c*) degradation of the neighbouring parenchyma cells of the vascular bundles (VB) (Medeira and Guedes 1989)

and Guedes (1989) studying the cause of bud drop in December common in apricot trees found that the first signs of the damage are occlusions in the vascular bundles of the receptacle and there are alterations of pith tissues at the insertion of the flower primordia. The cell wall of the parenchyma adjoining the vascular bundles is disrupted (Fig. 2.20*a, b, c*). The damage of the tissues in the flower primordia already to be found in October in the apricot according to Medeira et al. (1991) and it is due to plasmodial organisms. The site of the symptoms and their further spread is the same as has been described by Bubán et al. (1982). The formation of necroses in the flower primordia of the apricot has also been reported by Gülcan and Askin (1990).

2.4. Development of the Sexual Organs of the Flower

2.4.1. Formation of the Stamen and Pollen

The development of the stamen begins with the appearance of the *anther* primordium. The elongation of the filament happens a little before the anthesis. The anther primordium after it appears consists of the epidermis and the primary *archesporium* inside the epidermis. The periclinal division of the cells below the epidermis produces the primary parietal layer. Within this layer is the secondary archesporium. Two further divisions of this kind result in the two intermediate cell layers and the endothecium. The periclinal division of the secondary archesporium cells leads to the tapetum whose cells generally have multi nuclei related to their increased physiological activity (Terzyski and Stefanova 1981). The proteins synthesized by the tapetum along with lipids and carotenoids are injected into the wall of the pollen or are bound to its surface (Heslop-Harrison 1975) and they have a role in fertilization.

The site of pollen formation is in the archesporium which in the walnut is already present during the summer (Terzyski and Stefanova 1981). In stone fruit species formation of archesporium takes place before the end of dormancy at low temperatures. At high temperatures, the cells degenerate (Cociu and Bumbac 1973). In sour cherry and in peach the archesporium may form before the winter, too (Toptsiski and Mihaylov 1975).

In the course of pollen formation, called *microsporogenesis*, the pollen mother cells form by the separation of the archesporium cells (Fig. 2.21*a*) and the double division of mother cells leads to pollen tetrades (Fig. 2.21*b*). With the first meiotic division the chromosome number is halved and the 1 n haploid chromosome number characteristic for the gamete (Fig. 2.21*c*) results.

Microsporogenesis occurs in pear cultivars in the beginning of March or earlier but it takes place when the average temperature over five to seven days is higher than 12 °C (Mittempergher et al. 1965). In peach the threshold value for microsporogenesis is 4–6 °C (Toptsiski and Mihaylov 1975) and this is why microsporogenesis in peach can occur already in January (Cociu and Bumbac 1973). Microsporogenesis of the apricot is three–five days before bud swelling in January or February (Cociu and Bumbac 1973), in the gooseberry four weeks before flowering or in March (Toptsiski and Mihaylov 1975), and in plum toward the end of March (Cociu and Bumbac 1973). As for cherry, the critically high temperature that disturbs meiosis according to Lenander (1962 cit. in Whelan et al 1968) is 17–18 °C.

Meiosis itself in the apricot takes place at the end of February (Surányi 1977*a*), in the walnut in proterandric cultivars at the end of March, in proterogynic cultivars three weeks later (Terzyski and Stefanova 1981), and in the black currant at the time of bud break (Zeller 1968). No correlation was found between the time of flower bud formation or flowering and microsporogenesis. Yet it is well known that after the start of microsporogenesis the frost resistance of the buds decreases significantly. Therefore it is useful to know whether a given peach cultivar is early or late in completing microsporogenesis. For example, 'Springtime' and 'Elberta' completes microsporogenesis early, the 'Redhaven' is moderately early, while 'July Elberta' is late (Toptsiski and Mihaylov 1975).

The pollen becomes liberated with the opening of the theca of the anther, and opening of theca has a diurnal rhythm. Opening in sour cherry takes place slowly at about 9–12 hours. At around 11 hours it is spasmodic and remains intensive until 14 hours (Nyéki

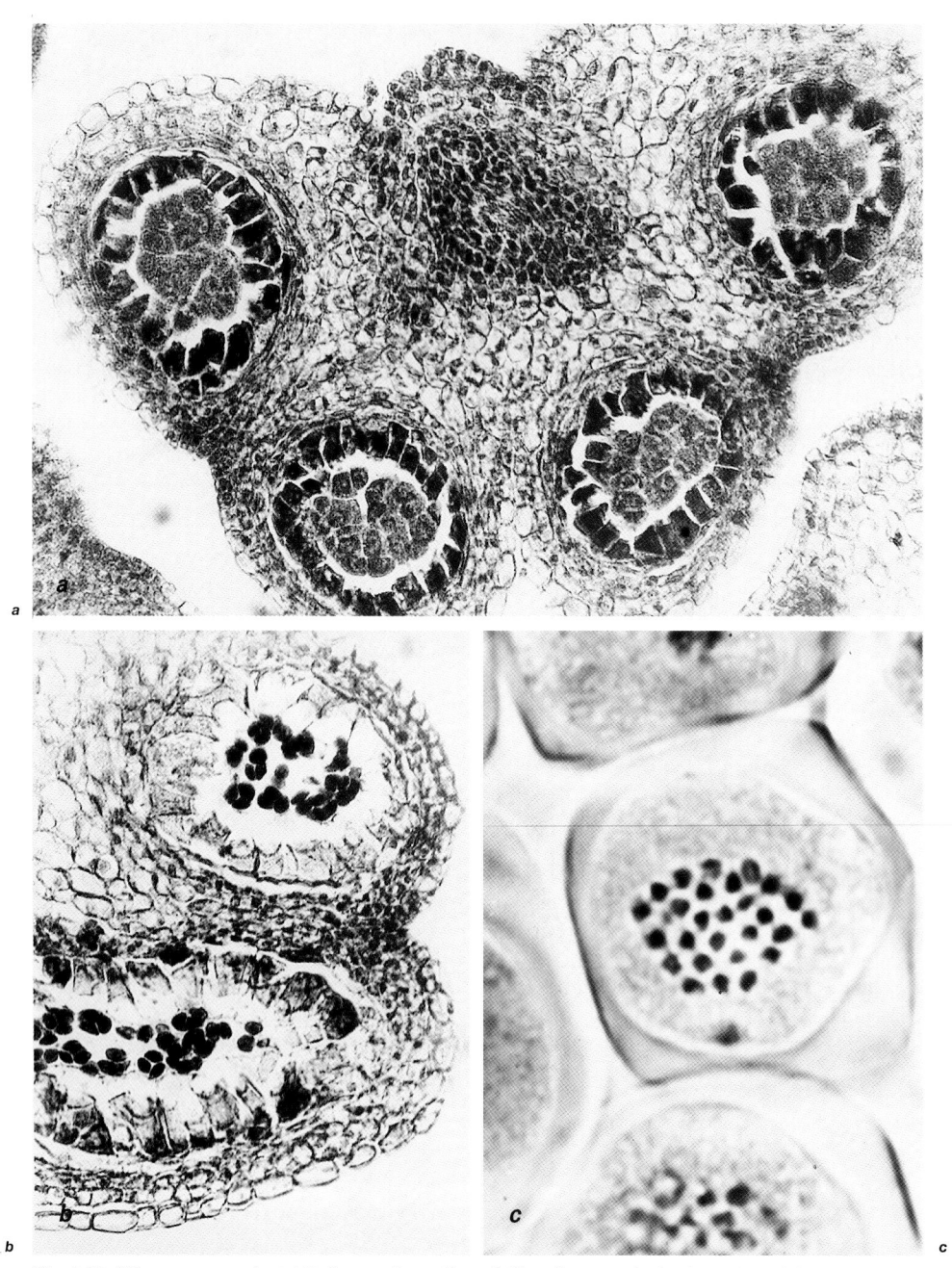

Fig. 2.21. Microsporogenesis. (*a*) Pollen mother cells and (*b*) pollen tetrades in the anther of the apple flower (after Bubán 1980*a*). (*c*) Meiosis in the pollen mother cells of strawberry (after Simon 1980, cit. Bubán 1980*a*)

1976*a*). The *dehiscence* of the anther in the apricot depending on the climate lasts for two-five days and within a given flower from half to two days (Brózik et al. 1978).

As for the sour cherry cultivar 'Pándy', there is a close correlation among dehiscence of anthers, appearance of stigma secretion and periodicity of nectar secretion (Orosz-Kovács et al. 1989). The peaks of pollen shedding usually are one hour before the maximums of nectar production occurring every four hours in flowers of apple trees (Orosz-Kovács 1990).

There may be 400 pollen forming in the anther of a plum cultivar such as 'Victoria', but in apple cultivars pollen grain number may be on average 5000 per anther (Stott et al. 1975, cit. in Bubán 1980*a*). It may happen in pear trees that the anther bursts before the anthesis (Modlibowska 1945), or the pollen germination begins before there is dehiscence of the anther as reported by Koul et al. (1985) in five of 70 apple cultivars and in several almond cultivars.

The *pollen* is a microspore of 1 n chromosome number whose nucleus divides with mitosis providing a larger vegetative and a smaller generative cell. They together constitute the pollen grain, i.e. male gametophyte, or microgametophyte. Then the generative cell divides again before pollen germination. In this way the hitherto bicellular pollen becomes tricellular (Russell 1991). However, the bi- and trinucleate terms are also common. The internal wall of the pollen, the intine, is produced by the haploid microspore (Heslop-Harrison 1975), the external wall, the exine, is composed of sporopollenin, the oxidative polymer of carotenoid and carotenoid esters (Brooks and Shaw 1968 cit. in Ende 1976). The exine of the mature pollen has an internal, nexine, and an external, sexine layer of particular structure. In certain plum cultivars the exine does not evolve. In such cases the young microspore degenerates on account of the splitting of the membrane (Pejkic and Popovic 1973). The shape of the pollen and the design of its surface enable the identification of species or even cultivars (Schwerdtfeger 1978).

In some case the pollen, developed in the lateral flowers of the shoots of the apple cultivars, does not have germination pores or it has less than the usual three (Zeller 1960*c*). There are no germination pores in 6–7% of pollens of the 'Starkrimson Delicious' apple cultivar but lack of germination pores can reach 16% in the cultivar 'Golden Delicious'. Another difference is that the G_1 phase is characteristic of the generative cell nucleus of the 'Starkrimson Delicious' pollen (DNA content/nucleus 1C, being the DNA level just after the second meiosis), while the G_2 phase is characteristic of the 'Golden Delicious' pollen (Avanzi et al. 1980). The pollen of 'Starkrimson Delicious' has well-developed mitochondria and a striking endoplasmic reticulum, this latter in the pollen of the 'Golden Delicious' is less developed while the dictyosomes are conspicuous (Beilani and Bell 1986). The size of the pollen in diploid apple cultivars is 20–40 microns and the percent of germination is usually 70–90%. The size of the pollen of the triploid apple cultivars shows a wide range of variation and the germination is a less than 25%, more often only 5–10% (Stott 1972).

The hormones in the pollen play a role in the *pollen tube growth* and/or in the *germination* capacity of the mature pollen. The important hormones are e.g. indoleacetic acid and the gibberellins (Vasil 1974). The hydrolytic enzymes such as basic phosphatase, ribonuclease, esterase and amylase are located in the intine, mostly under the germination pore (Knox and Heslop-Harrison 1969, cit. in Kapil and Bhatnagar 1975). There is more amino acid in the pollen and in the pistil in apple inflorescences consisting of more flowers comparing to those with less flowers (Ohno et al. 1961). Pejkic (1973) found higher nucleic acid and protein contents in the mature pollen of the sour cherry than in the earlier phases including meiosis, and the tetrade stage. The most important reserve of the

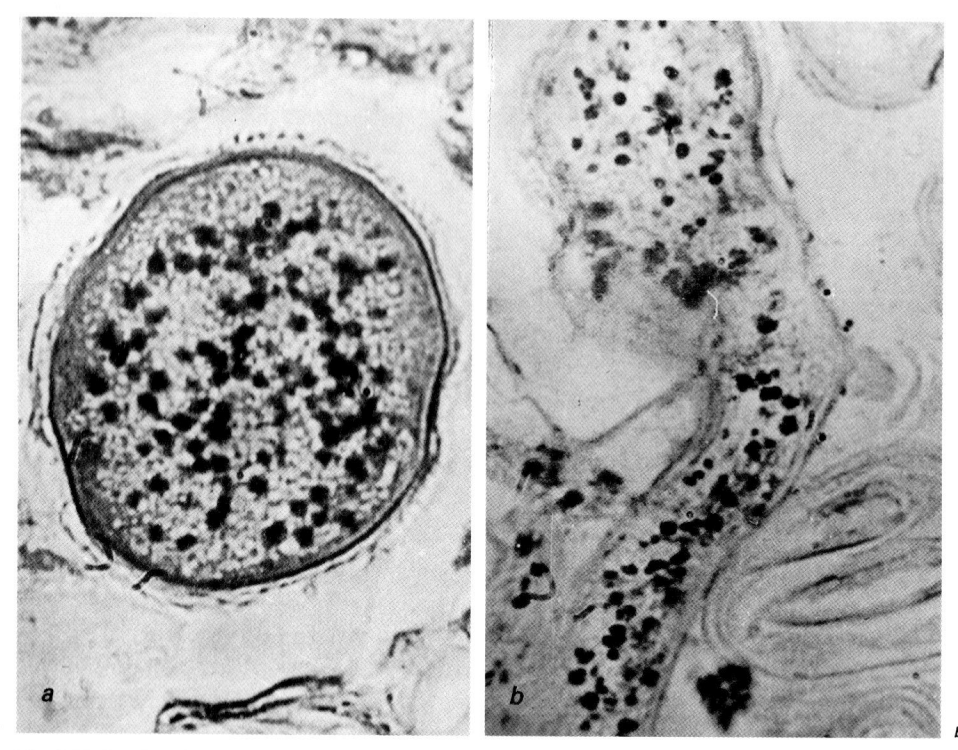

Fig. 2.22. Starch grains stored in the sour cherry pollen. (*a*) Stored grains in the pollen; (*b*) after pollen germination starch grains migrate into the pollen tube (after Stösser 1980*a*)

cherry and sour cherry pollen is starch which moves into the pollen tube after pollen germination (Fig. 2.22*a, b*; Stösser 1980*a*). The reserves of the pollen are only sufficient for the initial growth of the pollen tube (Braun et al. 1986). Later growth is ensured by the intercellular materials of the transmitting tissues of the style, which are rich in carbohydrates.

At least one of the basic causes of male sterility and abnormal pollen development is the presence of lethal pollen factors, e.g. in crab apple, wild pear, or in 15 diploid cultivars among 300 apple cultivars examined (Linder 1974) and in quince (Pejkic and Dokic 1968). In plum cultivars male sterility may be of biochemical physiological origin (Pejkic 1968, Pejkic and Popovic 1973), or may be caused by the overdevelopment of the tapetum layer to the detriment of the sporogenic tissues (Cociu and Bumbac 1968). It has also been observed that during the separation of the tetrades their plasma contents are resorbed (Cociu and Bumbac 1973). As for male sterility in the apricot, the faults are in microsporogenesis, i.e., the insufficiency of post-meiotic tetrade cytokinesis, irregular exine formation and abnormal tapetum development (Medeira and Guedes 1989; Fig. 2.23*a, b, c*) are determining factors in causing male sterility. According to a more recent finding (Medeira and Guedes 1990), during the course of microsporogenesis protein synthesis is prematurely blocked and there is a rapid decrease in esterase enzyme level following the second meiosis. This is probably related to the too early degeneration of the tapetum.

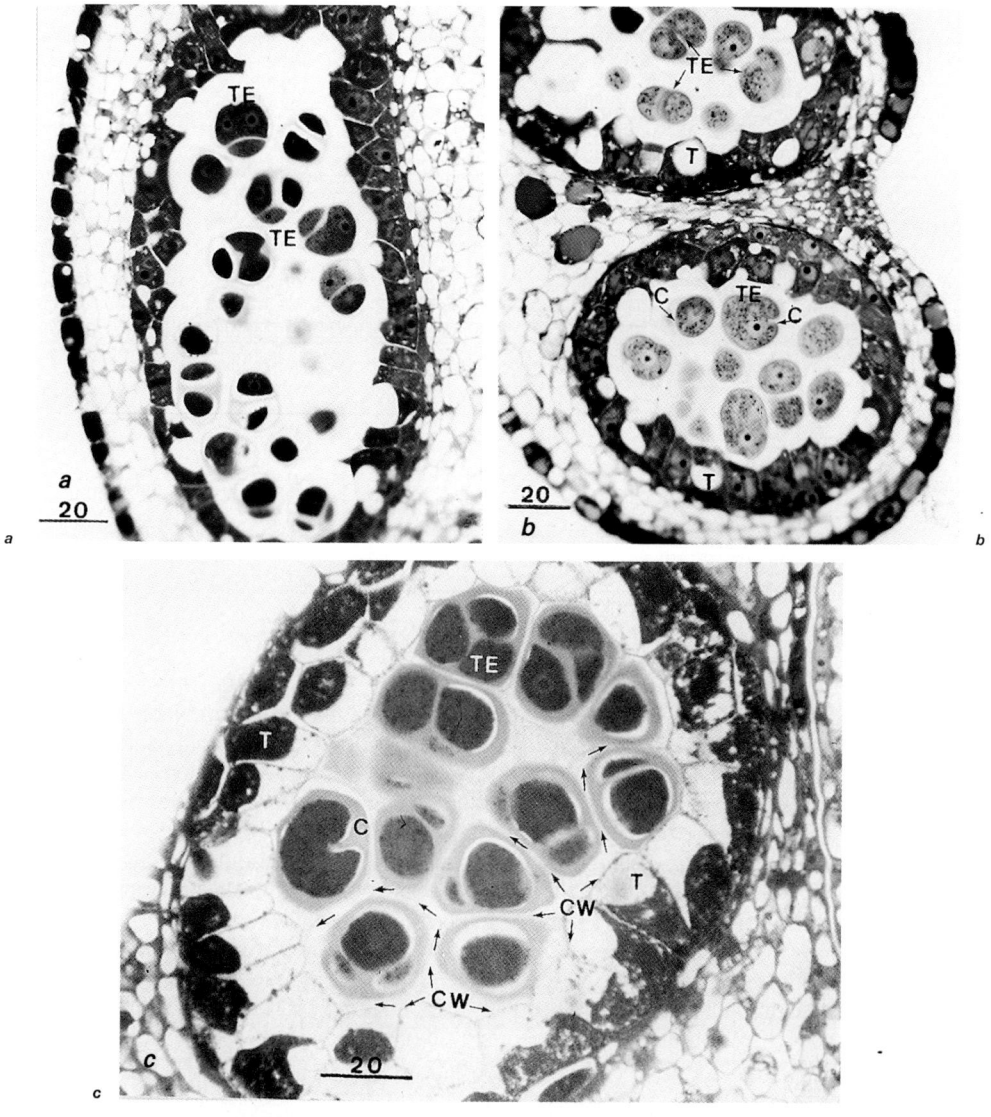

Fig. 2.23. Phenomena accompanying the partial male sterility of the apricot. (*a*) The variability of pollen tetrade formation; (*b*) the tapetum cells are vacuolized, the tetrade cytokinesis is incomplete or does not occur; (*c*) degeneration of the tapetum and tetrade cells within their cell walls. C = callose; CW = cell wall; T = tapetum; TE = tetrade (after Medeira and Guedes 1989)

A peculiar case of male sterility in 'Spencer's seedless' apple cultivar cannot be attributed to the above causes. In some apple cultivars, including 'Spencer's seedless' (Fig. 2.24), the flowers have no petals and along the numerous styles only remnants of the anther can be found. After appropriate pollination we induced fruit with normal seeds.

37

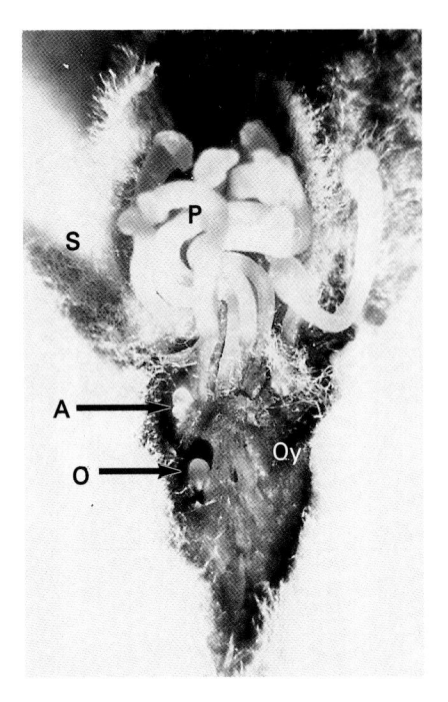

Fig. 2.24. Flower of the 'Spencer's seedless' apple cultivar. A = remnants of anther; O = ovule; Oy = ovary; P = pistil; S = sepal (Bubán 1978, unpubl.)

Chemicals of plant protection in fruit growing practice may cause problems in pollen germination and pollen tube growth. In studying the effect of pesticides only suitable and advised methods (Church et al. 1983a) will give realistic results. According to the traditional *in vitro* method, such as pollen germination on agar medium, fungicides mixed into agar drastically inhibit the pollen germination and pollen tube growth in apple and pear, but they inhibit pollen tube growth only if the fungicide is sprayed onto the agar surface (Marcucci and Filiti 1984). Despite the negative results of *in vitro* examinations, the *in vivo* testing in the orchard showed that the majority of fungicides did not influence the germination capacity of the pollen released from anthers opening two hours after spraying and did not influence the fruit set of the apple. In peach, however, it is advisable to be cautious (Manandhar and Lawes 1980). In any case there is less risk when instead of the high volume spray low volume techniques are used (Church et al. 1983b).

2.4.2. Structure of the Pistil

The pistil is that part of the flower in which the primary function of the flower, namely the fertilization, is realized. The "floral fertility" is represented by the maximum fruit setting percentage (Williams 1970a). Still it is not at all contradictory that, from a practical point of view, it is not necessary that fertilization should take place in all flowers of the tree. In apple trees 5% fruit set is sufficient (Modlibowska 1945) for a commercial crop. However, it may be problem when 1–49% of the flowers are abnormal on peach trees. It means that the pistil may be malformed, there may be necroses on other parts of the flowers and only 8–40% of apparently normal pistils are receptive (Crossa-Raynaud et al. 1984, Martinez-Tellez et al. 1982). The number of abnormalities is small in apricot and this is the reason why the statement could be made in reference to apricot: "No correlation was found between the aborted ovaries and productivity" (Quarta and Bunalti 1983). There are numerous comprehensive works on this subject (Nyéki 1989).

The generally known components of the pistil, gynoecium, collectively the stigma, style and ovary, are genetically determined with respect to morphology and size. In addition, the morphology also has functional significance. For example, in sour cherry the wide ovary influences fruit set favourably, while the long style detrimentally (Nyéki and Tóth 1976).

On the surface of the stigma there are papillae serving to receive the pollen. Their surface is covered with a layer of protein, so-called pellicle (Ende 1976). Cuticula cannot be found at least on the unicellular papilla of cherry flowers (Stösser 1980a). The papilla of the apple flowers one–two days after anthesis lose their turgidity and collapse, but this does not affect pollen germination (Braun 1984; Fig. 2.25a, b). The degradation of the papilla in pear and in stone fruit species is likewise rapid and accelerates after pollination (Cresti et al. 1985, Braun and Stösser 1985). According to Orosz-Kovács et al. (1992),

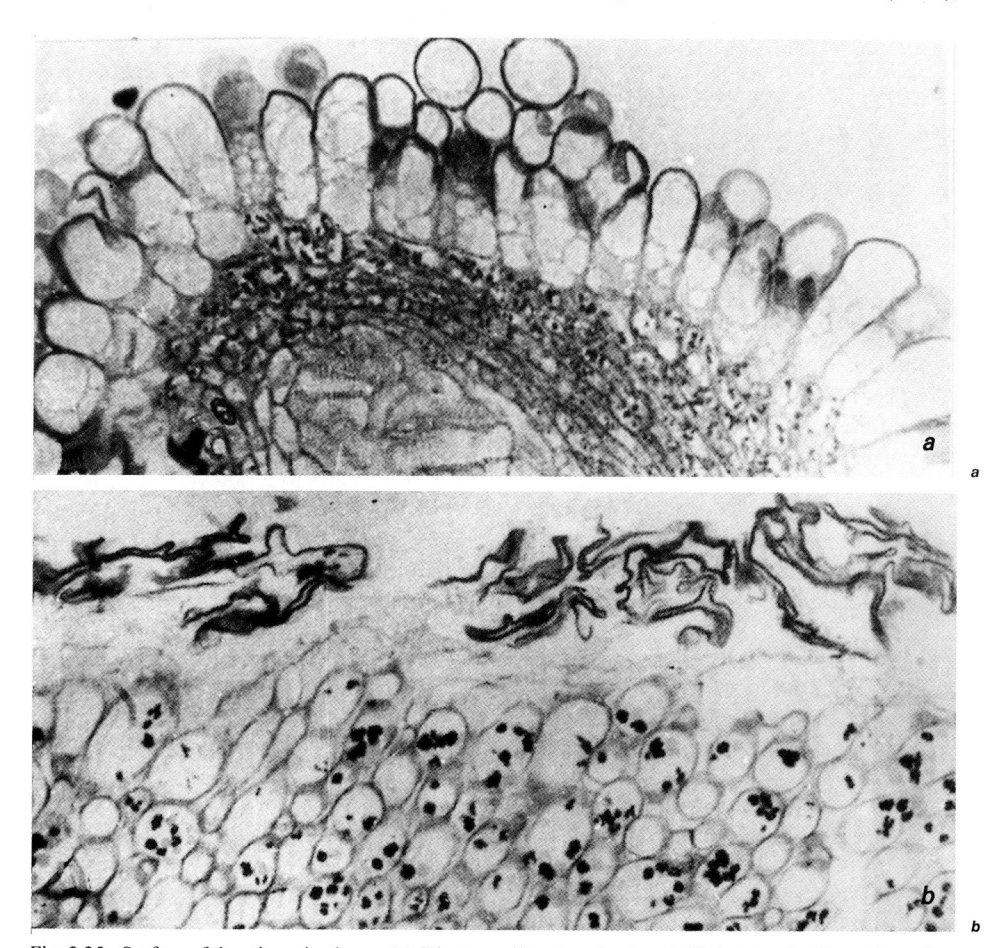

Fig. 2.25. Surface of the stigma in cherry. (a) Stigma papillae at anthesis and (b) their total collapse three days later (after Stösser and Anvari 1983)

the rapid papilla degradation is a process associated with dichogamy in sour cherry flowers. It may be mentioned here that the number of the papillae, their size, and the amount of style secretion, depending on the gametophytic or sporophytic type of the incompatibility system all may vary from species to species (Heslop-Harrison 1975).

The moist stigma, producing abundant secretion, is characteristic of the fruit species belonging to the Pomoideae and Prunoideae subfamilies. The appearance of the secretion indicates the receptivity of the stigma. The exudate is often emulsified with lipid droplets and the viscosity is ensured by the mucopolysaccharides. The secretion contains also sugars, phenolic compounds, free amino acids and various enzymes (Heslop-Harrison 1975, Kapil and Bhatnagar 1975). In sour cherry, flowers have a daily rhythm of the secretion activity. The maximum secretion occurs at eight–ten hours, there is no exudate production between 13–15 hours, but around 16 hours exudate is again produced (Nyéki 1976a). It has recently been reported (Orosz-Kovács et al. 1992) that stigmatic secretion activity has a rhythm of 12 hours in the dichogamous flowers of sour cherry. However, there is a rhythm of six hours in the homogamous ones. Both types of rhythms coincide with the daily rhythm of nectar production. In apricot, depending on the cultivar, there is secretion for one–four days but this only means half a day within any given flower (Brózik et al. 1978). The secretion may appear on the stigma before anthesis. Early secretion occurs in almond in traces and in apricot abundantly but in other fruit species it does not occur (Cresti et al. 1985). There is more secretion on the stigma of the gooseberry flower at full bloom than at the time of anthesis (Jefferies et al. 1982a).

The style in dicotyledonous plants is generally closed, in the monocotyledonous ones it is open because they have a style canal (Vasil 1974). The length of the style in the unopened bud of the pear is only half to one–third of its size at bloom (Hiratsuka et al. 1985). The size for the apricot is 17.5 mm on the average but flowers of those cultivars have the highest number of functioning stamen in which the pistil is the shortest (Surányi 1977a). The 12.8 mm long style of the almond after pollination grows to 17.2–21.8 mm in 120 hours (Godini 1981). Twin pistils may occur in peach and in almond flowers. In flowers of the *Amygdalo persica* hybrid there may even be 2–16 pistils and these after appropriate pollination produce three–eight fruits(!) from a single flower, according to Surányi (1978b).

The internal zone of the style which connects the stigma and the ovary is the conducting tissue which may be called stigmatoid or connecting tissue (Ende 1976) but most often designated as transmitting tissue (Stösser and Anvari 1983). It was a finding of historical importance when Amici (1824) (cit. in Vasil 1974) postulated that the transmitting tissues have a nutrient role when the pollen tube grows through the style. The transmitting tissue of the style in the apple flower is composed of cells and intercellular materials. The cells are abundant in cytoplasm, have numerous large endoplasmic reticulum cisterns, and are characterized by large Golgi vesicles rich in carbohydrates. The intercellular materials are probably the secretions of these cells, mainly carbohydrates and proteins, with lipids and pectin hardly present (Cresti et al. 1980; Fig. 2.26a, b). Saccharose is to be found in the highest proportion in the style of almost all apple cultivars, with fructose and glucose also present, while sorbitol is more characteristic of the pear (Braun 1984). The presence of starch is not characteristic of pome fruits (Braun 1984, Braun et al. 1986). It also accumulates, however, in the transmitting tissues of the style of the cherry. In cherry, it is in the highest amount during anthesis and it degrades within four–six days (Stösser and Neubeller 1980; Fig. 2.27a, b). This process is followed by the disintegration of the cells of the transmitting tissues (Braun and Stösser 1985; Fig. 2.28a, b) which, just as the degradation of inter-

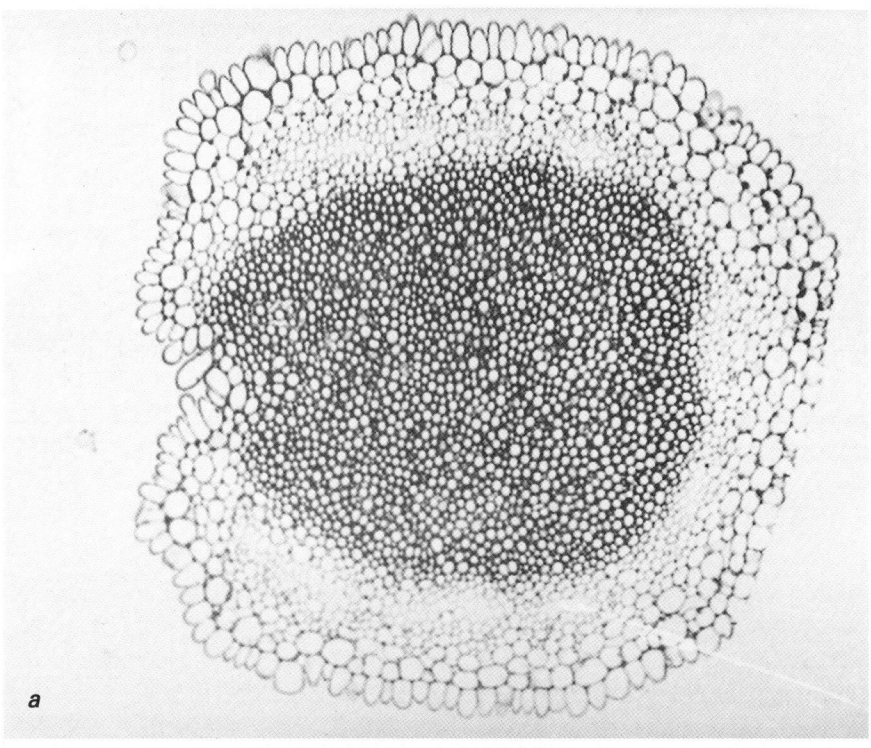

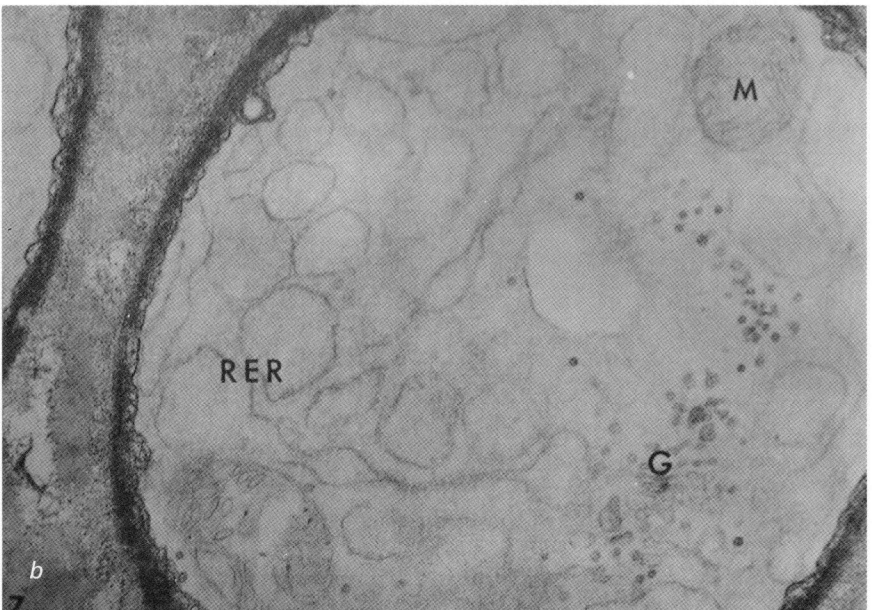

Fig. 2.26. Polysaccharides in styles of apple. (*a*) Staining proving the presence of polysaccharides in the intercellular material of the transmitting tissues in the style of the apple; (*b*) the more intensive reaction of cell walls. G = Golgi bodies; M = mitochondria; RER = rough endoplasmic reticulum (after Cresti et al. 1980)

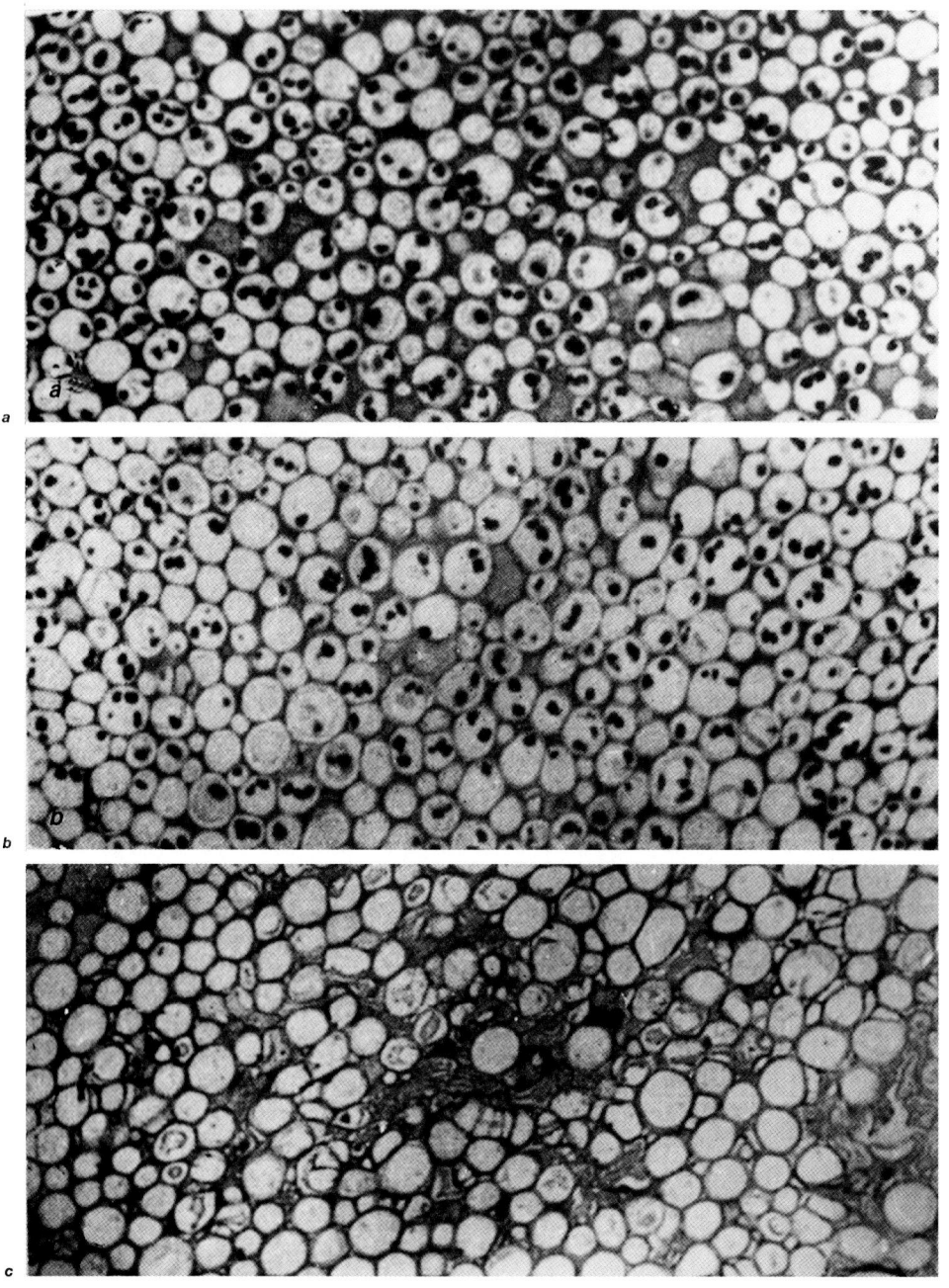

Fig. 2.27. Starch grains in the transmitting tissues of the style in a cherry flower (*a*) at anthesis, then (*b, c*) three and ten days later (after Stösser and Anvari 1983)

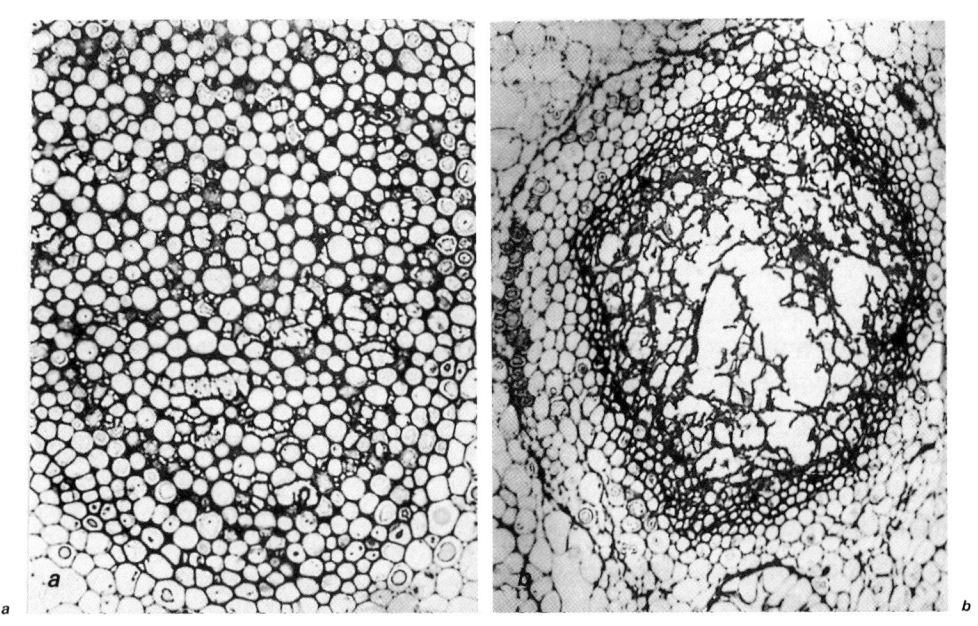

Fig. 2.28. Disintegration of the transmitting tissues of the style of the apple. (*a*) Initial and (*b*) total disintegration one and five days after pollination, respectively (after Braun and Stösser 1985)

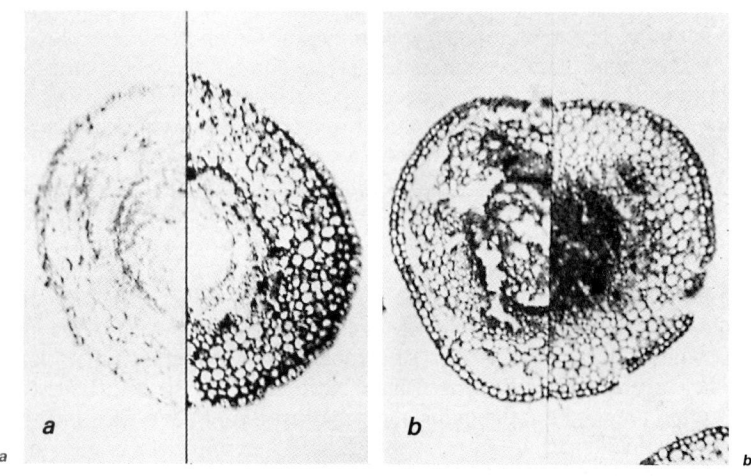

Fig. 2.29. Activity of the enzymes in the style of the cherry. (*a*) Peroxidase activity after heat inactivation (left half) and intact activity (right half); (*b*) acid phosphatase activity at anthesis (right half) and 17 days later (left half) (after Stösser 1983*a*)

cellular materials, is faster after pollination than in unpollinated styles. The first investigation of the peroxidase and acidic phosphatase enzyme's activity in fruit trees was carried out by Stösser (1983*a*). He found a parallelism between acidic phosphatase enzyme activities of the style (Fig. 2.29*a, b*) and the degradation of the transmitting tissues. Furthermore, in the pistil of oriental (Japanese) pear following cross-pollination the amount of soluble proteins temporarily declines but after 96 hours soluble proteins are more than 50% higher than after self-pollination (Hiratsuka and Tezuka 1980).

We have already discussed the size of the style but we also have to note that from the number of the stamen and the total length of the pistil a quotient can be formed (piece/mm = Q) that "is suitable for characterizing the extent of self-fertility", i.e. "the value of the quotient is the numerical expression of sexual correlation" (Surányi 1978*a*). The smaller this quotient in the flowers of apricot, cherry, sour cherry, plum, peach and almond cultivar the greater the self-fertilization capacity (Surányi 1977*a, b*; 1978*a, b*).

As opposed to the majority of fruit species, the walnut trees being monoecious plants have unisexual flowers. The bracts, the two bracteoles and the four sepals of the female inflorescence are grown together almost entirely in their length with the ovary. Since the style in walnut is very short the separation of certain flower parts is hardly possible (Sartorius 1990). The female flowers, are solitary or consist of two to five flowers varying with *Juglans* species (Sartorius et al. 1984).

The most important parts of the ovary are the ovule and the embryo sac forming within the ovule. These will be discussed in a separate chapter.

2.4.3. Development of the Ovule and Its Longevity

The ovules are located on the placenta, in stone fruit trees their number is two, in pome fruits ten, and in *Ribes* species the multiple of this. The female flower in the walnut has one ovule. The apple ovules appear first at the end of winter, or early spring (Horavka 1961) as undifferentiated meristem protrusion but the whole structure only forms before anthesis (Fig. 2.30*a, b*). The nucellus (macrosporangium) is larger in the terminal than in the lateral flowers of the apple inflorescences. The nucellus is also larger in trees on M9 rootstock than on seedling rootstocks and this is favourable from the point of view of fruit set (Marro and Lalatta 1978, Marro 1976).

Although, this is related to stone fruit species (Schauz 1989) it is relevant that seven to ten cell rows of the basal part of the nucellus have a role in the transport of the assimilates and obviously supplying the embryo with nutrients. There is a relationship between the development of the nucellus and the development of the vascular bundles in the funiculus. The apple cultivars producing weak fruit set have lesser developed vascular bundles (Lalatta et al. 1978*a*, Deveronico and Marro 1982) than those setting more fruit. The early abortion of young fruits is attributable to the insufficient vascular bundle connection between the funiculus and nucellus (Simons and Chu 1968). The ovules in flowers forming in axillary position on the shoots of apple are smaller and the integuments do not cover the nucellus (Zeller 1960*c*). These flowers are less fertilizable. It is interesting that in the flower of the 'Lodi' apple cultivar there are twice as many ovules and so 20 seeds may develop in a fruit (Simons 1974).

In the ovary of stone fruit species there are always two ovules but only one is capable of fertilization. The ovule capable of fertilization is called the primary or functional ovule, the secondary ovule is mostly underdeveloped (Eaton 1959*b*). In the pistil of 'Windsor' cherry cultivar twin ovules occur in 5–12% of cases, in such cases two nucelli become differentiated in the ovule primordium which may have joint integuments or both may have them individually. In such case there are two micropyles (Eaton 1959*a*) as has been reported also in the plum (Stösser and Hartmann 1982).

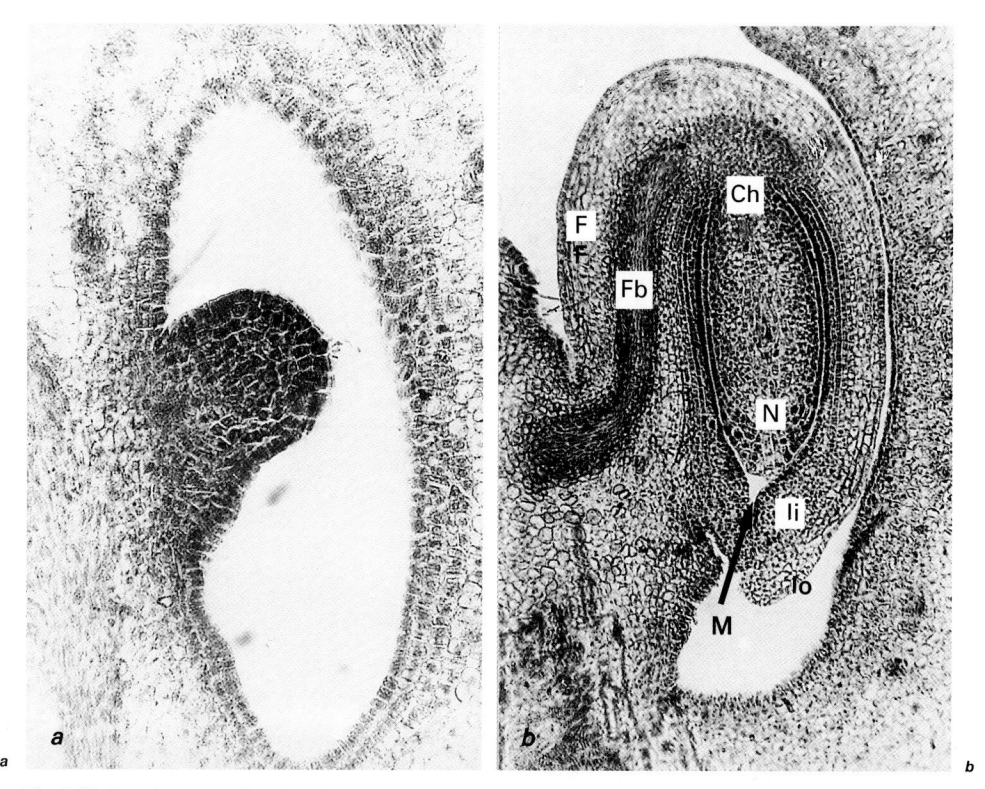

Fig. 2.30. Development of ovule primordium. (*a*) At the end of winter; and (*b*) before anthesis in apple. Ch = chalaza; F = funiculus; Fb = funicular bundle; Io an Ii = external and internal integuments; M = micropyle; N = nucellus (Bubán 1979, unpubl.)

In the ovule of the walnut—unlike most fruit species—there is only one integument surrounding the nucellus (Fig. 2.31; Sartorius 1990). Next to the ovule there is another formation which can be considered as the rudimentary external integument, but rather it is accepted as an outgrowth from the placental axis, and referred to as evagination (Catlin and Polito 1989). With respect to other fruit species, the obturator is a similar peculiar morphological structure which in the sour cherry ovary grows out of the placenta (Anvari and Stösser 1978*b*) but it may derive from the funiculus or from the integument also (Kapil and Bhatnagar 1975). The obturators guide the pollen tube (Marro 1982). In the sour cherry the pollen tube grows in the narrow canal between the obturators towards the micropyle (Anvari and Stösser 1978*b*).

The most decisive factor in realizing fertilization is the *longevity of the ovule*. The pollen germination and the growth of the tube to the base of the style may occur late even during the second week following anthesis when the transmitting tissues and papilla of the stigma have long been degraded (Stösser and Anvari 1983). However, this late there is hardly a chance of fertilization. Therefore, in a "strong flower", according to Williams (1965), the receptivity of the stigma, growth of the embryo sac and cell division in unfertilized ovules take longer time. All these conduce to the result: the fertilization ability of the ovule is twice as long as in the normal flower. In the apple tree ovule longevity

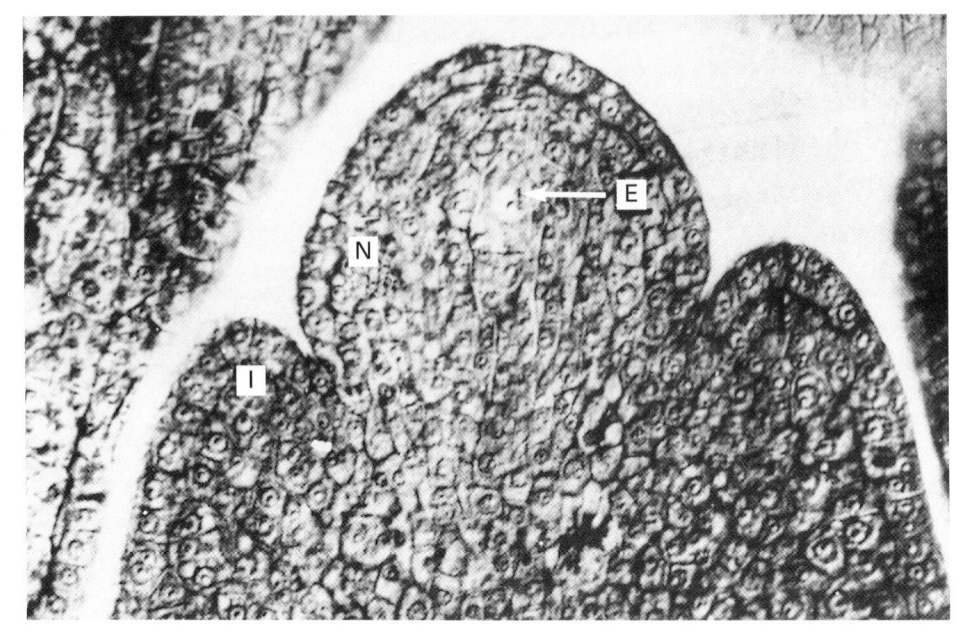

Fig. 2.31. Ovule of the walnut nine days before anthesis. E = mother cell of embryo sac; I = integument primordium; N = nucellus (after Sartorius 1990)

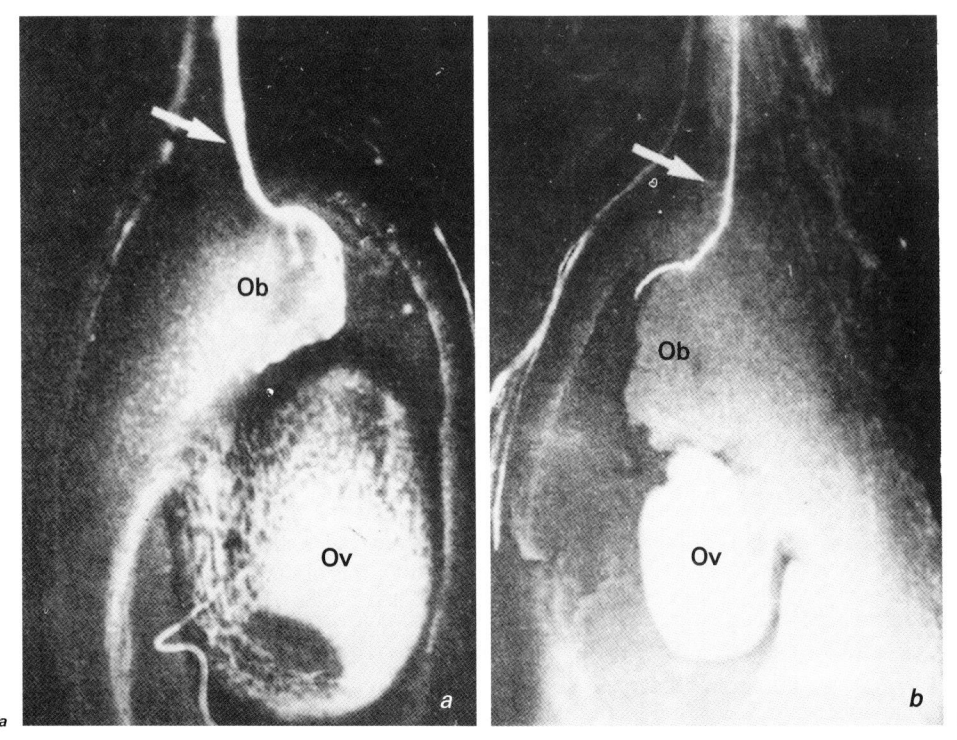

Fig. 2.32. Growth of pollen tube in cherry (arrow). Growth in the ovary of the cherry flower along the obturator towards the ovule. (*a*) Viable ovule shows little fluorescence; (*b*) the non-viable, senescent, ovule shows intensive fluorescence. Ob = obturator; Ov = ovule (Stösser and Anvari 1983)

may be 11–12 days at 11 °C average temperature (Child 1967). Functioning of pistils in sour cherry exists for two to three days (Nyéki 1976*a*) but by the time of anthesis 2–25% of the ovules may be degenerated (Bartz and Stösser 1989). Depending on the cultivars, ovules show fluorescence indicating inactivation three to six days after anthesis (Stösser and Anvari 1978; Fig. 2.32*a, b*). The ovules of cherry flowers, depending on the cultivar and the temperature, are viable for one to five days (Postweiler et al. 1985) or even 13 days (Guerrero-Prieto et al. 1985). The receptivity of the ovule in peach flowers drops quickly but the decline is slower in the male sterile cultivars than in the androfertile cultivars. This difference was also observed in *Prunus mahaleb* and *P. cerasifera* (Crossa-Raynaud et al. 1984).

2.4.4. Organization of the Embryo Sac

The embryo sac is the female gametophyte mentioned also as ginogamete or ovicells (Bagni and Gerola 1978). Its formation—the macrosporogenesis—begins with the appearance of the mother cell of the embryo sac, or mother cell of macrospore which is larger in size and is located in the middle or apical part of the nucellus (Fig. 2.33*a*). The mother cell of the embryo sac is the result of two steps of meiosis producing a haploid generation, the tetrade cells, which are located linearly or quadrantly. In the triploid *apple cultivars* the quadrant arrangement is more common (Abou-El-Nasr and Stösser 1989; Fig. 2.33*b, c, d*). The tetrade cell nearest to the micropyle enlarges, this is the mononucleated, or primary, embryo sac and the other three tetrade cells degenerate. After the first division of the nucleus of the primary embryo sac, followed by further enlargement and further division, the embryo sac of eight nuclei forms (Fig. 2.34*a*). In diploid apple cultivars the embryo sac of four and eight nuclei, respectively forms two days before anthesis mainly in the terminal flowers of the inflorescences. In triploid apple cultivars such embryo sac forms only at anthesis or one day later.

At the time of the anthesis of the *pear*, Lalatta and co-workers (1978*a*) observed still an early stage of tetrade cells. Pejkic and Jovovic (1968) reported that at anthesis the frequency of the embryo sac of eight nuclei is 63 and 82% in trees on seedling and quince rootstock, respectively. In *sour cherry* trees two to three days before anthesis there is only a mononucleate primary embryo sac but by flowering the complete structure of eight nuclei embryo sacs evolve (Anvari and Stösser 1978*b*). In *apricot* flowers the embryo sac may develop from any of the tetrade cells, therefore, several embryo sacs of various genetic constitution may be present in the ovule (Eaton and Jamont 1965). According to Martinez-Tellez and co-workers (1982), ten days before flowering of the *peach* there are only macrospore mother cells and the development of the embryo sac is completed only after anthesis. Others (Crossa-Raynaud et al. 1984) have come to the same conclusion.

Altogether seven species of the American blackberry (*Rubus* sp.) were studied by Pratt and Einset (1955) with respect to macrosporogenesis. In the nucellus of young ovule the subepidermal cells with periclinal division produce parietal (external) and sporogenous cells. The sporogenous cells will form the megaspore mother cells of which one becomes organized after meiosis into the embryo sac (Fig. 2.35*a, b, c*). The development of the embryo sac and the reduction of the number of the chromosomes occur regularly in *Rubus allegheniensis* (diploid) but meiosis and the reduced embryo sac is much less common in polyploid species. The sexual reproduction of the *Rubus spectabilis* described by Virdi and Eaton (1969) agrees with that of *R. allegheniensis* described above.

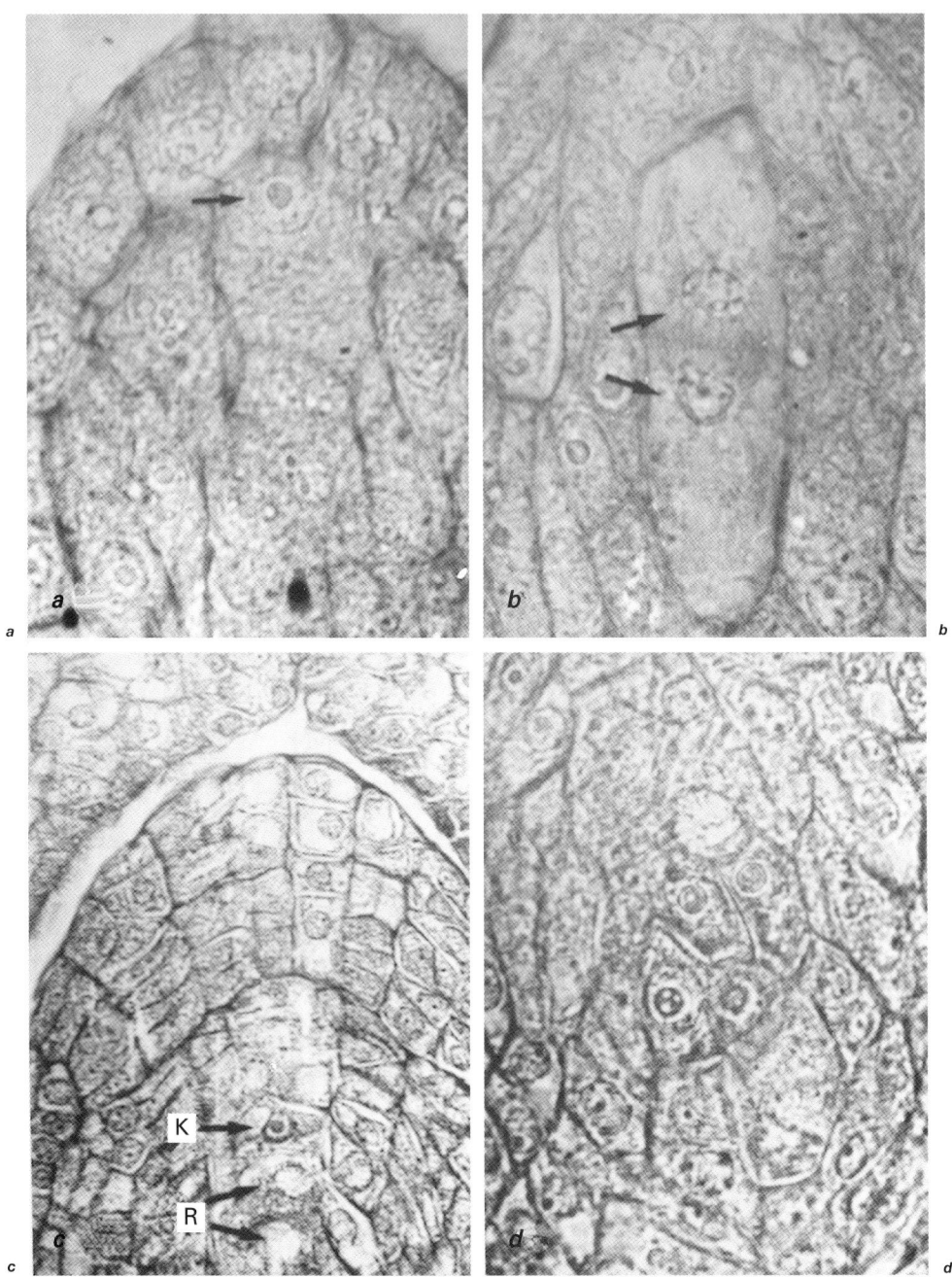

Fig. 2.33. Macrosporogenesis in the flowers of apple two weeks before anthesis. (*a*) Macrospore mother cell in the nucellus of the Boskoop (triploid) cultivar; (*b*) nucellus of two cells in the ovule of the 'Jonagold' (triploid) cultivar (arrow); (*c*) the linear arrangement of macrospore tetrades and (*d*) quadratic arrangement in 'Cox Orange' (diploid) cultivar. K = macrospore cell nucleus; R = remnants of degenerating tetrade cells (after Abou-El-Nasr and Stösser 1989)

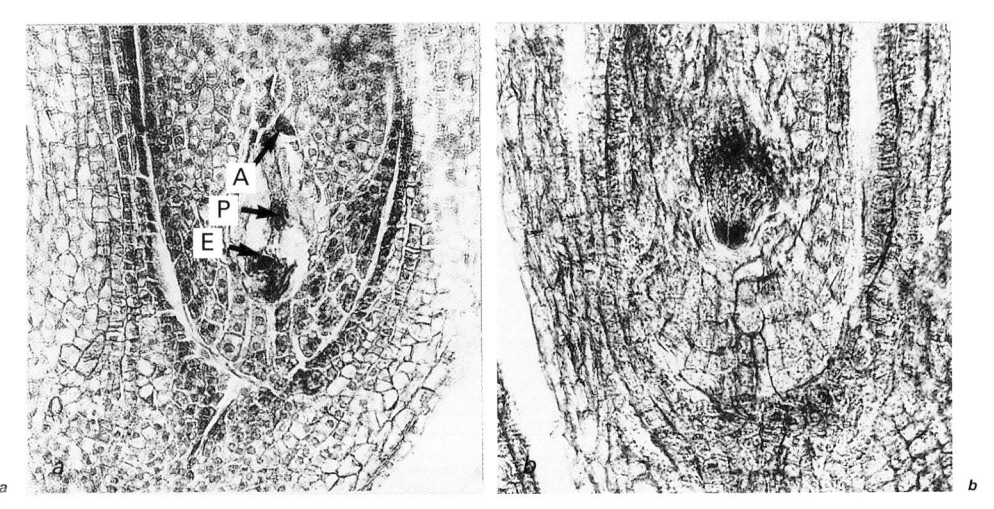

Fig. 2.34. (*a*) Normal and (*b*) twin embryo sacs with proliferating nuclei in the nucellus of the apple flower. A = antipodal cell nuclei; E = egg cell apparatus; P = polar cell nuclei (Bubán 1979, unpubl.)

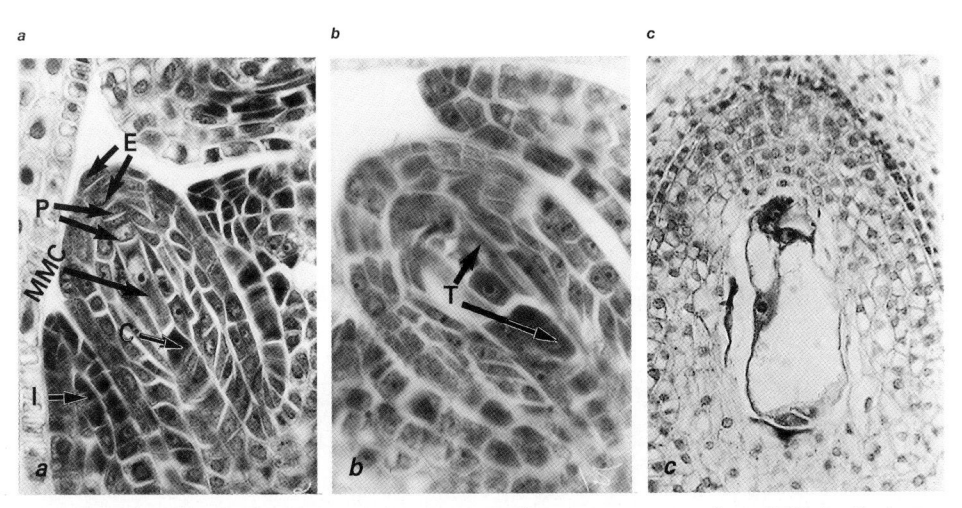

Fig. 2.35. Macrosporogenesis in *Rubus* species. (*a*) *R. canadiensis*, macrospore mother cell; (*b*) *R. allegheniensis,* macrosporogenesis; (*c*) *R. canadiensis*, embryo sac; egg cell, two synergids, one of both polar nuclei and three antipodal cell nuclei. C = chalaza cell; I = integument; MMC = macrospore mother cell; P = parietal cells; T = macrospore tetrade (after Pratt and Einset 1955)

The formation of the embryo sac mother cell in the ovule of the *black currant* takes place three to five days before anthesis (Zeller 1968; Fig. 2.36).

In the female flower of the *walnut* 20 days before anthesis there is only a primordium of the nucellus (Sartorius 1990) but ten days later the mother cell of the embryo sac can be recognized (Fig. 2.37), from which a linear tetrade will be formed (Fig. 2.38*a, b*). Of the tetrade cells always the nearest towards the chalaza (the lowest in the figure) will be embryo sac. It is considered an irregularity when two cells further develop among the tet-

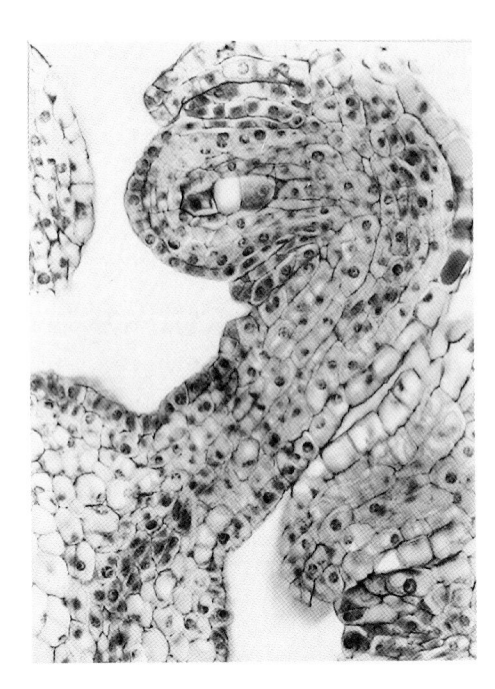

Fig. 2.36. Formation of the embryo sac mother cell in the black currant three to five days before anthesis (after Zeller 1968)

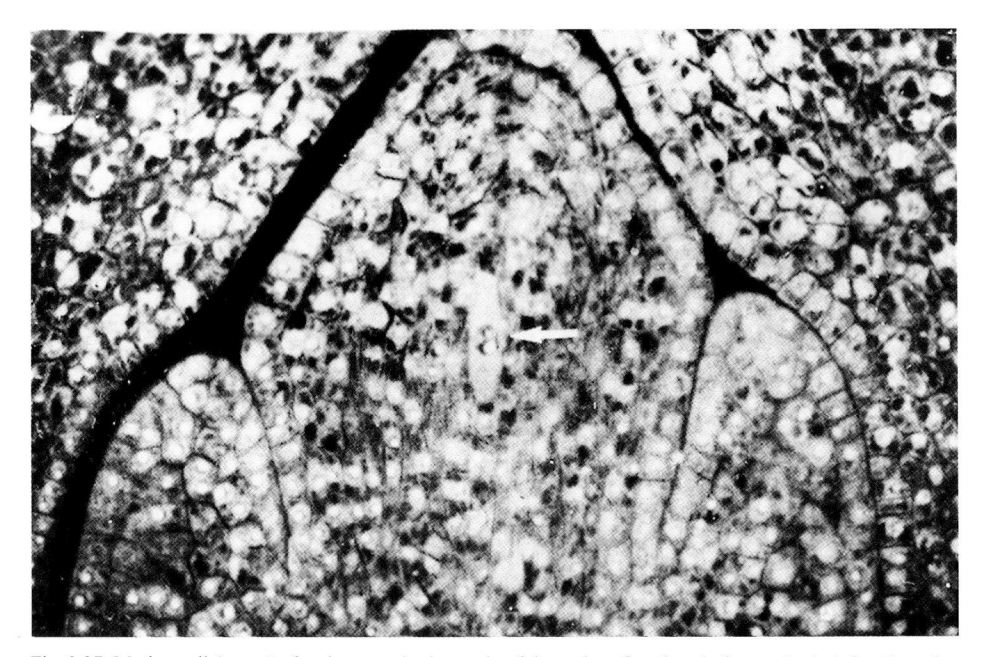

Fig. 2.37. Mother cell (arrow) of embryo sac in the ovule of the walnut five days before anthesis (after Sartorius 1990)

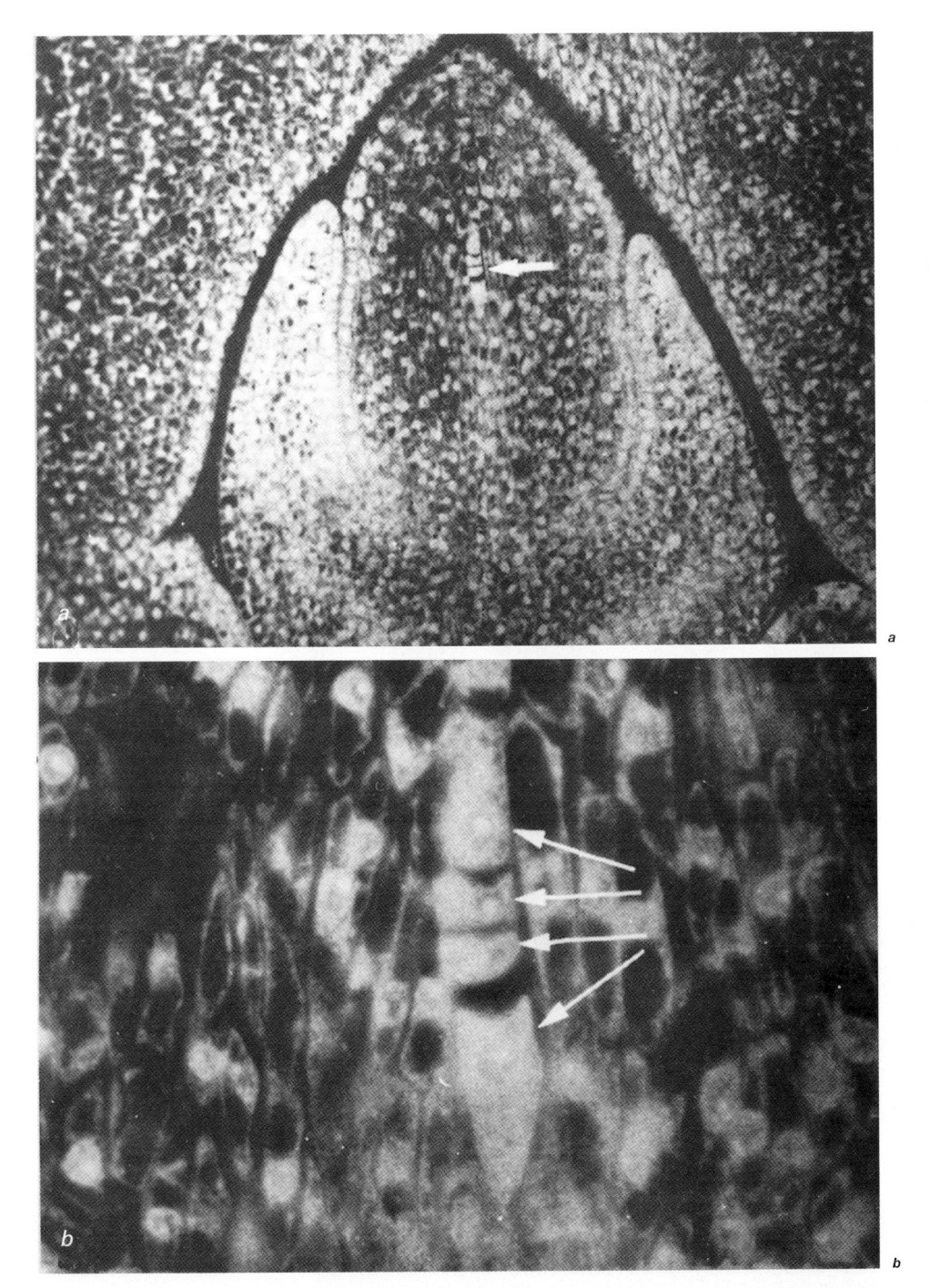

Fig. 2.38. Linear tetrade (arrows) in the nucellus of walnut three days before anthesis. (*a*) View of ovule; (*b*) linear arrangement of four cells of the tetrade (after Sartorius 1990)

51

Fig. 2.39. Two macrospores (arrows) existing in one nucellus of walnut at anthesis (after Sartorius 1990)

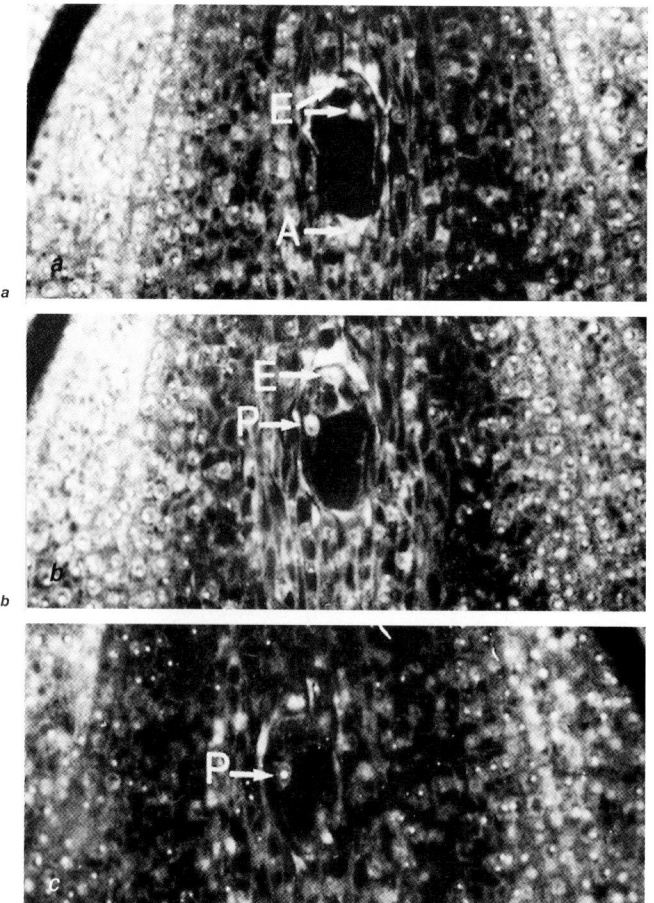

Fig. 2.40. Arrangement of eight nuclei in the embryo sac of the walnut nine days after anthesis (demonstrated by three sections: *a, b, c*). A = antipodals; E = nucleus of the egg cell; P = polar nuclei (after Sartorius 1990)

Fig. 2.41. Embryo sac in the nucellus of the ovule of the sour cherry. (*a*) Premature and (*b*) mature (after Furukawa and Bukovac 1989)

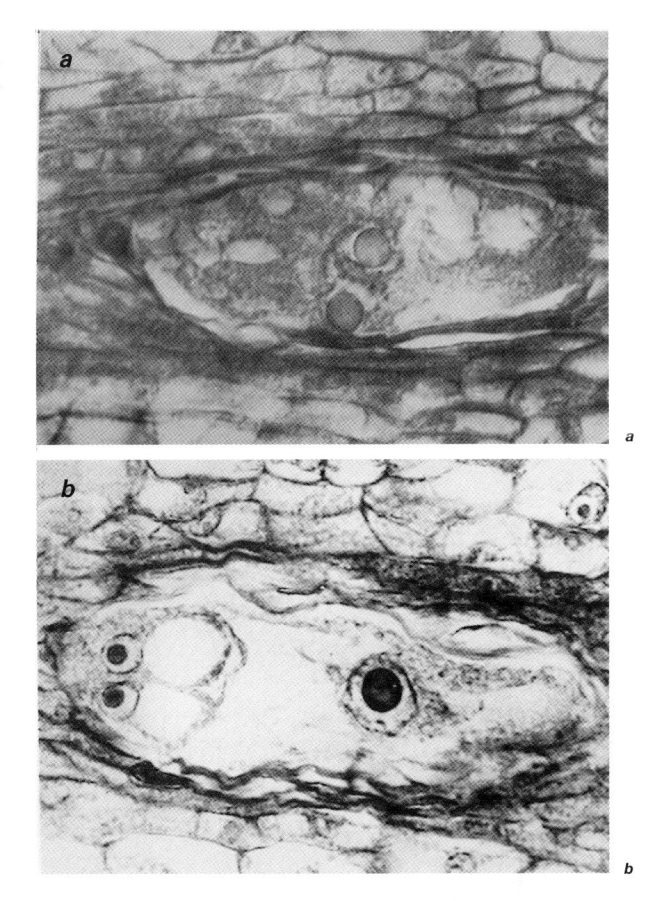

rade cells (Fig. 2.39). At anthesis there is only a binucleate embryo sac that completes its development in the next six to nine days (Fig. 2.40*a, b, c*).

The well-known function of the *egg cell* settled in the micropylar end of the mature embryo sac is sexual reproduction. The cytoplasm of the two *synergids* next to the egg cell is strongly polarized, i.e. it is quite dense towards the micropyle pole, while on the opposite side there is a vacuole (Marro 1982). The synergids are rich in mitochondria and in dictyosomes, their endoplasmic reticulum is well developed and they are important in guiding the pollen tube and opening the tube (Kapil and Bhatnagar 1975). This can be realized by chemotropic substances derived from the filiform apparatus of synergids, or the enzymes released from the synergids may induce the micropyle tissues to produce chemotropic substances. The two polar nuclei in the case of the sour cherry (Anvari and Stösser 1978*b*) migrate to the middle but they do not fuse into a secondary embryo sac. This is only realized in the mature embryo sac (Furukawa and Bukovac 1989; Fig. 2.41*a, b*). The fusion of polar nuclei in cherry can also be observed shortly before fertilization or during fertilization (Stösser and Anvari 1978). On the chalazal pole of the embryo sac three *antipodal cells* can be found whose role has not been clarified.

Two embryonic sacs may form in the nucellus (Fig. 2.34*b*). The frequency of such formation in 'Golden Delicious' and 'Yellowspur' apple cultivars are 5% and 7%, resp.

(Deveronico et al. 1982). The multiple embryo sacs in apricot in some cases fuse (Eaton and Jamont 1965). When there are two embryo sacs in the plum ovule their ratio without egg cell is over 60% (Pejkic 1969). The frequency of the two embryo sacs in sour cherry is 2–18% (Pejkic 1971). The appearance of two embryo sacs in walnut (Sartorius 1990) and *Rubus spectabilis* (Virdi and Eaton 1969) is also reported.

As we have found for the whole ovule, the *longevity of the embryo sac* is vital from the point of view of fertilization. The rate of degeneration of embryo sacs is inherently more in the triploid than in the diploid apple cultivars and more in lateral than the terminal flowers (Milutinovic 1973, 1975). The longevity of the embryo sac is longer in the terminal flowers of the apple inflorescence than in the lateral flowers (Marro and Lalatta 1978) and it may be doubled with nitrogen supply at the end of summer (Williams 1963, 1965).

The longevity of the female gametophyte in flowers of pear trees on quince and seedling roostock is 192 and 96 hours, respectively. In these flowers at anthesis, the corresponding rate of the unfunctional embryo sacs is 5% and 19%, respectively (Pejkic and Jovovic 1968). In the flowers of pear trees the degeneration of the embryo sac in Poland is much faster than in Italy (Callan and Lombard 1978). In plum trees a similar rootstock effect has been found (Pejkic 1969). In the sour cherry embryo sac, depending on the year, degeneration starts 9–11 days or 18–22 days after anthesis. First the nuclei of synergids degenerate followed by the egg cell and last by the polar nuclei (Anvari and Stösser 1978*b*). In the cherry the nuclei of the unfertilized embryo sacs die in a similar order beginning two weeks after anthesis (Stösser and Anvari 1978). When two (twin-) flowers form in the bud of the almond the frequency of the sterile flowers is 51% but in solitary flowers only 19%. The frequency of the sterile and twin flowers increases among the late opening flowers (Socias i Company and Felipe 1987).

3. Morphological Characteristics of Flowers and Fruits
E. Dibuz

3.1. General Characteristics of Flowers of Fruit Species

Here we summarize the morphological features influencing flowering, pollination, fertilization and fruit set the most.

Temperate zone fruit species belong to relatively few taxonomic units. We review them here giving the characteristic flower formula for certain families (A = androecium, Ca = calyx, Co = corolla, G = gynoecium, P = perigonium). The number of components for each characteristic is indicated below. The characteristics of fruit species belonging to a given family are shown in the figures:

subclass : *Hamomelididae*
order : Fagales
family : Corylaceae $A_{(3-14)}$; $G_{\overline{(2)}}$ (Fig. 3.1)
Fagaceae $P_{4-8}\ A_{4-50}$; $P_{3+3}\ G_{\overline{(6)}}$ (Fig. 3.2)

order : Juglandales
family : Juglandaceae $P_{3-5}\ A_{3-40}$; $P_4\ G_{\overline{(2)}}$ (Fig. 3.3)
subclass : *Dilleniidae*

order : Ericales
family : Ericaceae $Ca_{(4-7)}$; $Co_{(4-5)}$, $A_{8-10}\ G_{(4-5)}$
subclass : *Rosidae*

order : Grossulariales (Saxifragales)
family : Grossulariaceae $Ca_5\ Co_5 A_5 G_{\overline{(12)}}$ (Figs 3.4 and 3.5)

order : Rosales
family : Rosaceae
subfamily : Maloideae $Ca_5 Co_5 A_{10+5+5}\ G_{5-1}$ (Figs 3.6–3.8)
Rosoideaea $Ca_5 Co_5 A\ \infty\ G_{4,5}\ \infty$ (Figs 3.9–3.11)
Prunoideae $Ca_5\ Co_5\ A_{10+10+10}\ G_1$
$Ca_5\ Co_5\ A_{10+5+5}\ G_1$ (Figs 3.12–3.16)

order : Cornales
family : Cornaceae $Ca_{4-5}\ Co_{4-5-0}\ A_{4-5}\ G_{\overline{(2)}}$

order : Elaeagnales
family : Elaeagnaceae $Ca_{2-4}\ Co_0\ A_4\ G_{\underline{1}}$

subclass : *Asteridae*
order : Dipsacales
family : Caprifoliaceae $Ca_{(5-4)}\ [Co_{(5-4)}\ A_{4-5}]\ G_{2-5-8}$

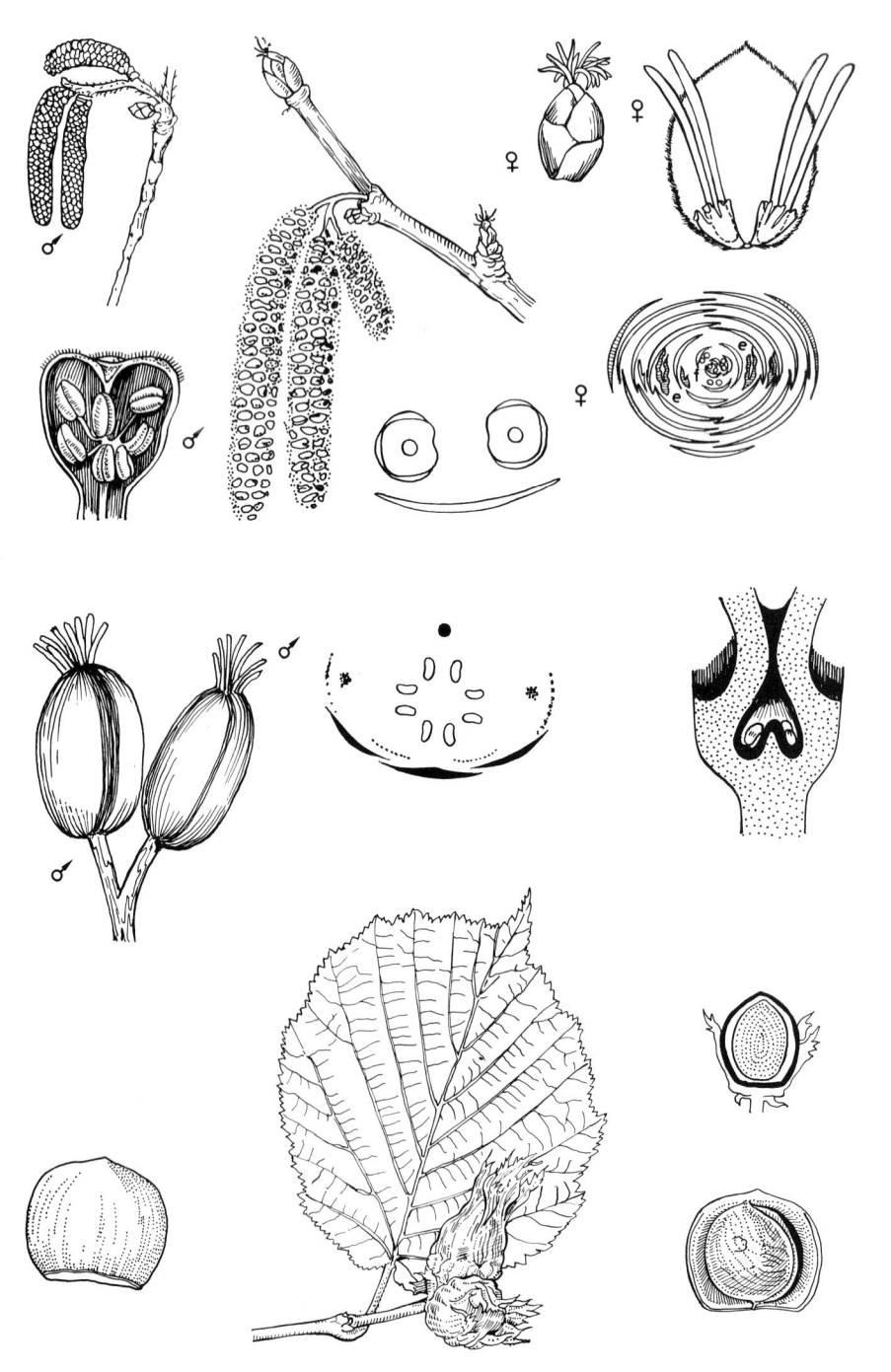

Fig. 3.1. Flower and fruit of hazelnut (*Corylus avellana*) (compiled after Hortobágyi 1962, Bordeianu et al. 1963, Kárpáti and Terpó 1968, Kárpáti 1969, Siegfried 1973, Winter et al. 1974, Horn 1976 and Terpó 1987)

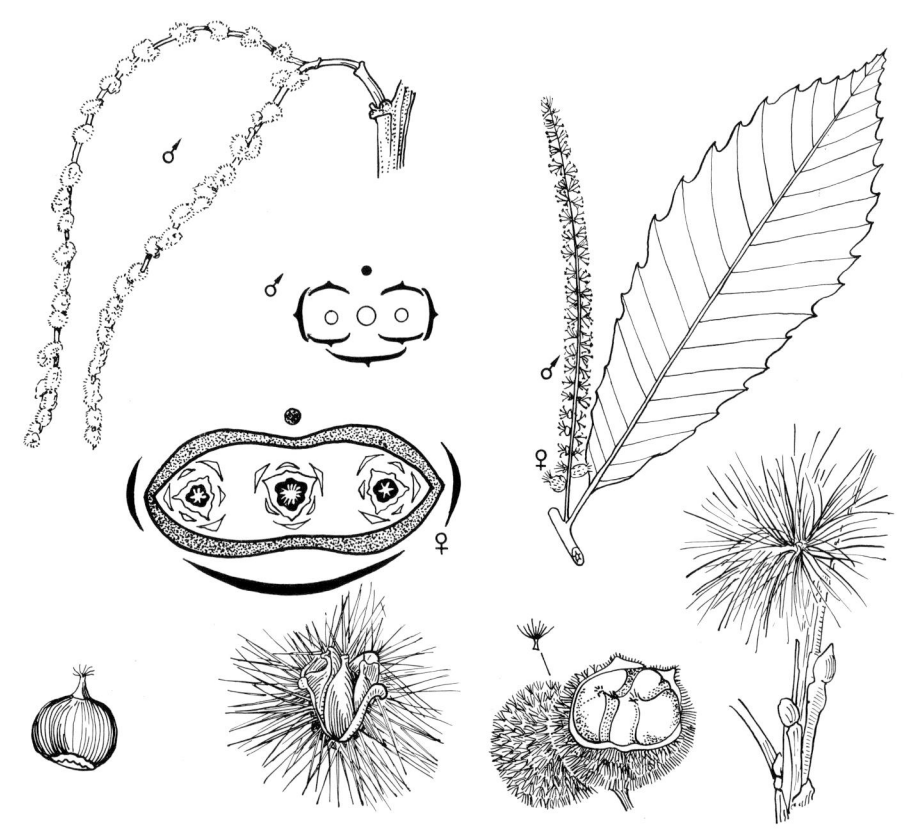

Fig. 3.2. Chestnut flower and fruit (*Castanea sativa*) (compiled after Soó 1965, Kárpáti 1969, Jacob et al. 1981, Csapody and Tóth 1982 and Terpó 1987)

Male inflorescences of fruit species are mainly actinomorphs having radial symmetry; weakly zygomorphic flowers only occur in blueberry (Porpáczy 1987) and black elder (Jakob et al. 1981, Terpó 1987). In black elder the whorls of five in the flowers remained even though the scattered leaf position on the vegetative shoots became aligned diagonally (Sárkány 1969).

In hazelnut, catkins develop in the flower buds which are in the bud stage 4–6 mm thick and 20–30 mm long (Horn 1976). The bracts tightly touch each other and only loosen at blooming, then the length of the catkin may reach 50–150 mm. Pedicle of catkins are shorter toward the end of the shoot. On the top quarter of the short shoots catkins have short pedicles and towards the apex of the shoot they do not have pedicles at all. On the double node branchings (dichasium) of the catkin inflorescence, one to three male flowers can be found. The flower pedicles are pre-leafed (Hortobágyi 1968) and their apical parts are free, the rest grows to the bracts. Their joint role is to support the catkin but they are not considered as flower parts.

The female inflorescence containing 4–16 simple female flowers develops inside the mixed buds and during blooming only the crimson stigma stretches out from the end. In the axillary point of bracts of the inflorescence there are two flowers (Horn 1976). According to

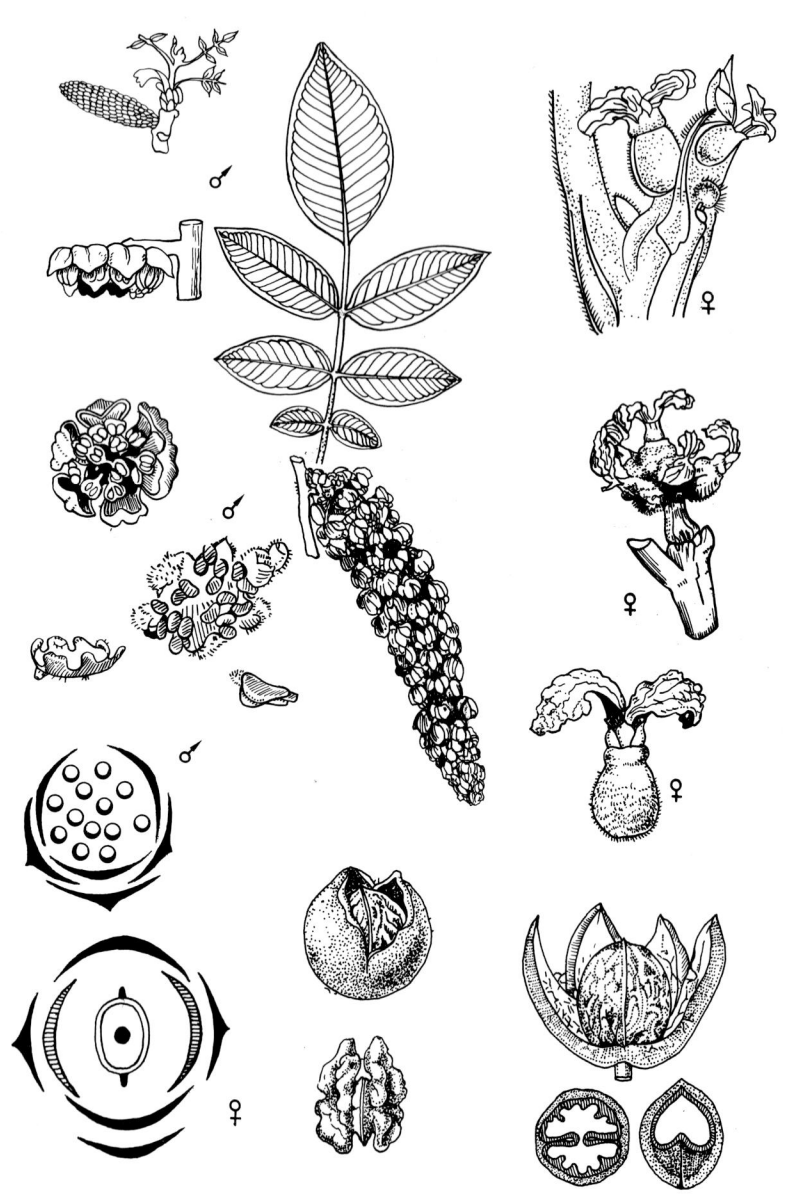

Fig. 3.3. The walnut (*Juglans regia*) flower and fruit (compiled after Bordeianu et al. 1963, Soó 1965, Kárpáti and Terpó 1968 and Terpó 1987)

Jacob et al. (1981) the bracts also take part in forming the shuck protecting the nut. Others (Hortobágyi 1968, Horn 1976) attribute the origin of the shuck to the bracts found on the flower pedicle surrounding the flower. The shuck is bell or jug-shaped, and fimbriate.

The less widespread hazelnut species, such as *Corylus pontica, C. colurna, C. maxima*, have similar flower structures. Filbert is covered with a shuck, pecan with a hull, walnut with a husk, and chestnut with a bur.

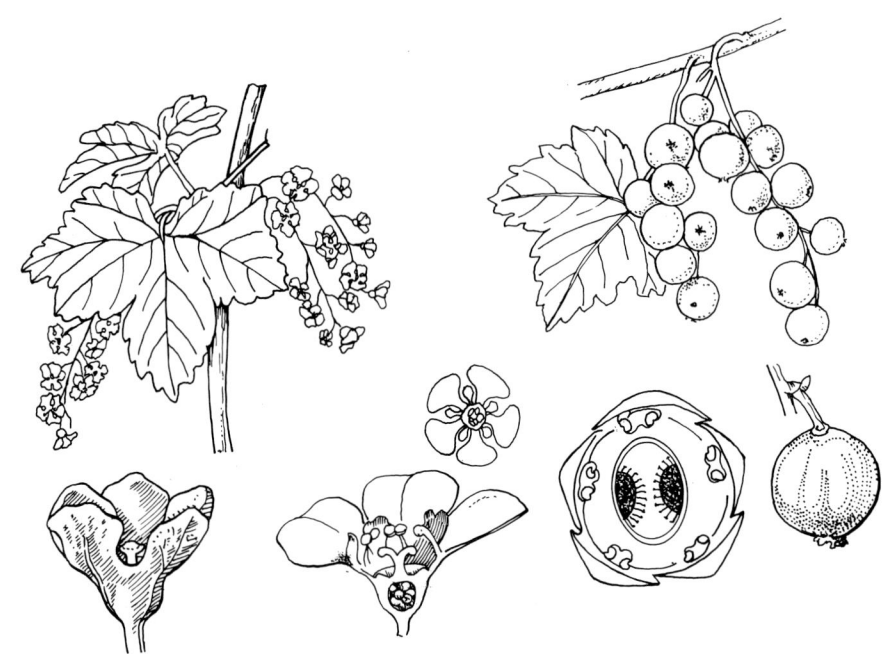

Fig. 3.4. Red currant flower and fruit (*Ribes rubrum*) (compiled after Kárpáti 1969 and Dibuz, unpubl.)

In chestnut, the structure of the catkins is similar to that of the hazelnut. However, the male flowers in the inflorescence of the hazelnut are heterotactic and open in double cymose. In contrast, in chestnut, the catkin develops in mixed buds in lateral position mainly in the central region of the shoots. The catkins developing in the axils of leaves are upright 15–20 cm long, and 1½ cm thick (Szentiványi 1976b). At the base of bracts of the inflorescence three male flowers develop representing a double cymose.

In the apical part of the shoot there are heterotactic, double cymose inflorescences in the axils of smaller leaves which appear elongated (Terpó 1987). On the lower branches of the inflorescence three female flowers in each dichasium can be found sorrounded with bracts. On the middle and top branches of inflorescences there are male flowers.

In the terminal mixed buds double cymose inflorescences develop generally with one to ten female flowers but according to Szentiványi (1976b) often 15 flowers can also be found. In the female inflorescence and in the female flowers of mixed inflorescences together with the development of seed, from the hypsophylls, a thorny, closed shuck also develops.

In single sex chestnut flowers it often happens that the undeveloped parts of the other sex appear (Jacob et al. 1981). Similar to the sweet chestnut, the flower structure of other species in cultivation (*Castanea dentata, C. crenata, C. molissima*) is similar.

In walnut, the male flowers develop as simple catkins. The tepals characteristically grow together with the hypsophylls and bracteoles of the inflorescence (Hortobágyi 1968). Others (Szentiványi 1976a) consider that the male flower of the walnut is an uncovered catkin-bud which clings without peduncle from the point of attachment. It may even reach 10–15 cm in length when developed.

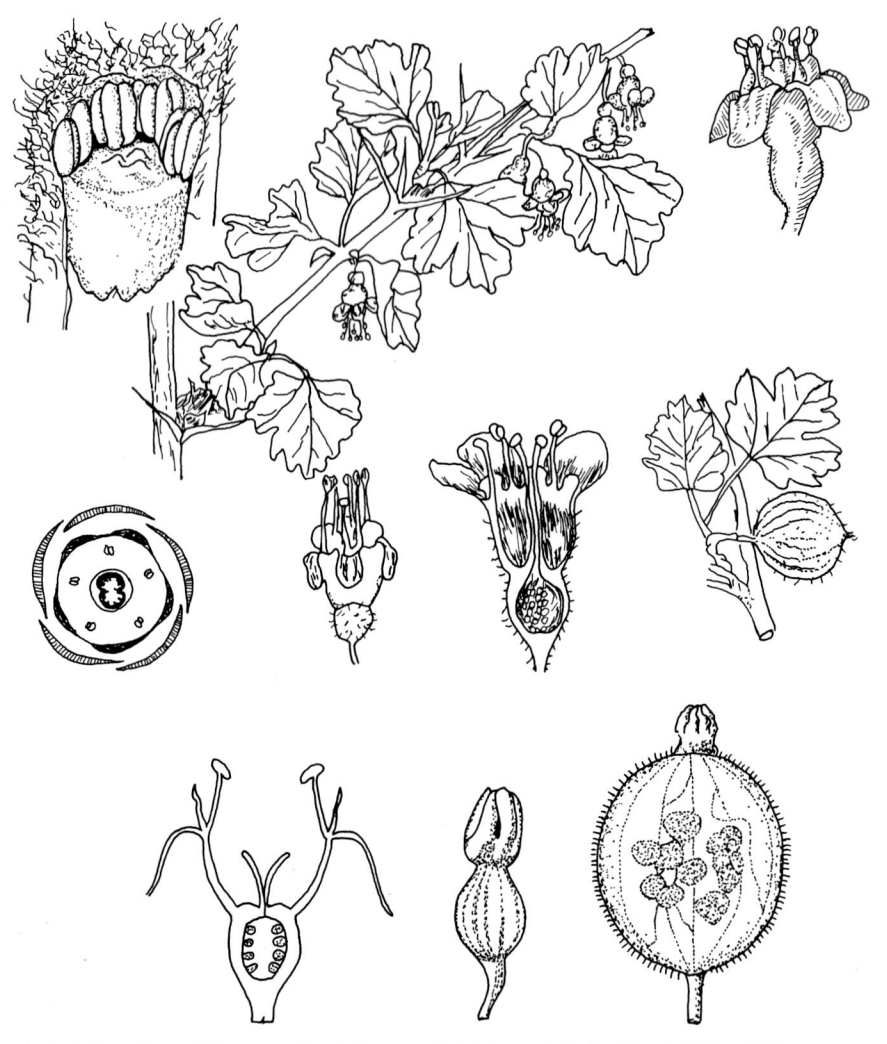

Fig. 3.5. Gooseberry (*Ribes uva-crispa*) flower and fruit (compiled after Terpó 1987 and Dibuz, unpubl.)

The female flowers of the walnut develop in mixed buds in groups of one to four (Terpó 1987), rarely as 5–15 flower strings (Szentiványi 1976*a*). The four united sepals participate with the two bracts in forming the green hull (Hortobágyi 1986).

The highbush blueberry's (*Vaccinium corinbosum*) racemose inflorescence is mainly in terminal position. Fifty to eighty flowers may occur in one inflorescence.

In the *Ribes* species, in the mixed buds, there is a negative correlation between the leafy part of the shoot and the size of the bunch flowers forming in the apex. The simple racemose inflorescence in the red currant has the most flowers (generally 10–35). In the black currant 2–15 flowers are characteristic and the flower number in the gooseberry is even less (one to three).

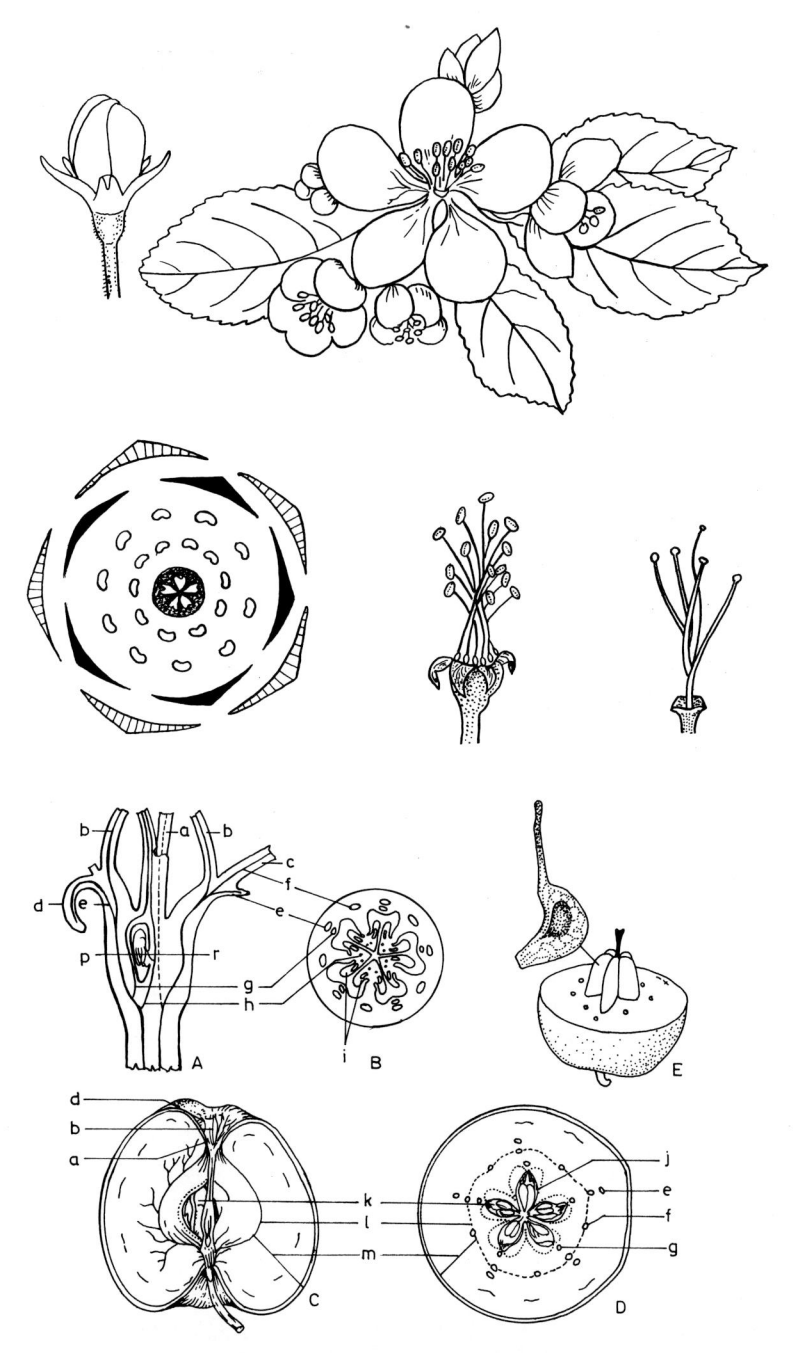

Fig. 3.6. The flower and fruit of the apple (*Malus domestica*). A = flower long-section; B = cross-section of the receptacle along the P-r line of Fig. A; C = fruit long-section; D = fruit cross-section; E = opened endocarp; a = style of flower; b = stamen; c = corolla; d = calyx; e = vascular bundles of the calyx; f = vascular bundles of the petals; g = dorsal vascular bundles of the carpels; h = ventral vascular bundles of the carpels; i = ovule; j = seed; k = endocarp; l = external abutment of the core; m = dimension of the receptacle wall (compiled after Kárpáti 1969 and Terpó 1987)

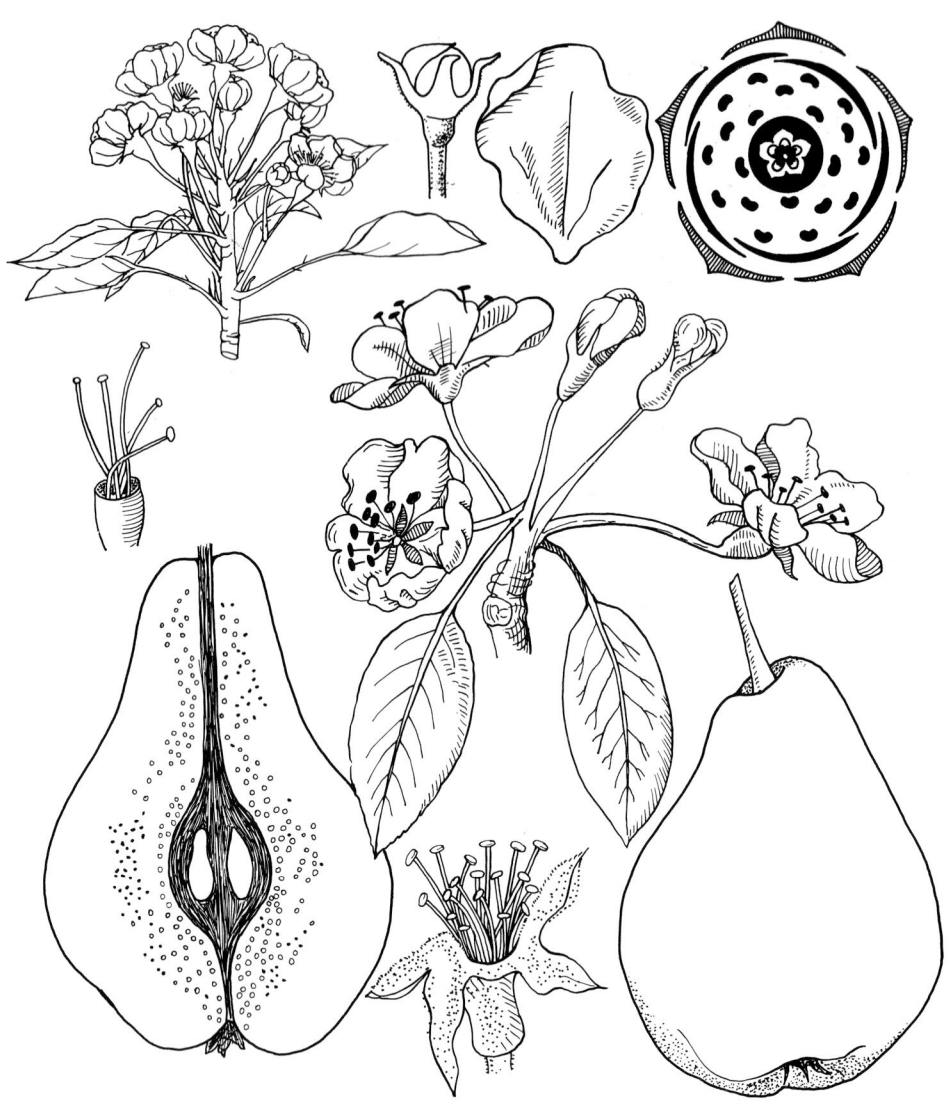

Fig. 3.7. The flower and fruit of the pear (*Pyrus communis*) (compiled after Hortobágyi 1962, Kárpáti 1969 and Dibuz 1989, 1991)

Ribes rubrum is the collective name of the red currant (Kárpáti and Terpó 1968). The parent of most currant cultivars is *Ribes silvestre* (western red currant). It is characteristic of this species that a quint-angled ring can be found in the receptacle and the flower appears in the pseudo berry fruit. In the *Ribes spicatum* (northern red currant) this outstanding ring in the receptacle and on the pseudo berry is missing. The *Ribes petraeum* (subalpine red currant) is striking with its red flower.

The apple and pear with respect to their flower type and the number of flowers per inflorescence show wide deviations. This has importance to the duration of bloom and to the order the flowers open. In the cymose apple inflorescence the opening order in the in-

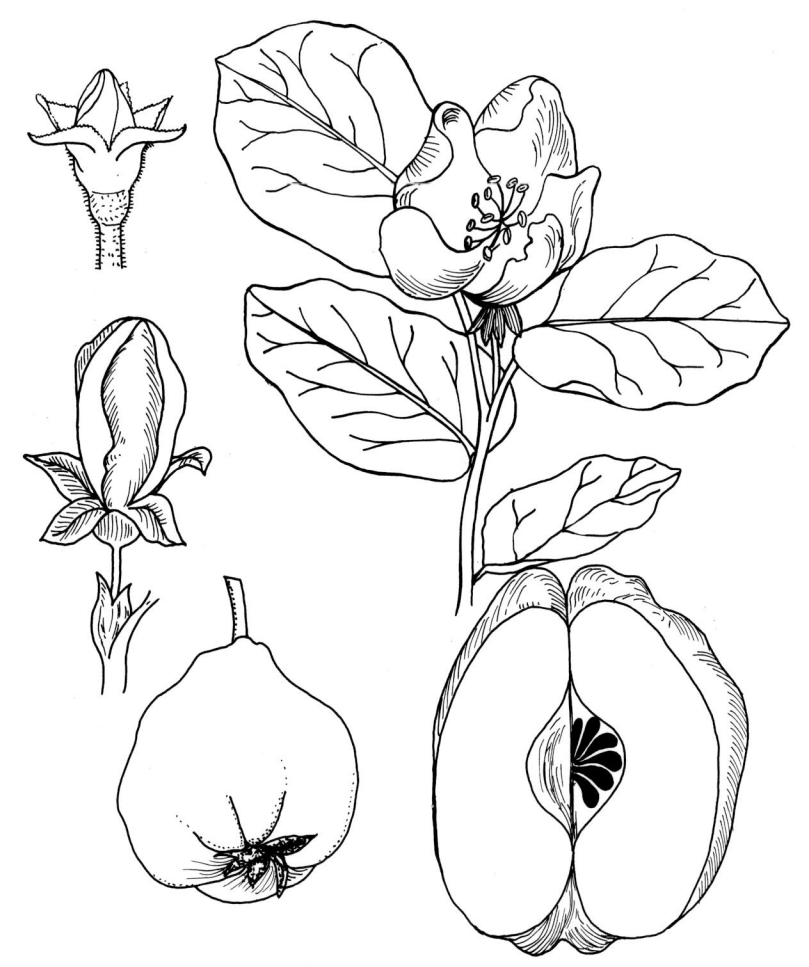

Fig. 3.8. The flower and fruit of the quince (*Cydonia oblonga*) (compiled after Hlava et al. 1983, Terpó 1987 and Dibuz, unpubl.)

florescence is always centrifugal. In the corymbose inflorescence of the pear it is mostly acropetal. However, Dibuz (1991, 1993) observed that centrifugal opening order also occurs as a divergent form in pear cultivars. This indicates that we cannot generalize.

The quince and the medlar have single flowers. In *sorbus* more flowers can be found in apical than in axillary position (Porpáczy 1987) such varied characteristics of the inflorescence re-inforce the multitude of flower organization of the Pomoideae subfamily.

Among all the fruit species mentioned here only the strawberry produces its flowers on a scape. In the strawberry flower the sepal and the corolla are verticillate, the stamen and the pistil are scattered (Sárkány 1969). The flower peduncle, after transforming its bracteole, forms the so-called external calyx-leaves (calyx-orbicular). In other cases the inflorescence consists of bracts that are foliage-like. Terpó (1987) has described this in the everbearing type *Fragaria vesca* var. *semperflorens*. The cymose inflorescence flow-

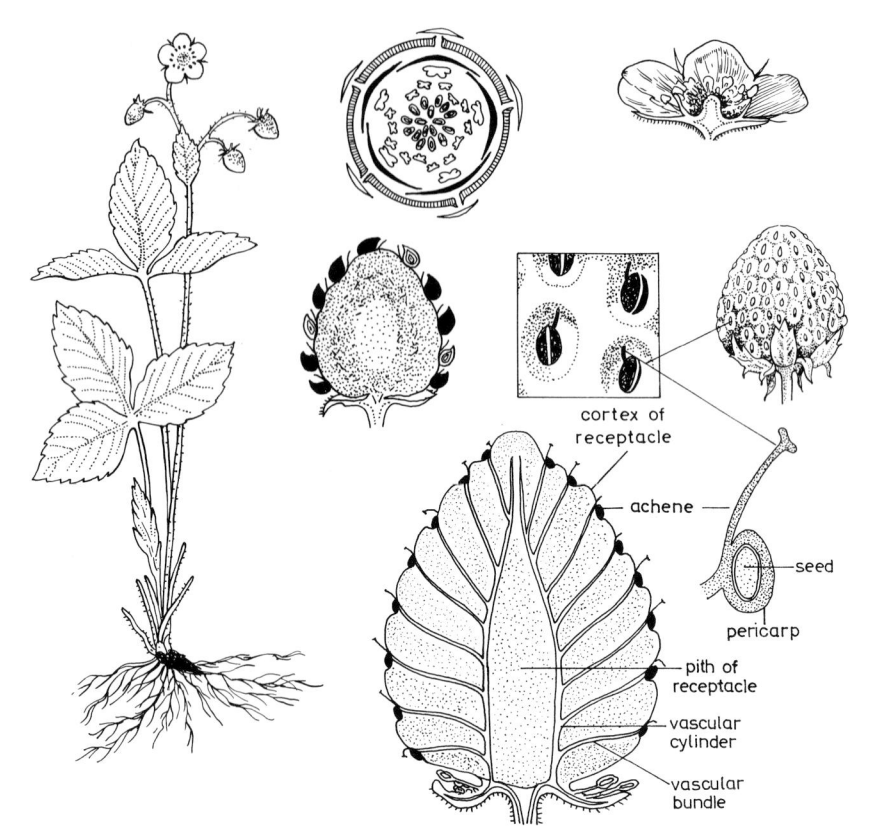

Fig. 3.9. The flower and fruit of the strawberry (*Fragaria magna*) (compiled after Soó 1965, Wilson and Loomis 1967, Papp 1984 and Terpó 1987)

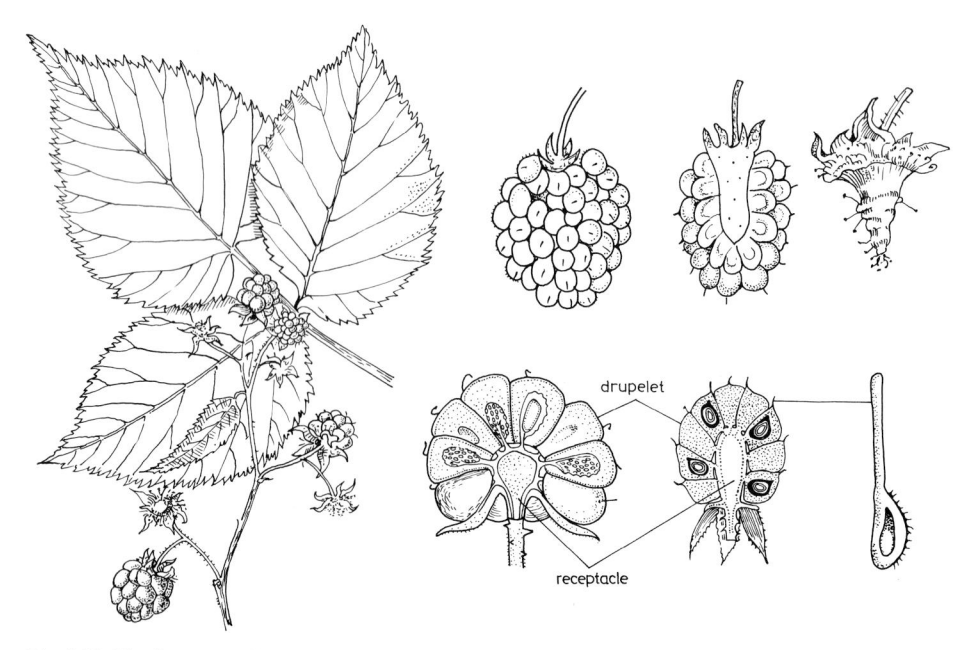

Fig. 3.10. The flower and fruit of the raspberry (*Rubus idaeus*) (compiled after Wilson and Loomis 1967, Kárpáti 1969, Winkler et al. 1974, and Hortobágyi 1986)

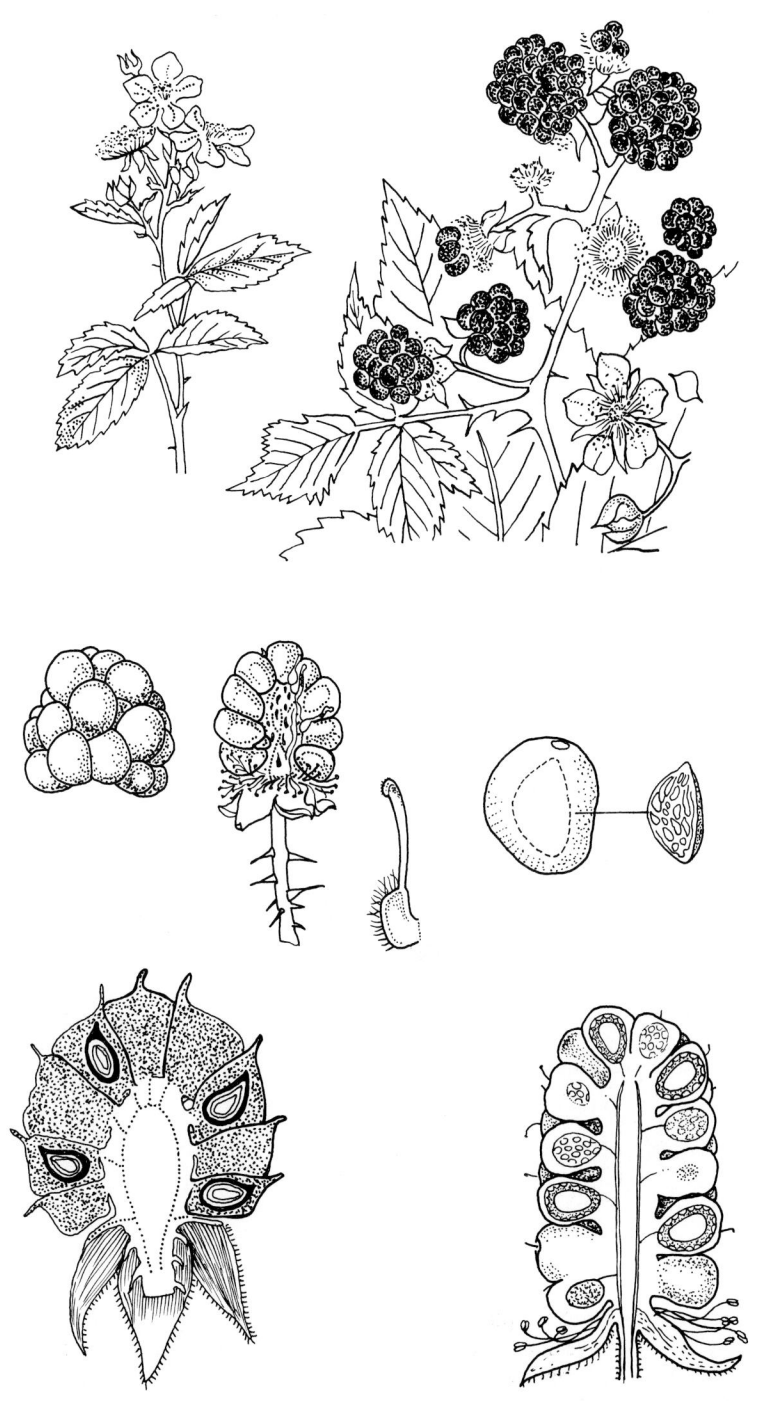

Fig. 3.11. The flower and fruit of the blackberry (*Rubus* sp.) (compiled after Soó 1965, Wilson and Loomis 1967, Papp 1984 and Terpó 1987)

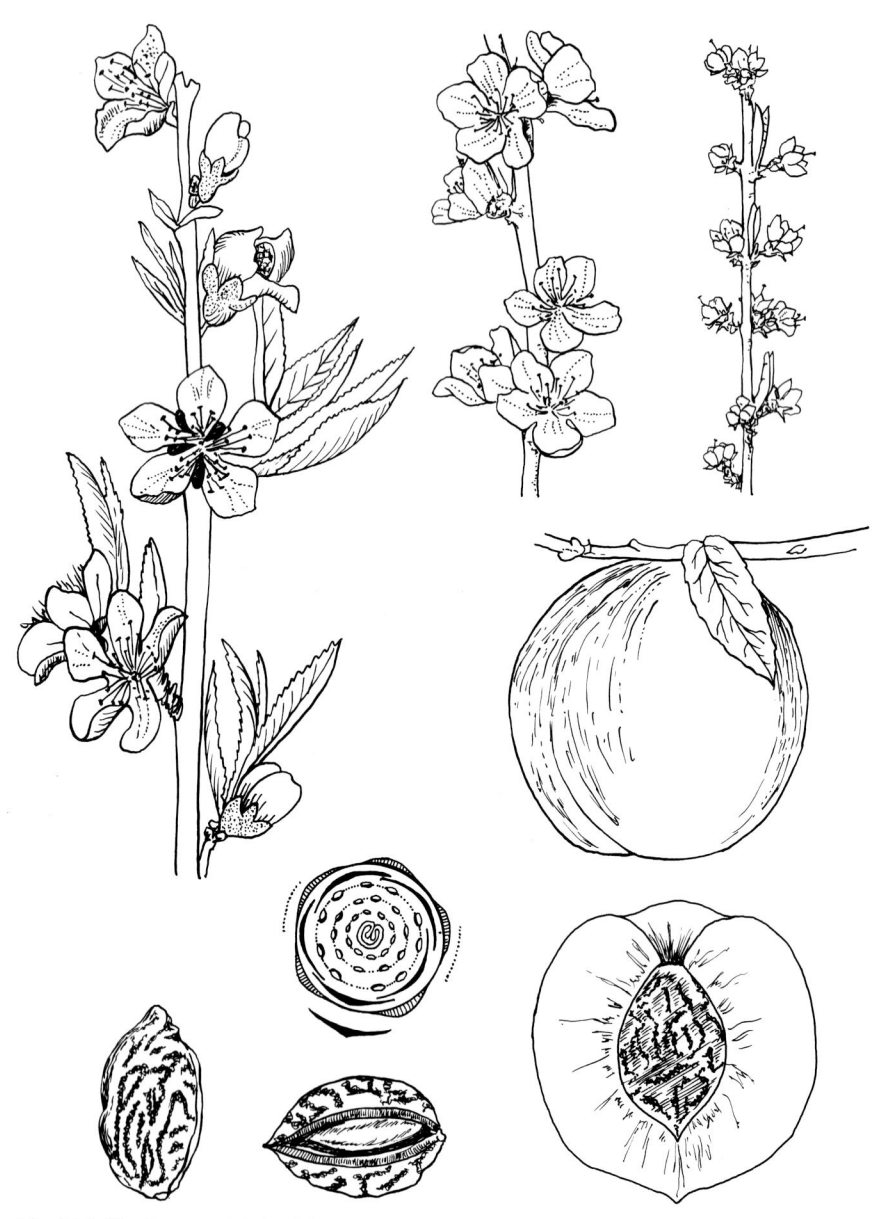

Fig. 3.12. The flower and fruit of the peach (*Persica vulgaris*) (compiled after Ninkovski 1989 and Dibuz, unpubl.)

ers are male in the diploid cultivars, in the case of hexa- and octoploids they are dioecious or both (Szilágyi 1975). The male flowers are always larger and more striking. Szilágyi (1975) reported that *Fragaria chiloensis* are dioecious, the *Fragaria vesca* and *Fragaria viridis* are monoecious and unisexual. The latter are considered by Kárpáti and Terpó (1968) to be dioecious just as the *Fragaria moschata*.

Fig. 3.13. The flower and fruit of the almond (*Amygdalus communis*) (modified after Sárkány and Szalai 1966)

The inflorescence of the raspberry and the blackberry develops in mixed buds. The branching of the inflorescence and its degree of development shows a tight correlation with the development of the vegetative shoots (Rünger 1971). There are larger and more prominent flowers in the inflorescence of blackberry than in the simpler inflorescence of the raspberry. The hypsophylls in the inflorescence of the raspberry and the blackberry are often foliage-like (Mohácsy et al. 1965). The lower flowers of the inflorescence can

Fig. 3.14. The flower and fruit of the apricot (*Armeniaca vulgaris*) (compiled after Terpó 1987 and Dibuz, unpubl.)

be found in the axils. Here we should mention that the *Rubus idaeus* (raspberry) is a collective name because raspberry cultivars also, belong to *Rubus strigosus* (drooping inflorescence) and *Rubus occidentalis* (black raspberry) (Terpó 1987). In *R. occidentalis* the flowers occur in packed umbrella-like inflorescence (Mohácsy et al. 1965).

Blackberry cultivars originated from several species *Rubus laciniatus, R. loganobaccus, R. intermis, R. procens, R. caesius* and their collective name is *Rubus fructicosus* (Terpó 1987). When considering the variable morphology of the flower of the *Rubus* species it is important to consider the origin of the cultivars. Especially we have to take into account that the crossing of the raspberry and the blackberry subgenera is also common. A good example of this is *Rubus mohacsyanus* (*R. idaeus* ×*caesius* ×*loganobaccus*).

The flowers of the rose depending on the species are single or appear as bunched inflorescence.

The flowers of the peach and the almond appear singly and axillary (Rünger 1971). The flower buds located in groups are generally next to each other but in the peach Terpó

Fig. 3.15. The flower and fruit of the plum (*Prunus domestica*). A = stone development; B = stone fruit long-section; C = ripe fruit; a = endocarp; b = nucellus; c = embryo sac, d = seed; e = exocarp; f = mezocarp; g = cavity of the stone; h = fruit stem; i = belly suture of the fruit; j = pistil-point of the fruit (compiled after Hortobágyi 1986 and Terpó 1987)

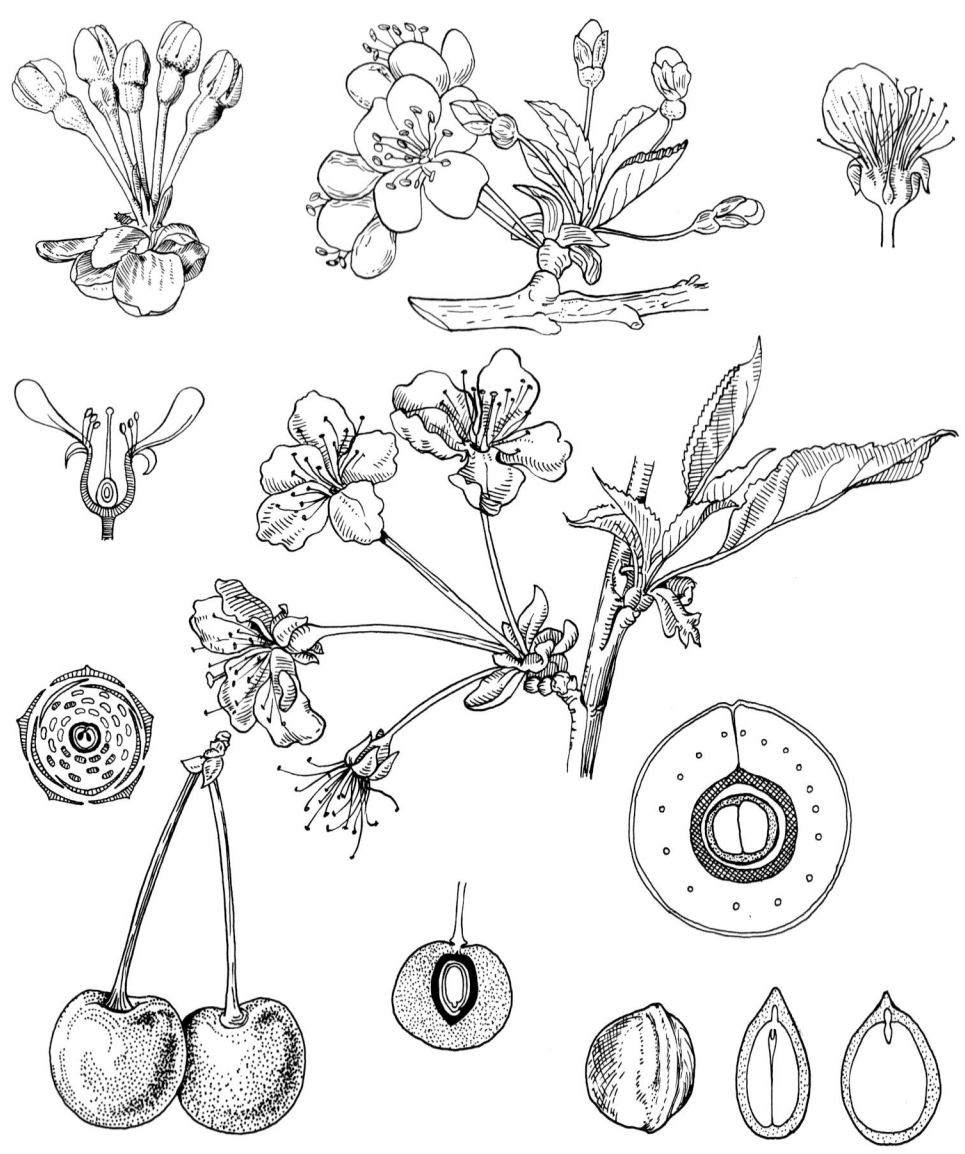

Fig. 3.16. The flower and fruit of the cherry (*Cerasus avium*) (compiled after Sárkány and Szalai 1966, Hortobágyi 1968, Kárpáti 1969 and Winter et al. 1974)

(1987) mentions that the flowers may be above each other. The rose-flower shape and the bell-flower shape are both common in peach. In the nectarines (Ninkovski 1989) the rose flower is more common. Pejovics (1976) does not distinguish between the frequency of the "rose" flower and the "star" flower.

The reduction of the complexity of the flower is well illustrated in *Prunus* (Jacob et al. 1981): the elongated inflorescence is common in *Padus avium*; a short bunch in *Cerasus mahaleb*; a one-to-five flowered cymose in *Cerasus avium* and *Cerasus vulgaris*; and

Table 3.1. The Characteristics of the Flowers of Fruiting Plants (Compiled after Dibuz)

Morphological characters	Hazelnut	Chestnut	Walnut	Cranberry	Red currant	Black currant	Gooseberry	Apple	Pear	Quince	Medlar	Black sorbus	Strawberry	Raspberry	Blackberry	Rose	Peach	Almond	Apricot	Plum	Cherry	Sour cherry	Cornel-berry	Sea buckthorn	Black elder
Flowers in compound bud	X	X	X			X	X		X	X	X	X			X	X			X	X	X	X	X		X
Flowers in floral bud	X	X	X	X	X	X	X	X	X	X	X	X		X	X	X	X	X	X	X	X	X	X	X	X
Flowers on short shoot, terminally	X	X	X	X	X	X	X	X	X	X	X	X		X	X								X	X	
Flowers on short shoot, laterally	X	X	X	X	X	X		X	X	X	X	X				X	X	X	X	X	X	X	X	X	
Flowers on long shoot, terminally	X	X	X			X		X	X	X		X		X	X	X	X	X	X		X	X	X		
Flowers on long shoot, laterally	X		X		X	X		X	X							X	X	X						X	
Flowers alone																									
alone—axillary					X	X	X		X					X	X										
in raceme	X	X	X																					X	
in spike	X	X																							
in ament	X	X																							
in cyme								X					X	X	X	X				X	X	X			X
in umbel or dichasium	X	X	X	X	X	X	X	X	X	X	X	X	X	X	X	X	X	X	X	X	X	X	X		X
Bisexual flowers																									
Diclinous and monoecious flowers	X	X	X																						
Diclinous and dioecious flowers		X																						X	
Flowers that can be heteromorphic													X												
Hypsophyll can be found on the axil of inflorescence	X	X	X		X	X	X	X	X		X	X		X	X	X				X	X	X	X	X	X
Hypsophyll cannot be found on the axil of inflorescence	X	X															X	X	X						
Hypsophyll on the peduncle				X	X	X	X	X	X	X	X	X		X	X	X				X	X	X	X		X
No hypsophyll on the peduncle					X	X	X						X	X	X										
Receptacle is: scooped, flattened, plate-shaped				X											X										
Receptacle is: convex													X	X	X								X		
Receptacle is: concave								X	X	X						X							X	X	
Receptacle is: tubular																									
Number of flowers in the inflorescence: 1–3	X	X	X		X	X	X	X	X				X	X	X		X		X	X	X	X	X	X	
Number of flowers in the inflorescence: 4–10	X	X	X	X	X	X	X	X	X				X	X	X	X		X		X	X	X		X	
Number of flowers in the inflorescence: 11–20				X											X										
Number of flowers in the inflorescence: 21–50												X													
Number of flowers in the inflorescence: many	X	X	X																						X

single flowers in *Persica vulgaris, Armeniaca vulgaris, Amygdalus communis, Prunus domestica* and *Prunus spinosa.*

The tiny flowers of cornel-berry (*Cornus mas*) can be found in a small umbrella (Soó 1965, Hortobágyi 1968). Terpó stresses the occurrence of the hypsophylls in the inflorescence in contrast with Siegfried (1973) earlier observations. According to Porpáczy (1987) in sea-buckthorn the corolla is missing in the tiny insignificant flowers and the receptacle is tube-shaped. The flowers only become differentiated on new shoot this is why they only grow on the shoots. The male flowers are brown in colour the females are yellow. The bisexual flowered cultivars were described two decades earlier (Burmistrov 1972) but their prominence will be expected in the future.

The black elder produces its flowers in composite cymose (Soó 1965), 200–10,000 flower per inflorescence can be expected (Porpáczy 1987).

The further study of the general morphological signs of the flowers and their categorization was made possible with extensive source works and own observations from which Table 3.1 was prepared.

3.2. Characteristics of the Perianth, Androecium and Gynoecium

The characteristics of the perianth (Table 3.2) the androecium (Table 3.3) and the morphological signs of the producing region (Table 3.4) are summarized for fruiting species. The majority of fruit-bearing plants have a double (heteroclamideus) perianth. Homoiochlamideus perianth can be found in the dioecious flowers of chestnut, hazelnut, walnut and sea buckthorn. The shape of the sepals and the degree of intergrowth show great variations according to the species (Szabó 1980). It is characteristic for the *Vaccinium myrtilus* (whortleberry), *V. vitis-idaea* (red whortleberry) *V. macrocarpon* (American cranberry) *V. corymbosum* (swamp blueberry) and *V. angustifolium* (lowbush blueberry) cultivars to have fleshy calyx, and its function in the fruit-organization.

In the *Ribes* species the sepals have merged and are tube-like (Kárpáti and Terpó 1968). It should be mentioned that *Ribes aureum,* a species used as rootstock has petaloid calyx. (Terpó 1987). In gooseberry, Harmat (1987) has described the sepals as egg-shaped and the majority of them bend back and rise above the petals.

In pomaceous fruits the calyx is also significant. This is especially striking in medlar (*Mespilus germanica*) where the huge sepals remain on the fruit.

We referred earlier to that in the double calyx of the strawberry the external circle is of bract origin. This is why the calyx is striking. The shape of the calyx is characteristic of the species and even of the cultivars. The calyx with projecting sepals is striking in *Fragaria moschata* and in *Fragaria vesca.*

The rose is significant with its fringed elongated sepals. The calyx of the cherry bends back and have intact edges. In the sour cherry edges of sepals are serrated. The red calyx of the apricot also folds back.

In androgynous fruit species the flowers without perianth (apochlamydeus) occur rather as an irregularity. However, in the diclinous flowering species the perianth is often missing (achlamydeus flower) or is strongly reduced. The tepal in the hazelnut and the chestnut is green. The tepal in the walnut is generally coalescent. This is rarely the case in the hazelnut and the chestnut (Terpó 1987). The tepals in the walnut have a reproductory role as well.

In a large number of fruit species belonging to the Rosales order a secondarily multiplied group of stamens (secondary polyandria) located in whorls can be observed

Table 3.2. The Organization of the Flower Circles and the Characteristics of Perianth in Various Fruit Species (Compiled after Dibuz)

Morphological characters	Hazelnut	Chestnut	Walnut	Cranberry	Red currant	Black currant	Gooseberry	Apple	Pear	Quince	Medlar	Black sorbus	Strawberry	Raspberry	Blackberry	Rose	Peach	Almond	Apricot	Plum	Cherry	Sour cherry	Cornel-berry	Sea buckthorn	Black elder
Number of flower circles: 2	X	X	X																					X	X
Number of flower circles: 4				X																			X		
Number of flower circles: 5					X	X	X	X	X	X	X	X	X	X	X	X	X	X	X	X	X	X			X
Number of members in flower circles: many																									
Number of members in flower circles: identical				X	X	X	X	X	X	X	X	X	X	X	X	X	X	X	X	X	X	X	X	X	X
Number of members in flower circles: different			X																						
Symmetry of flowers: actinomorphic				X	X	X	X	X	X	X	X	X	X	X	X	X	X	X	X	X	X	X	X	X	X
Symmetry of flowers: zygomorphic				X	X	X	X	X	X	X	X	X	X	X	X	X	X	X	X	X	X	X	X	X	X
Perianth single sex				X	X	X	X	X	X	X	X	X	X	X	X	X	X	X	X	X	X	X	X	X	X
Sepals: free					X	X	X	X	X	X	X	X	X	X	X		X	X	X	X	X	X	X		X
Sepals: attached					X	X	X	X	X	X	X	X	X	X	X		X	X	X	X	X	X	X		X
Sepals in single circle				X	X	X	X	X					X	X			X	X	X	X	X	X	X		X
Sepals in double circle				X	X	X	X						X												
Sepals: symmetrical				X	X	X	X	X	X	X	X	X	X	X	X	X	X	X	X	X	X	X	X		X
Sepals: petaline, showy																									
Sepals: strongly reduced					X	X	X																X		
Sepals: fall				X	X	X	X	X	X		X	X	X	X	X		X	X	X	X	X	X	X		X
Sepals: take part in fruit formation								X	X	X	X	X											X		
Corolla: free petals								X	X	X	X	X	X	X	X	X	X	X	X	X	X	X			
Corolla: fused petals				X	X	X	X																X		X
Petals: white				X	X	X	X	X	X		X	X	X	X	X			X	X	X	X	X	X		X
Petals: reddish-white					X	X	X	X	X								X	X							
Petals: green, greenish-white				X	X	X	X																		
Petals: yellow																X				X			X		
Petals: yellowish-white		X																							
Petals: reddish-green					X		X																		
Perianth unisexual	X	X	X																					X	X
Petals: in one circle	X	X																						X	X
Petals: in two circle																									
Petals: free	X	X																						X	X
Petals: fused	X	X																						X	X
Perianth may be missing	X	X																							

Table 3.3. The Characteristics of the Androecium in the Flowers of Fruit Species (Compiled after Dibuz)

Morphological characters	Hazelnut	Chestnut	Walnut	Cranberry	Red currant	Black currant	Gooseberry	Apple	Pear	Quince	Medlar	Black sorbus	Strawberry	Raspberry	Blackberry	Rose	Peach	Almond	Apricot	Plum	Cherry	Sour cherry	Cornel-berry	Sea buckthorn	Black elder
Filaments are: long	X	X				X	X		X	X	X	X	X	X	X	X	X	X	X	X	X	X	X	X	X
short, sessile			X	X	X	X	X	X																	
Filaments remain: long, on fruit as well	X	X	X	X	X	X	X		X	X	X	X	X	X	X	X	X	X	X	X	X	X	X	X	X
for short period			X	X	X	X	X	X	X	X	X	X	X	X	X	X	X	X	X	X	X	X	X	X	
Filaments are: non-branching	X																								
branching																									
Stamens grow on corolla	X	X	X																						X
Connective: small surfaced																								X	
large surfaced			X	X	X	X	X	X	X	X	X	X	X	X	X	X	X	X	X	X	X	X	X	X	X
On the anther: there are no extensions			X		X	X	X	X	X	X	X	X	X	X	X	X	X	X	X	X	X	X	X	X	X
extensions are anther-like				X																					
extensions similar to a flock of hair	X																								
The dehiscence of anther is: longitudinal	X			X	X	X	X	X	X	X	X		X	X	X	X	X	X	X	X	X	X	X	X	X
through apical pores																									
From the anthers: pollen scatters freely	X	X	X		X	X	X	X	X	X	X		X	X	X	X	X	X	X	X	X	X	X	X	X
pollentetrades scatter				X																					
The colour of anthers is usually yellow	X	X	X	X	X	X	X	X		X	X		X	X	X	X	X	X	X	X	X	X	X	X	
The colour of anthers can be other than yellow									X			X													X

Table 3.4. Characteristics of Pistil Gynoeceum in Flowers of Fruit Species (Compiled after Dibuz)

Morphological characters	Hazelnut	Chestnut	Walnut	Cranberry	Red currant	Black currant	Gooseberry	Apple	Pear	Quince	Medlar	Black sorbus	Strawberry	Raspberry	Blackberry	Rose	Peach	Almond	Apricot	Plum	Cherry	Sour cherry	Cornel-berry	Sea buckthorn	Black elder
Gynoecium: monocarp	X	X	X	X	X	X	X						X	X	X		X	X	X	X	X	X	X	X	X
polycarp								X	X	X	X	X	X	X	X	X	X	X	X	X	X	X	X	X	X
Pistil: monomer	X							X	X	X	X	X	X	X	X	X									
dimer			X		X	X	X																X		X
trimer			X																				X		X
polymer		X		X																					
Ovary: upper position	X	X	X	X	X	X	X						X	X	X		X	X	X	X	X	X	X		X
medium position													X			X									
lower position					X	X	X	X	X	X	X	X											X	X	
Marginal placentation																							X		
Wall placentation																							X		
Angular placentation	X	X																							
Axial placentation								X	X	X	X	X					X	X	X	X	X	X		X	
Angular-axial placentation																									
Basal placentation			X													X									
Ovule: straight	X																								
turned		X																							
curved				X	X	X	X	X	X	X	X	X	X	X	X	X	X	X	X	X	X	X	X	X	X
Stigma: papillous, warty																	X	X	X	X	X	X	X	X	X
villous	X		X																						
fluffy							X			X	X					X									
branching extends from closed flowers	X		X														X						X		
Flower: with 1 stigma	X	X		X								X												X	
with 2 stigmas					X	X	X						X	X	X	X									
with 3–5 stigmas	X	X	X		X	X	X	X	X	X	X		X	X	X										
Ovule in ovary: 1																	X	X	X	X	X	X	X	X	
2								X	X	X	X						X	X	X	X	X	X	X		
more									X	X															

75

(Jacob et al. 1981). The length of such stamens may vary according to the whorl the stamens are located in.

The stamen in the cranberry flowers is located in two circles freely (Porpáczy 1987). Cornel-berry has an episepal stamen circle (Terpó 1987). In the walnut the number of the stamens may show great deviations (in the axillary branches of the catkin) (Szentiványi 1976b).

The connective between the anthers generally has a larger surface in insect-pollinating species. The flexibility of the filaments of stamens may vary according to species or cultivars as well as the space between the stamens. This is important from the point of view of effective insect pollination.

The spiral arrangement of the pistil in Rosaceae is generally characteristic (Sárkány 1969). This arrangement even remained where the other components of the flower are verticillate (raspberry, strawberry). The evolutionary development is toward reduction of the carpel number and coalescing of various parts of the flower (flower axis, tepal, sepal, etc.). This view is underlined by the large number of species which form pseudo fruit organized from flower parts to form edible fruit-like structures.

The flower may be monocarp (with a single pistil) or polycarp (several pistils are present). In a polycarpic species with the coalescence of the carpels we get coenocarpic fruit, whereas when carpels remain independent we speak of apocarpic fruit (Hortobágyi 1986). In certain species the placentation and the ovule location may vary.

We can find two ovules in the double-segmented carpel of the hazelnut, but generally only one develops into a seed. In chestnut six carpels can be found per space in the flowers with two clinging nucelli (one ovary forms from three carpels).

In the dimer ovary of the *Ribes* species many ovules with large surface placentae form. In contrast with this the members of genera *Fragaria*, *Rubus* and *Rosa* have monomerous ovaries with one to two ovules. Among the rosaceous types there may be more than two ovules in apple and pear, while there are strikingly many in quince.

In the fruit-producing plants, discussed here, always real fruit develops when ovaries are located in medium position, while when ovaries are located in the upper and lower positions generally pseudo fruit is produced. Exceptions are the hazelnut, chestnut and cornel-berry.

The dimerous ovaries (hazelnut, walnut, red currant, black currant and gooseberry), the trimerous ovaries (elderberry, rarely the walnut) and the polymerous ovaries (chestnut, cranberry, occasionally black elder) produce coenocarpic fruits, the other fruit species are apocarpic.

The gradually narrowing part of the ovary continues in the style and ends with the stigma. The location of the stigma, its shape, and organization may vary but they have a common feature in that their papillar epidermis is strikingly filled with cytoplasm. In the insect-pollinated species secretion promotes pollen adhesion. In the wind-pollinated species for assuring pollen adhesion large-surface stigmas develop. Nectar forms in the cavities of the carpel at the base of the petals in nectaries which have papillary cells in the surface of glandular tissues.

3.3. Important Characteristics of Fruits

Real fruit is rare in edible fruit-producing species. Pseudo fruits occur in many kinds of variations (Table 3.5).

Hazelnut can be found either singly or in multiples (two to eight) on the fruiting shoot

Table 3.5. Types of Fruit in Fruit Species (Compiled after Dibuz)

Morphological characters	Hazelnut	Chestnut	Walnut	Cranberry	Red currant	Black currant	Gooseberry	Apple	Pear	Quince	Medlar	Black sorbus	Strawberry	Raspberry	Blackberry	Rose	Peach	Almond	Apricot	Plum	Cherry	Sour cherry	Cornel-berry	Sea buckthorn	Black elder
Real stone fruit																	X	X	X	X	X	X	X		
Real nut	X	X																							
Blackberry fruit (group of drupes on a receptacle)														X	X										
Strawberry fruit (achenes on a receptacle)													X												
Rose hip (achenes in a fleshy fruit)																X									
Pseudo berry																									
Stone accessory			X																						
Pome fruit								X	X	X	X	X													
One fully developed seed in a fruit	X	X	X														X	X	X	X	X	X	X	X	
Two fully developed seeds in a fruit		X						X																X	
Three to ten fully developed seeds in a fruit				X	X	X	X	X	X	X	X	X													X
Many fully developed seeds				X	X	X	X			X			X	X	X	X									

77

(Horn 1976). Within a nut only one seed develops which completely fills the cavity. The endosperm is missing. We consume the two large fleshy cotyledons. The exocarp is entirely sclerenchymatized. The chestnut is organized from several carpels. The extent of the vagination of the internal endocarp and the twin nuts are important morphological characters of the chestnut (Szentiványi 1976*b*).

In the pseudo fruit of the walnut the ovary has a single cavity and within the ovary a single ovule develops. The involucre consists of fused tepals, bracts and bracteoles. The involucre is adnate to the pericarp. From these develops the green hull (Hortobágyi 1986). Jacob et al. (1981) considers that the external hull is of cupule origin. The hard shell of walnut is also formed by the sclerezation of the endocarp.

The pseudo berry of *Vaccinium* is formed from the lower positioned ovary onto which the fleshy sepals grow (Terpó 1987). We also find pseudo berries in the *Ribes* species but here the nonfleshy calyx residuum remains on the fruit and only the receptacle surrounding the ovary becomes fleshy. The fruit of the *Ribes grossularia* has brittle hairs while that of *Ribes uva-crispa* is soft-haired. Such hairs are present earlier on the receptacle of the ovary (Kárpáti and Terpó 1968).

Mention should be made here of yosta [(*Ribes nigrum* ×*Ribes divaricatum*) × (*Ribes nigrum* ×*Ribes uva-crispa* cv. 'Grüne Hansa')] and rikö (*Ribes nigrum* cv. 'Silvergieter' ×*Ribes uva-crispa* cv. 'Lady Delamore'). The yosheline is hybrid of *Ribes nigrum* cv. 'Silvergieter' ×*Ribes uva-crispa* cv. 'Grüne Riesenbeere'. These hybrids of black currant × gooseberry unite the properties of the parents but their fruit resembles that of the gooseberry (Harmat 1987).

In pome fruit the receptacle and the bottom of sepals besides the pistil take part in producing the pseudo fruit. The remains of stamen do not have a role in developing the pome but they can be found on the fruit. Terpó described (1987) that the walls of cores may be soft and tunicated in pear, quince and sorbus or parchment-like in apple or sclereid, hard in medlar. In protecting the membranous core a layer of stone cells also helps especially in the pear (Dibuz 1989, 1993).

In rose achenes are completely surrounded by the fleshy receptacle and together they turn into a rose-hip. In this case the ovary is in a lower position. The strawberry fruit forms from the upper apocarpic pistil region by the achenes occurring on the external surface of the receptacle which is suitable for human consumption. Achenes are either completely outstanding or are located in cavities (Szilágyi 1975). Outstanding achenes are characteristic for *Fragaria chiloensis*, achenes located on cavities for *F. virginiana* (Terpó 1987). According to the latter author, the achenes of *Fragaria vesca* var. *eflagellis* do not have a continuation.

The blackberry likewise develops from the apocarpic gynoeceum. In this species a group of drupelets are located on the surface of the fleshy receptacle. The external tissues (exocarp) of drupelets become fleshy and suitable for human consumption. The internal wall of the drupelet (endocarp) remains hard. Thus the receptacle takes part in the organization of the blackberry drupelet but it is smaller and becomes less fleshy than in the strawberry. In the raspberry we do not consume the receptacle because the mature fruit separates either in whole or part fruit from the receptacle. In some current cultivars of blackberry and raspberry-blackberry we cannot pull off the group of fruits from the receptacle. Such cultivars are difficult to harvest. In the raspberry on the terminal of drupelets the residua of stigmas can also be found (Mohácsy et al. 1965).

The real fruit of stone fruit has a membraneous exocarp and a fleshy mesocarp. In almond the mesocarp cracks upon ripening and the stone seed is liberated. The stone then cracks and the seed is suitable for consumption. In the other stone fruits the mesocarp is consumed. The endocarp below becomes hard from sclerenchymatization. Together with the seed this gives the hardness or the so-called stoniness (Jakob et al. 1981, Terpó 1987). The shape of the stone, its surface and crackability are characteristic morphological indications of plum types. In the cornelberry also a stone fruit develops from the lower ovary. In the case of cornel, it has not been properly clarified whether, in contrast with the perigynous ovary of plum cultivars, the receptacle takes part, to some extent, in forming the fruit in addition to the pistil.

The fruit of sea buckthorn appears to be a stone fruit (pseudodrupa) but in reality the fleshy receptacle of the single-seeded nut and often the fleshy calyx form the fruit (Terpó 1987). The origin of the fruit in the relatively recently cultivated black elder is controversial. Siegfried et al. (1973) reported that in the black elder a multiseeded stone fruit develops. Porpáczy (1987) also mentions stone fruit, while Terpó (1987) refers to a pseudo berry as a fruit. In our opinion, the pyrenarium develops from the perigynous pistil region, the organization of the fruit resembles a rose hip, but in the black elder the fleshy receptacle surrounds the stone fruits (the endocarp forms a separate stone shell around every seed).

4. Flowering

M. Soltész

4.1. Time of Bloom

4.1.1. Time of Bloom and Its Succession in Fruit Species

Bloom of various species occurs at different times. Bloom starts early spring with hazelnut and following it there is no period until fall when we could not find a flowering fruit-producing species. The knowledge of bloom time of the species, and of cultivars within the species, is important for assembling the various cultivar combinations that give the best pollination possibilities, or decrease the loss caused by frost in the orchard.

Information of bloom phenology is also important in existing orchards. The knowledge of time of bloom is essential for production and plant protection technology, for determining the best time for bee placement in the orchard for pollination, and for estimation of the best possible date of harvest.

The time of bloom is a species characteristic. However, the beginning of bloom and the length of the bloom period of a given species also depend on the comparison that we may make with other cultivars in addition to ecological factors that may influence bloom.

The use of specific cultivars may change the order in which species flower in case of species that flower close to each other. For example, the pear and apple bloom order may change in a given orchard if we use early blooming apple and late blooming pear cultivars. The order then becomes apple and pear instead of pear and apple.

The bloom order may also change if we consider cultivars that span the entire period of bloom for the species. Maliga (1946) mentioned that in the southern and northern parts of the U. S. the bloom order of peach, pear and apple shows major differences. In Europe the common bloom sequence is peach, pear and apple. In Hungary, according to Okályi and Maliga (1956), the bloom order of stone fruits is: apricot, peach, cherry, sour cherry and plum. We summarized the relative bloom order of fruit-producing species in Hungary in Fig. 4.1.

Hungary is located in the middle of Europe between the 16° and 23° parallels east, and the 45° and 49° latitudes north. The majority of orchards are located between the 46° and 48° latitudes. The orchards are between 100 and 300 metres above sea level. In the various species there is no correlation between the time of bloom and the time of maturity of the fruit. In other ecological conditions the same species may flower at different times and the period from bloom to maturity of fruit may also be modified.

One can classify the bloom time of species according to the classifier. Maliga (1946) considered peach as "early" blooming species and apple as a "mid-period" blooming species. Brózik (1975) considered peach as a "mid-season" blooming species and apple as a "late" blooming species. The blooming sequence described by these two authors is also different. The description of blooming sequence of Brózik (in Brózik and Nyéki 1975) is given in Fig. 4.1. In the bloom order the late blooming rose occupies a special place: it flowers through several months and its fruit matures through several months. Therefore, it cannot be easily classified in the order.

With the exception of hazelnut, the short bloom time is characteristic of early blooming species. In years when bloom is late the duration of bloom increases in all

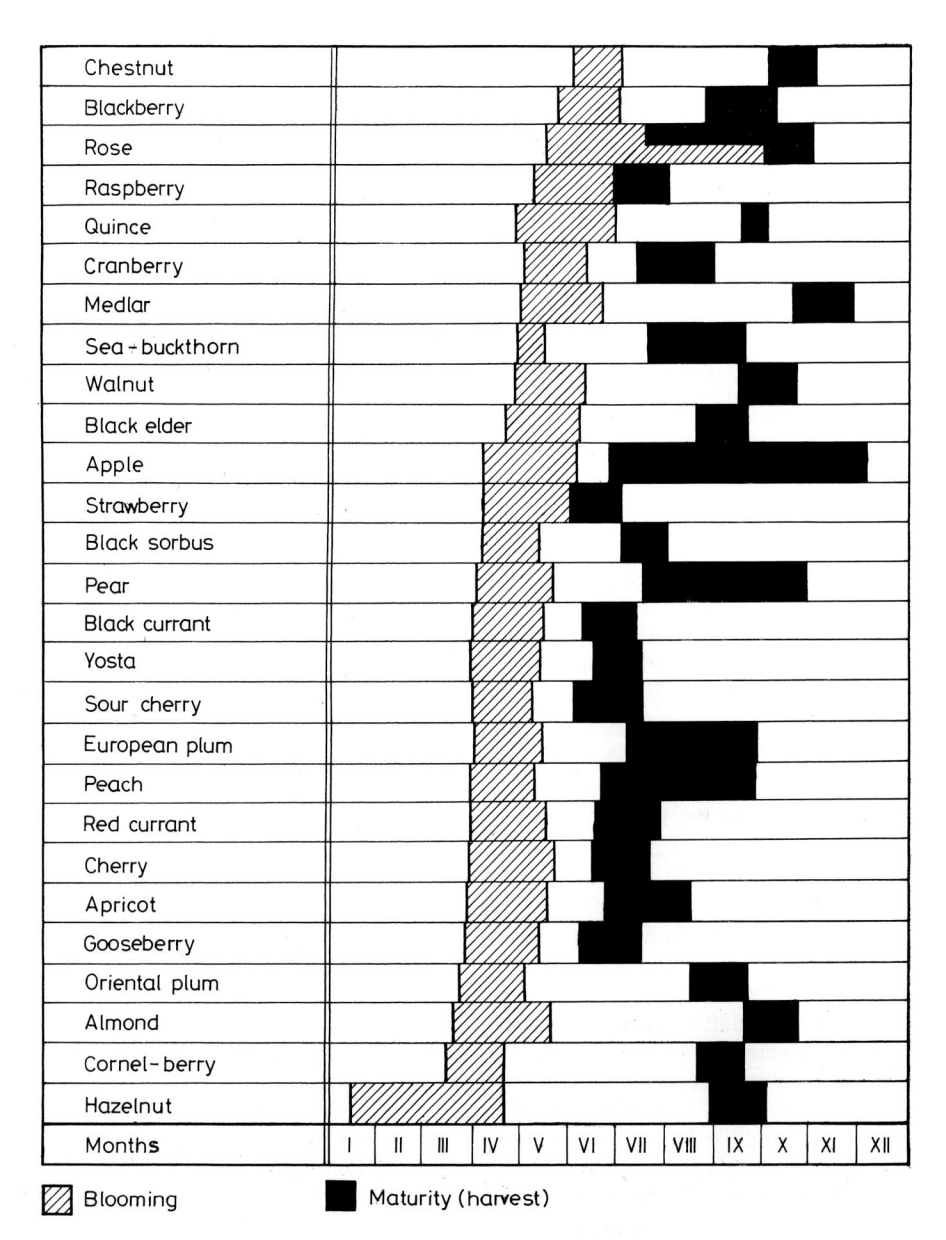

Fig. 4.1. The blooming and fruit harvesting period of temperate zone fruit species in Hungary (compiled after Soltész)

species but especially in the early blooming species. The increase in bloom can be as long as five days (Table 4.1). We summarized the change in time and length of bloom of species from data collected for several decades at the same location (Fig. 4.2).

The hazelnut and cornel bloom before the appearance of the leaves, and flowers and leaves appear at the same time in sea buckthorn. Peach and apricot usually bloom before

81

Table 4.1. The Length of Blooming of the Fruit Species in Days in an Early and in a Late Blooming Year (after Brózik 1975)

Fruit species	In early blooming year	In late blooming year
Hazelnut female flowers	79	60
Hazelnut male flowers	40	15
Almond	10	4
Gooseberry	20	15
Apricot	7	11
Cherry	15	6
Red currant	25	16
Peach	4	4
Plum	10	11
Sour cherry	17	6
Black currant	25	21
Pear	16	12
Strawberry	40	38
Apple	19	16
Walnut female flowers	25	30
Walnut male flowers	30	15
Medlar	10	21
Quince	14	12
Raspberry	33	32
Chestnut	12	14
Average	23	18

the leaves appear, but in occasional years blooming may occur after the leaves are out. The change in bloom from before to after the appearance of the leaves is also characteristic of early blooming *P. domestica* cultivars and oriental plums. The relative time of bloom compared to the appearance of the leaves is dependent on the place of origin of the species, the degree of completion of chilling requirement and the prevailing spring weather.

For some later blooming species the extended bloom period is characteristic. These species are, in addition to rose, strawberry, raspberry and elderberry. The bloom period is even more extended in everbearing strawberry and in raspberry if we also consider the fall producing raspberry types.

The bloom time of cultivars greatly depends on their origin. This is especially applicable to red and black currant. According to Oprea and Palocsay (1977), cultivars that are descendants of *R. sativum* bloom earlier than those that are descendants of *R. petraeum*. The bloom period of plum is changing. The use of the oriental plum cultivars influences the bloom time of this fruit greatly.

The phenology of bloom is also of significance in fruit breeding. It is not only important that we know the bloom time of new cultivars, but changing the bloom time also has its place in breeding. The interest is increasing in late blooming cultivars that may avoid late freezes. Pear is a good example how bloom time is influenced by the genetic origin of the cultivar. According to Westwood (1975), the sequence of bloom of species in *Pyrus* is as follows: *P. kochney*, *P. ussuriensis*, *P. calleryana*, *P. communis*, *P. cordata*, *P. longpipes*, *P. regelii*, and *P. nivalis*. In the research of Faust et al. (1976) *P. calleryana* bloomed 16–21 days before 'Bartlett' (Williams), a *P. communis* cultivar. Seedlings of P.

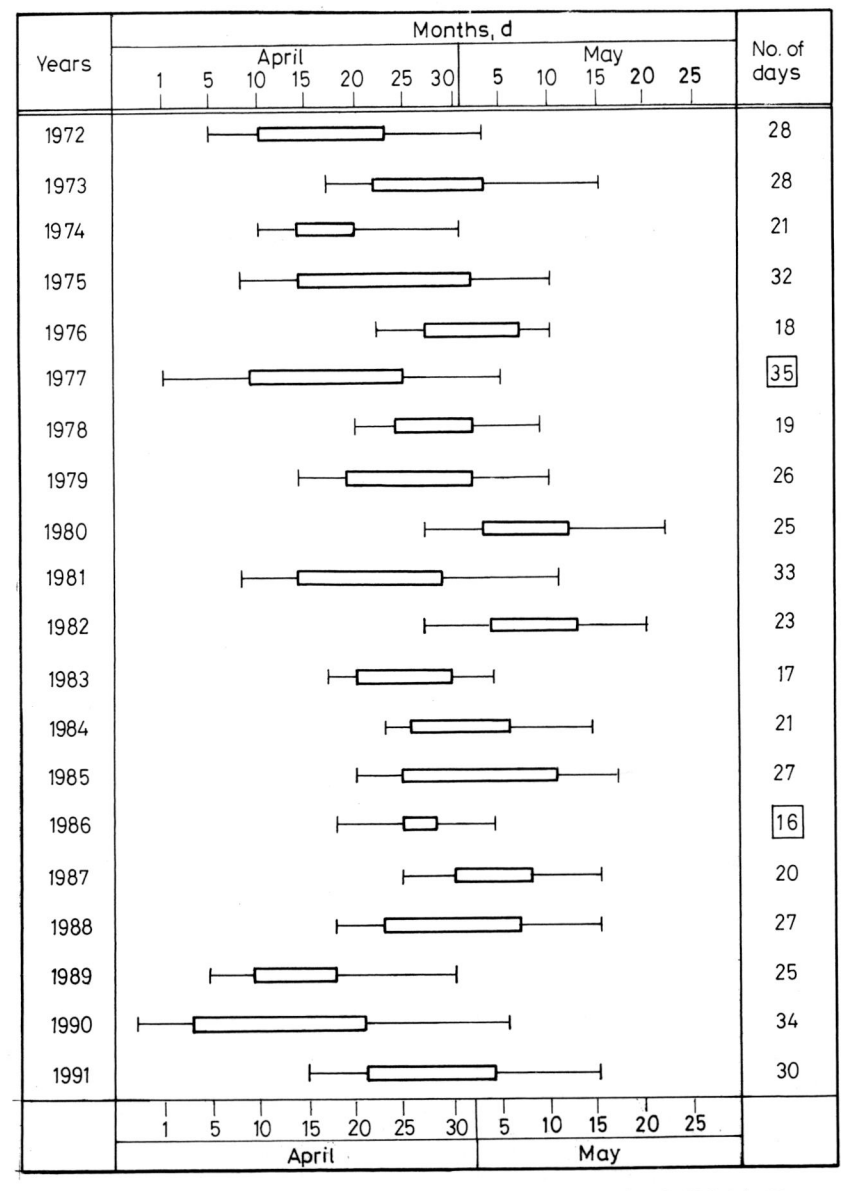

Fig. 4.2. Bloom period of apple cultivars between 1972 and 1991 at Kecskemét–Helvécia, Hungary (after Soltész 1992)

calleryana ×*P. communis* flower 5 to 12 days before 'Bartlett'. Flowering is even closer to 'Bartlett' in seedlings of (*P. calleryana* ×*P. communis*) ×*P. communis* and commences within zero to five days before 'Bartlett' ('Williams' Bon Chretien'). A selection of *P. saricularis bulbiformis* (*Pyrus* ×*Sorbus*) flowers after other *Pyrus* species.

4.1.2. Beginning of Bloom in Various Cultivars

Phenophases of bloom are: the beginning of bloom, full bloom and end of bloom. We can observe the beginning of bloom with the greatest precision. According to Schmidt (1954), the beginning of bloom expresses the best the genetic differences among the species. It is still difficult to compare data in the literature, because various authors designate the beginning of bloom according to varying amounts of open flowers. Data collected in Table 4.2 clearly indicate these differences.

In the beginning of bloom of cultivars there can be major differences according to location and years. In addition to genetic differences, the most often occurring date for beginning of bloom reflects well the effects of the location and agrotechnique. The actual date of the beginning of bloom is only of practical importance for the location where the observation was made, or at identical ecological location. This is underlined by the observation of Kronenberg (1985) at 119 locations in Europe where he found a difference of 80 days in the beginning of bloom of apples in the same year.

Data available in the literature clearly indicate that there are major differences in the beginning of bloom also at the same location. In apple, Ivan et al. (1981) observed four days of fluctuation in the beginning of bloom. Others found similar differences for apple.

Table 4.2. The Estimated Time of the Beginning of Flowering in Each Fruit Species

Species	The beginning of flowering	Author/Year
Valid for all species	The first flowers are open The first flowers opened, others continue 10% of the trees has at least 1 flower open 1–5% of flowers are open 10% of flowers are open* 12–15% of flowers are open	Maliga (1946), Soenen et al. (1978) Soltész (1992) White (1979) Nyéki (1989) Wociór et al. (1976) Faust (1989)
Apple	5% of flowers are open* 10% of flowers are open*	Soyanov and Gormevsky (1984), Palara et al. (1985) Faedi and Rosati (1975), Marro and Lalatta (1982)
Pear	The first flower has opened	Brown (1943)
Cherry	The first flowers are open 5% of flowers are open* 10% of flowers are open*	De Vries (1967) Ryabov and Ryabova (1970), Brózik (1969) Christensen (1974)
Sour cherry	The first flowers are open 10% of flowers are open*	Blasse (1964) Vasiliev and Rodeva (1978)
Plum	10% of flowers are open* 25% of flowers are open	Tóth (1967) Vondracek (1975), Bellini and Bini (1978)
Hazel	20% of flowers are open*	Modic (1974)

*The value given shows the average number of flowers that are open at most, but lower number is also possible on individual trees.

Rasmussen (1984) described 19 days, Brown (1940) 23 days, Soltész (1982a) 27 days and Reichel (1964a) 28 days. In pear, Sherman and Janick (1964) described an eight-day difference in the beginning of bloom, Soenen et al. (1978) 13 days, Braniste and Amzár (1964) and Vondracek and Kloutvor (1976) both 22 days, and Schnelle (1955) 30 days. The largest difference in pear was described by Brown (1943) concerning the early and changeable blooming cultivar, 'Brockwort Park', which fluctuated in the beginning of bloom 46 days. In quince, Maliga (1966b) described one to two days differences, Angelov (1966) two to four days and Brózik and Nyéki (1975) three to five days.

In apricot, Szabó and Nyéki (1989) described a fluctuation of eight days in the beginning of bloom. In peach, Szabó and Nyéki (1989) gave two days of fluctuation and for plum Levitzkaya and Kotoman (1980) reported ten days for the same location. Cherry has a more labile beginning of bloom among the stone fruits. For cherry Christensen (1974) reported eight days of fluctuation for the beginning of bloom, Stancevic (1967) reported 11 days, Christensen (1970) 14 days and Basso and Natali (1972) 15 days.

According to Porpáczy (1987) the fluctuation in the beginning of bloom in black currant was six days and in strawberry fluctuation was 30 days (Szilágyi 1975). In chestnut the largest difference in the beginning of flowering was 20 days (Solignat 1973).

4.1.3. The Relative Bloom Time of Cultivars

A number of factors influence the course of bloom (Fig. 4.3). This is the reason why the relative bloom time was introduced several decades ago (Chittenden 1911). The relative bloom order determines the relationship of bloom of a given cultivar compared to others. As a basis of comparison one should choose a cultivar with relatively stable bloom. This can be the earliest blooming cultivar, or any other cultivar in the succession. A relative order for apple cultivars, using the early blooming 'Red Astrachan' is given in Table 4.3. The relative blooming succession makes bloom data useful at locations other than they have been collected. The usefulness of the relative blooming order has been debated.

There are two approaches to this. One theory states that the relative bloom order of the cultivars in different years is constant. This was established for apple, by Brown (1940), Maliga (1961), Neumann (1962), and Rasmussen (1984); for pear by Brown (1943) and Vondracek and Kloutvor (1976); for cherry by De Vries (1968), Stancevic (1967), Mattusch (1970), Christensen (1970), Ryabova (1970), and Krapf (1972); and for plum Tóth (1957), Bellini and Bini (1978), and Levitzkaya and Kotoman (1980). Another group of researchers concluded that there can be quite significant deviations in the year to year time of relative bloom of the cultivars. Beakbane et al. (1935), Gautier (1971) and Faust (1989) refer to the instability of relative bloom time. In apple, Grubb (1949) and Soltész (1982a, 1985, 1992); in pear, Chollet (1965) and in the sour cherry Kobel (1954) gave data which support the fact that there is no stable bloom order which would be constant from one year to the other. Grubb (1949) found that the differences between the cultivars remained. Soltész (1982a) pointed out that the uncertainty of the relative bloom time in various habitats is even more pronounced than the instability at the same site over the years.

We cannot do away with the concept of relative bloom order. However, Table 4.3 confirms that we should not aim for an order which will always be constant but for one which will give the expected order with the greatest probability. These estimated values may also have inaccuracies but they have a wide application.

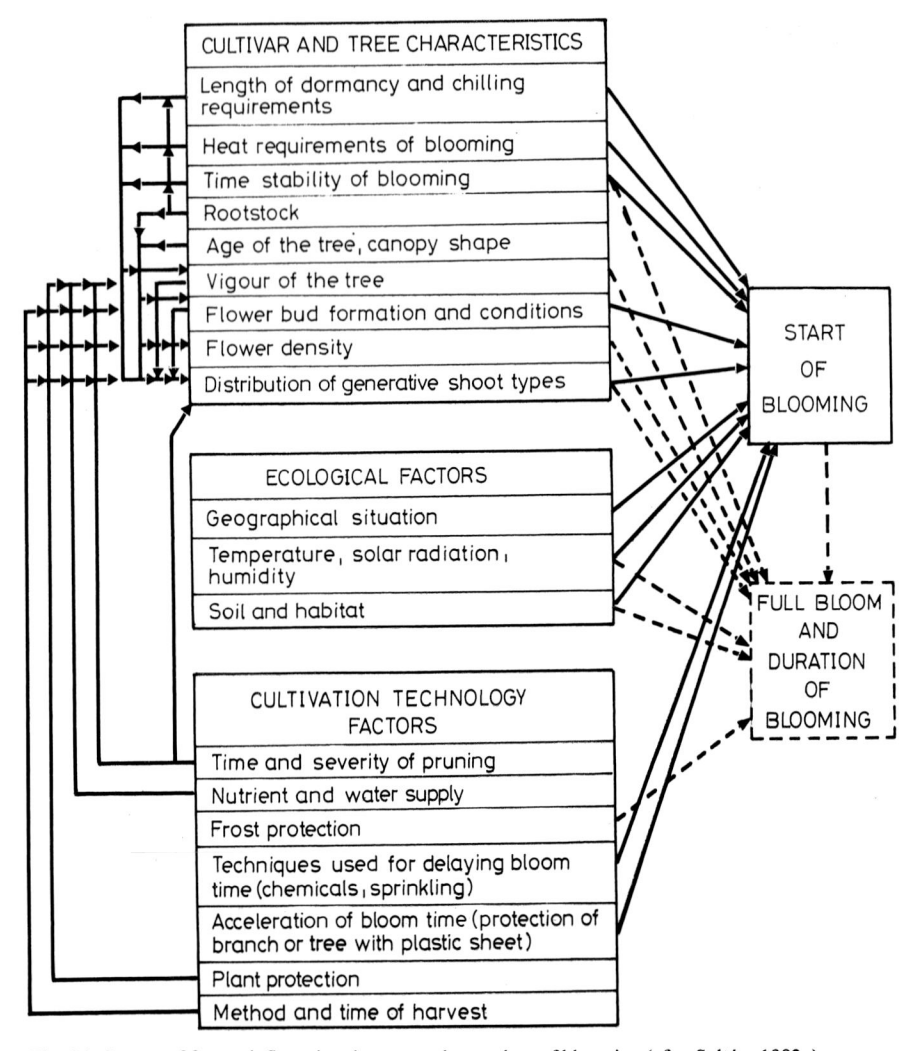

```
CULTIVAR AND TREE CHARACTERISTICS
Length of dormancy and chilling requirements
Heat requirements of blooming
Time stability of blooming
Rootstock
Age of the tree, canopy shape
Vigour of the tree
Flower bud formation and conditions
Flower density
Distribution of generative shoot types

ECOLOGICAL FACTORS
Geographical situation
Temperature, solar radiation, humidity
Soil and habitat

CULTIVATION TECHNOLOGY FACTORS
Time and severity of pruning
Nutrient and water supply
Frost protection
Techniques used for delaying bloom time (chemicals, sprinkling)
Acceleration of bloom time (protection of branch or tree with plastic sheet)
Plant protection
Method and time of harvest

START OF BLOOMING

FULL BLOOM AND DURATION OF BLOOMING
```

Fig. 4.3. System of factors influencing the start and procedure of blooming (after Soltész 1982*a*)

These findings agree with those of Rawes (1922), Crane (1927/28), Crane and Brown (1954) as well as those of Potter (1956) claiming that determination of a theoretically possible relative order requires many years of observations.

Investigations have shown (Soltész 1992) that from the point of view of stability of relative bloom time we may use the apple cultivars. The flowering order of 'Red Astrachan', 'Delicious', 'Early McIntosh', 'Golden Delicious', 'McIntosh' cultivars can be considered relatively stable whereas this cannot be said for 'Gravenstein', 'Early Red Bird' and 'James Grieve' cultivars.

It may be supposed that where the start of relative bloom could be shown, the observations were not continued from the first open flowers but only later (Soltész 1992). Where the method applied had the condition that at most 10% could be the ratio of the

Table 4.3. Relative Succession Order of Beginning of Blooming of Apple Cultivars (MMI, Helvécia, Hungary, 1975–1984) (after Soltész 1985)

Relative succession order according to 10 years' average	Species	1975	1976	1977	1978	1979	1980	1981	1982	1983	1984	Average difference
		8/4	22/4	1/4	20/4	14/4	27/4	12/4	27/4	18/4	23/4	
					Difference from 'Red Astrachan' (in days)							
1	Red Astrachan	1	0	3	0	0	0	0	0	0	0	0.0
2	Early Red Bird	1	0	3	0	0	0	0	0	−3	0	0.1
3	Red Gravenstein	−1	2	0	2	0	2	0	4	−2	1	0.8
4	Gravenstein	1	0	0	3	0	3	0	0	0	−1	0.8
5	Alkmene	5	1	4	2	3	4	0	4	0	1	2.4
6	James Grieve	11	1	11	1	3	5	4	4	3	2	3.8
7	Peasgood's Nonsush	9	3	6	4	3	4	0	4	3	2	3.8
8	Merton Worcester	9	3	7	3	4	4	1	4	3	2	4.0
9	Staymared	7	3	8	2	3	6	2	6	3	2	4.2
10	Carola	9	3	9	3	4	5	0	4	4	2	4.3
11	Jonathan	10	2	7	4	3	6	1	7	2	2	4.4
12	Granny Smith	9	3	8	4	6	5	0	7	3	2	4.7
13	Maigold	9	3	7	3	7	6	2	7	4	2	5.0
14	Winter Banana	12	4	8	4	4	5	2	7	2	2	5.0
15	McIntosh	9	3	8	4	7	6	1	7	2	4	5.1
16	Jonadel	11	3	6	5	6	6	1	7	4	5	5.4
17	Delicious	12	5	9	4	4	9	2	8	3	2	5.8
18	Redgold	12	3	8	4	7	9	1	7	2	5	5.8
19	Cortland	11	4	11	4	5	7	1	7	3	6	6.1
20	Lobo	10	3	12	4	7	10	1	7	3	6	6.3
21	Golden Delicious	14	3	8	6	9	10	2	7	4	5	6.8
22	Starking	13	6	11	5	7	9	2	7	3	5	6.8
23	Early McIntosh	14	4	10	5	7	8	3	7	3	8	6.9
24	Splendor	15	6	10	4	8	10	3	9	4	5	7.4
25	Golden Spur	14	6	12	6	9	0	6	7	4	6	7.9
26	Reinette musquée	17	10	11	4	7	8	3	10	4	8	8.2

Note: Start of blooming = first flowers have opened.

opened flowers at the start of bloom, and the time until the 10% opening could dissolve the smaller deviations.

Despite the fact that the time of bloom is genetically determined there are factors that influence bloom and the genetically determined bloom date cannot be expected to be expressed exactly. It may be concluded that we have to pay more attention to the factors influencing the bloom date because this is the only way we can achieve a more precise estimate for the date of blooming.

4.1.4. Full Bloom of Cultivars

We consider the full bloom time the phase during the course blooming when the flowers are open in the greatest number. Within this the time the exact point full bloom is the day when the proportion of the open flowers is the greatest. The full bloom time is generally not more than a day, the full bloom period may be several days depending primarily on the method how we indicate the full bloom time (Table 4.4).

The full bloom date of the chestnut cultivar (Fig. 4.4) will be shown as the simplest model. Plotting the flowering as shown in apple (Fig. 4.5) and in walnut (Fig. 4.6) allows the observer to pinpoint the full bloom time from the steepness of the curves.

Table 4.4. The Estimated Time of Full Bloom of Each Fruit Species

Species	Period of full bloom	Author/Year
Valid for all species	More than 50% of flowers have opened	Weger et al. (1940), Nyéki (1980, 1989), Soltész (1982a, 1987, 1992)
Apple	25–75% of flowers are open Petal fall has begun, 60% of flowers have opened 80% of flowers have opened The first fall of petals	Soyanov and Gormevsky (1984) Marro and Lalatta (1982) Palara et al. (1985), Redalen (1980) Faedi and Rosati (1975b)
Pear	More than 25% of flowers have opened More than 50% of flowers have opened More than 85% of flowers have opened	Vondracek and Kloutvor (1976) Brown (1943), Brózik and Nyéki (1971) Faust et al. (1976)
Cherry	25–75% of flowers are open 50–75% of flowers are open 75% of flowers are open 90% of flowers are open	Ryabov and Ryabova (1970) Albertini (1981) DeVries (1968), Brózik (1969) Christensen (1974)
Sour cherry	25–75% of flowers are open More than 50% of flowers have opened 75% of flowers are open	Ryabov and Ryabova (1970) Wociór et al. (1976) Brózik (1969)
Plum	70% of flowers are open 80% of flowers are open and petal fall has begun	Tóth (1967) Bellini and Bini (1978)
Hazel	More than 50% of flowers have opened 80% of flowers are open	Modic (1974)

Fig. 4.4. Bloom time of Hungarian chestnut cultivars (after Szentiványi 1980)

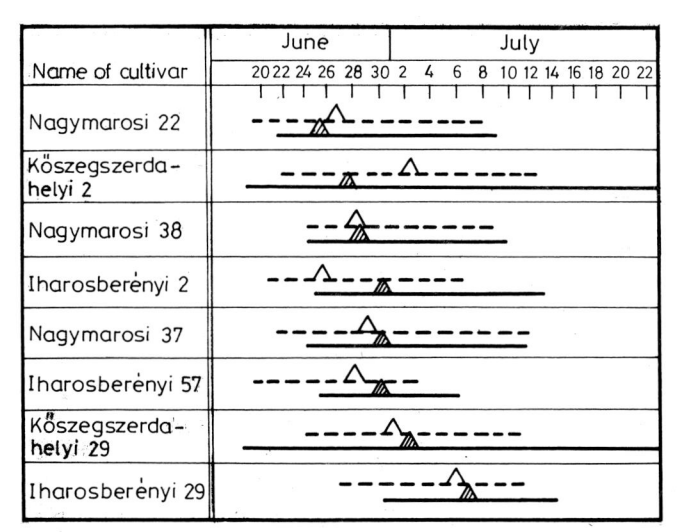

Name of cultivar	June 20 22 24 26 28 30 2	July 4 6 8 10 12 14 16 18 20 22
Nagymarosi 22		
Kőszegszerda-helyi 2		
Nagymarosi 38		
Iharosberényi 2		
Nagymarosi 37		
Iharosberényi 57		
Kőszegszerda-helyi 29		
Iharosberényi 29		

Flower type:

male △   female ▲

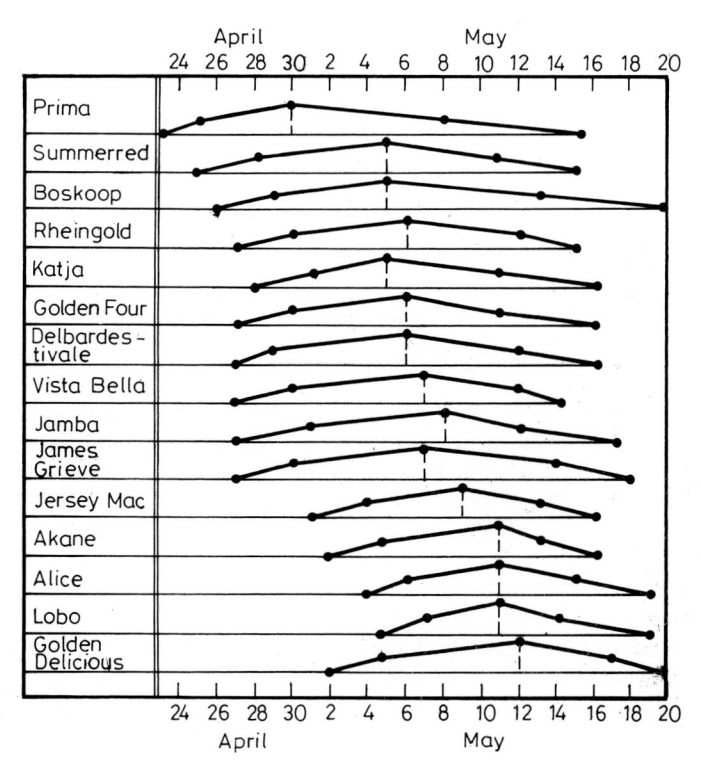

Fig. 4.5. Blooming of apple cultivars in 1982 in Oppenheim, Germany (after Steinborn 1983)

89

Fig. 4.6. Bloom time of Hungarian walnut cultivars (after Szentiványi 1980)

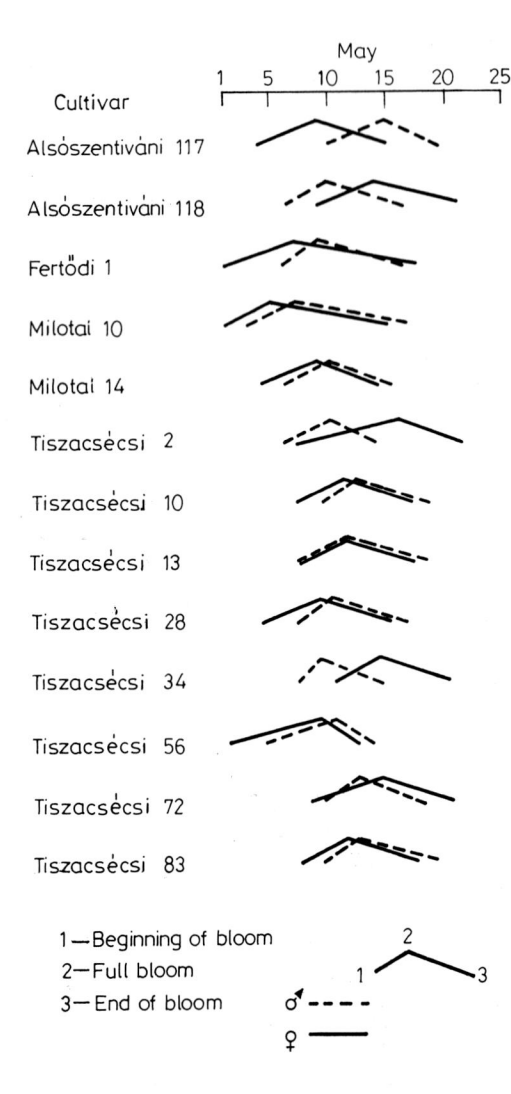

1 — Beginning of bloom
2 — Full bloom
3 — End of bloom

♂ - - - -
♀ ———

From year to year there may be rather large deviations in the full bloom time. Brózik (1969) observed three to eight days difference in the full bloom date of the 74 'Pándy' sour cherry clones and two to ten days difference in the 23 'Cigánymeggy' selections. DeVries (1967) found 7–14 days difference in the early and late blooming cultivars with respect to the full bloom date of 27 sour cherry cultivars.

Wociór et al. (1976b) found three to six days deviation in the full bloom date of eight sour cherry cultivars. Brózik and Nyéki (1975) demonstrated a difference of three to seven days in the full bloom date of the quince cultivars. The characteristics of the full bloom time of stone fruit species have been summarized by Nyéki (1989) in Table 4.5. The full bloom period deviated five days or less in sour cherry and non-melting flesh peaches, as well as in the free-stone peach. In the European plum, nectarine, apricot, cherry and oriental plum cultivars the deviation was more than five days. In the early

Table 4.5. The Characteristics of Full Bloom* of the Stone Fruits (after Nyéki 1989)

Fruit species	Differences between the full bloom of earliest and latest cultivars (in days)	Length of full bloom of species (in days) Average	Length of full bloom of species (in days) Extreme and values	The average coincidence** of full bloom of cultivars (in days)
Cherry	2–9	2.7	1.0–4.0	2.7
Sour cherry	1–4	1.7	1.3–1.9	1.3
European plum	2–6	4.4	3.0–5.8	1.8
Oriental plum	5–9	5.7	4.3–7.3	2.7
Apricot	1–7	4.5	3.9–5.6	3.0
Peach	2–6	6.4	5.5–7.1	3.7
Average of stone fruits	–	4.2	–	2.4

*Full bloom = more than 50% of flowers are open on trees; **Among cultivars to be pollinated and pollen donors

blooming seasons the average day of the full bloom occurs relatively early, that is, the full bloom date is closer to the start of blooming. Of the stone fruits, sour cherry, oriental plum and peach fluctuated more in the full bloom while in the case of the apricot and the cherry the full bloom date was more stable.

The stone-fruit species are grouped according to the length of the full bloom time when more than 50% of flowers are open, by Nyéki (1989) as follows:

1. Short (three days less): sour cherry, cherry;
2. Medium (three to five): European plum, apricot;
3. Long (five days more): oriental plum, peach.

Soltész (1992) systematized the apple cultivars on the basis of their full bloom times. The two main types are those which have one maximum peak and two maximum peaks. The full bloom time may be: (1) of identical distances from the start and end of bloom (e.g. 'Red Astrachan', 'Starking'); and (2) the same distance with respect to the beginning of blooming (e.g. 'Idared', 'Jonathan', 'Jonagold'); or (3) to the end of blooming (e.g. 'Golden Delicious', 'Red Rome'). In the case of suitable spur formation and weather types 2 and 3 may form two maximum peaks of flower opening. In both cases generally a larger and a smaller flower opening can be observed. In the 'Idared', 'Jonathan' and 'Jonagold' the second smaller peak is created by the flowers formed on the axial part of the shoots. In 'Golden Delicious' and 'Red Rome' cultivars the reverse is true, the flowers created by the smaller flower opening peak precede the large maximum peak. These observations apply to bearing trees. In young trees there are further changes in the full bloom time.

The one maximum full bloom time lasting for several days occurs in those cultivars where flowers form on short spurs. In the majority of the cultivars, in seasons which ensure prolonged blooming, we may find two maximum peaks. Other authors give the time of the full blooming separately for flowers developing on shoots and on spurs and in these cases the curves for flower opening only have one peak (Roemer 1968–70).

There are not many studies on the relative full bloom time. This has been observed in the cherry and the apple. In the cherry Hooper (1932) and Beakbane et al. (1935) estab-

Table 4.6. Relative Order of Full Bloom of Apple Varieties (MMI, Helvécia, Hungary, 1975–84) (after Soltész 1985)

Relative order (10 year average)	Cultivar	1975	1976	1977	1978	1979	1980	1981	1982	1983	1984	Average difference
					\multicolumn{7}{c}{Difference from 'Red Astrachan' (days)}		
1	Red Astrachan	16/4	28/4	9/4	25/4	19/4	5/5	15/4	4/5	20/4	27/4	0.0
2	Early Red Bird	6	1	8	−1	2	−2	0	−1	1	0	1.4
3	Red Gravenstein	4	2	7	0	4	0	0	0	0	1	1.9
4	Gravenstein	6	−2	8	1	4	2	2	1	1	2	2.2
5	Alkmene	12	0	4	1	6	3	3	2	1	4	3.4
6	McIntosh	5	2	9	4	5	3	3	3	1	4	3.8
7	Carola	6	4	11	2	6	3	0	3	4	4	4.3
8	Merton Worcester	7	4	7	4	5	3	4	4	4	5	4.4
9	James Grieve	13	4	13	0	4	3	2	1	2	3	4.4
10	Peasgood's Nonsush	10	4	6	6	7	3	2	2	4	4	4.6
11	Granny Smith	7	4	10	3	9	4	3	3	3	4	4.9
12	Cortland	8	3	11	5	7	5	2	4	2	7	5.3
13	Maigold	8	6	11	3	7	4	3	3	3	8	5.3
14	Jonathan	10	3	8	6	6	4	4	4	3	5	5.3
15	Early McIntosh	12	4	13	4	7	4	4	4	3	7	5.5
16	Delicious	9	6	10	4	6	6	3	4	3	7	6.3
17	Starking	9	6	9	6	8	6	7	4	5	7	6.4
18	Winter Banana	12	6	12	6	9	6	5	4	5	6	6.4
19	Staymared	11	6	10	7	9	3	5	3	4	5	6.5
20	Redgold	11	6	11	6	7	5	5	5	2	6	6.6
21	Lobo	8	6	16	6	9	6	7	5	4	7	6.7
22	Jonadel	13	7	10	6	9	4	6	3	4	8	7.2
23	Golden Delicious	11	8	11	7	10	6	5	5	4	8	7.2
24	Splendor	12	8	14	6	7	5	6	4	5	7	7.4
25	Golden Spur	13	8	14	7	13	4	12	5	5	7	8.1
26	Reinette musquée	16	9	14	7	11	7	9	4	5	8	8.5
								12	9	6	8	9.9

lished that the relative order also exists in the full bloom time. However, this was not confirmed by Stancevic (1967).

In the apple, Soltész (1985) was the first to investigate the relative full bloom time. Of the several hundred cultivars investigated we will analyse the full bloom time of the 26 in which the relative blooming period was compared with that of 'Red Astrachan' (Table 4.6). The relative order of the full bloom is even more unstable than that of the relative beginning of blooming. The investigations have continued, but the situation has not changed (Soltész 1992). Over 20 years the same relative order has never been repeated in the full bloom time. Furthermore, there was not one case when the start of blooming and the relative order of the full bloom period were synchronous. The relative full bloom values are hardly appropriate for comparative investigation of the blooming date of cultivars.

4.1.5. Duration of Bloom in Various Cultivars

Bloom time is that interval of time which begins with the anthesis of the first flower and ends with the last. There is a difficulty in analysing the bloom time given in the literature because there is no uniform interpretation for the end of bloom (Table 4.7)

Bloom date for fruit species shows great deviations (Table 4.8). The possible role of the season in bloom was compiled for two differently blossoming apple cultivars by Soenen et al. (1978) and can be studied with the help of Table 4.9. The duration of bloom was influenced by the seasons in the two cultivars differently. This correlates with the experiences of Roemer (1970) Childers (1975), Blasse (1976) and Soltész (1982a) an apple; of Brózik (1972) in cherry; and Wociór (1976b) in sour cherry. In other words, the longer bloom time does not follow automatically an early blooming period cultivar group, or early blooming year.

Table 4.7. The Estimated Time for the End of Blooming of Fruit Species

Species	End of blooming	Author/Year
Valid for all species	Pistils are not functioning, the spread of pollen has finished 90% of flowers have fallen 95–100% of flowers have fallen	Soltész (1992) Soenen et al. (1978) Faust (1989), Nyéki (1989)
Apple	The beginning of petal fall	Palara et al. (1978)
Pear	90% of flowers have fallen	Brown (1943), Brózik and Nyéki (1971)
Cherry	90% of flowers have fallen 95% of flowers have fallen	Christensen (1974) Ryabov and Ryabova (1970)
Sour cherry	90% of flowers have fallen	Vasiliev and Rodeva (1978)
Plum	70% of flowers have fallen 80% of flowers have fallen	Tóth (1967) Bellini and Bini (1978)

Table 4.8. The Average Length of Blooming of Cultivars in the Fruit Species

Species	The average length of blooming (days)	Author/Year
Apple	7–11	Badescu et al. (1981)
	7–15	Way (1971)
	13–22	Soltész (1982a)
range* = 11–17	15–20	Fritzsche (1972)
Pear	5–18	Brózik and Nyéki (1975)
	6–12	Stancevic (1972)
	10–16	Brown (1943)
range = 9–16	15–20	Fritzsche (1972)
Quince	6–20	Angelov (1966)
Cherry	5–18	Brózik and Nyéki (1975)
	6–14	Maliga (1952)
	7–10	Srivastava and Singh (1970)
	11–18	Christensen (1970)
range = 8–15	13	Stancevic (1967)
Sour cherry	5–14	Brózik and Nyéki (1975)
	5–12	Vasiliev (1978)
	8–12	Redalen (1984b)
range = 7–13	9–13	Gozob et al. (1979)
Peach	3–10	Nyéki and Brózik (1975)
	8–10	Fideghelli and Capellini (1978)
range = 6–10		Moretti et al. (1962)
Almond	10–15	Brózik (in Nyéki 1980)
Apricot	3–18	Brózik and Nyéki (1975)
	6–14	Szabó and Nyéki (1989)
range = 6–15	10–12	Fideghelli and Capellini (1978)
Plum	3–12	Levitzkaya and Kotoman (1980)
range = 5–16	7–20	Bellini and Bini (1978)
Raspberry	25–30	Kollányi (1990)
Black berry	21–23	Terettaz (1978)
Red currant	15	Blasse and Hoffman (1989)
Black currant	14–21	Fernquist (1961)
	15–20	Porpáczy (1987)
Gooseberry	10–15	Harmat (1987)
Black sorbus	15–20	Porpáczy (1987)
High cranberry tree	14	Porpáczy (1987)
Chestnut	12–21	Breviglieri (1951)

* Calculated range from the author s data

Table 4.9. The Average Meterological and Blooming Data of a 20-Year Long Research (1958–77) in Gorsem with 'Boskoop' and 'Golden Delicious' Apple Cultivars (after Soenen et al. 1978)

Cultivar	Characteristic for the year	No. of year	No. of days from bud burst till the beginning of bloom	The beginning of bloom day/month	The end of bloom day/month	The length of bloom days	The sum of daily maximum temperatures °C from bud burst till bloom	The sum of daily maximum temperatures °C during bloom
Boskoop	Early blossoming	9	47.5	22/4	4/5	16.1	563.8	261.6
	Late blossoming	11	45.7	2/5	17/5	14.6	526.2	247.8
	All years	20	46.5	28/4	11/5	15.3	543.1	254.0
Golden Delicious	Early blossoming	14	45.3	26/4	11/5	15.6	591.3	264.0
	Late blossoming	6	43.8	6/5	24/5	18.0	538.6	300.7
	All years	20	44.8	28/4	15/5	16.3	549.8	275.0

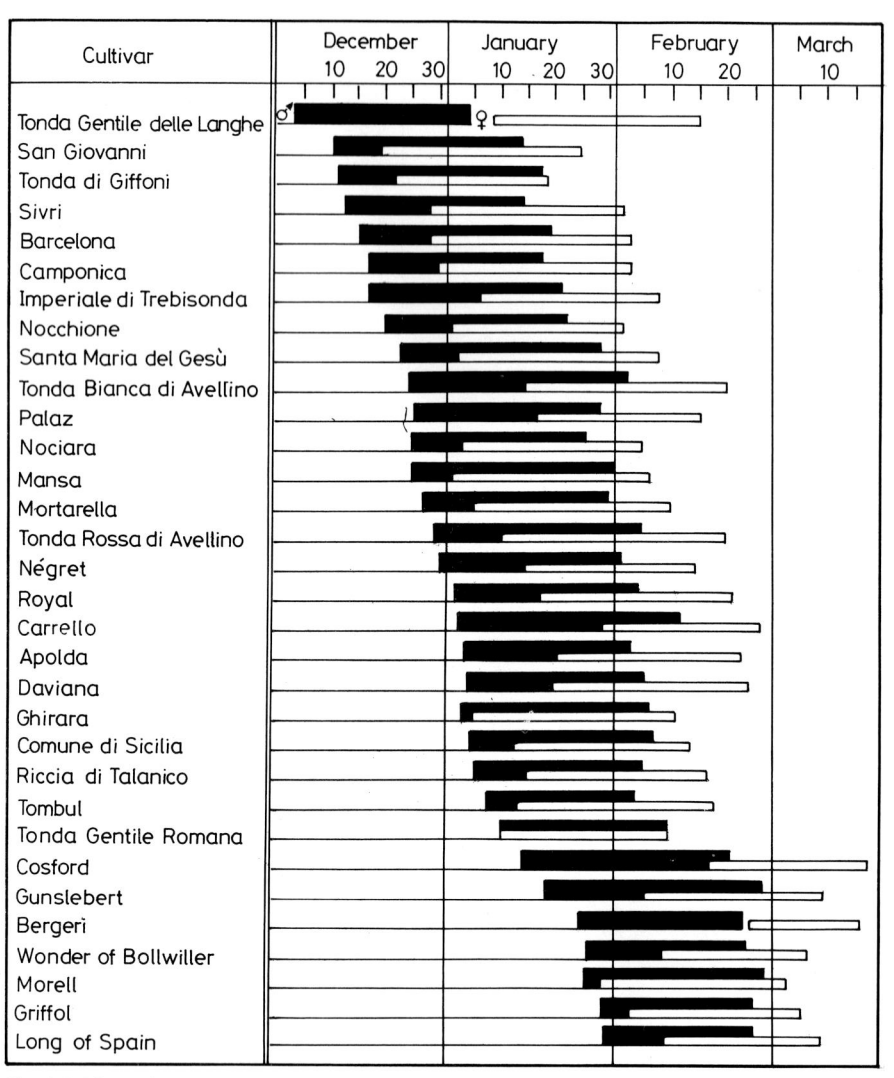

Fig. 4.7. Bloom time of male and female flowers near Rome (average of observations of years 1975 to 1980, after Manzo, P. and Tamponi, G., cit. by Jona 1986)

In contrast, Schmidt (1954) emphasized that an early blooming year increased the length of bloom. Ruggiero (1955, cit. in Porpáczy 1964) and Brózik and Nyéki (1975) found this generally true, and Reichel (1964b) reported this for apple, and Tóth (1957) found it in plum.

It is obvious from what has been discussed that there is a great need to study the relationship between duration of bloom and the season. This was pointed out with respect to the apple by Brown (1940) and Crane and Brown (1942).

In hazelnut cultivars not only the bloom time is different in the male and female flowers but so is the bloom duration (Fig. 4.7). It is striking that the duration of bloom in diclinous flowers is not influenced by the relative bloom time nor by dichogamia.

4.2. Groups of Blooming Time and the Extent of Joint Blooming

It has become widespread practice to group cultivars according to blooming time to establish cultivar groups that flower jointly. There may be several ways to do this which will be presented in the section on floral phenology.

The number of the blooming time groups used in pome fruit species was summarized in Table 4.10. Several researchers use three or four groups about equally while the use of five and six groups occur more rarely. The number of the time groups deviates considerably according to countries. There is a tendency that in colder climate investigation-sites use less groups.

It has been demonstrated in the apple, that five years are needed for bearing trees to group a cultivar definitely into a time group. In those cultivars which produce flowers on short shoots this period can be also accepted for young trees. Cultivars that flower on long shoots about six to eight years are needed to establish a definite grouping (Soltész 1987). In 119 apple cultivars only seven could be grouped constantly into the same blooming group during the observation time. Only many years of investigations will give the blooming time group in which flowering occurs with the greatest frequency.

In stone fruit species three blooming-time groups are most frequent (Table 4.11). Nyéki (1989) found great deviations in the groupings of cherry cultivars. In sour cherry five bloom-time groups have been recommended for countries with extreme climates. The cherry and sour cherry cultivars have been listed under eight bloom groups by Maliga (1953a) primarily because of the unstable blooming cultivars. Nyéki (1989) also found a lot of discrepancy in the grouping by authors of the sour cherry.

In plum, the cultivars are mainly grouped into three or four bloom-time groups. For example, the 'President' cultivar is described as a very early cultivar by a CTFL (1972) report. Nicotra et al. (1976) and Belmans (1986) have listed it under early while Krapf (1976) and Kellerhals (1986) considered it a medium early type.

Table 4.10. The Number of Blooming-Time Groups of Pome Fruits Based on the Literature

No. of blooming-time groups		Fruit species/Author
3	Apple	Vukolova (1960), Gautier (1971), Way (1971), Branzanti et al. (1974)
	Pear	Andreies et al. (1981)
	Quince	Brózik and Nyéki (1975)
	Medlar	Brózik (in Brózik and Nyéki 1975)
4	Apple	Maliga (1962), Silbereisen and Sherr (1969), Krapf (1976), Soltész (1982a, 1986, 1987), Gautier (1983)
	Pear	Brózik and Nyéki (1971), Nyéki (1980), Lombard-Hull and Westwood (1980)
5	Apple	Christensen (1977), Le Lezec et al. (1981)
	Pear	Vondracek and Kloutvor (1976)
6	Apple	Way (1973)
	Pear	Gautier (1974)

Table 4.11. The Number of Blooming-Time Groups of Stone Fruits Based on the Literature

No. of blooming-time groups		Fruit species/Author
3	Cherry	Hooper (1932), Beakbane et al. (1935), Crane and Brown (1937), Stancevic (1967), Götz (1970), Ryabova (1970), Christensen (1971), Cociu et al. (1981)
	Sour cherry	Blasse (1964), Stancevic (1969), Götz (1970), Ryabov and Ryabova (1970), Radulescu (1971)
	Apricot	Layne (1967), Kostina (1970), Brózik (in Brózik and Nyéki 1975), Brózik Jr. (in Nyéki 1980), Nyujtó et al. (1982, 1983), Lamb and Stiles (1983)
4	Cherry	Kamlah (1928a,b) Maliga (1952), Stösser (1966b, 1979), Brózik (1971), Krapf et al. (1972), Stösser and Neidhart (1975), Ryabova (1977), Brózik and Nyéki (in Nyéki 1980), Ivan et al. (1981)
	Apricot	Maliga (1966a), Fideghelli and Monastra (1978)
	Plum	Bellini and Bini (1978), Fideghelli and Monastra (1978)
	Almond	Brózik (in Nyéki 1980), Kester (1981)
5	Cherry	Christensen (1973, 1977), Albertini (1981), Kellerhals (1986), Trefois (1986)
	Sour cherry	Maliga (1953a), Brózik (1969), Nyéki (1974), Kellerhals (1986)
	Apricot	Monastra et al. (1984)
	Plum	Bellini (1973), Baldini and Scaramuzzi (1980), Kellerhals (1986)
6	Cherry	Caillavet (1973)
	Plum	Vondracek (1975)

In dioecious species the bloom-time groups have to be given separately for female flowers and male flowers on account of the major dichogamy. Only in this way has the blooming-time groups informative value. Jona (1986) formed five bloom groups in the hazelnut for female and also for male flowers. The bloom times for the different sexes were grouped similarly. This, of course, does not mean that homogamy is possible because during the bloom time there is the difference in calendar time. (Table 4.12). Of seven cultivars that had early female bloom three also had early male flowering, two medium early and one medium male flowering. The two medium-early flowering female cultivars had no male flowers during the same period and there were no late flowering females but there were six cultivars that were grouped into late flowering when male flowering was concerned.

The cultivars belonging to the same bloom-time group have a good chance to cross-pollinate each other. The cultivars belonging to the neighbouring bloom-time groups can also come into account when establishing an orchard of mixed cultivars. In the latter case it is especially important to clarify the bloom times so as to ensure appropriate pollen supply.

According to Kamlah (1928b), in years with cool spring, when suddenly warm weather occurs, the early and late blooming cultivars may coincide. In Fig. 4.8 we illustrate how the apple cultivars representing four bloom groups bloomed during the course of ten years. Based on all cultivar combinations, the average joint blooming is a good es-

Table 4.12. The Division of Hazelnut Species in Each Blooming-Time Group Relating to Pistillate and Male Flowers of a Different Blooming Time (after Jona 1986)

Female group designation	Female group time and No. of cultivars	Male group designation	No. of male blooming cultivars in each group
A*	Early: 6	A	3
		B	2
		C	1
		D	–
		E	–
B	Medium early: 2	A	–
		B	2
		C	–
		D	–
		E	–
C	Medium time: 8	A	1
		B	–
		C	7
		D	–
		E	–
D	Medium late: 7	A	–
		B	–
		C	4
		D	3
		E	–
E	Late: 10	A	–
		B	–
		C	1
		D	3
		E	6

*A to E means time of blooming for both male and female flowers.

timate for establishing an orchard of different bloom-time groups. From the point of view of joint blooming we consider it acceptable if the average of four cultivars with joint blooming reaches 40%.

Soyanov and Gormesky (1984) consider that if the full bloom of apple cultivars coincides for two to three days then there is no problem with pollen supply. Nyujtó (1958) claimed three such days were needed for the 'Pándy' sour cherry. Brózik (1971) determined in the cherry that out of a 10–14 days blooming at least four to six should be coincidental. Nyéki (1989) found that in stone fruits three days were acceptable during the full bloom time. In plum, Tóth and Surányi (1980) found three to four days to be necessary.

Ryabov and Ryabova (1970) in their joint blooming scale had three grades: full bloom is not coincidental, the full bloom overlap does not reach 50% and the coincidence of the full bloom of cultivars is above 50%.

The warning of Way (1971) is essential in reference to grouping cultivars. If apple cultivars are grouped into three blooming groups in the northeastern U. S. the early and

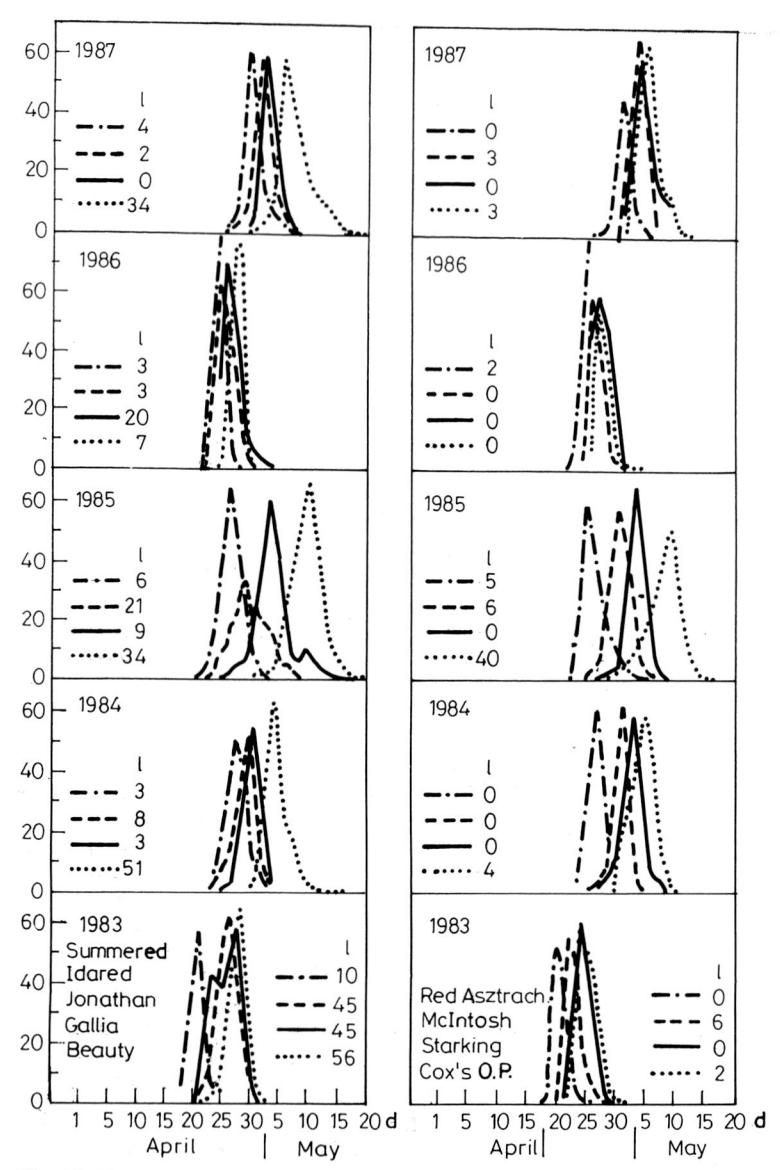

Fig. 4.8. Changes in joint blooming of apple cultivars belonging to different time groups in the course of ten years (after Soltész 1992) ...... early, ------ medium early; —— medium late; ∘∘∘∘∘∘ late; l = ratio of flowers on long shoots

late blooming cultivars only once did not flower at the same time during a ten-year period, but this is not the case in the southern states.

Often two cultivars are necessary to cover the bloom period of the main cultivar. Two pollen giving cultivars are used to pollinate the triploid 'Mutsu' apple cultivar in Hungary. Their bloom times are illustrated in Fig. 4.9. For the 'Pándy' sour cherry the pollen giving cultivars blooming at two different times increase the availability of pollen. It can

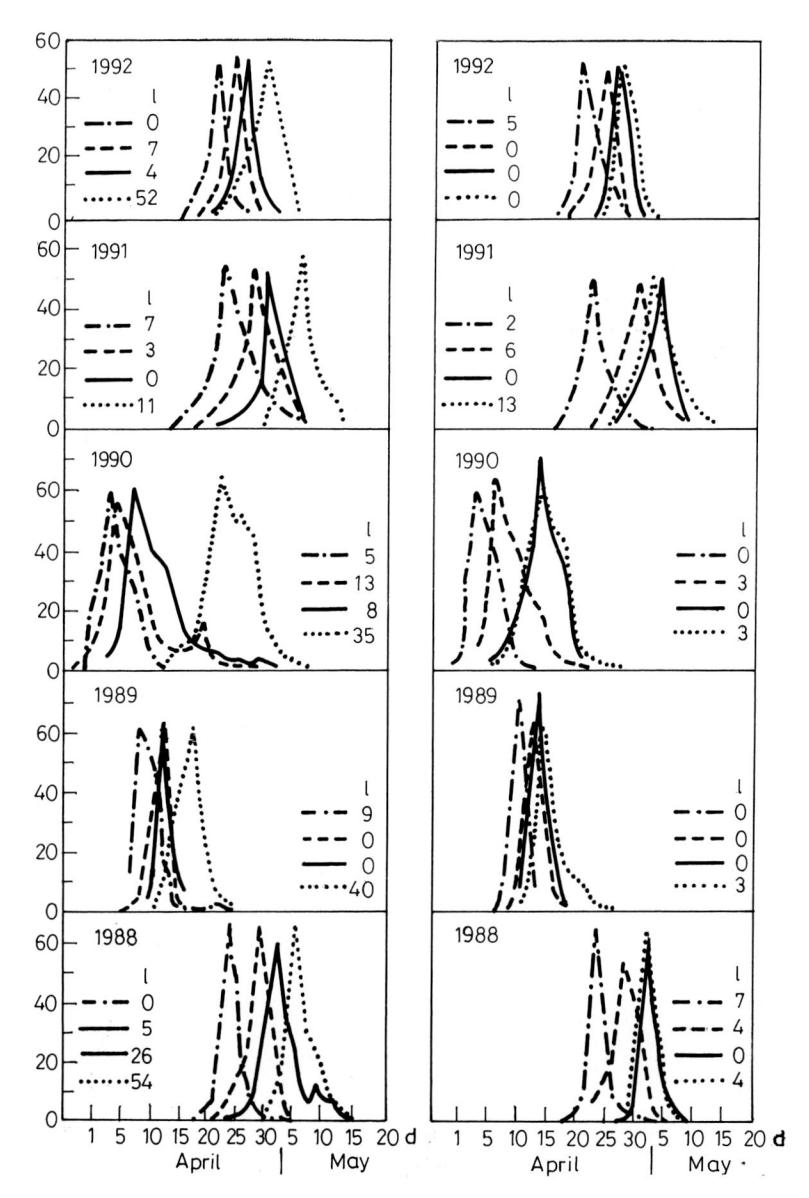

Fig. 4.8.

be seen well in Fig. 4.10 that the self-sterile 'President' plum and the joint blooming of three-pollen-donor varieties varies according to the years.

Nyéki (1989) considers the overlap in the full bloom time may have three possibilities:

1. The full bloom times do not overlap;
2. Blooming periods partly overlap (there are one to four days deviation in the full bloom of certain cultivars).

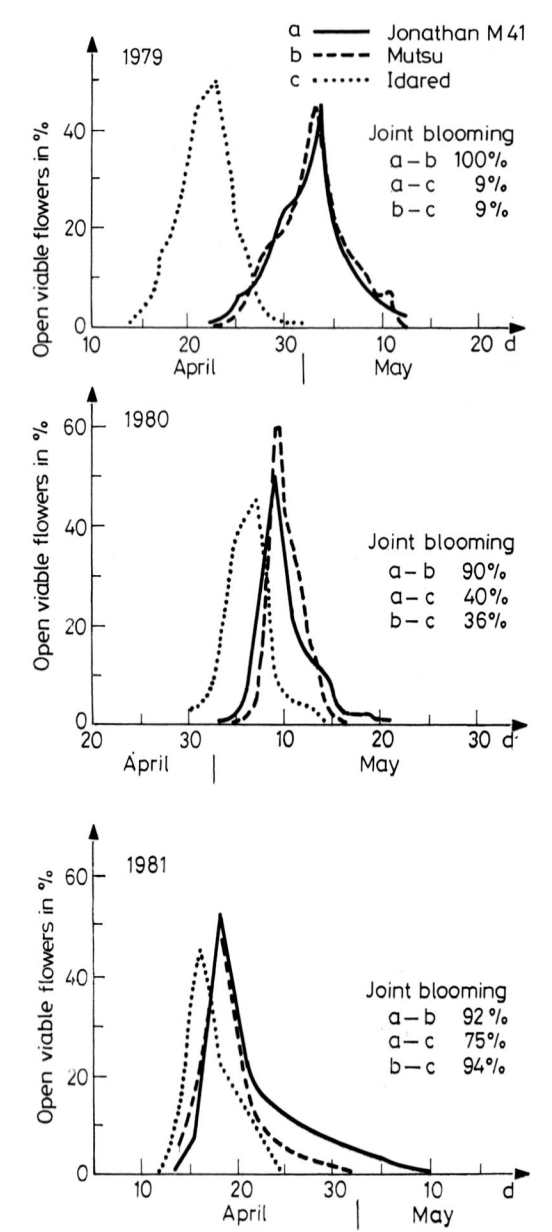

Fig. 4.9. The 'Mutsu' apple cultivar and two pollen donors joint blooming between 1979 and 1981 (after Soltész 1992)

3. The full bloom times coincide (there is only one-day difference in the full bloom of certain cultivars).

To establish the combinations the least favourable seasons of blooming should be the guideline.

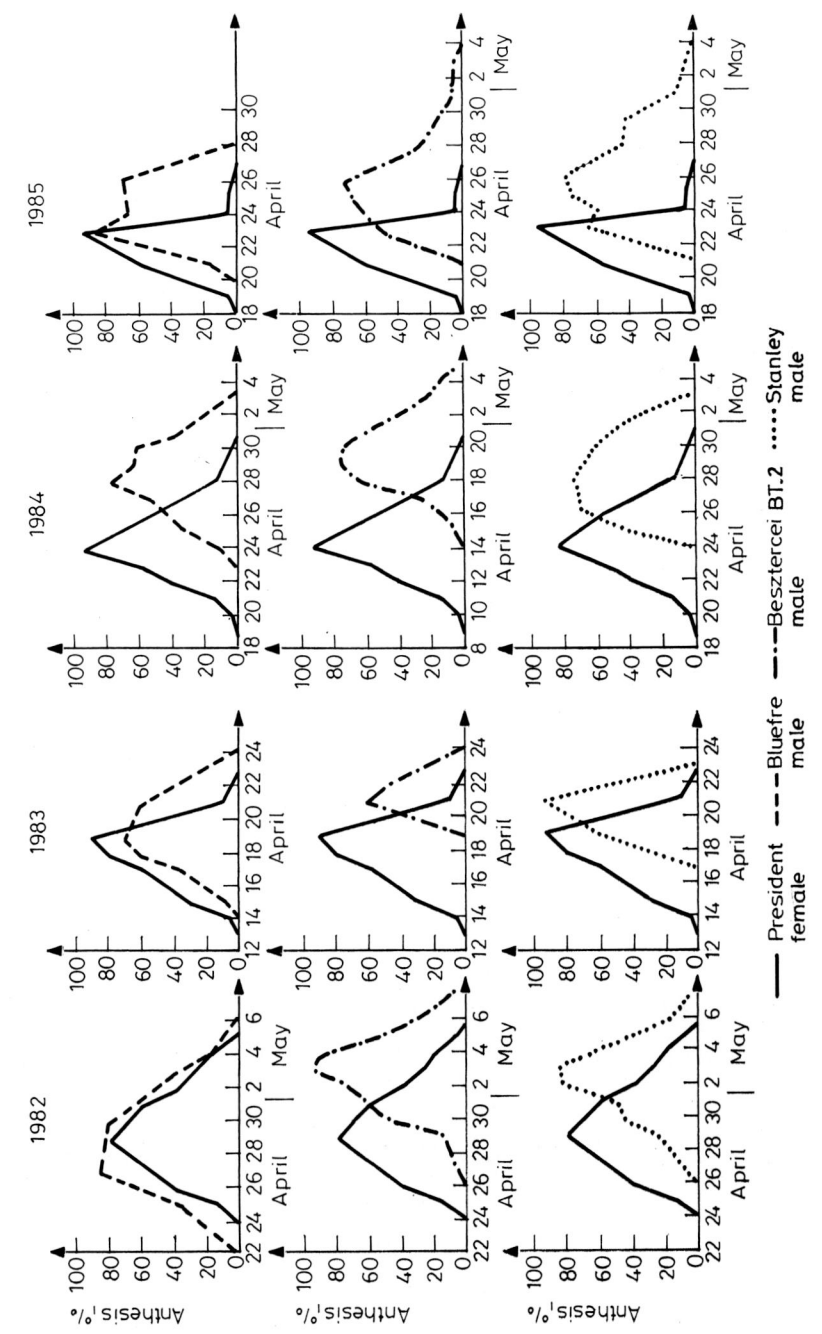

Fig. 4.10. Extent of joint blooming of self-sterile 'President' plum cultivars with pollen donor cultivars (after Nyéki 1989)

4.3. Factors Influencing the Beginning and the Course of Bloom

4.3.1. Length of Dormancy and Requirement for Cold

The length of the dormancy period and cold requirement are genetic characteristics of temperate fruit species. For the process of dormancy to be completed appropriate cold effect is required by the plants. Thus the cold keeps them in dormancy which immediately ceases if the given heat threshold level for the fruit species is surpassed.

The cold requirement is met at a temperature range of 0 to + 7 °C (Chandler et al. 1937 cit. by Faust 1989; Lamb 1948). Chandler (1942) found that if the cold effect is not realized then the bloom time is delayed. In apricot in addition to late flowering a number of the flowers drop as reported by several researchers (Crossa Raynaud 1955, Breviglieri 1958, Tabuenca 1968, Legave 1978).

Table 4.13. Chilling Requirement for Ending the Dormancy in Various Fruit Species

Fruit species	Hours of cold*	Author/Year
Apple	200–2000	Chandler et al. (1937, cit. by Faust 1989)
Pyrus communis	1000–1500	Lombard et al. (1980)
		Gautier (1981)
Asian pears	800	Lombard et al.
Quince	50–400	Chandler et al. (1937, cit. by Faust 1989)
Cherry	1000–1300	Tabuenca (1972)
	1000–1700	Bargioni (1982)
	500–1300	Chandler et al. (1937, cit. by Faust 1989)
Sour cherry	1000–1700	Bargioni (1982)
	600–1500	Chandler et al. (1937, cit. by Faust 1989)
Apricot	500	Tabuenca (1972)
	250–900	Chandler et al. (1937, cit. by Faust 1989)
Peach	650–800	Fideghelli and Capellini (1978), Weinberger (1956), Tabuenca (1965)
	200–1000	Chandler et al. (1937, cit. by Faust 1989)
European plum	700–1700	Chandler et al. (1937, cit. by Faust 1989)
Japanese plum	500–1500	Chandler et al. (1937, cit. by Faust 1989)
Hazelnut female flower	600–800	Bergamini and Ramina (1968)
male flower	350–600	
Cranberry	1300–1400	Mainland (1985)

*Chilling requirement under +7 °C for ending dormancy

Griggs (cit. by Soltész 1982a) found that high chilling-requiring pear cultivars, when planted in California, reacted differently to the mild winters because the required cold was lacking. The low chilling-requiring 'Bartlett' (Williams) pear bloomed at the usual time while the high chilling-requiring 'Winter Nellis' cultivar bloomed very late. Bargioni (1979) experienced the same in the cherry.

Overcash and Loomis (1959, cit. by Rom and Arrington 1966) pointed out the different cold requirements of apple cultivars. When the winter was mild, cultivars requiring

Table 4.14. Dates on Which the 1000 Hours Below 7 °C are Reached in 119 Places in Europe (after Kronenberg 1979

Place	Date (day/month)	Place	Date (day/month)	Place	Date (day/month)
Marseille	12/3	Ambérien	22/1	Stettin	2/1
Perpignan	8/3	Chatillon	22/1	Halle	2/1
Split	3/3	Cologne	22/1	Göteborg	1/1
		Constanta	22/1	Hamburg	1/1
Pescara	28/2	Odessa	22/1	Potsdam	1/1
Ancona	28/2	Bucharest	17/1		
Biarritz	28/2	Limoges	17/1	Biglandsfjord	30/12
Cherbourg	23/2	Edinburgh	15/1	Odense	30/12
Brawdey	21/2	Claremorris	15/1	Lärwick	29/12
Cognac	20/2	London	15/1	Multdorf	27/12
Dinard	19/2	Slåttersy	15/1	Warsaw	27/12
Brest	19/2	Stavanger	15/1	Karup	27/12
Valley	11/2	Eelde	15/1	Allenstein	26/12
Valentia observatory	11/2	Eindhoven	15/1	Kiev	26/12
Venezia	5/2	Trappes	15/1	Kironograd	25/12
Mount Batten	4/2	Prague	12/1	Bronnosund	25/12
Carcassone	3/2	Stornoway	12/1	Jönköping	22/12
Stalin	2/2	Chaermont	12/1	Stockholm	22/12
Tiree	1/2	Skagen	12/1	Karlstad	22/12
		Karlsruhe	9/1	Riga	22/12
Toulouse	30/1	West Raynham	9/1	Tartin	22/12
Nantes	30/1	Wahnsdorf	8/1	Leningrad	22/12
Poitiers	30/1	Dublin	8/1	Velikie Luke	22/12
Bordeaux	30/1	Aberdeen	7/1	Kwisk	22/12
Felixstone	27/1	Oban	7/1	Zürich	22/12
Rennes	25/1	Bremen	7/1	Kråkenes fyr	22/12
Shannon	25/1	Trier	7/1	Uppsala	16/12
Belgrade	24/1	Frankfurt a/M.	7/1	Turku	16/12
Milan	24/1	Emden	7/1	Tampere	15/12
Cardiff	24/1	St. Etienne B.	6/1	Gorky	14/12
Malin Head	22/1	Exeter	6/1	Donnarvet	10/12
Acklington	22/1	Hantsholm	6/1	Voshiy Voloshek	10/12
Blackpool	22/1	Bydgoser	5/1	Vallersund	7/12
Spurn Head	22/1	Nancy	4/1	Smolensk	7/12
Pershore	22/1	Wick	4/1	Voronez	6/12
Dungeness	22/1	Kiel	4/1	Nyland	4/12
Den Helder	22/1	Friedrichshafen	3/1	Lillehammer	3/12
De Bilt	22/1	Freistadt	3/1	Kajaaini	2/12
Valkenburg	22/1	Osnabrück	3/1		
Zeebrugge	22/1	Caen	3/1	Gunnarn	17/11
Lille	22/1	Hela	2/1		
Dunkirk	22/1				

high chilling bloomed later. Loewel (1966 cit. by Soltész 1992) reported the delayed blooming of 'Glockenapfel'. Rapillard (1981) wrote about the 'Gala' apple cultivar having a short dormancy and so its heat requirement was low.

Bargioni (1982) is of the opinion that if cherry and the sour cherry do not receive sufficient chilling more serious damage can be caused by the delay in bloom and its associated side-effects than frost during blooming. Bargioni (1979) stressed the especially high cold requirement of the cherry. Philp (1947) listed the cherry according to their requirements for cold.

Guerriero (1982) found that the apricot of continental origin had longer dormancy periods. Following a mild winter, because of the lack of the cold effect, flower buds fall before anthesis. In such years, there are many malformed flowers and blooming is late and irregular. Because of the late bud break, buds will be vegetative. On the same tree small fruits, the size of a nut, may be found together with flowers.

Griggs (cit. by Soltész 1982a) found in apple and in almond that after mild winters the difference between the times of bloom of given cultivars increases. Lalatta (1982b) observed that if bloom was late and chilling requirement was not met then bloom was late also in pear, peach and cherry as well as in hazelnut.

Szilágyi (1975) expressed the opinion that in June bearer strawberries the cold requirement may vary.

The cold requirements in hours are listed in Table 4.13 for various species. In Table 4.14 the data of Kronenberg (1979) are presented giving the dates when the cold hours below 7 °C reach 1000 for 119 sites.

A detailed summary has been given by Faust (1989) related to chilling requirement and dormancy. He also mentions that the greatest problem in growing temperate zone fruits in warmer climatic regions is the dormancy and the lack of appropriate chilling effect. However, greater emphasis has to be given in the future to the study of the correlations between dormancy and blooming. The chilling requirement and its fulfilment may also influence the heat accumulation necessary for the start of bloom (Bargioni 1982). Knowledge of the exact end of dormancy may help in the estimation of heat requirement (Swartz et al. 1984). The view that cold effect has no role in the bloom time, only accumulated heat is important should also be mentioned (Overcash 1965).

Investigations carried out on the apple by Plancher (1985) revealed several problems. In his view, measuring the cold requirement of fruit trees is complex, the scatter of the data is wide and the seasonal effect is strong. When estimating the amount of cold the evaporation loss cannot be taken into consideration since instead of the temperature of the tissues we measure the temperature of the air.

4.3.2. Calculation of the Amount of Heat Necessary to Start and Sustain Blooming

The bloom time of fruit species and cultivars is a stable genetic property which is manifested only after accumulation of a certain amount of heat estimated from a threshold level. The role of the temperature in beginning of blooming in fruit species has been recognized a long time ago (Ziegler 1879, Herbst and Weger 1940, Brown 1943, Sisler and Overholser 1943, Kobel 1954, Schmidt 1954, Tamás 1959, Efimov 1963, Vitanov 1964, Sherman and Janick 1964, Ryabova 1970, Solignat 1973, Andreies et al. 1981).

The actual bloom date of the cultivars and the beginning of relative blooming depend on the realization of the given heat requirement. However, other factors influencing the beginning of bloom prevent a simple linear relationship between bloom time and temperature requirement characteristic of the cultivar.

The changeable weather preceding bloom hampers the accumulation of heat to be put into effect to start the blooming of the cultivar. Ryabova (1970) attributed 90% of the effect of heat just before beginning of bloom to be responsible for anthesis and estimated 10% as a characteristic of the cultivar. Tóth (1957) and Nyéki (1975) likewise stressed the weather effects rather than the cultivar characteristics.

The amount of heat requirement for blooming and its estimation is affected by the starting temperature that is taken into consideration. The heat threshold level or effective minimum is the temperature value where the environmentally imposed dormancy (ecodormancy) is interrupted and the heat accumulation may start.

Generally +5 and +6 °C are given as heat threshold values. In addition to species listed in Tables 4.15 and 4.16 we also mention +6 °C for red and black currant (Blasse and Hofmann 1989), +5 °C for red currant, black currant and black sorbus (Porpáczy 1987) and for raspberry (Kollányi 1990).

Earlier in some fruit species temperatures above +6 °C were used, but Kobel (1954) regarded this disputable. Adapting Kobel's opinion, it appears that it was not a bad decision to take the meteorological 0 °C as a basis for the termination of ecodormancy and not to restrict it to a single value but to a range between 0 and +6 °C. However, when estimating heat unit a definite value has to be used, therefore this value should be uni-

Table 4.15. Amount of Heat Required for Pome Fruits to Begin Blooming

Fruit species	Date of start month/day	Method of temperature calculation	Temperature threshold value, °C	Temperature amount, °C	Author/Year
Apple	February 1	A	0	513–639	Sisler and Overholser (1943)
	January 1	A	6	87–144	Schmidt (1954)
	January 1	B	6	550–720	Tamás (1959)
	February 1	A	0	437	Vitanov (1963)
	January 1	B	6	4040	Horney (1965)
	Sprouting	A	0	373.0	
			5	195.8	Timon (1979)
	First warming up 5 °C	A	0	415–460	Sinkova (1970)
	First warming up 5 °C	A	5	185±10	Vrublevsky (1970)
	January 1	A	0	537	Babóné (1974)
	Sprouting	A	0	366	Babóné (1974)
	March 1	A	0	320–460	Soltész (1987)
Pear	January 1	B	6	3013–3440	Herbst and Weger (1940)
	January 1	B	6	3435	Schossig (1959)
	February 1	A	0	389	Vitanov (1963)
	January 1	B	6	3340	Horney (1965)
	January 1	B	6	341–3539	Nyéki (1973)

A = sum of daily averages above threshold; B = sum of hourly temperatures above threshold

Table 4.16. Heat Requirements of Stone Fruits for the Start of Bloom

Fruit species	Date of start month/day	Method of temperature calculation	Temperature threshold value, °C	Temperature amount, °C	Author/Year
Peach	February 1	A	0	299	Vitanov (1963)
	March 1	A	0	143–388	Bulatovic (1959)
	Sprouting	A	0	224. 8	Timon (1970)
			5	108. 7	
Apricot	March 1	A	5	284	Kobel (1954)
	34 to 38 days before flowering	A	6	141±11	Nyújtó and Tomcsányi (1959)
	February 1	A	0	248	Vitanov (1963)
	30 days before flowering	Max.	5	230–301	Bespechalnaya (1970)
	January 1	Max.	0	342	Molnár and
	December 16		0	366	Stollár (1971)
Sour	Sprouting	A	5	135	Kirakosian and Beketovskaya (1970)
cherry	Sprouting	Max.	5	177–207	Efimov (1963)
	January 1	A	0	328–417	Apostol (1976)
Cherry	February 1	A	0	315	Vitanov (1963)
Plum	February 1	A	0	399	Vitanov (1963)
	January 1	B	5	3045	Horney (1965)
	Sprouting	A	0	271. 8	Timon (1970)

A = sum of daily averages above threshold; B = sum of hourly temperatures above threshold; M = sum of maximum temperatures

formly 0 °C. If the calculation would start at +6 °C then the temperature effects between 0 and +6 °C, that are important for the start of anthesis, would be left out.

Modic (1974) considered +4 °C in the hazelnut as the threshold value but he also experienced that bloom already began at 0 °C.

It appears from the data on pome and stone fruit species that there are rather large variations in temperature values depending on the source of information.

Herbst and Rudloff (1939), Herbst and Weger (1940), and Weger et al. (1940) using previous experience and detailed phenological investigations established the basis of heat estimation related to blooming. The starting point of January 1 used by them in their calculations has been taken over by others. This is acceptable in the species with short dormancy period (apricot, almond, hazelnut) but in the majority of the species it is too early. We have to find a starting date for the estimation of the heat units after endodormancy has ended. With this we do not want to exclude the temperature effects during endodormancy but they should not appear in the estimated total for heat requirement.

In defining heat requirement, it is actually better to use the total heat amount because this is what we measure and sum during the course of observations. The temperature requirements given in the literature considering the various calculation methods and the various conditions of data taking give little possibility for generalizations. The heat requirement values at the same investigation site are very scattered. After 20 years of investigations, Soltész (1992) found that in apple there is a loose relationship whichever method is used between the amount of heat received and blooming.

One of the goals for calculating heat requirement is to forecast the bloom data. It is somewhat of a paradox that forecasting would be important if it is given earlier but early estimates reduce the reliability of the data.

It is worth considering White's (1979) work who using temperature data for 53 species worked out a bloom data forecast. Fruit growers, especially with respect to the early blooming species, are not in a position to use the method considered to be best, because they cannot use marker plants for forecasting bloom data.

Newer observations support the fact that the later the blooming of the fruit species the heat calculation gives the less reliable result, and it is especially important to take into account other temperature factors (night minimum, day maximum temperature and their changes) and climatic features. The data show that the bloom period of the apple is the turning point in this respect.

Herbst and Weger (1940) investigating the pear found that blooming closely followed the estimated heat value. Pearce and Preston (1954) did not find this in apple. After the estimated heat amount was reached the start of bloom depended primarily on how the night temperatures varied. If by the middle of April the received heat does not enable early anthesis then the night temperature has a larger role than the further accumulation of temperature values.

This is supported by Soltész's (1992) data which indicate that during the early blooming years the temperature requirements were smaller. In contrast, Simidichev (1971*b*) found in the apple that there was more need for heat in years when blooming occurred earlier.

Schossig (1959) found in pear that in the case of late blooming there was relatively more need for heat. The deviation from the average heat requirement in the pear amounted only to ±1 day.

Balchin and Pye (1950, cit. by Soltész 1992) gave more precise blooming forecasting with minimum temperatures than the daily maximum or average temperature. Lake (1956) stressed the importance of the minimum temperature in chestnut. In sour cherry, Efimov (1963) mentioned the advantages of maximum temperature values. Without further examples it can be emphasized that species flowering at different time require different methods of estimation for establishing reliable heat requirement and for forecasting bloom data. White (1979) reported methods from which one can select and modify as needed.

There is a close relationship between daily heat value, the active heat requirement and the process through the various phenophases. After reaching the necessary heat level a temperature stimulus is needed so that anthesis may begin. The level and nature of this heat stimulus may vary according to the fruit species.

Generally, the earlier the fruit species blooms after reaching the necessary heat amount the sooner bloom begins even at lower temperature. In these the heat amount can be estimated with the usual methods and reliable results can be obtained.

The heat stimulus necessary to accelerate anthesis according to Sinkova (1970, cit. by Soltész 1992) is at least 10 °C. In Kobel's view (1954), in apricot, the bloom following the accumulation of the appropriate heat amount takes place if the daily temperature reaches 9.5 °C.

During the period from bud burst to start of bloom there is smaller deviation in the number of days and the temperature values than in the calendar for the start of bloom. This deviation is bigger in species which have later blooming.

The process of bloom itself is also influenced by the temperature. After anthesis has commenced a meteorological stimulus needs time to take effect. Apple cultivars react to the temperature stimulus in two days according to Hoffmann (1962). In sour cherry, Nyéki (1974c) found the reaction time to be one to two days and it was the same in cherry (Nyéki et al. 1974). Kozma's work (1950) is worthwhile studying for fruit growers with respect to the temperature induced blooming periodicity of the grape.

Nyéki (1974c) in sour cherry demonstrated that flower anthesis pauses at daily 5 °C temperature, and rising temperature up to 16–17 °C is stimulative with respect to the rate of blooming. The process of bloom is rapid and spasmodic at temperatures above 20 °C. In apple, Hoffmann (1962) found that the close correlation between blooming and the temperature occurs at between +6 and 15 °C. Below +6 °C the temperature of the previous days has no effect. Above 25 °C temperature hardly has a stimulative effect.

Temperature greatly affects the duration of bloom. In sour cherry, Kirakovsian and Beketovskaya (1970) demonstrated the combined effect of temperature and precipitation on the duration of blooming. Nyujtó (1980) found that the duration of bloom in the apricot at 20 °C maximum was six to seven days while at 12–17 °C the time doubled to 12–14 days. Oprea (1981) observed that the duration of bloom of apple depending on the temperature varied between 5–20 days. Rasmussen (1984) also emphasized the role of temperature in the process of blooming.

Soltész (1986) considers that there is a tight correlation between temperature and bloom dynamics in apple but it is difficult to estimate it on account of the flowers formed on the shoots. He demonstrated that the year has a decisive role with respect to the bloom duration which depends partly on the temperature and partly on the ratio of the flowers formed in axial buds of the shoots.

Mention should be made of the effect of temperature below 0 °C on the process of blooming. Efimov (1963) found that in the sour cherry this only had an effect if it damaged the flowers. Ryabova (1970) reported that in cherry below 0 °C it had no effect. Tóth (1967) established a decrease in the bloom duration due to spring frost in the plum.

The freezing of the apical flowers of the apple inflorescence leads to a more rapid anthesis in the other flowers (Babaleanu 1938). Whichever flower is affected by the frost its bloom time is reduced (Soltész 1982a, 1992). Higher temperature speeds blooming and the difference decreases between bloom time among the cultivars (Krapf et al. 1972).

There is a further complication in the relationship between bloom time and temperature. Within a species cultivars react differently to temperature. In sour cherry, the differences in the bloom time among the cultivars can be explained by the heat requirements of the cultivars (Nyéki 1974c). In the apple, Tamás (1959) demonstrated the heat requirements of bloom-time groups of cultivars but the basic cultivar and its mutants may have deviations. Schossig (1959) pointed out the major differences in the pear cultivars. Tóth and Surányi (1980) indicated that the plum has a relatively low heat requirement but within this low requirement there is still difference in the bloom times for *P. cerasifera*, *P. salicina* and *P. domestica* cultivars.

Trees on various rootstocks react differently to the temperature influencing blooming because their heat requirements changed (Tamás 1959). Babóné, cit. by Brózik and Nyéki (1975) showed that in apple trees on stronger growing rootstock bud burst starts later, but because this is so the time between bud burst and bloom falls to a warmer period and the lag is made up. Thus the actual time of flowering is not influenced by the rootstock.

Sinkova (1970, cit. by Soltész 1992) reported on deviations in heat requirements according to location in apple. The temperature necessary for certain phenophases to be accomplished depend also on the external environmental factors. The cultivars with higher heat requirements react more sensitively under unfavourable temperature conditions. This is what Lake (1956) experienced in 'Cox's Orange Pippin' apple cultivar.

Finally, heat requirements and temperature effects connected with other climatic factors should be discussed. These are considered as individual factors whereas their main effects are indirect and they mainly influence the temperature conditions before and during blooming.

Solar radiation may have a role in influencing the time of blooming, but its direct effect cannot be separated from the temperature effect (White 1979). The direct role of solar radiation in moderating the heat threshold level may be considerable. At 0 °C it may interrupt the environmentally imposed dormancy. The radiation energy at the absorption site transforms into heat energy and the plant parts which are exposed to solar radiation may have higher temperatures than the surrounding atmospheric temperature (Herbst and Weger 1940). In trees that have small canopies (slender spindle, super spindle, etc.) and in fruit species with anthesis before development of foliage we can expect the influence of radiation in increasing the temperature of the tissues. Therefore, in such situation it is especially justified to use lower heat threshold levels.

The amount of precipitation has an effect on the beginning and the course of blooming. Safer and Bespechalnaya (1971, cit. by Soltész 1992) showed in the apricot the bloom-delaying effect of 57–87 mm precipitation which occurred 30 days before bloom. After a longer period without precipitation daily doses of 12 mm precipitation favourably influenced the rate of blooming in apple while greater precipitation than this delayed blooming (Hoffmann 1962). Delay of blooming with sprinkling is based on this observation applied successfully.

Möhring (1942) pointed out the effect of soil temperature on bloom time. It is still disputed whether the soil temperature has a direct influence or not on flowering or its role is limited to warm the air immediately above the soil.

In the literature the effect of June temperature is often mentioned as a factor in determining the time of next year's flowering. It has not been determined whether the June heat effect influences the time and extent of flower bud formation or the effect may be direct. Pearce and Preston (1954) were the first to report that in apple the hot June delayed blooming the next years.

In any case, McLagen (1933) investigated the effect of multiple years on flowering. His finding with respect to the pear would be worth considering generally. The temperature of three seasons affects the beginning and the course of blooming. The temperature during previous spring may influence petal fall thus affecting the beginning of flower bud formation. The temperature during winter before blooming may influence the course of dormancy, and the temperature during the spring may provide the needed heat unit for the beginning of bloom.

4.3.3. The Role of Geographical Latitude and Height Above Sea Level

There are contradictory opinions concerning the effect of geographical location on blooming. Roemer (1970) reported that the geographical location and the habitat have a greater role in regulating flowering than any other factors that may occur at the given location. Bellini and Bini (1978) in observations on plums stressed that the effects of the geographical latitude and the height above sea level are small in comparison to the effect of atmospheric temperature.

Of course both, Roemer and Bellini and Bini, may be right. We have to take into consideration how far the two sites are that are compared. It may be that in one habitat the ecological variability from year to year is greater than the difference between the two geographically close locations.

The following examples will demonstrate that latitude and height above sea level can produce differences in bloom time. The two factors, temperature and geographical location, cannot really be compared since the geographical location manifests its effect through the changes in temperature.

In habitats which are distant from each other (in geographical latitude, and in height above sea level) the differences are striking with respect to the bloom time. In habitats close to each other these two effects may overlap. Often, the favourable geographical location allows the occurrence of temperature values which give the same heat units that are not indifferent from the point of view of bloom time.

Hoffmann (1887) compared the bloom time in 333 geographically varied sites in apple and in 262 locations in pear. Progressing northwards one degree of latitude, produced four to six days delay in bloom in the apple and four or five days in the pear. One hundred metres increase in height above sea level meant a two days delay in flowering in the apple and three days in the pear.

Phillips (1922), who is the most quoted in this topic, established that the start of bloom time was delayed by four to six days at every degree of latitude northward and every 33–34 m above sea level meant one day delay.

According to Blasse (1964), relatively small difference in geographical location caused several days in the start of bloom in 13 sour cherry cultivars. Radulescu (1971) observed major differences in blooming of 26 sour cherry cultivars in four habitats.

Way (1974) compared cherry-tree blooming in the northeast and western parts of the U.S. In New York state the cultivars started to bloom at almost identical times within four to five days. In the western states there are great differences in bloom time of the same cultivars.

Krapf et al. (1972) found that the cherry completed blooming faster at higher locations and the simultaneous blooming among the cultivars was common. Nagasawa et al. (1972) observed the start of bloom of 'Golden Delicious' apple cultivar at 340, 480, 770 and 850 m heights above sea level. He found a difference of 14 days in the bloom time between the highest and lowest orchards.

Ryabov (1975) observed that the further north the peach is cultivated the smaller the difference in the time of blooming of the cultivars. In apricot, 1° change in latitude, on average, caused three to four days of difference in the start of bloom.

Soltész et al. (in Nyéki 1980) compared the bloom time of 18 apple cultivars in two identical habitats but the orchards were located 1° latitude apart. The relative start of bloom of cultivars was only partly identical, but it did not change to the extent that any cultivar have bloomed earlier at 1° latitude further north than at the more southern location. At habitats further north the bloom occurred closer to each other, the bloom duration declined but the simultaneous blooming was more common especially in 1977 (Fig. 4.11).

Bargioni (1982) reported that in Italy the blooming of the cherry was significantly influenced by the geographical location. Lalatta (1982a) found that in apple blooming there was a 20-day difference due to a change of 250 m above sea level.

Cultivar	April 5–30 / May 5–10	Relative rank (o)	Relative rank (●)
Red Astrachan		1	2
Stark Earliest		2	4
Éva		3	5
Close		4	9
Gravenstein		5	3
Ceglédi piros		6	6
Idared		7	8
Melba		8	7
Jonathan		9	10
Early Red Bird		10	1
Jonadel		11	16
Jonared		12	11
Starkrimson Delicious		13	14
Staymared		14	12
Golden Delicious		15	13
Starking		16	15
Winter Banana		17	18
Golden Spur		18	17

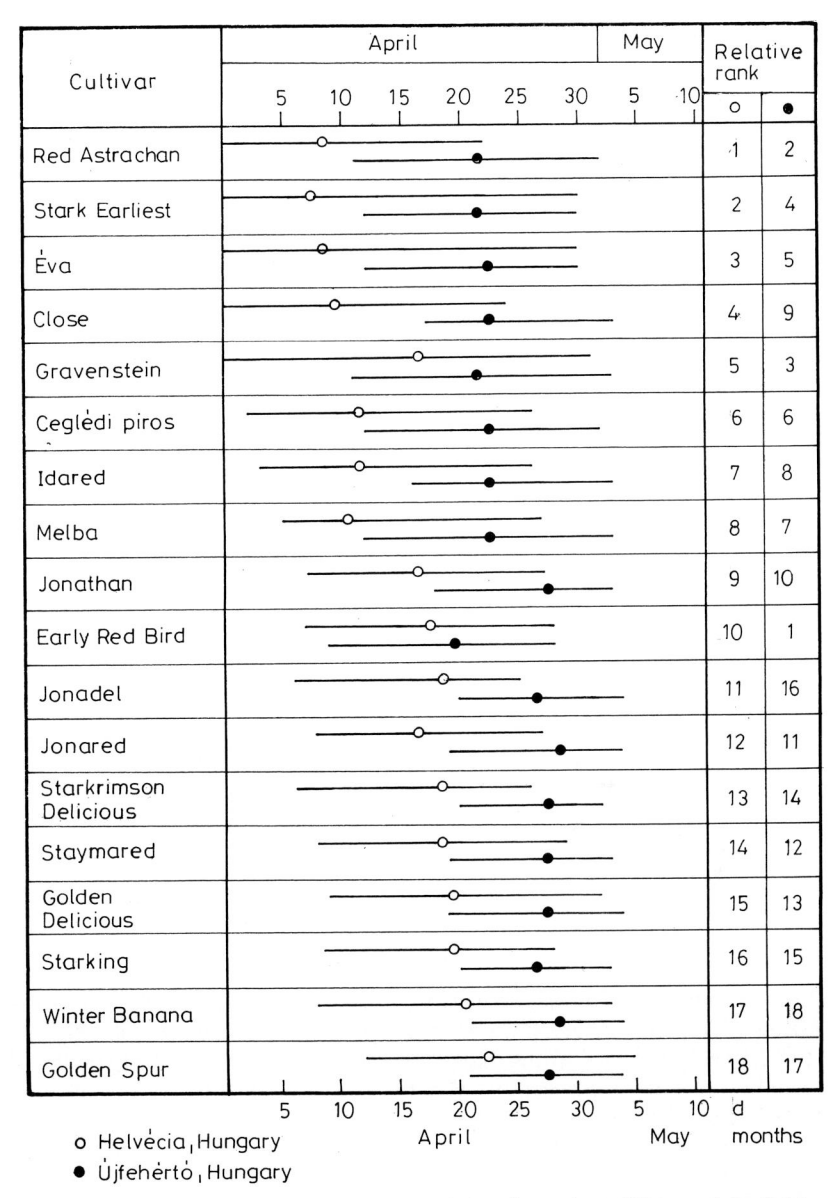

o Helvécia, Hungary
● Újfehértó, Hungary

Fig. 4.11. Bloom time of apple cultivars in two apple growing regions of Hungary (after Soltész et al. in Nyéki 1980)

Faust (1989) established that the further north the cultivar is located the later its blooming. Later blooming causes compacted flowering far apart further south and compacted flowering causes overlap between cultivars that flower.

However surprising it may be, the experts living in the northern hemisphere investigated the latitude effect generally speaking only northwards. Only Griggs (cit. by Soltész

113

1982*a*) mentioned that the almond bloomed earlier in the northern part than in the southern part of California. Almost the same can be said for the height above sea level progressing upwards. The question has never been raised what happens if we proceed in the other direction. Smith (1970) raised the issue that species (cultivars) having high chilling requirements become damaged at site with mild winters. Contrary to expectation, their blooming does not occur earlier but due to the inadequate chilling effect, blooming is late.

The 'Cafona' apricot cultivar was surprisingly late in blooming in south Italy by several days compared to the colder Central Italian region which has a later spring. This is attributed by Guerriero (1982) to the lack of appropriate cold effect.

There is plenty of information that after the chilling requirement satisfied additional cold decreases heat requirement. This means that in northern, cold locations, when the air warms, usually in early May, blooming is explosive because heat requirement is relatively small after prolonged cold exposure. Cherries and apples bloom only days apart. Perhaps cherry blooms May 5, apple May 10. In the south where chilling is satisfied but extra cold is not imposed blooming is more spread. Cherries may bloom March 25 to April 1, apples April 24 to May 1.

Further south, when chilling is not sufficient, blooming is very late, slow and sporadic, and lasts for weeks, only a few flowers (1% – 5%) are open on a given day.

We have to count on increased damage of cultivars requiring greater chilling, if they are grown on habitats with milder winter (Faust 1989).

Kronenberg (1985) on the basis of a survey illustrated the effect of geographical latitude and height above sea level on the blooming of the apple cultivar. In Northern Europe the bloom of the 'Boskoop' begins when summer cultivars in the southern countries are about to ripen. Kronenberg (1985) claimed that the geographical situation of 'Boskoop' and 'Golden Delicious' apple cultivars give a rather varied picture with respect to blooming. Analysing the data it appears that among the 119 investigation sites in 87 sites 'Golden Delicious' blossoms later than the 'Boskoop'. The colder the climate is in these regions the closer the two cultivars get to each other in blooming. At 14 investigation sites the two cultivars blossomed together. Surprisingly, at the warmest 18 sites—primarily at the warmest South European locations–'Golden Delicious' blossomed before 'Boskoop'. Compared to 'Boskoop' 'Golden Delicious' bloom time of deviated form –21 to + 31 days.

Lalatta and Sansavini (1983) have proposed and are expecting the apple cultivation in the mountain regions of Italy to increase. They remarked that bloom time in the mountain region is a much more complicated question than just a matter of the effect of altitude on temperature. In fruit-growing mountain regions, besides the temperature fluctuations, humidity changes significantly. The technological elements (size of the tree, shoot growth, etc.) also change.

4.3.4. Cultivar Properties Influencing Bloom

The geographical origin of fruit species and cultivars determines the beginning bloom. *Prunus cerasifera* originating from Asia Minor and the Balkan region and the *Prunus triflora* group, originating from Eastern Asia, blossom early often already in March. The *Prunus domestica* group members accounting for the most valuable prune cultivars generally blossom in April (Schaer 1952). The flower type also determines the start of bloom. Ryabov (1957) observed that the rose-flowered peach cultivars generally flower earlier than the bell-flowered ones.

The black currant derived from the continent (Poland, Russia, Siberia) are earlier blooming types than the Atlantic cultivars (English, Dutch, French, German). These properties are inherited (Plancher and Dördechter 1983).

One can select clones and spur variants with properties such as ripening time, better fruit quality as well as favourable growth characters. The species could be grouped into three categories with respect to bloom time of their mutants. The first category contains those in which there are no mutants and they are walnut, chestnut, peach, almond, hazelnut. The second group comprises those in which the bloom time of the mutants greatly deviates from the basic cultivar and includes cherry, sour cherry and apricot. The third group contains those cultivars of the species apple and pear where very many mutants can be found but the genetic variations did not influence bloom time.

In almond, grouped into the first group, mutants of 'Tardy Nonpareil' flower 10–14 days later than the basic cultivar (Kester 1981).

In the apricot Nyujtó (1978) found three bloom time groups of 'Magyar kajszi' clones.

Maliga (1953a) compared the full bloom time of selected clones of 'Pándy' sour cherry. Brózik (1969) found three to eight days deviation in full bloom in the clones of 'Pándy' sour cherry and two to ten days in the 'Cigány' sour cherry cultivar types and one to six days deviation in the clones of the 'Germersdorfer' giant cherry.

The 'Compact Lambert' blooms later than 'Lambert', the basic cherry cultivar in contrast what is expected from its growing type. From this Lapins and Schmidt (1976) concluded that there can be a mutation which can occur not only by changing growth type but blooming only.

Tóth (1967) found two types of blooming in French prunes (Agen). 'Agen type 1' blooms in mid-period whereas 'Agen type 2' blooms late. Soltész (1990, unpublished) found that mutants of 'Bartlett' (Williams) pear do not change in their time of bloom. Similarly, no change in time of bloom was observed in mutants of apple cultivars of 'Jonagold' and 'Delicious' (Albertini 1980, Soltész 1992). In spur-type apples half to one day delay was observed (Soltész 1992). This is somewhaat less than what Palara et al. (1983) established (one to two days).

Silbereisen (1970) found significant differences in bloom time among the 'Jonathan' clones in some years. Soltész (1992) experienced no more than one-day difference in 30 'Jonathan' mutants. Faedi and Rosati (1974a) reported three to four days earlier blooming in 'Blackjon' a 'Jonathan' mutant. Cassani (1981) found surprisingly great differences in the bloom time of two mutants of 'Golden Delicious'. 'Goldensheen 1960' bloomed eight days later than the basic 'Golden Delicious' while in the 'Goldenir INRA Sel. P8–R12 A17' mutant the difference in bloom occurred 15 days later.

Judging from the literature, we can establish that the climate (weather) has various effects on the cultivars. Nyéki (1990) distinguishes the following:

a) Stable blooming cultivars depend less on the weather;
b) Labile blooming cultivars are strongly influenced by the weather.

Wociór (1976a, b) found the rapid bloom rate to be a property of the sour cherry cultivars. According to Brózik and Nyéki (1980), the 'Cigány' sour cherry types are characterized by fast blooming. Nyéki (1989) considered the 'Pándy' sour cherry to be a labile blooming cultivar.

Nyéki (1973a) observed 'Beurre Hardy', 'Olivier de Serres', 'Beurre Diel', 'Madame Favre' and 'Curato' pear cultivars to have labile bloom times. Chollet (1976) considered 'Comice' also a labile pear with respect to blooming time.

According to Brózik and Nyéki (1975), contradicting other data the 'Conference' pear cultivar has a characteristically short bloom time. Chollet (1976) regarded the pear to be a typically fast blooming type species. This species has a tendency to have second blooming, especially the early bearing and fertile cultivars. Second blooming is considered disadvantageous because it increases the infection by *Erwinia amylovora*, the bacteria causing fire blight.

Cociu and Tudor (1981) consider short bloom time of 'Prima' and 'Elita 20' apple cultivars and the long bloom time of 'Spartan' and 'James Grieve' to be a stable cultivar property. In contrast, in Rasmussen's opinion (1984), the early and medium early blooming apple cultivars are labile in respect to bloom time.

A special clone effect has been reported by McGranahan and Forde (1985) in the walnut. The bloom time is greatly influenced by the age of the tree from which the scion was originated.

The relationship between bloom and maturity time is raised not from the aspect how bloom time is necessary for estimation of harvesting time but the reverse: namely, how maturity time, considered as a cultivar property, is correlated with bloom time (Soltész 1992).

Without drawing far-reaching conclusions, it should be mentioned that Badescu et al. (1981) and Soltész (1992) reported on a loose correlation between maturity and bloom in the apple. Lamb (1966) and Ivan et al. (1981*a, b*) reported the same in cherry, Cociu (1981), Albertini (1980) and Stancevic (1969) in sour cherry, while Vondracek (1975) in plum. Terrettaz (1978) reported that in the blackberry the duration of bloom is not related to the extended maturity time. Kollányi (1980) found a tight correlation between these two phenological characteristics in raspberry.

Most researchers found no correlation between bloom time and maturity time:

in apple: Reichel (1964*a*), Way (1971), Faedi and Rosati (1975*b*) Mullins and Lockwood (1979);
in pear: Baldini (1949);
in cherry: Kobel (1954) Ryabova (1970, 1977), Christensen (1970), Sansavini et al. (1981);
in plum: Tóth (1967),
in red and black currant: Blasse and Hoffmann (1989).

Ryabova (1977) noted that the early maturing cultivars do not necessarily flower early but in a given year the early blooming is generally followed by early maturity. Plancher and Dördechter (1983) did not find any correlation between the bloom time and maturity in black currant cultivars originating from the continent of Europe while the correlation was more characteristic of the Atlantic cultivars.

The majority of researchers found a close correlation between bud burst and bloom. Let us mention a few exceptions first. Vondracek (1972) did not find any correlation between these phenophases in sour cherry. Terrettaz (1978) reported in the blackberry that there was a looser relationship between bud burst and bloom time than in bloom and maturity time. Plancher and Dördrechter (1983) found a correlation in black currant between bud burst and bloom in the Continental cultivars, however, this was not the case in the Atlantic cultivars.

The researchers who found a close correlation between the bud burst and bloom are:

in apple: Schmidt (1940), Murawski (1959), Reichel (1964*a*), Visser and Scharp (1967), Schwartz et al. (1984), Soltész (1992)

in pear: Thibault (1979)
in cherry: Vondracek (1969), Bodi (1981a)
in plum: Vondracek (1975).

Volorina (1978) found a correlation between the cherry and sour cherry with respect to anthesis and pollen grain starch synthesis. Dale's (1988) investigations did not show a close correlation between the shoot growth of the raspberry and blooming. Visser and Scharp (1967) found a correlation between the time of leaf fall and the extent and time of bud burst.

According to Bellini (1980), in walnut bud burst is not correlated with the capacity of how the cultivars are able to flower in the axial buds.

The time of bloom is influenced by the site of flower formation, the type of bearing shoot and the structure of the inflorescence. These show great deviations in the cultivars of certain species.

Numerous factors may influence flower formation in the lateral buds of long shoots in apple. However, the cultivar property is decisive. According to Silbereisen and Scherr (1969), the formation of flowers in lateral bud, is a characteristic of 'Golden Melon', 'Rome Beauty' and 'Staymared' cultivars while it does not apply to 'Delicious'. Silbereisen (1970) considered formation of lateral buds a characteristic of 'Jonathan' apple cultivar but he found deviations among its mutants. Branzanti et al. (1974) found the tendency to flower on long shoots to be characteristic of about 40% of apple cultivars.

There is also a difference in the flower-forming tendency in the apical buds and the apex of long shoots of the apple cultivars (Soltész 1986). It should be realized that the cultivar characteristics are strongly influenced by other factors of flower formation, therefore, it is rather difficult to give correlations which would be valid for cultivars everywhere. In addition, there is no close correlation between the flowering tendency and extent of flower formation on long shoots.

In the late blooming walnut the tendency to form flowers in lateral buds of long shoots is the most important property, for raising the reliability of bearing fruit. In the apple, which blooms earlier than the walnut, the flowers formed on the long shoots are unable to fill this role because their viability is much lower for fruit set than the flowers on short shoots.

The length and nature of the bearing shoots mostly influences the stability of bloom time in the apple, the annual deviation of joint bloom time of given cultivars and the process of bloom itself. In these the quantity of flowers formed in the lateral buds on long shoots has a decisive role. It is important to know what to expect in this respect in certain cultivars.

On the basis of flowering tendencies on long shoots, Soltész (1986, 1989) has grouped apple cultivars into four groups:

a) There is negligible or little tendency in young and old trees such as 'Delicious' as well as in mutants of 'Cox's Orange Pippin', 'Reine des Reinettes', ('Benoni', 'McIntosh' and 'Red Resistant'.

b) The flowering tendency of long shoots is great in the first three years and then drastically reduced in cultivars such as 'Red Astrachan', 'Ceglédi Piros', 'Fertődi Téli', 'Idajon', 'Gloster', 'Jerseyred', 'London Pippin', 'Maigold', 'Mutsu', 'Spigold' and 'Staymared'.

c) Flower forming tendency in young and old trees is significant and relatively stable on long shoots such as 'Cheasepeake', 'Gallia Beauty', 'Kecskeméti Vajalma' and 'Red Rome'.

d) Tendency for young and old trees to form flowers on long shoots but the ecological and cultivation factors make it uncertain. Such cultivars include 'Golden Delicious', 'Idared', 'Jonagold' and 'Jonathan'.

Roemer (1968, 1970) compared the bloom times of short and long shoots. If we evaluate the results of short and long shoots together it appears that maximum apical blooming occurred in 'Golden Delicious' and 'James Grieve' cultivars but the bloom duration increased only in the 'James Grieve' cultivar. In the 'Cox's Orange Pippin' cultivar the later blooming on long shoots not only extended the bloom time significantly but also resulted in two peaks of blooming.

Cultivar characteristics can also be indicated by how quickly the anthesis on long shoots follows that on short bearing shoots. Within this, it is important to know in what region of the long shoots do the lateral flowers form. Do they form close to the apex or further away from it (Soltész 1992). First on long shoot the apical flowers are opening and for the last time the lowest lateral flowers of the shoot blossom.

What part of the tree produces the flowers also have a role in stone fruit species in determining the time of blooming. The difference between the time of blooming in cultivars having spurs and cultivars producing lateral flower is the largest in the apricot (Berezenko 1963, Tarnavschi et al. 1963, Branzanti et al. 1965). The appearance of flowers in lateral buds of long shoots has also a decisive effect on delaying the bloom time of the walnut (Bellini 1980).

In the raspberry cultivars, the position of the bunch on the bearing shoot is very important. Most flowers are on the lower part of the branch but these are the least developed. Flowers arranged in the middle of the shoots are the most fertile. Kollányi (1990) studied bloom time in four raspberry cultivars at four different heights on the bearing shoots. He found a divergence in the anthesis order. The averages of four cultivars were as follows:

Number of inflorescence from the apex	Anthesis time using the apical flower as a basis (May 15), days
1 – 3	0
7 – 9	+3.7
4 – 6	+4.5
10 – 12	+5.5
13 – 15	+6.2
16 – 18	+7.0

In the black sorbus the inflorescence developing in the lateral buds of the shoots contain less flowers and, according to Porpáczy (1987) they may vary depending on the cultivar. Lateral flowers may flower faster, but regardless of fast flowering they still significantly extend the whole bloom time of the plant.

The structure of the inflorescence may influence the process of blooming and this may cause important deviations in bloom time among cultivars. In the apple, the centrifugal anthesis order of the inflorescence is a rather stable property and the change indicates irregularity. This may vary in the cultivars. Williams (1975*b*) experienced in the 'Cox's Orange Pippin' cultivar that in certain years the apical flowers were not the first to open in the inflorescence. In the apple the duration of anthesis of inflorescence, the duration of flowering for individual flowers and the rate of anthesis of apical and lateral flowers are characteristics of the cultivars (Soltész 1992). We have found a slower blossoming rate in 'Alkmene' in 'James Grieve' and 'Julyred' inflorescences than in other cultivars.

The order of anthesis of inflorescences was observed by Dibuz (1993) in 77 pear cultivars. The acropetal order, generally described in the literature, is characteristic of 58% of the cultivars, 23% of the cultivars have a centrifugal order and in 19% the divergent form is characteristic. We can count on acropetal order primarily in those cultivars where the number of the flowers per inflorescence is more than five to seven.

Generally, in the inflorescence of the black sorbus cultivars there is centripetal order of bloom (Porpáczy 1987). In the umbelliflora of the cornel-berry centripetal flowering is also the typical form.

In the *Rubus* species the different structure of inflorescence of the raspberry and blackberry types also has an influence on the time of bloom (Nybom 1986). In the raspberry that produces the fruit only once apical flowers of the inflorescences open first (Kollányi 1990). The bud below the apical flower lags in development and the other flowers open regularly progressing towards the base. On the basal part of the compound cyme inflorescence, on the pedicle of the first order flowers second- and third-order flowers are located whose blooming essentially takes place later. The number of flowers per inflorescence (5–15) is property of the cultivar influencing the duration of blooming. The cultivar 'Malling Jewel' is characterized by few flowers, moderate number of flowers is characteristic of 'Malling Exploit' and large number of flowers is characteristic of 'Norfolk Giant'.

The flowers of black raspberry cultivar can be found in an umbel in black raspberry, and the centripetal anthesis is characteristic, in contrast, of red raspberry types. In these cultivars the duration of blooming is shorter because of the simpler inflorescence and the faster anthesis order.

The cultivar characteristics also have a role in the raceme inflorescence of the red and black currant. There are 13 to 15 flowers in a raceme of the red currant depending on the cultivar with an acropetal anthesis order. According to Porpáczy (1987), 5–20 flowers can be found in the black currant inflorescence. In black currant in some cultivars there are side branches on the raceme. These side branches flower together with the apical flower of the raceme, which is the last flower to open of the raceme in an acropetal order. The different structure of the inflorescences results in a one or two days shorter blooming time in red currant having more flowers in the inflorescence than that of the cultivars of the black currant.

4.3.5. The Effect of Cultivation on Bloom Time

The literature in this field has become evermore extensive on this subject and it would be impossible to cite all of it. Therefore, we shall refer to several characteristic examples to point out two essential questions.

Before an orchard is established, apart from a knowledge of floral phenology, we have to take into the consideration the rootstock, the material to be planted and the bloom time of the planned system since these cannot be changed once the orchard is established. It should also be emphasized that by taking into account the bloom time when setting up an orchard we can forget about this factor later because there is no cultivation intervention by which we do not influence the time of blooming either directly or indirectly.

There are numerous examples in the literature referring to the effect of rootstocks. Often, the effect of the rootstock is oversimplified as growth force and it is not considered as a factor in coexistence with the other influences of the rootstock.

Several researchers (Lucke 1959, Reichel 1964*a,* Roemer 1968–1970 Church et al. 1983*c,* Williams and Church 1983) were not able to demonstrate the effect of the rootstock during the bloom period of the apple cultivars. Reichel (1964*b*) referring to the complexity of the rootstock effect reported that it had no direct effect even at the bud differentiation time. Baboné (1974) showed that weakly growing rootstocks induce earlier bud burst.

There have been numerous reports on the effect of the rootstock on the bloom time of apple trees (Gardner et al. 1949, Tamás 1959, Cummins and Norton 1974, Brózik and Nyéki 1975, Crabtree and Westwood 1976, Steinborn 1985). The majority of authors claim that the early blooming is due to weak growth of the rootstock (Gardner et al. 1949, Visser 1973, Crabtree and Westwood 1976, Steinborn 1985). Cummins and Norton (1974) consider that numerous flowers formed already in the nurseries in the lateral buds of the shoots in 'Idared' on M 9 rootstocks.

Tamás (1959) explained the earlier blooming on weak rootstock to be due to the lower heat requirement. It should also be taken into account that the rootstocks can change the cold requirement of dormant trees. Young and Werner (1985) found that among the examined apple rootstocks the one with the highest cold requirement was the MM 106, the least cold requiring was M 7 and the rest, M 111, M 26, and M 9, fell somewhere in between.

Soltész (1982*a*, 1992) found that in the apple a certain tendency can be expected with respect to the effect of the bloom time but this can only be determined reliably if we examine the rootstock–scion combination. The combined effect of the different growth strengths of the rootstocks and the scion cultivars are additive in formation of the bearing surface which manifests itself in the strength of the spurs and how great the differences are in flower formation on the bearing shoots. If the genetic properties of a cultivar do not allow the flower formation on long shoots then the rootstock cannot modify this significantly. Therefore, the effect of rootstock can be demonstrated on those apple cultivars in which the nature of the bearing shoots and the tendency for flowering is variable.

In pears, because of the existence of only a few rootstocks, it is relatively easy to assess the effect of the rootstock on blooming time. Brózik and Nyéki (1975) did not find the effect of rootstock in the bloom of the pear. Ghena and Braniste (1978) experienced pear blooming one to three days earlier on quince rootstock and attributes this to the weaker growth of the cultivars on quince emphasizing the rootstock–scion combination.

No correlation was found by Griggs et al. (1972) between the vigour and bloom time of 'Bartlett' (Williams) pear trees. They supposed that if there was any rootstock effect on bloom time that had to occur through the changed cold requirement during dormancy.

Sansavini et al. (1980) also investigated the bloom time of 'Bartlett' (Williams) pear trees on different rootstocks. There was no tight correlation between vigour of the rootstock and the vigour of the scion with respect to bloom time. The start of bloom was affected by interstocks and virus infection. Cultivars on rootstock which were infected with virus showed later blooming.

There is less data regarding the effect of rootstocks on the bloom time in stone fruits.

In the cherry, Capucci (1959) and Bargioni (1982) found that trees grafted on mahaleb flowered four to five days earlier. According to Willings' (1960) data, in 14 cherry cultivars blossoming on mahaleb rootstock caused one day earlier blooming compared to wild cherry rootstock. Srivastava and Singh (1970) claim we have to count on the earlier flowering effect in rootstocks. Nyujtó (1960, 1967) also observed that bloom of cultivars was one to two days earlier on mahaleb than on wild cherry or on wild sour cherry rootstock. According to Blauhorn and Schimmelfleng (1986), the later blooming of the 'Pándy' sour cherry trees having on wild sour cherry rootstock was favourable for fruit set.

Bellini and Bini (1978) mention the rootstock effect in plums. Lalatta (1982b) claimed that the plum rootstock weakens the growth of peach trees, which intensifies fruit set. This is attributed to several causes: fertilizability is increased, the effective pollination period is longer and fruit drop is less. However, bloom time was not influenced significantly by the rootstock.

In stonefruit species there is a predominance of rootstock grown from seed so it is worth mentioning that in the blooming in orchards the bloom time heterogeneity is high (Capucci 1959).

The bloom of the walnut scion cultivars is influenced by the age of the tree providing the scion and the rootstock age. McGranahan and Forde (1985) observed that the age of the tree providing the scion had a greater effect than that of rootstock. Goldwin (1990) warned that greater attention has to be paid to the properties of the trees providing the propagation material.

We have to admit that we know very little regarding rootstock effect on flowering and time of blooming. Mention will be made of a problem which has been raised by Soyla and Ludders (1988) in apple but which also has general implications. Often, we attribute an effect to the rootstock, whereas only the deviating susceptibility of the rootstocks to factors, such as salinity of the soil, or calcium content, is seen as rootstock effect, making itself felt in the fruit tree. Increasing the vigour of the trees obtained by propagating trees by tissue culture, flower formation and bloom time may have to be somehow regulated (Rosati and Gaggioli 1987).

Apart from many harmful effects of virus infection, the bloom time is also influenced by viruses. Virus-free propagating materials are therefore essential. Unfortunately, virus infection may ensue later also. The delay in blooming due to virus infection was reported in the cherry and plum (Desvignes and Savio 1970), in peach (Marenaud and Desvignes 1965, Savio 1970) and in the sour cherry (Vértessy and Nyéki 1974).

Savio (1970) found bloom in peach trees infected with necrotic ring virus to be two to five days later than on virus-free trees. According to Vértessy and Nyéky (1974), the bloom time in the 'Montmorency' sour cherry cultivar inoculated with ringspot virus was three to nine days later than on healthy trees. In 'Pándy' clone 48 sour cherry the virus infected trees bloomed one week earlier.

The direct effect of the cultivation system, such as canopy shape and trunk height, is difficult to demonstrate since it is hardly possible to discriminate it from the rootstock-applied agro- and phytotechnique and other cultivation effects. Lalatta (1982b) emphasized the indirect effect of the shape of the canopy in peach and sour cherry explaining the effect by radiation effectiveness influencing bloom time through flower bud formation.

The different cultivation systems in peach (Williams and Corton 1990) and in cherry mainly influence the density of the flowers (Miller et al. 1990), thereby having an indirect effect on blooming.

The cultivation system used determines the size of the tree. On smaller trees (plants) the bloom is faster and earlier because of the more favourable temperature conditions. In larger trees, the extension of bloom time is more common. When radiating frost occurs during blooming the bloom may separate more in the different levels of the canopy. Roemer (1970) found that frost in the early phase of bloom in the apple can postpone the blooming in the upper part of the canopy by one week. In frost-exposed sites, Lake (1956) reported major delays in bloom in dwarf apple trees. However, this is probably the consequence of winter damage.

The bloom duration of 'Idared' apple cultivar and the later blooming on free spindle is greater than on narrow spindle, and we have to take this into consideration if we wish to cultivate them together with later blooming cultivars (Soltész 1992).

More flower buds are formed in the lateral buds of long internode shoots on "small trees" of gooseberry than in bush cultivation (Harmat 1987). The lateral flowers of long shoots, which are less common in bush cultivation than in plants grown on trunks, have anthesis earlier in the basal part of the shoots and later in the apical part of the shoots. The trunk of the tree-type gooseberry flowers three to four days earlier but the blooming is more extensive.

The size of the plants in any cultivation system is not constant and it changes with the age of the trees. The larger, older trees flower earlier (Grubb 1949). This has been confirmed by Krapf et al. (1972) in the cherry. Soltész (1992) found the same for apple. In young apple trees blooming may start one to two days later but the full bloom time—connected with bearing shoot characteristic of the cultivar—may induce larger than one or two days of change.

Blooming within a tree does not take place simultaneously. It starts in the external part of the canopy then the internal part follows. Those parts of the canopy that are exposed to radiation bloom earlier (Fritzsche 1972). In the upper part of the tree there is also earlier blooming (Seaton and Kremer 1939, cit. by Soltész 1992). The beginning bloom on the southern side of the tree may be 1. 6 days earlier than on the northern side (Budig 1960).

In Nyujtó's view (1980), there is no difference in bloom time in young apricot trees among various parts of the canopy. On older apricot trees there may be two to seven days difference in flowering within the canopy. In the sour cherry very often the flowering began on the lower part of the canopy (Nyujtó 1980).

In all kinds of cultivation systems flower density, type of bearing shoot and its distribution are important from the point of view of bloom time. Since pruning and other phytotechniques have the greatest impact these give the most striking effect on bloom time.

Roemer (1970) observed in 'Cox's Orange Pippin' cultivar that, on the one-year-old long shoots, the flower formation is closely connected with the two-year-old branch flowering tendency. Flowers will only occur on long shoots if the extent of flowering on two-year-old spurs is 61%. Soltész (1982a) found a similar tendency in 'Gravenstein', 'Staymared', 'Ingrid Marie', 'McIntosh' and 'Red Astrachan' cultivars.

The flowers that are borne on long shoots significantly extend the bloom time of apple trees. Lucke (1959) was the first to report that the flowering is determined by the type of the shoot. The blooming order is as follows: flowers open on spurs, smooth bearing spur, then apical inflorescences of long shoots and finally lateral inflorescences of the shoots. Soltész's (1982a) investigations confirmed this order and, to a lesser extent, it applies to the different aged spurs also.

The blooming order of the bearing part of the pear and the black sorbus is similar to that of the apple. In black sorbus, Porpáczy (1987) observed that bloom began when the shoots reach 10 to 15 cm.

According to some earlier observations (Bouché-Thomas 1953, Engel 1960), there may be a difference of five to ten days between the apical inflorescence of short and long shoot. The apical bud of long shoot opens later. Fritzsche (1972) established that the flowers developing in the lateral buds of long shoots only begin to open when those on the short bearing parts have bloomed. Later this was not confirmed as generally applicable.

These different bloom times indicated by Fritzsche (1972) were only experienced in those years of early blooming when a lasting cold spell at the beginning of bloom inhibits further anthesis. In regular flowering processes the anthesis takes place, depending on the cultivar, continuously in the bearing shoots. In years of rapid blooming the anthesis may overlap in the short and long shoots so that the difference between anthesis of apical inflorescences of the two types of shoots may not be more than one day, and this deviation in the lateral inflorescence of long shoots may only be four to five days (Soltész 1992).

A comparison was made of the bloom time of five- and ten-year-old trees on trees on the same rootstock, standing next to each other in a year when the weather did not prevent the prolonged continuous blooming (Fig. 4.12). The differences of trees to flower in the full bloom period, the duration of blooming and the joint blooming could be attributed to the flowering tendency of long shoots. The effect of the age of trees with respect to the tendency to flower on long shoots can be traced in the various years (Soltész 1986).

In stone fruits, Berezenko (1963), Tarnavschi et al. (1963) and Branzanti et al. (1965) observed in the apricot that blooming took place two to three days later on the long shoots derived from the spring shoot growth, while in the second phase of shoot growth flowering was five to six days later than on the short bearing spurs. Tóth (1957) estab-

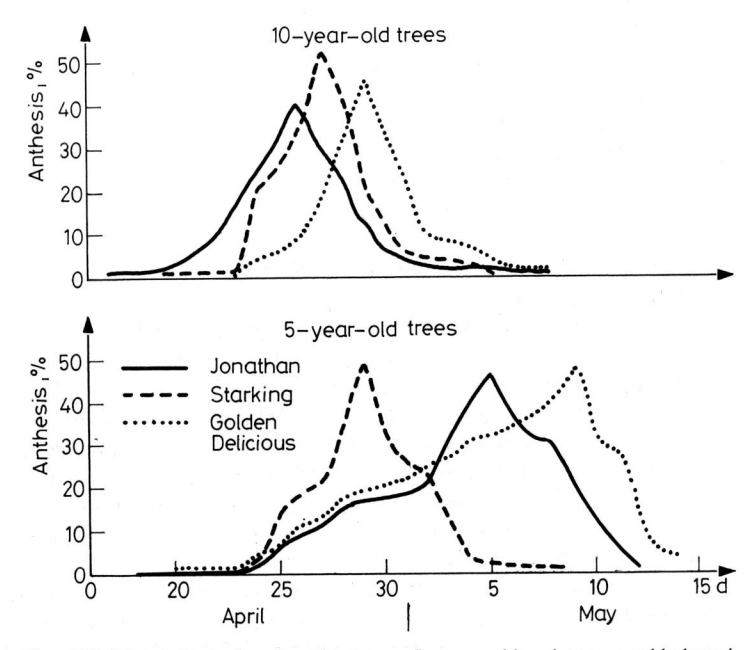

Fig. 4.12. Bloom time of apple cultivars on five-year-old and ten-year-old slanted hedgerow trees in 1979 at Kecskemét-Helvécia, Hungary (after Soltész 1982a)

Table 4.17. Effect of Time of Pruning on Time of Full Bloom in Cultivated Highbush Blueberry in Relation to Full Bloom Date of Unpruned Control (after Gough 1983)

Time of pruning	Mean No. of days after full bloom of unpruned control		
	1979–80 season	1980–81 season	1981–82 season
September	4	3	3
November	1	1	1
February	1	1	0

lished a similar order in the plum between the short bearing spurs and long bearing shoots. In peach, the location of the flowers on the shoots influences the time of bloom. At pruning we not only set the level of fruit production but we also change to a small extent the blooming duration of the shoots and the entire tree (Spence and Couvillon 1975).

Studying the pruning on the time of blooming in apple, it turned out that it had no effect on the bloom time in the year of pruning. In the following year there was a difference between the pruned and unpruned trees with respect to blooming, but not with severity of pruning. The time of pruning had a greater influence on the amount of flowers formed on the long shoots, and in this way affected the process of blooming (Lalatta 1982a).

The effect of pruning time on blueberry can be seen in Table 4.17. Early pruning delayed bloom more than late pruning.

According to Soltész (1992), the change in the shoot binding in the apple was primarily influenced by the flower density. It had no significant effect on bloom time. Flower density is also changed by ringing of apple trees (Rünger 1971). Thus ringing may change bloom time. Pisani (1962) established that elimination of leaves and shoots did not influence the bloom time of the apple and pear.

Irrigation has various effects. In apple it promotes the flowering tendency of long shoots (Streitberg 1975) if other conditions are met. Irrigation was used successfully before blooming in delaying bloom time. The uptake of nitrogen during the previous summer influences the beginning of blooming and its course in apple (Hill-Cottingham and Williams 1967).

Bellini and Bini (1978) mention fertilization, shading defoliation and pruning as factors influencing blooming.

Growth-regulators are widely used to regulate flower bud formation. Their application could always have indirect effects on the bloom time. We shall only refer to a few growth-regulators which had special modifying effects on bloom time.

Sullivan and Vidmayer (1970, cit. by Soltész 1992) reported on the flower-delaying effect of *Alar*. Champagnat et al. (1986) found that ethephon accelerated the flower bud burst and also resulted in earlier bloom time in apple. Crisosto et al. (1990) established that when ethephon was applied at the time of 10% leaf fall it delayed flowering by 13 to 16 days in plum. In peach the flower-delaying effect was achieved, but it had an unfavourable after-effect. (Durner and Grianfagna 1990).

Williams and Wilson (1970) by applying heat during frost at blooming brought flowering earlier by two days. With present energy prices we cannot expect to be able to heat orchards but earlier blooming of pollen donor trees can be achieved with two to three days if the cultivar to be pollinated has no other way of being pollinated. Small pollen donor trees and certain blooming branches can be covered by plastic sheets temporarily. This method was applied in an apple orchard by Soltész (1982a), and the later blooming pollen donor cultivars could be forced to flower earlier by three to ten days depending on the time of covering with plastic sheets and the temperature conditions.

In strawberry cultivation this method of covering is generally used in other early fruits, such as peach and cherry not so often. In the strawberry, covering the soil influences bloom time through heating the soil and the plants. Pritts (1989) demonstrated that the covering with transparent plastic sheets resulted in earlier blooming than with the black plastic sheets.

Larsen and Fritts (1987) studied the blooming of regrafted apple trees. The beginning of bloom, its duration and the tendency of flowering on long shoots greatly depended on the age of the regrafted tree.

4.4. Methods for Observing and Evaluating Phenology of Bloom

4.4.1. Determining the Time of Blooming

Estimation is the simplest way to determine the beginning and the end of blooming and the time of full bloom. In certain years this is a rapid method requiring limited labour input. However, collecting definite information characteristic of the cultivar requires observation in most cases for ten years (Maliga 1980). If one observes a cultivar for a long enough period then the relative position of bloom, the average length of bloom, the time of full bloom and the group where the cultivar belongs, as far as blooming is concerned, are all determined. For using a uniform technique, we recommend that the beginning of bloom is the day when the first flowers open, the end of the blooming is when no functioning flower is left on the tree, even the last flowers had opened and completed their blossoming.

One can compare the bloom time of cultivars by "line diagrams" by recording the date of blooming or by utilizing the relative differences between cultivars (Silbereisen 1982). By recording the relative differences between cultivars, one can use the collected data also to characterize the course of the blooming. However, in this case one has to record the distribution of cultivars that show full bloom daily. Recording the dates only has the disadvantage that it is difficult to select cultivars blooming together for choosing the best pollen donors. This method is also unsatisfactory for analysing the course of blooming and determining the factors responsible for influencing the course of blooming.

A more precise method is the analysis of the rhythm of blooming. This method determines the newly opened flowers daily. This can be done in two ways: (1) estimating the number of flowers open every day on a number of sample trees, or (2) by designating a sample limb on the trees, counting the flowers on this limb before the bloom begins and during the bloom to determine how many flowers are open. We use the blooming rhythm analysis if we want to obtain information on flowering of a large number of cultivars at the same time. The information value of data collected by the blooming rhythm analysis greatly depends on the collector's experience. This is extremely important when the flower number is estimated. When estimating the flowers the tree must be approached every day from the same side. For counting flowers on a sample branch one has to select a branch with at least 100 flowers. The sample branches must be in similar locations, at similar height and receiving about equal radiation. If there are few flowers on the tree it is advisable to count all the flowers. Estimation of flowers is difficult on trees which have few flowers and estimation on trees with few flowers will result in inaccurate data.

Determination of the dynamics of blooming, collecting data on numbered flowers, is presently the most accurate method for collecting data on blooming. Unfortunately, it can only be used with a limited number of trees at the same time. This method allows collection of data on all phases of bloom phenology and requires a large number of experienced persons if one wants to study details of floral phenology on several cultivars at the same time. This method may provide data on such details as the functioning of stigmas, changes in the rate of shedding of pollen, or the period of pollen shedding. With the help of this method we can provide information on the number of flowers opened daily, or the number of flowers completing their blooming daily. In other words, we can provide information on the dynamics of blooming.

Observation of dynamics of blooming with accuracy can only be done on numbered individual flowers. The method was developed by Herbst and Rudloff (1939) and was introduced by Weger (1940). Since then several investigators have used it. In sour cherry, Nyéki (1974c) used it and developed it further. With this method several smaller branches are designated per tree. In round canopy trees it is advisable to select branches located in all four directions. In rectangular canopy trees it is satisfactory to select branches on both long sides of the tree. If one has to select only one branch per tree that branch should be on the side of the tree that receives the most radiation.

Observations about the state of the flowers, opening of the flower, number of open flowers, falling of petals, etc. must be counted daily. We do not recommend observing flowers between 12 and 16 h during the warmest period of the day. When the flower opens the petals separate, the stigma is visible and green and the anthers are visible. In the open flower the petals are extended, the stigma is shiny and covered with secretion. The end of bloom is signified by an "old" stigma which does not secrete any more, and by that falling of the petals has started. The functioning of stigma and the dehiscence of pollen can be determined daily or hourly.

Marking the individual flowers can be done before the blooming begins. This way we can make ready more branches for observation. One can also attach the marking labels at the time when observing the opening of the flowers. This way the marking and the recording of flower opening occur at the same time. We note here that the marking of individual flowers also aids in following insect pollination, changes in the floral morphology, such as flattening of stamens, changing in the colour of petals, and determination of the time of fertilization.

4.4.2. Blooming Phenograms

Illustrating the flower counting results on phenograms was introduced by Herbst and Rudloff (1939). Phenograms can be made for various purposes. Figure 4.13 has four variations of blooming phenograms. Figure 4.13a and b are valuable for determining the blooming dynamics, whereas c and d are more useful for deciding the degree of joint blooming. Phenograms illustrated in Fig. 4.13c and d are produced mostly from data obtained by observation of blooming rhythm.

With the help of blooming phenograms we can determine accurately the various phases of blooming, the length of the blooming period, or the rate of flower opening or

Fig. 4.13. Method of depicting the anthesis of flowers (after Soltész et al. 1980)

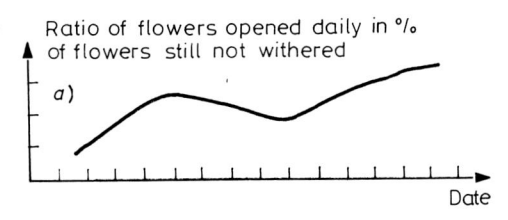

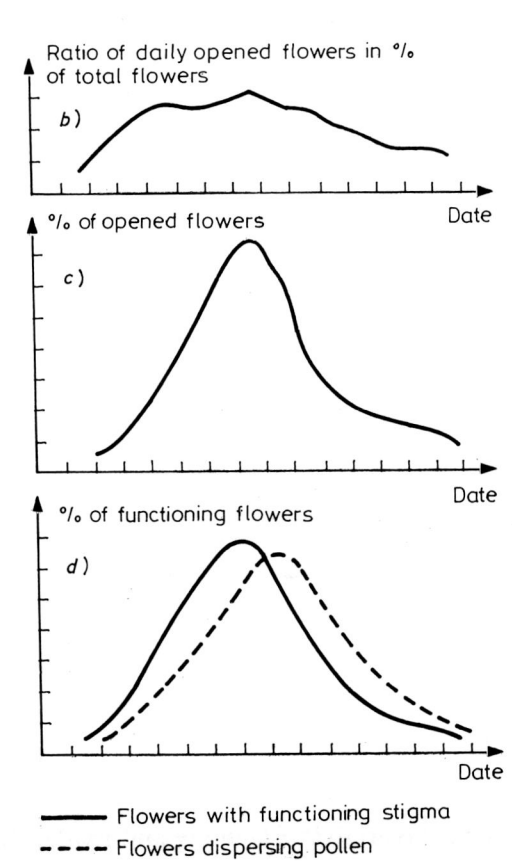

a) Ratio of flowers opened daily in % of flowers still not withered — Date

b) Ratio of daily opened flowers in % of total flowers — Date

c) % of opened flowers — Date

d) % of functioning flowers — Date

——— Flowers with functioning stigma
- - - - Flowers dispersing pollen

completion of bloom (Fig. 4.14). The phenogram is also useful for determining the time of full bloom. It is advisable to consider the time period as full bloom when the percentage of open flowers is above 50 (Herbst and Rudloff 1939; Roemer 1968, 1970; Nyéki 1989). If on a phenogram we plot the opening of flowers and separately the end of blooming, where the two lines intersect the 50% line is the period of full bloom.

In stone fruits Nyéki (1989) distinguishes four types of blooming phenograms: (1) Flat rate of bloom. The largest rate of flower opening is under 35%; (2) Single peak phenogram, with a short full bloom period; (3) Single peak phenogram, with full bloom that lasts for several days; (4) Phenogram with two maximum peaks. Soltész (1992) modified this classification for apple by using less than 50% values for the first group.

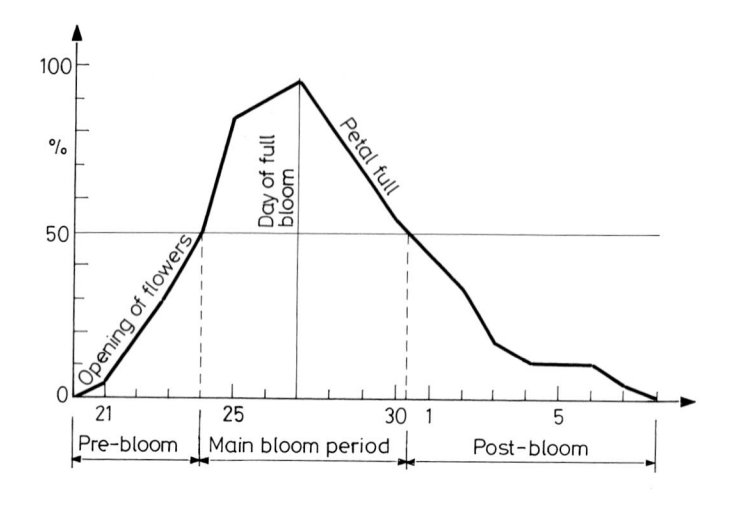

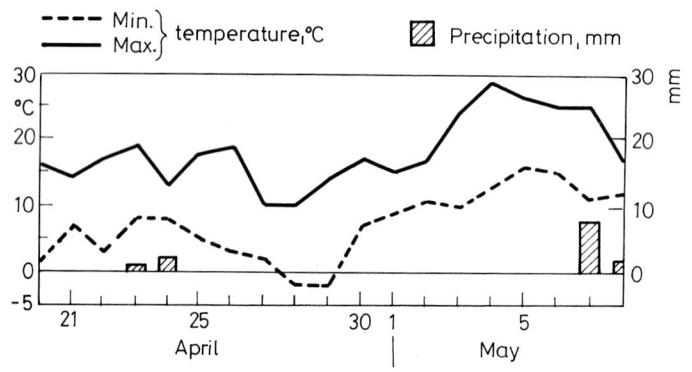

Fig. 4.14. Blooming produce of 'Stanley' cultivars in Csany in 1984 (after Nyéki 1989)

4.4.3. Overlap in Blooming and Evaluation of Joint Blooming

The overlap in blooming of cultivars can be determined in various ways. The simplest way is to overlay the diagrams of blooming to determine how much overlap is in blooming between two cultivars in relation to the full bloom time. Only limited information can be gained from this since this method does not give the rate of flower opening. Another method is to determine the overlap in the full bloom time. For this we can use either the data obtained from flower rhythm or flower dynamic methods. This gives us either the overlap in the data of full bloom or in the overlap in the full bloom period. We can work out different scales for overlap in various species. Nevertheless, three categories are generally correct for every species. These are: the full bloom time of cultivars does not overlap, partially overlaps, or completely overlaps.

The overlap in blooming can be best illustrated with bloom phenograms (Fig. 4.15). We express the overlap in percent of blooming found under the curve. Nyéki (1989) in

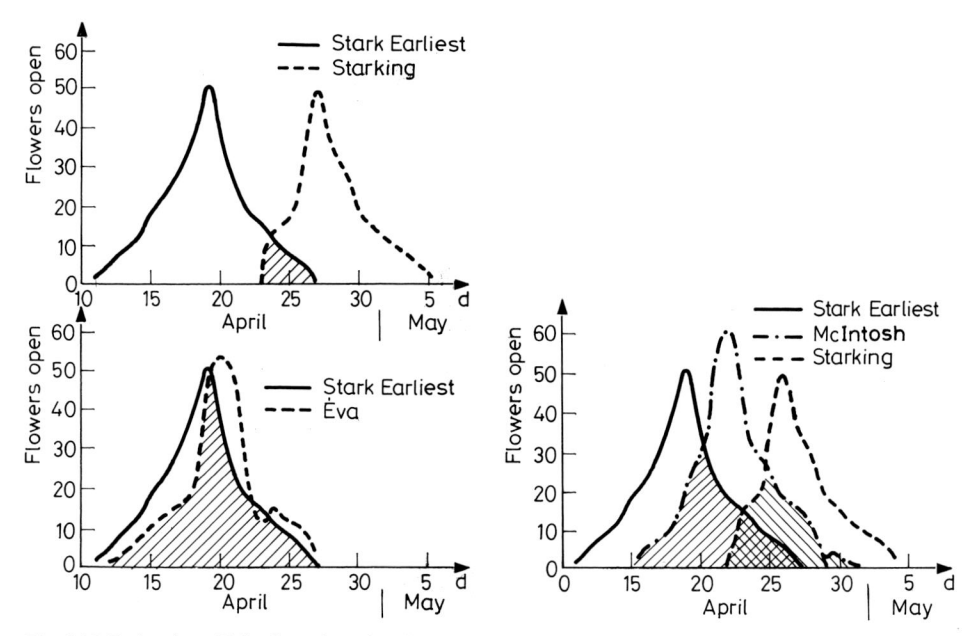

Fig. 4.15. Estimation of joint flowering of apple cultivars (after Soltész 1982a)

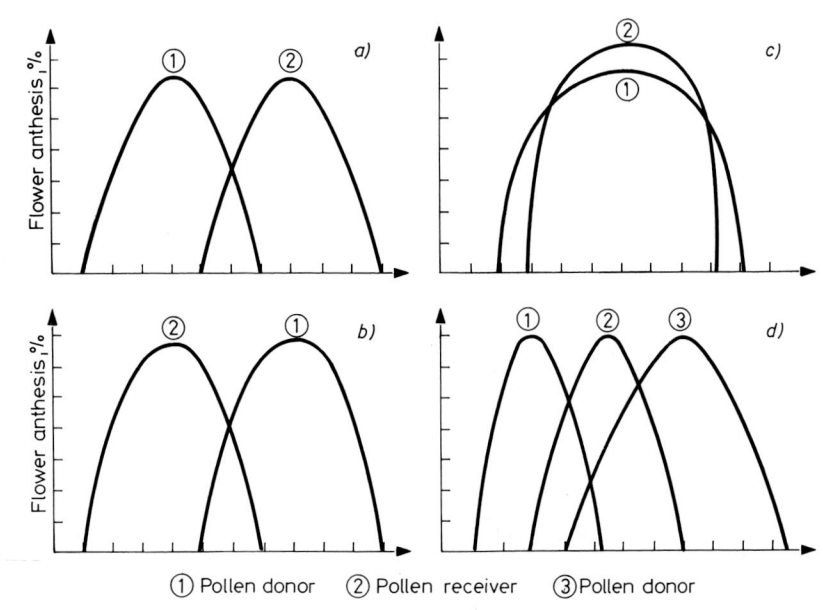

① Pollen donor   ② Pollen receiver   ③ Pollen donor

Fig. 4.16. Overlapping types of blooming periods (after Nyéki 1989)

129

Table 4.18. Designations of Bloom-Time Groups 4 to 6 in the Individual Fruit Species*

	A	B	A	B	A	B	A	B	A	B
1	early	apple, pear	very early	apricot	early	cherry	early	plum	medium-early	pear
2	medium-early	cherry	early		mid-season	almond	medium-early		mid-season	
3	medium-late	chestnut	mid-season		late		mid-season		late	
4	late	plum	late		very late		late		very late	
1	early	apple	very early	cherry	early	pear	very early	apple	very early	pear
2	medium-early	cherry	early	sour cherry	medium-early		early	plum	early	cherry
3	mid-season	apricot	mid-season		medium-late		mid-season		medium-early	
4	medium-late	plum	late		late		medium late		mid-season	
5	late	almond	very late		very late		late		medium late	
6		hazelnut					very late		late	

*After authors of previous tables; chestnut: after Solignat (1973); hazelnut: after Jona (1986); A = group names; B = fruit species

stone fruits and Soltész (1992) in apples used planimeter to determine the area of overlap under the curve. Today this can be done with computerized techniques. The types of overlap in blooming are illustrated in Fig. 4.16. In addition to the percentage of overlap, we need to consider the total number of flowers that bloom together. This is applicable to both, the entire cultivar blooming overlap and the blooming overlap on the various bearing shoot types.

4.4.4. Classifying Cultivars into Blooming Groups

For classifying the cultivars into blooming groups we can use either the beginning of bloom or the full bloom times. The full bloom time gives more reliable information and is used more widely.

For classifying a cultivar into a category most accurately all important factors influencing blooming, such as age of the trees, length of the observation, weather, etc. need to be considered. We especially must be careful in using information obtained on young trees when making final classification of a cultivar. It is also advisable to use data for classification from years when the blooming was farthest apart among the cultivars. In considering cultivars with different chilling requirements we have to choose years when the highest chilling requiring cultivar was satisfied and the lowest chilling requiring cultivar was not injured.

The presence of known cultivars in the collection that have a stable early or stable late blooming helps in the classification of others. With the help of such stable blooming marker cultivars we can divide the blooming period into groups as far as blooming is concerned. The width of the group in days is influenced by the speed of blooming and the weather conditions which may change yearly. The larger the number of groups the longer period of observation is necessary to classify a cultivar in the proper group accurately.

There is no agreement in the literature for naming the blooming groups. If there are only three groups there is no problem naming them. They are early, mid-season and late, and there is no need for variations. Using four, five or six categories the naming shows a different picture (Table 4.18). In many cases, considering the classification of different authors, the group of a given cultivar may not be different, but the name of the group is different thus it appears that the cultivar is classified differently.

It would be desirable to name the categories uniformly. Since this probably will not happen, it is advisable to accept Christensen's (1973) classification who used the following groups for cherry: 1 very early; 3 early; 5 mid-season; 7 late; 9 very late. In addition to the name, it would be more precise to use the number assigned to the blooming group. Christensen (1973) used the uneven numbers for the groups and reserved the even numbers for the intermediate types.

Finally, we have to mention a few problems associated with working with blooming. Floral phenology does not lend itself to semantics. This is especially true for variation of data between or within years, and variation between sample trees or sample branches. Blinova and Sedov (1976) and Rasmussen (1984) recommended that, in addition to the averages of sample trees, the multiyear characteristics of blooming also need to be analysed. It is very important to situate the research conducted on blooming of cultivars in homogeneous conditions. Publishing blooming results without giving details of the conditions in which the data were obtained greatly narrows the usefulness of the information.

5. Receptivity of Sexual Organs

M. Soltész, J. Nyéki and Z. Szabó

5.1. Stigma Receptivity

The receptivity of the stigma means that the stigma is able to ensure that pollen grains are secured on its surface; assure the fast germination of pollen and facilitate the quick start of pollen tube growth through the style. Soon after completing this task, the stigma turns brown and gradually loses its receptivity. This process is partly affected by the speed of pollen tube growth toward the embryo sac. High temperature results in the failure of the reception of further pollen grains because of the quick senescence of the stigma, while low temperature increases the possibility of multiple pollination.

"Receptivity of the stigma" and other terms are used in the literature interchangeably. Among the most common expressions are: "maturity of stigma", "active stage of stigma", "sexual maturity", and "viability" of the stigma.

There are a lot of arguments against the substitution of "viability" for "receptivity" because the former, but not the latter, also includes the role of the style. The viability of the style is still required for fertilization even after the stigma has lost its receptivity. The demand for the viability of the style may depend on the particular fruit species and cultivars.

Apples and *sweet cherries* are characterized by a faster pollen tube growth in comparison to *sour cherries* and especially to *plums*. A slow pollen tube growth is characteristic of the *plum*, as is often reported (e.g. Lee and Bünemann 1981, Stösser 1984). In other words, there is a significant difference between the period of the viability of the stigma–style complex and the period of the receptivity of the stigma surface. The difference is even more evident when a hexaploid plum cultivar receives pollen from a diploid cultivar (Kursakov and Dubovitzkaya 1974).

Although in insect-pollinated flowers, secretory activity increases the receptivity of the stigma, viability and receptivity, these two terms do not mean the same. The stigma may be pollinated even if it is not covered with stigmatic secretion, but its surface is only glossy, though in this case it is undoubtedly more difficult for pollen grains to adhere.

The effective pollination period (EPP) is very short. Pollination after the EPP is useless because the pollen tubes cannot reach the embryo sac in time. The EPP often coincides with the receptive period of the stigma but differences may also take place. If the EPP is shorter than the receptive period of the stigma then a concentrated and well-timed pollination is required. In other cases, the theoretically possible EPP will be shortened by the early loss of receptivity.

The EPP does not necessarily mean the period after anthesis (even if stigmas are receptive). According to Williams et al. (1984) and Williams and Brain (1985), the period just after anthesis is not the most suitable for pollination in 'Cox's Orange Pippin' apple cultivar. But the period four to five days later is satisfactory.

Eaton et al. (1968) observed that in *raspberry*, pollination two days after anthesis yielded the best results. Another study by Eaton (1959b) illustrates the opposite case: stigmas of 'Windsor' sweet cherry are receptive for more days but only the pollination

Table 5.1. Receptive Period of Stigmas in Pome Fruits, Chestnut, Walnut and Currant

Species	Duration of receptive period (day), end values and maximums	Author (year)
Apple	1–4	Soltész (1992)
	1–5	Davary and Nyéki (1990)
	2–4	Nyéki and Ifjú (1975)
	2–5	Soltész (1982a)
	4–6	Sandsten (1909)
	7	Kurennoy (1968)
	9–10	Greznitshenko (1969)
Pear	1–3	Nyéki (1973a)
	2–5	Nyéki and Ifjú (1975)
	2–7	Gautier (1983)
	9–10	Greznitshenko (1969)
Quince	6	Angelov (1966)
Currant	5–6	Wellington (1921), Sokolova (1951)
Chestnut	30	Shimura et al. (1971)
Walnut	1	Kuppuswami (1954)
	3–5	Forde and Griggs (1972)
	4–6	Germain (1975)
	5–7	Germain (1973)
	1–8	Nedev and Stefanova (1979)
	5–8	Szentiványi (in Nyéki 1980)

performed on the first day of anthesis will result in a good fertilization. Toyama (1980), however, achieved satisfactory results on the last (fifth) day of the receptive period in 'Lambert' sweet cherry.

Hartman and Howlett (1954) reported the EPP of 'Delicious' apple to be much shorter than the receptive period of its stigma. In 'Comice' pear, both periods are very short (Gautier 1974). In 'Conference', the EPP is ten days; but unfortunately, it is halved by the shorter receptivity of the stigma, reported by Gautier (1974).

Finally, it is to be noted that the receptive period of the *plum* stigma was observed to be longer than its blooming period by Levitzkaya and Kotoman (1980).

Secretory activity generally begins after anthesis but receptivity can be observed before this time. *Sweet cherry* stigmas were reported to be receptive three to seven days before anthesis by Tukey (1933). Smykov (1974) claimed that *apricot* stigmas were receptive three to four days before anthesis. In *plum*, two days of prior receptivity was reported by Knuth (in McGregor 1976) and one day by Randhawa and Nair (1960). In contrast, *sweet cherry* stigmas are suited for pollination only after anthesis (Stösser 1966a).

Receptivity before anthesis is of practical importance in the case of artificial pollination and cleistogamy. Tables 5.1 and 5.2 provide a comparison of available data concerning the receptivity of stigma. Data refer to the number of the days counted from anthesis.

Table 5.2. Receptive Period of Stigmas in Stone Fruits

Species	Duration of receptive period (day), end values and maximums	Author (year)
Sweet cherry	1–2 2 2–5 4–6 4–7 5	Srivastava and Singh (1970) Bargioni (1978) Nyéki (1974c) Brózik (1972) Toyama (1980) Guerrero-Prieto et al. (1985), Lombard et al. (1983)
Sour cherry	1 (some hours) 1–2 1–6 2–5 2–6	Nyujtó (1970) Nyujtó (1966), Pejkic (1972) Nyéki (1974), Ifjú (1975) Nyujtó and Banainé (1974) Brózik and Nyéki (in Nyéki 1980)
Plum	1–5 3–4 4–5 4–6 3–7 4–8 7–12	Szabó (1989) Nyéki and Ifjú (1975) Randhawa and Nair (1960) Dorsey (1919), Bellini and Bini (1978) Levitzkaya and Kotoman (1980) Lee and Bünemann (1981) Stösser (1984)
Apricot	1 1–4 4–7	Heideman (1984) Brózik et al. (1978) Toyama (1980)
Peach	1–4 7–12	Nyéki and Brózik (in Nyéki 1980) Toyama (1980)
Almond	4–8	Gagnard (1954)

Table 5.3. Receptive Period of Stigmas in Apple* (after Soltész 1992)

Stigma receptivity (day)	1981	1982	1983	1984	1985	1986
Range	1.8–3.2	1.7–2.8	1.4–2.4	1.9–2.9	2.2–3.5	0.9–1.9
Means of all cultivars	2.5	2.2	2.0	2.3	2.8	1.4
Means of diploid cultivars	2.4	2.1	1.9	2.2	2.7	1.3
Means of triploid cultivars	2.9	2.6	2.1	2.9	3.2	1.5

Stigma receptivity (day)	1987	1988	1989	1990	Average
Range	0.8–1.8	1.1–1.7	0.9–1.7	1.3–2.6	0.8–3.5
Means of all cultivars	1.4	1.4	1.3	2.0	1.9
Means of diploid cultivars	1.3	1.4	1.2	1.8	1.8
Means of triploid cultivars	1.5	1.4	1.5	2.6	2.2

* King flowers of spur flower clusters

Table 5.3 provides detailed data, concerning *apple*, indicating that yearly weather fluctuations influence receptivity of stigmas more than the characteristics of apple cultivars. Nevertheless, there are discernible differences between diploid and triploid cultivars.

The receptivity of *sour cherry* flowers has been evaluated in terms of both self-fertility and bloom time (Table 5.4). In *plum* cultivars, receptivity is measured in hours rather than days (Table 5.5). As for *apricot*, the period of receptivity clearly shows a yearly variation (Table 5.6). The data of two years for *peach* (Table 5.7) demonstrate that

Table 5.4. Receptive Period of Stigmas in Sour Cherry (after Nyéki 1974)

Characteristics of stigma receptivity period	1974	1973	1974	1972-1974
Range (day)	3.3–4.1	1.9–2.8	1.6–3.3	1.6–4.1
Means of cultivars (day)	3.6	2.3	2.5	2.8
Means of self-sterile cultivars (day)	3.8	2.3	2.7	2.9
Means of self-fertile cultivars (day)	3.5	2.2	2.1	2.6
Means of early blooming cultivars (day)	3.3	2.7	2.8	2.9
Means of late blooming cultivars (day)	3.8	2.1	2.1	2.7
Beginning of bloom (date)	April 7	April 20	April 4	–
End of bloom (date)	April 24	May 7	April 14	–
Duration of bloom (day)	18	18	11	16
Sum of daily mean temperatures during bloom (°C)	222	250	112	128

Table 5.5. Receptive Period of Stigmas in Oriental and European Plum Cultivars (after Szabó 1989)

Characteristics of stigma receptivity period	Oriental plums 1984	Oriental plums 1985	European plums 1984	European plums 1985
Range (hour)	35–51	25–29	27–36	20–29
Means of cultivars (hour)	45	27	33	26
Means of early blooming cultivars (hour)	42	29	33	27
Means of late blooming cultivars (hour)	51	25	31	25
Beginning of bloom (date)	April 14	April 18	April 20	April 19

Table 5.6. Receptive Period of Stigmas in Flowers of Apricot Cultivars (after Nyéki and Szabó 1986, unpublished)

Cultivar	1976	1977	1978	Average
Magyar kajszi	·	3.0	1.0	2.5	2.2
Ceglédi óriás	2.5	1.0	2.0	1.8
Kecskeméti rózsa	3.0	1.0	2.5	2.2
Kécskei rózsa	2.0	1.0	4.0	2.3
Mean	2.6	1.0	2.7	–

Stigma receptivity (day)

Table 5.7. Receptive Period of Stigmas in Peach (after Nyéki and Szabó 1986, unpublished)

Characteristics of stigma receptivity period	1983	1984
Range (hour)	12–24	34–41
Mean of cultivars (hour)	19	35
Mean of early blooming cultivars (hour)	19	34
Mean of late blooming cultivars (hour)	19	36
Mean of freestone cultivars (hour)	12	34
Mean of clingstone cultivars (hour)	22	38
Mean of nectarine cultivars (hour)	23	34

bloom time does not play a determining role in receptivity, nor do the differences among the cultivar groups.

Weather of the given year affects receptivity the most and especially the temperature and precipitation during bloom are determinant. Nyéki (1974c) observed in *sour cherry* that the receptive period was only two to three days long at a mean daily temperature of 15 °C. At 4 to 12 °C, the length of this period doubled. High temperature (25 °C) is very unfavourable in 'Pándy' *sour cherry* (Nyujtó 1970, Nyujtó and Banainé 1974, Wociór 1976a, b). At this temperature, the receptive period lasts only a few hours.

Toyama (1980) also emphasized that high temperature plays the crucial role in decreasing the receptivity in *apricot, sweet cherry* and *peach*. Szabó (1989) observed in *plum* that temperature rising from 10 °C to 13 °C would halve the receptive period.

Low temperature and rainfall also drastically shortens receptivity, as is demonstrated by Williams (1975a, b) in *apple*.

In insect-pollinated fruit species, stigmas secrete sticky substances. Weather has a crucial influence upon this activity, hence we should enter into details here.

Nyujtó and Banainé (1974) argued that in *sour cherry*, stigmatic extracts would appear on the stigma surface 9 to 13 hours after anthesis. Anthesis in the morning can shorten this period to one to five hours (Nyéki and Ifjú 1975).

The daily course of secretory activity is undoubtedly a function of temperature, as is demonstrated in *sour cherry* (Fig. 5.1). The maximum secretory activity can be observed between 6 and 10, then the intensity of secretion decreases between 11 and 12, it completely ceases between 13 and 15, and a discernible activity can be observed again after 16 h. This process is shorter in self-fertile cultivars due to the greater and earlier chance of fertilization.

Nyéki (1989) could not observe secretory activity in most *apricot* flowers. The stigma surface is being very rough and covered with pollen grains which makes observation difficult. At 8 h no stigmas display secretory activity, then there occur a larger and a smaller maximum at 10 h and at 16 h, respectively. The activity reaches an end at 18.

Stigmatic secretion plays an important role in pollen adhesion and tube development. It is a sticky and sweet medium consisting of various components (nutritives and hormones) with slightly acidic pH. Secretion prevents the stigma from drying out and regulates the osmotic pressure of pollen grains. Secretory activity indicates the receptivity of the stigma. This process is influenced by weather.

Stigmatic secretion rapidly disappears at high temperature and low humidity. Dry and windy weather has the same effect. Secretory activity lasts longer and the extract remains

Fig. 5.1. Receptivity of stigmas in 'Érdi bőtermő' sour cherry cultivar (Érd-Elvira 1974 in Hungary) (after Nyéki 1974c)

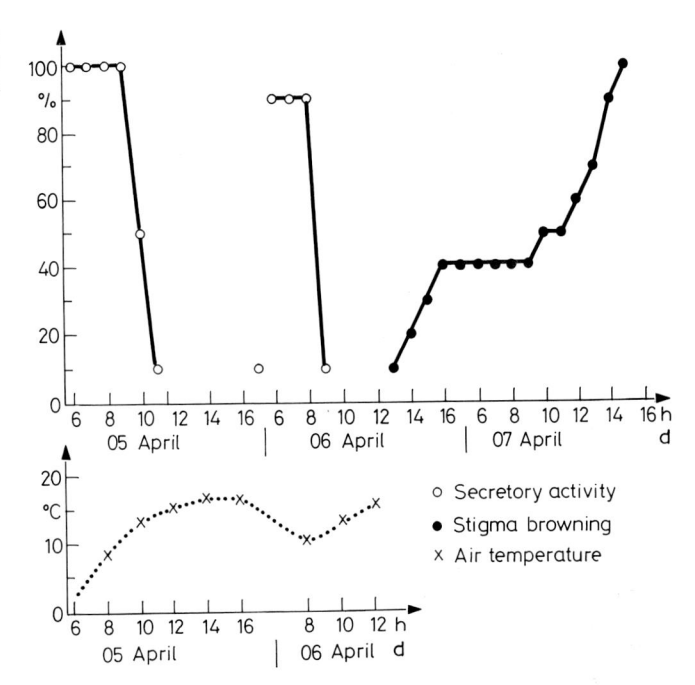

○ Secretory activity
● Stigma browning
× Air temperature

on the stigma surface for the entire day under humid and cool circumstances. Short and weak rain is not harmful but strong rainfall removes secretion once and for all.

An importance is attributed to papillae on the stigma surface by Watanabe (1984) and Cresti et al. (1985). Studies made by Stösser and Anvari (1983) with *sweet cherry* and by Braun and Stösser (1985) with *apple* led to the opposite conclusion. They consider the influence of stigmatic secretion and the turgescence of papillae negligible in the development of pollen tubes.

Characteristics of a given species or cultivar also have an effect on the receptivity of the stigma. Lemaitre (1978), for instance, observed an extremely short period of receptivity in *strawberries* having long-lasting blooming. Soltész (1992) found a year by year smaller receptivity in some diploid apple cultivars ('Red Astrachan', 'Alkmene', 'Cox's Orange Pippin', 'Red Delicious'). A study carried out by Nyéki (1974c) showed that 'Pándy 48' was the best 'Pándy' sour cherry clone in respect of receptivity. Among *oriental plums*, reported by Szabó (1989), 'Shiro' had a great deviation in receptivity. Among the European cultivars, 'President' can be characterized the same way. The receptivity period is consistently the shortest in 'Methley' oriental plum. In 'Marcona' almond receptivity cannot be demonstrated because the stigma has already been pollinated in the balloon stage before anthesis, as was reported by Gagnard (1954).

Some other factors affecting stigma receptivity are as follows:

In *apple*, shoot type and the arrangement of flowers in an inflorescence have a significant effect on receptivity. The receptivity period may be decreased by 50% in lateral inflorescences located on long shoots compared to those on short spurs (Soltész 1992). Lateral flowers of cyme inflorescence show the same tendency to a lesser degree (Lalatta 1982a, Soltész 1992).

In *sour cherry*, stigma receptivity is lower in the outer part of the canopy and on the basal part of shoots as well as in the flowers of short spurs (Ifjú in Brózik and Nyéki 1975). The receptivity period increases in direct proportion to the number of flowers in the inflorescence. Irrigation was observed to have similar effects whereas sod yields the opposite results. Position effect of flowers in *plum* canopy was also reported by Szabó (1989).

The structure, activity and receptivity of stigmas in wind-pollinated fruit species differ from those of the species discussed above and, hence it is summarized separately.

In *walnut*, the bifurcate (two-branched) stigma is situated at the top of the pistillate flower without stamen. Branches open and spread only when the stigma is receptive. The effective pollination period coincides with the stigma receptivity. There is an increasing possibility for pollination when the stigma branches open and begin to spread. The EPP ends when the spreading stigma branches reach an angle of 45° (Schuster 1961, Germain et al. 1973, Germain 1975).

During spreading stigmas are deeply wrinkled, glittering and greenish-yellow. Though their tissues are full of water, there appears a material rich in lipids on their surface (Szentiványi in Nyéki 1980). When the stigmas turn brown or their inclination exceeds 45°, there is no further possibility for pollination.

The *chestnut* has compound heteromorphic inflorescences containing pistillate, staminate and bisexual inflorescences, originating from the axillary buds under shoot tips. The double cymose female inflorescences are situated in the lateral branches of the compound inflorescence, with male inflorescences above them. After the petal fall, the male inflorescences either fall down or remain on the fruit and dry (Szentiványi in Nyéki 1980). The bisexual inflorescences, which sometimes occur in the area that separates male and female inflorescences, have a weak receptivity and are aborted soon (Solignat 1973).

The receptivity period begins in independent female inflorescence and the female branches of heteromorphic inflorescences when all the stigmas have been completely developed. The receptivity of stigmas waiting for each other needs to be considered together. The pollination period may last for a month (Shimura et al. 1971).

Hazelnuts are characterized by a peculiar stigma position as well as a special mechanism supporting pollination (Szentiványi in Nyéki 1980). If stigma tips lose their receptivity as a result of frost, their task will be taken over by the lower style regions due to the continuous style growth.

Another feature of *hazelnuts* is that the length of shoots determines viability and receptivity of female flowers. Proper receptivity requires a shoot length of 15 to 25 cm, the shorter length applies to old plants and the longer length to young ones.

5.2. The Period of Pollen Shedding

Open pollen sacs are an indication of the sexual maturity of anthers. Pollen shedding from mature anthers may begin and exceptionally may finish before anthesis. In most fruit species, however, the dehiscence of anthers only takes place after anthesis. Smykov (1974) observed in *apricot* that 15 days before anthesis pollen was "mature" and some anthers dehisced and shed their pollen. The dehiscence of anthers was reported by Brózik et al. (1978) to vary from half to two days in different *apricot* cultivars and to start in popcorn stage before anthesis. In 'Ceglédi óriás' in 1977, dehiscence had already finished in one-fifth of flowers by the time of anthesis. These flowers are cleistogamic. The dehiscence of anthers lasted longer than the secretory activity of stigmas.

In *peach*, anthers dehisce at the time when flowers opened or after opening (Martinez-Tellez et al. 1982). Therefore, peach flowers cannot be pollinated in the balloon stage, that is, cleistogamy is excluded. Nyéki (1989) argued that the dehiscence of anthers began zero to four hours after receptivity of stamens.

In the 'Pándy' *sour cherry* cultivar this process begins one to two hours after opening of flowers (Maliga 1944).

Nyujtó and Banainé (1974) demonstrated, in different *sour cherry* cultivars anthers dehisce 13 to 44 hours after opening of flowers. The minimum temperature required for dehiscence is between 13 and 15 °C.

Wociór's (1976b) detailed studies on different *sour cherry* cultivars showed that anther dehiscence required 4 to 28 hours from opening of flowers. This period was 4 to 24 hours for those flowers that opened in the morning, while it took 23 to 28 hours for those opened at 14 h.

In *plum*, pollen shedding was observed to begin two days after anthesis (Randhawa and Nair 1960). According to Szabó (1989), this period lasts for 0 to 30 h and there is no great difference between the *oriental* and the *European* cultivars.

Church et al. (1983c) emphasized that anther dehiscence in *apple* may vary to a great extent, depending on the annual weather. The dehiscence of the first anthers requires 0 to 2.4 days from opening of flowers but it also depends on the cultivar. Soltész (1992) stated that the dehiscence is 0 to 2.5 days and attributed a greater role to the annual weather than to cultivar characteristics. In two of ten years pollen shedding began at the time of flower opening in several cultivars. Earlier observations of Soltész et al. (in Nyéki 1980) suggest that high temperatures may result in pollen shedding immediately before the flowers opened.

The dehiscence of anthers begins in *gooseberry* when sepals completely bend down towards peduncle. The direction of dehiscence is always acropetal (Harmat in Nyéki 1980).

The time of pollen dehiscence and the period of pollen shedding in fruit species are summarized in Table 5.8.

The beginning of pollen dehiscence and the duration of pollen shedding concerning *apple* are summarized in Table 5.9. The effect of the years is significant. Pollen shedding periods vary from 0.9 to 6 days during the course of the ten years investigated. Church et al. (1983c) reports five to eight-day-long pollen shedding periods in two years while in an earlier work Williams (1970a, b) reports not more than two-day-long pollen shedding periods.

Soltész (1992) examined the daily course of pollen shedding in *apple*. His experiments showed that it was the temperature that determined the quantity of dehisced anthers and the daily maximum always fell between 11 and 15 hours.

The main features of pollen shedding in *sour cherry* are summarized in Table 5.10. Nyéki et al. (1974) observed an induced periodicity of pollen shedding, which is similar to that of anthesis. The dehiscence of anthers is continuous the whole day. This process is slow between 9 and 11 h, reaches a maximum at 11 h and remains intensive until 14 h (Fig. 5.2). Ninety percent of the anthers open on the first day. The rate of anthers dehisced is over 10% per hour during the most intensive noon period (Nyéki 1974c). The daily peak of pollen shedding is between 8 and 12 h in *sweet cherry* (Percival 1965).

Dehiscence of anthers in *oriental* and *European plums* are given in Table 5.11. Szabó (1989) observed the dehiscence of anthers to be continuous the whole day, but its level was low till 11 h. The intensive phase of pollen shedding begins at 12 h in *Japanese plums* and two hours later, at 14 h, in *European plums*. It is the first day of pollen shed-

Table 5.8. Duration of Anther Dehiscence and Pollen Shedding in Different Fruit Species

Species	Duration (day)	Author (year)
Apple	1–4	Davary and Nyéki (1990)
	1–5	Soltész (in Nyéki 1980)
	1–6	Soltész (1992)
	5–7	Church et al. (1983)
Pear	1–5	Nyéki (1973a)
Sweet cherry	1–5	Nyéki (1974b)
Sour cherry	1–5	Nyéki (1974c), Nyéki et al. (1974)
	40 hours	Nyujtó and Banainé (1974)
	3	Wociór (1976b)
Plum	1–5	Szabó (1989)
	1–6	Bellini and Bini (1978)
Peach	1–3	Nyéki and Brózik (in Nyéki 1980)
Almond	2–6	Gagnard (1954)
Apricot	1–2*	Brózik (in Nyéki 1980)
Currant	from anthesis to petal fall	Baldini and Pisani (1961), Porpáczy (1987)

* Considering only the period after anthesis

Table 5.9. Duration of Anther Dehiscence and Pollen Shedding in Apple* (after Soltész 1992)

Year	Beginning of anther dehiscence after anthesis day — Range of cultivars	Mean	Duration of pollen shedding (day) — Range of cultivars	Mean
1981	1.5–2.2	1.8	2.5–5.4	4.0
1982	1.4–2.5	1.7	2.3–4.6	3.7
1983	1.1–2.1	1.4	1.5–3.7	2.4
1984	1.6–2.0	1.8	2.4–6.0	4.3
1985	1.5–2.4	1.9	2.8–5.9	4.2
1986	0.0–1.3	0.4	1.0–2.6	1.8
1987	0.0–1.5	0.5	0.9–2.8	1.7
1988	0.5–1.2	0.9	2.1–4.8	2.9
1989	0.6–1.4	0.9	1.6–4.1	2.6
1990	1.2–2.3	1.8	2.1–5.3	3.4
1981–90	0.0–2.5	1.3	0.9–6.0	3.1

* King flowers of spur flower clusters

Table 5.10. Duration of Anther Dehiscence in Sour Cherry Cultivars (after Nyéki 1974c)

Characteristics of anther dehiscence	1972	1973	1974	1972-74
Range (day)	1.7–2.3	1.8–2.6	1.1–2.1	1.1–2.6
Mean of cultivars (day)	1.9	2.1	1.8	1.9
Mean of self-sterile cultivars (day)	2.0	2.2	1.8	2.0
Mean of self-fertile cultivars (day)	1.7	2.2	1.8	1.8
Mean of early blooming cultivars (day)	1.7	2.6	1.9	2.1
Mean of late blooming cultivars (day)	2.0	2.0	1.7	1.9

Table 5.11. Duration of Anther Dehiscence in Oriental and European Plum Cultivars (after Szabó 1989)

Characteristics of anther dehiscence (hour)	Oriental plums	European plums
Range	46–54	95–117
Mean of cultivars	50	104
Mean of early blooming cultivars	50	117
Mean of late blooming cultivars	51	99

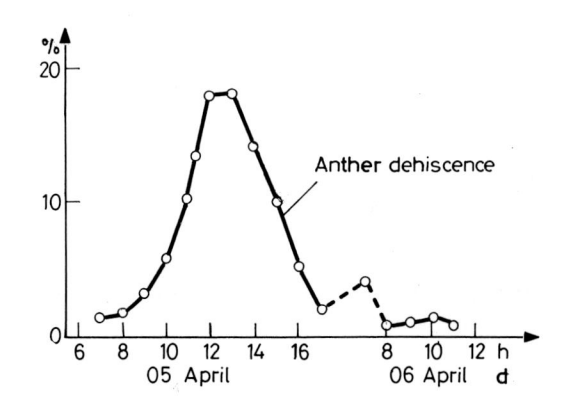

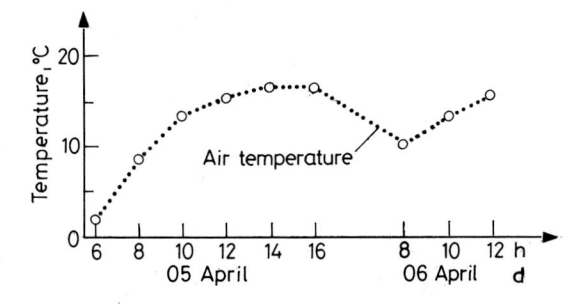

Fig. 5.2. Dehiscence of anthers in 'Érdi bőtermő' sour cherry cultivar (Érd-Elvira in Hungary) (after Nyéki 1974c)

ding that is the most intensive in years of rapid blooming; otherwise, it is the second day. There is pollen shedding also in the night to a lesser degree.

In insect-pollinated fruit species anther dehiscence is influenced mostly by the weather, specifically by temperature and precipitation.

Strong rainfall at a given temperature can significantly decelerate the dehiscence of anthers in *apple* (Lalatta et al. 1982). A foggy morning may result in a delay of a few hours in dehiscence (Soltész 1992). Williams et al. (1984) demonstrated that the anthers previously dehisced and then closed again under the influence of rain were difficult to separate from those not dehisced before. The considerable damage to the anthers forced them to close again which may entirely inhibit pollen shedding.

In every *sour cherry* cultivar, the average pollen shedding period was reported to be three days. In rain, pollen grains do not shed but swell up, and they begin to be shed again only a few hours after the rainfall. Heavier rain shortens the pollen shedding period by one day (Nyéki 1974c).

There is a strong correlation between the hourly rate of dehiscence and temperature, in *plum* (Szabó 1989).

Rate of dehiscence during the first day after anthesis

	under 14 °C	14 to 20 °C	above 20 °C
European cultivars	1 to 5%	2 to 10%	4 to 16%
Oriental cultivars	1 to 2%	2 to 10%	6 to 24%

The duration of pollen shedding is given for four Hungarian *apricot* cultivars in Table 5.12. In every year investigated, pollen shedding started in the balloon stage. The first peak is at 14 h on the first day. The second day after anthesis can be characterized by a strongly decreasing pollen shedding activity; only a small peak can be observed two hours after the last flowers have opened.

The dehiscence of anthers lasts for a short time in *peach* (Table 5.13). Comparisons between the data of two years indicate that in 1983 the rate of dehisced anthers was under 10% until 9 h with a daily maximum between 11 and 13 h. In the following year the rate of dehiscence reached a level of 10% two hours sooner and the daily maximum was also reached two hours sooner.

Table 5.12. Anther Dehiscence in Apricot Cultivars (after Nyéki and Szabó 1986, unpublished)

Cultivar	Average duration of pollen shedding (day)				Pollen shedding in balloon stage (%)			
	1976	1977	1978	Mean	1976	1977	1978	Mean
Magyar kajszi	2.0	1.0	1.5	1.5	69	100	15	61
Ceglédi óriás	1.0	0.5	1.5	1.0	53	100	17	57
Kecskeméti rózsa	1.5	1.0	1.0	1.2	48	100	6	51
Kécskei rózsa	1.0	1.0	2.0	1.3	39	100	25	55
Mean	1.4	1.2	1.5	–	52	100	16	–

Table 5.13. Duration of Anther Dehiscence in Peach (after Nyéki and Szabó 1986, unpublished)

Characteristics of anther dehiscence	1983	1984
Range (hour)	7–8	7–17
Mean of cultivars (hour)	7	12
Mean of early blooming cultivars (hour)	7	10
Mean of late blooming cultivars (hour)	7	13
Mean of freestone cultivars (hour)	7	10
Mean of clingstone cultivars (hour)	7	12
Mean of nectarine cultivars (hour)	7	13
Beginning of bloom (date)	April 5	April 18

In every insect-pollinated fruit species, the unfavourable effects of cool and rainy weather can be summarized as follows:

* Low temperature unfavourably affects the removal of pollen from anther (Nyéki 1974a,c).
* Rain (Dorsey 1919), but not high humidity (Percival 1955), inhibits anther dehiscence.
* Steady rain during bloom results in pollen grains to burst due to their high osmotic concentration (Ducom 1968).
* Rain washes away pollen from the dehisced anthers or prevents pollen from being shed from the wet anthers (Burchill 1963, cit. by Nyéki 1989).
* Pollen shedding starts again a few hours after rainfall.

It is instructive to cite here some statements of Ifjú (in Nyéki 1980) concerning the effect of other factors: a sod system shortens the pollen shedding period while irrigation increases it. The dehiscence of anthers begins later and lasts longer in the northern and inner parts of the canopy.

The following examples demonstrate the effect of cultivar characteristics on pollen shedding. According to Church et al. (1983), some *apple* cultivars are characterized by an earlier beginning of pollen shedding (e.g. 'Lord Lambourne' and 'James Grieve'). In 'Golden Delicious', however, anthers dehisce relatively late. Studies by Davary and Nyéki (1990) led to the same conclusion in 'Golden Delicious' and 'Jonathan'.

Rudloff and Schanderl (1950) stated that opening started in the inner whorl of stamens in *stone fruits*, whereas Nyéki (1974c) and Ifjú (in Brózik and Nyéki 1975) held the opening sequence to be cultivar-specific in *sour cherry* though they also considered opening from inner to outer whorl to be most common. Bargioni (1978), in *sweet cherry*, confirmed the opinion of Rudloff and Schanderl.

Pollen is easily shed from the pollen sacs of dehisced anthers in some cultivars preventing pollen grains from sticking together. The walls of anthers retain 50 to 80% of pollen grains; thus, they remain stuck in clumps on the surface of pollen sacs instead of being shed. Nyéki (1974c) found few pollen grains, stuck in clumps, prevented from being shed, in anthers of 'Pándy' *sour cherry*.

Sour cherry cultivars, which are characterized by spreading pollen shedding, are more likely to be pollinated successfully (Maliga 1980). Maliga observed in some *sweet cherry* cultivars that pollen appeared on the petals in little clumps after anthesis. Faccioli (1981) stated that this was an early anther dehiscence that characterized 'Bigarreau Moreau' sweet cherry.

Anthers of 'President' plum open early and intensively (Szabó 1989). In 'Stanley', however, this process begins later and lasts shorter. Cool and wet weather prevents a portion of anthers from dehiscing, especially in 'President' and 'Bluefre' plums. The dehiscence of anthers is more intensive in *oriental* cultivars than in *European* ones.

The dehiscence of anthers was observed to be significantly more intensive in clingstone peaches than in other cultivars in certain years (Nyéki 1989).

Anthers of 'Gorella' and 'Redgauntlet' strawberries are inactive at the beginning of bloom, and the vitality of pollen also reaches a satisfactory level only later (Lemaitre 1978).

The bloom period of male flowers in *walnut* is shorter than that of female flowers (Germain et al. 1975, Szentiványi in Nyéki 1980). Pollen grains appear at a catkin length of 2 cm. When the catkins become light green and hanging, the anthers draw apart, begin to turn yellow and start to dehisce from the base of the catkin shedding pollen. The daily peak of this process falls between 10 and 12 h. The empty anthers soon turn black. The shed pollen remains viable for two to three days.

In *chestnut*, appropriate anther development and pollen shedding is to be expected only in flowers with long filaments (Breviglieri 1951). The commercially valuable cultivars are characterized by a poor pollen production and shedding. The marone-type cultivars, that have large fruits, also have flowers with short or incomplete filaments.

Solignat (1973) and Szentiványi (in Nyéki 1980) provided a detailed description of the development of female flowers in *chestnut*. Secondary catkins develop which are fewer and smaller on shoot tips 15 to 20 days after the primary catkins have appeared. They begin to open when the primary catkins have completed opening. Pollen shedding is completed by the male flowers in compound heteromorphic inflorescences which are the latest to shed pollen. Thus, the pollen production of a tree lasts for an entire month due to the appearance of three waves of male inflorescences.

It was observed in *hazelnut* that up to 600 m above sea-level young catkins often shed pollen uselessly at the end of summer or in fall.

5.3. The Overlap of Pollen Shedding and Stigma Receptivity

Homogamy means that there is no difference in time between the receptivity of stigma and the dehiscence of anthers in a bisexual flower. In the case of dichogamy, the sexual organs of the flower will be mature in distinct points of time, excluding self-pollination.

Dichogamy has two forms. In one case, the period of anther dehiscence and pollen shedding precedes the period when stigmas are receptive (protandry). In the other case, stigmas have already been receptive when anthers begin to dehisce (protogyny).

Cleistogamy also occurs in fruit species. In this case, an obligatory self-pollination takes place in the closed flower before anthesis.

The development of the different organs in a flower proceeds from outside to inside. Hence, dichogamy (protandry) is natural because stamens become mature first (Emery 1877, cit. by Soltész 1992). According to Gardner et al. (1952), dichogamy is a system preventing self-fertilization. Bisexual plants show a lower degree of dichogamy, depending on species. Dichogamy is always partial in temperate zone fruits; but it is complete in tropical fruits (Barbier 1985).

It is important to know the degree of dichogamy and the factors affecting it, especially in the case of species requiring cross-pollination. Checking the overlap of bloom periods only provides some preliminary information in combining cultivars. Cultivars suitable to be pollinizers are those whose pollen shedding period coincides with the stigma receptivity period of the cultivar to be pollinated.

The importance of dichogamy is different in self-fertile and self-sterile cultivars, in bisexual and unisexual flowers, and in single flowers or inflorescences.

Apple flowers are characterized by protogyny (Maliga 1946, 1956a, Wilcox 1962, Porpáczy 1964, Soltész et al. in Nyéki 1980, Soltész in Nyéki 1980, Williams et al. 1984, Barbier 1985, Davary and Nyéki 1990, Davary 1992). The degree of protogyny is usually one to two days but even a period of five days is possible (Wilcox 1962). Recent investigations of Soltész (1992) have suggested that the degree of dichogamy (and the prevalence of its cultivar-specific features) in a given year depends on the prevailing weather.

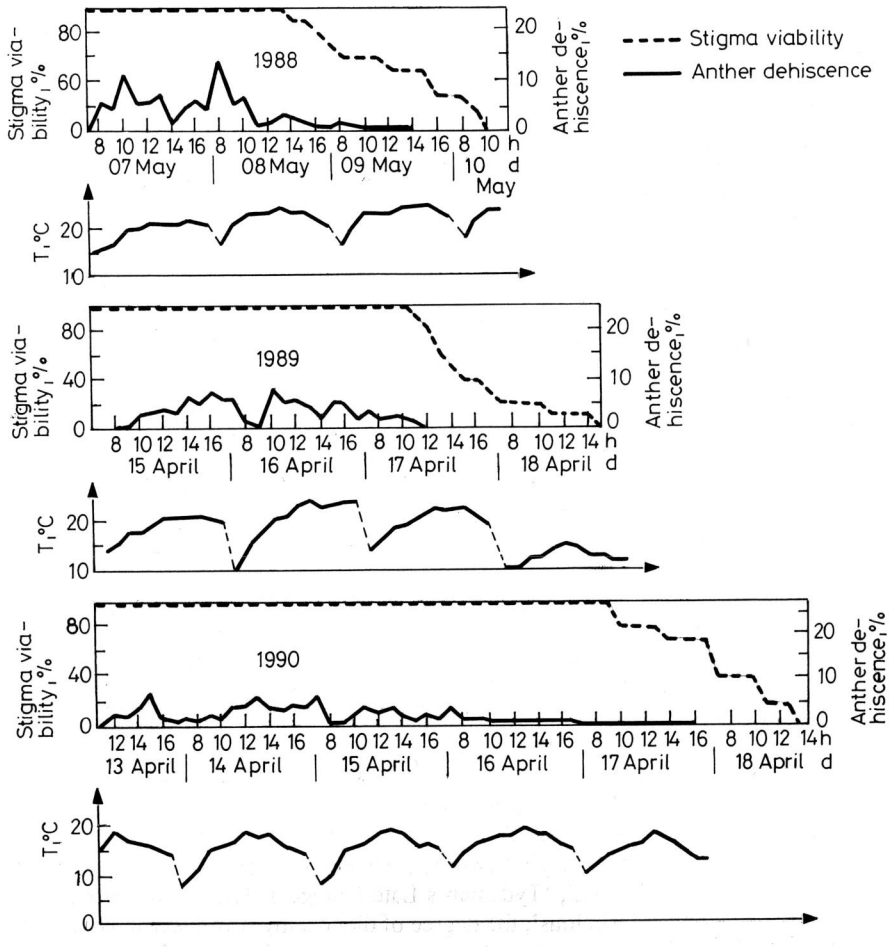

Fig. 5.3. Receptivity of sexual organs in flowers of 'Watson Jonathan' apple cultivar (Debrecen 1988 to 1990) (after Davary 1992)

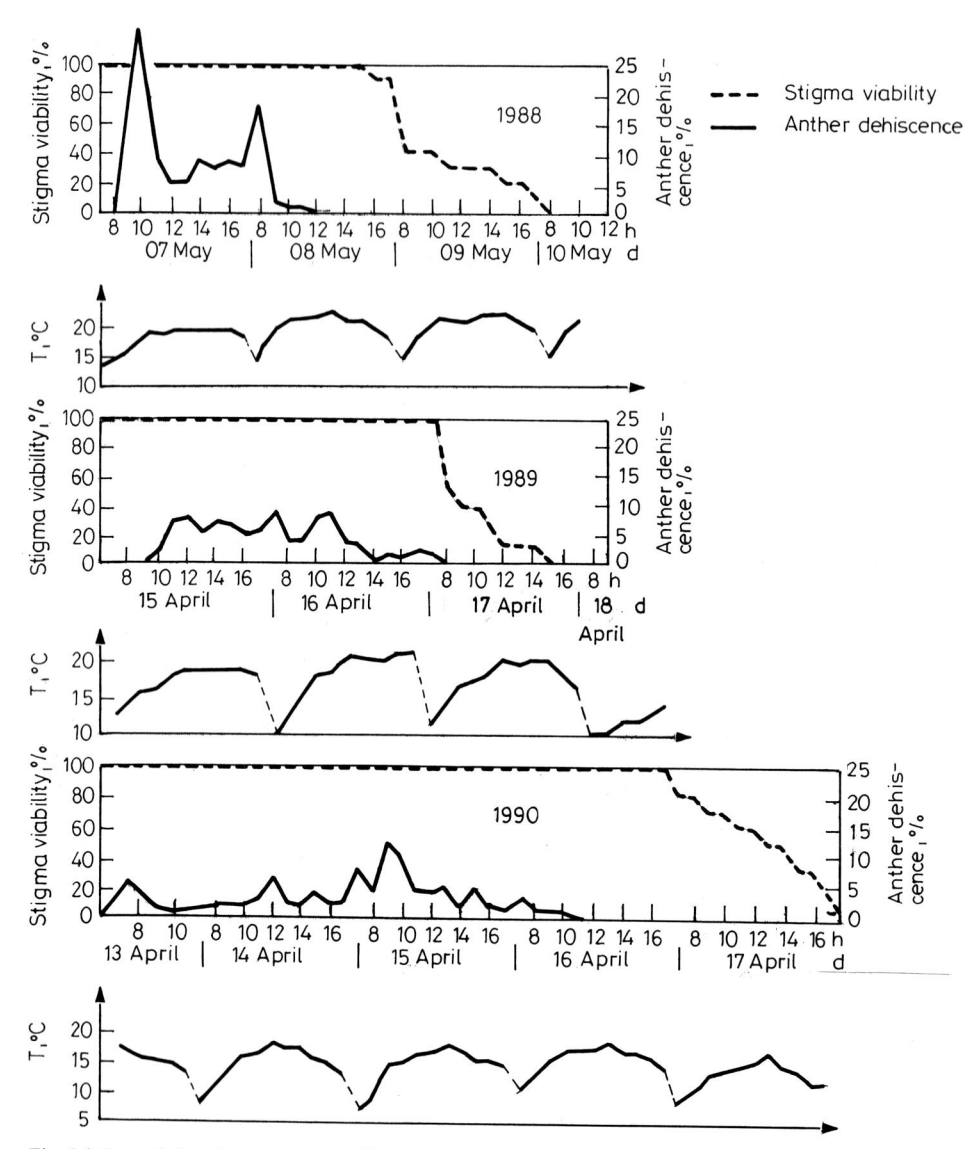

Fig. 5.4. Receptivity of sexual organs in flowers isolated by parchment bag in 'Watson Jonathan' apple cultivar (Debrecen 1988 to 1990 in Hungary) (after Davary 1992)

Precipitating anther dehiscence, high temperature will decrease the degree of dichogamy. Sometimes the maturity stage of the two sexual organs almost coincides in some cultivars ('Antonovka', 'Bramley's Seedling', 'Charden', 'Granny Smith', 'Gravenstein', 'Idared', 'Jerseyred', 'Mantet', 'Tydeman's Late Orange'). The age of trees also affects dichogamy. In 'Golden Delicious', the degree of dichogamy was lower in young trees.

Baldini (in Nyéki 1980) mentioned protandry in *apple*. Soltész (1992) observed in the years 1986 and 1987 that pollen shedding preceded the secretory activity of stigmas and started on the day of anthesis in a part of flowers in some cultivars ('Alkmene', 'Ceglédi piros', 'Close', 'Early Red Bird', 'Idared', 'James Grieve', 'Julyred', 'Merton Worcester', 'Spartan', 'Summerred'). Most of these cultivars are characterized by an early or mid-early flowering, supporting the statement of Williams et al. (1984) that a lesser degree of dichogamy is characteristic in early blooming cultivars.

The effect of temperature upon dichogamy is illustrated in Fig. 5.3. The role of temperature is elucidated by enclosing the flowers in parchment bag to influence their temperature (Fig. 5.4).

The *pear* is also characterized by protogyny (Maliga 1946, Nyéki 1973a). The data concerning *quince* in the literature are contradictory: Maliga (1946) regards it as protogynous, whereas Gardner et al. (1952) reports protandry.

Protandry is also mentioned in *sour cherry* (Maliga 1946), but in most studies protogyny (Mohácsy and Maliga 1956, Nyéki 1974c), or sometimes homogamy (Lott and Simons 1968, Nyéki 1974c), is argued for. Nyéki's (1974c) data are given in Table 5.14. Changes in the period of anther dehiscence highly influence the overlap of the activity of the two sexual organs. An overlap in the same flower is of importance only in self-fertile cultivars. Homogamy is striking in 'Pándy 279' sour cherry clone. Figure 5.5 illustrates

Table 5.14. Overlap of Stigma Receptivity and Pollen Shedding in Flowers of Sour Cherry Cultivars (after Nyéki 1974c)

Characteristics of receptivity of sexual organs	1972	1973	1974	Average
Beginning of anther dehiscence after anthesis (hour)	11	26	9	15
Overlap of stigma receptivity and anther dehiscence (day)	1.9	2.1	1.8	1.9
Homogamy (%)	20	67	58	48
Protandry (%)	0	32	13	15
Protogyny (%)	80	1	29	37

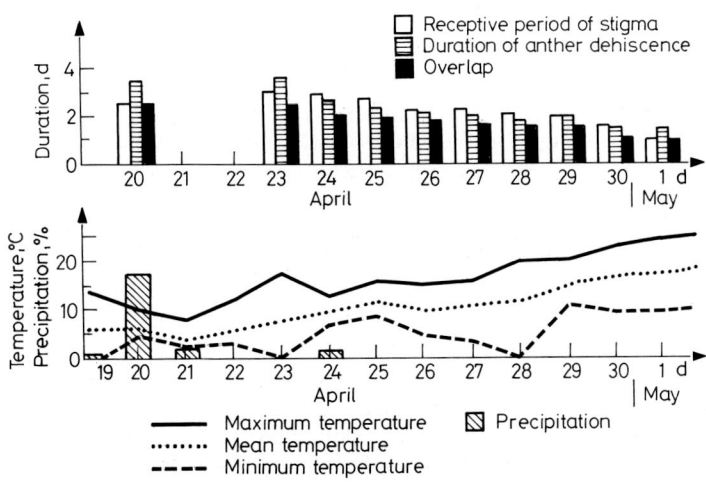

Fig. 5.5. Variability of receptivity period of sexual organs in flowers of sour cherry cultivars (Érd-Elvira 1973 in Hungary) (after Nyéki 1974c)

Table 5.15. Overlap of Stigma Receptivity and Pollen Shedding in Oriental and European Plum Cultivars (after Szabó 1989)

Characteristics of receptivity of sexual organs	Oriental plums	European plums
Beginning of anther dehiscence after stigma receptivity (hour)	7	7
Overlap of stigma receptivity and anther dehiscence (hour)	37	20
Homogamy (%)	3	20
Protogyny (%)	97	80

the effect of bloom time on the maturation of sexual organs and the overlap of their functioning in *sour cherry*.

The *sweet cherry* is similar to the sour cherry in respect of dichogamy. Some authors considered it protogynous (Stösser 1966a, Nyéki 1974c), but homogamy was also reported (Okályi and Maliga 1956, Nyéki 1974c).

The *plum* is protogynous according to most studies (Ewert 1929, Maliga 1946, Okályi and Maliga 1956, Tóth and Surányi 1980). Szabó (1989) confirms this opinion but also reports the occurrence of homogamy (Table 5.15). Homogamy may reach a higher rate in European cultivars. The overlap of the maturation and the activity of sexual organs (in a European cultivar) is illustrated in Fig. 5.6. The effect of years is remarkable in both cases. Figure 5.7, providing data on 'President' plum, serves as an illustration of the fact that individual flowers may differ so much that numerous flowers are to be investigated so as to discern cultivar characteristics.

The rate of dichogamy is also determined by the time and period of anther dehiscence in *plum*. Examining the yearly variation in dichogamy, it is important to consider temperature and other factors affecting the dehiscence of anthers and pollen shedding. The rate of homogamy may reach a level of 50% in 'President', while the majority of the cultivars can be characterized by protogyny (Szabó 1989).

Among stone fruits, the main features of the *apricot* are the high rate of protandry and the occurrence of cleistogamy. The rate of cleistogamy is highly influenced by the prevailing weather (Brózik et al. 1978, Nyujtó et al. 1979). Protogyny has never been mentioned in the literature. The proportion of protandry to homogamy depends on the weather of the given year (Table 5.16).

The overlap of the receptivity periods of the sexual organs are illustrated in *apricot* in Fig. 5.8. The temperature has also a strong effect on the activity of anthers in apricot. Warm weather decreases the degree of homogamy and increases protandry but trees are scarcely hindered from self-fertilization even under such circumstances.

Older cultivars used in *peach* production also showed protogyny, protandry and homogamy together (Mohácsy et al. 1967). Nyéki and Szabó (1986) carried out detailed studies with cultivar groups on homogamy and dichogamy (Table 5.17). Protandry was not found in any of the cultivars. The average degree of protogyny is not more than one to two hours, therefore we can consider homogamy as a rule in *peach*. The receptivity period of 'Springcrest' flowers is given in Fig. 5.9.

Almond is protogynous (Gagnard 1954). The *gooseberry* (Harmat 1987) and the *red currant* (Günthardt 1915, Porpáczy 1987) are also protogynous, but protandry or homogamy are characteristic of *black currant* flowers. In *blackberry*, homogamy and both forms of dichogamy are observable (Maliga 1946).

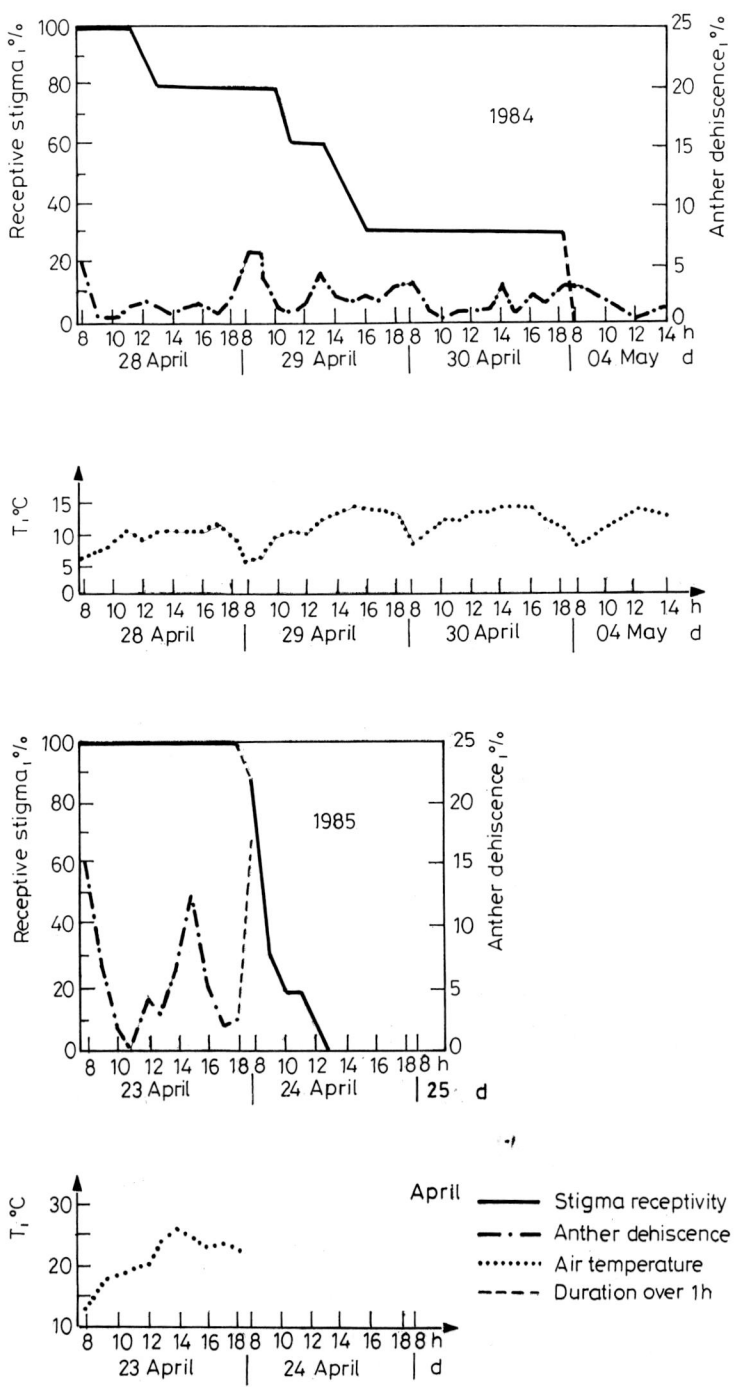

Fig. 5.6. Receptivity of sexual organs in 'President' plum cultivar (Siófok in Hungary) (after Szabó 1989)

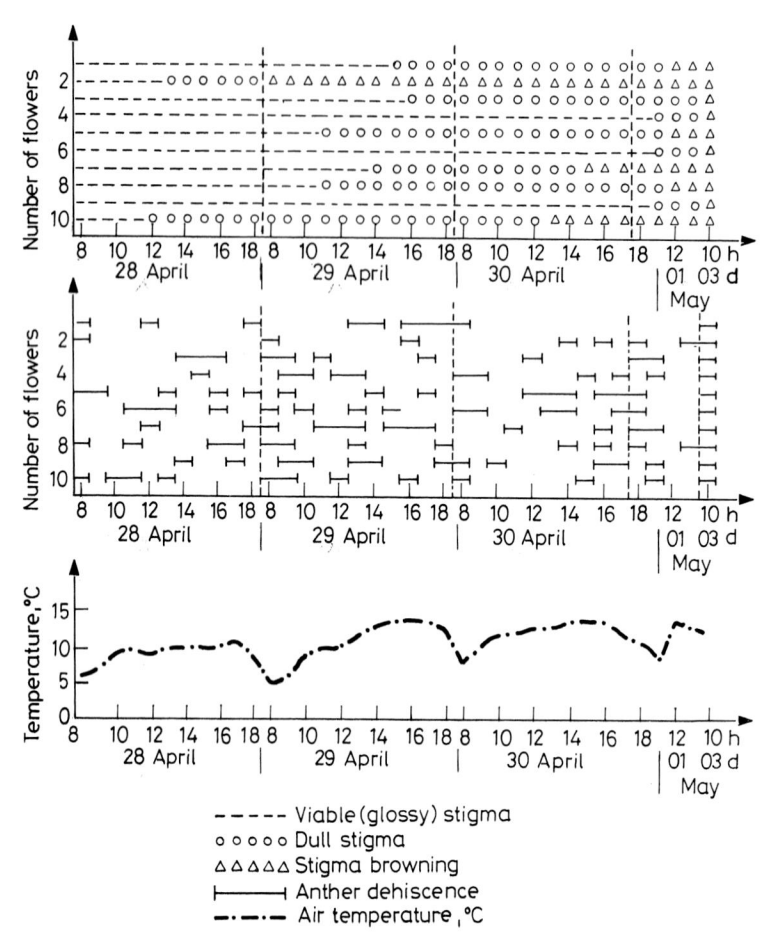

Fig. 5.7. The heterogeneity of receptivity in flowers of 'President' plum cultivar (Siófok in Hungary) (after Szabó 1989)

----- Viable (glossy) stigma
o o o o o Dull stigma
△ △ △ △ △ Stigma browning
├───┤ Anther dehiscence
─·─·─ Air temperature, °C

Table 5.16. Overlap of Stigma Receptivity and Pollen Shedding in Flowers of Apricot Cultivars (after Nyéki and Szabó 1986, unpublished)

Characteristics of receptivity of sexual organs	1976	1977	1978	Average
Time between anthesis and end of pollen shedding (day/flower)	1.2	0.5	1.6	1.1
Time between anthesis and end of secretory activity (day/flower)	2.6	1.0	2.9	2.2
Difference between the receptivity of two sexual organs (day)	1.4	0.5	1.3	1.1
Homogamy (%)	1.2	3.7	0.0	1.6
Protandry (%)	98.8	96.3	100	98.4

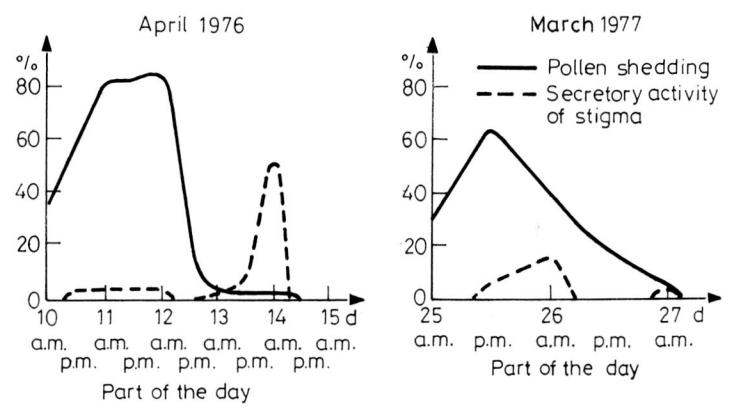

Fig. 5.8. Receptivity of sexual organs in 'Magyar kajszi C. 235' apricot clone (Budaörs in Hungary) (after Brózik in Nyéki 1980)

Protandry was reported by Gardner et al. (1952) in species such as *sea buckthorn*, *walnut* and *hazelnut*.

The definite dichogamy of *walnut* is often emphasized by several workers in this field (Krapf 1971, Germain et al. 1975, Krapf and Bryner 1977, Tsurkan and Pintya 1979, Szentiványi in Nyéki 1980, Yedrov et al. 1982, Kellerhals 1986). The development of male flowers is faster than that of female flowers in warm springs, increasing dichogamy. Cool conditions will result in a delay in the maturation of anthers and, consequently, protogyny. Protandry is characteristic of young trees to a greater extent. Young trees often produce either only male or only female flowers for a few years (Germain et al. 1975, Krapf and Bryner 1977).

Yedrov et al. (1982) reported the majority of *walnut* cultivars to be protandrous, but he also mentioned a lower rate of protogyny and an even lower rate of homogamy. The degree of protandry is 2 to 12 days, and that of protogyny is 4 to 13 days depending on the year. The weather in Hungary causes a fluctuation of 60 to 80% in homogamy (Szentiványi in Nyéki 1980).

Table 5.17. Overlap of Stigma Receptivity and Pollen Shedding in Flowers of Peach Cultivars (after Nyéki and Szabó 1986, unpublished)

Characteristics of receptivity of sexual organs	1983	1984
Beginning of anther dehiscence after stigma receptivity (hour)	1.0	1.7
Overlap of stigma receptivity and anther dehiscence (day)	7.3	18.9
Homogamy (%) mean of cultivars	45	13
mean of freestone cultivars	70	20
mean of clingstone cultivars	20	15
mean of nectarine cultivars	45	5
mean of early blooming cultivars	57	20
mean of late blooming cultivars	33	10
Protogyny (%) (mean of cultivars)	55	87

Fig. 5.9. Receptivity of sexual organs in 'Springcrest' peach cultivar (Siófok 1983 in Hungary) (Nyéki, original)

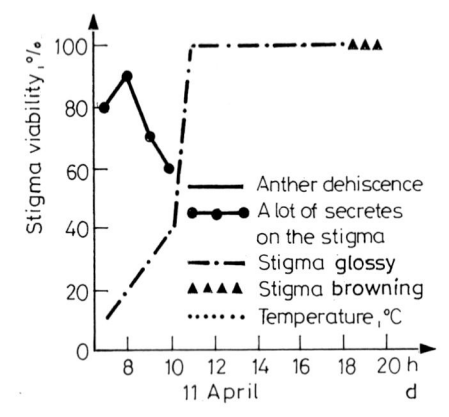

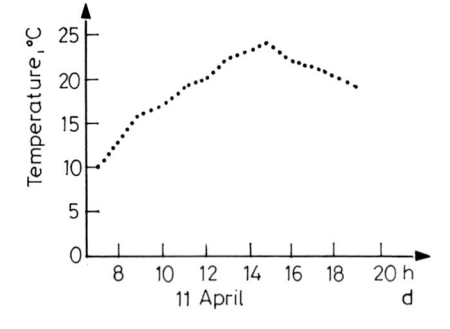

Breviglieri (1951) found a difference of five to six days in the maturation of sexual organs of *chestnut* trees, while this difference was much more (7 to 14 days) in the flowers themselves. Porpáczy (1964) reported double dichogamy. The bloom of male inflorescences precedes that of female inflorescences by eight to ten days, but the male inflorescences of compound heteromorphic inflorescences bloom last. None of the primary, secondary and tertiary flowers of male inflorescences bloom at the same time.

6. Pollination and Fertilization

6.1. Means of Pollination and Factors Affecting Pollination

J. Nyéki

Pollination or pollen transfer is a process during which "mature" pollen grains freed from pollen sacs and are transferred to the seed primordium of gymnosperms or the stigma of angiosperms by insects or wind.

Plants can be divided into two groups according to the type of pollen transfer: there are wind-pollinated (anemophilous) and insect pollinated (entomophilous) plants. The grouping of fruit species according to their pollen transfer is shown in Table 6.1.

6.1.1. Pollination Types

The pollen may come from the same cultivar, from another cultivar, or from the flower of another taxon.

Pollination types in fruit growing are: cleistogamy and chasmogamy.

Cleistogamy means pollination in the balloon stage, and it can be regarded as a natural self-fertilization (spontaneous autogamy). It is typical of species such as *apricot*, *quince* and *peach*.

In the case of chasmogamy, pollination takes place after anthesis. This is the most common pollination type in fruit species. Chasmogamy can be divided into two types: idiogamy (self-pollination) and allogamy (cross-pollination).

In idiogamy stigmas receive pollen from the same plant. In this case, genetically homogeneous gametes unite in the course of the process of fertilization. Idiogamy may take place in two different ways. In autogamy, the stigma receives pollen from the anthers of

Table 6.1. Grouping of Fruit Species According to Their Pollination Types

Insect-pollinated (entomophilous) species	Occurrence of cleistogamy	Wind-pollinated (anemophilous) species	Pollination by both wind and insects
apple	quince	walnut	chestnut
pear	apricot	hazelnut	
quince	peach	sea buckthorn	
sweet cherry			
sour cherry			
plum			
apricot			
peach			
almond			
gooseberry			
currant			
raspberry			
strawberry			

the same flower. In geitonogamy, pollen is transferred from other flowers of the same fruit tree. The pollen transfer itself may be spontaneous or artificial self-pollination. In the case of spontaneous self-pollination, the transfer of pollen to stigmas requires no carrier, neither insects nor wind. It is the morphology of flowers that makes self-pollination possible:

a. The direct contact of the anthers and the stigma facilitates natural autogamy.

b. The stamens are close to the stigma in the flower and transfer of pollen is naturally possible.

c. The anthers being situated above the stigma, pollen grains will fall down to the stigmatic surface due to gravitation.

In artificial self-pollination studies, pollen transfer is carried out by the pollen of the same cultivar and autogamy, geitonogamy, or clonal geitonogamy are rarely distinguished, but mentioned uniformly as self-pollination. Clonal geitonogamy, a special type of self-pollination, is also to be regarded as a kind of xenogamy. Different trees of the same cultivar pollinate each other in this case.

Flowers of most fruit trees are pollinated by pollen coming from flowers of other fruit trees, named cross-pollination (allogamy).

Xenogamy takes place when the flowers of two genetically different plants pollinate each other.

The protandry or protogyny of sexual organs discussed before (Chapter 5), may prevent a flower from being fertilized by its own pollen. However, this is only a superficial obstacle because flowers blooming at different times in a tree can pollinate each other, and the fertilization resulting from this pollination is equivalent to self fertilization.

In fertilization studies, xenogamy is regarded as pollination with the pollen of another cultivar (or species). Xenogamy is required in fruit species when:

a. the flowers are unisexual,

b. the receptivity period of male sexual organs and that of female sexual organs do not overlap each other (dichogamy),

c. their own pollen is sterile or self-incompatible with the stigma.

In search for pollinizers, cross-fertilization is carried out, and the choice of parents is conscious. In the case of cross-pollination, the stigmas of the given cultivar will be pollinated and hopefully fertilized with the pollen of another cultivar (species). In self-sterile cultivars, pollination is to be carried out without emasculation, but flowers of self-fertile cultivars are to be emasculated.

In most studies on cross-pollination a reciprocal cross is made. The cross partners play both the role of the female and the male partner. Therefore, the female entity of species *A* is pollinated by species *B* and vice versa.

6.1.2. Factors Affecting Pollination

The factors affecting pollination can be classified into the following groups:

1. weather (climatic) circumstances,
2. the quality and quantity of pollen, and its shedding from anthers,

3. the intensity of insect visitations and the activity of pollinators,
4. the length of bloom,
5. the composition, quantity, rate and placing of the pollinizers in the orchard.

Here we enter into the details of the weather conditions unfavourable to pollen transfer.

Temperature. Warm and dry conditions during bloom have a positive effect on pollination and fertilization.

The activity of bees, the most common pollinators, is the best between 17 to 22 °C. Temperatures below 10 °C slow or hinder insects in their work. In *sour cherry*, the following conditions are held to be favourable to pollination: the minimum temperature is over 4 °C for four to five days, the average temperature remains over 10 °C, the maximum temperature is between 18 and 20 °C, and the relative humidity of the air is under 50% (Parnia et al. 1979).

The self-fertile *sour cherry* cultivars will give a regular yield even if unfavourable weather conditions shorten the pollination period to one day during the bloom. The flower is highly sensitive to low temperature. During blooming temperatures between –1.5 and 0 °C may be critical. In the case of high temperature and low humidity rapid evaporation of stigmatic secretion will prevent pollen grains from adhering on the surface of stigmas.

Rain and humidity. Steady rain in the blooming period will hinder pollination. The rain prevents pollen grains from being shed from anthers, hindered from dehiscing, washes pollen out of the dehisced anthers, and dilutes and washes down stigmatic secretion from stigmas. The rain decreases the concentration of nectar as well which discourages insect visitation and, consequently, pollination. High humidity also prevents pollens from being shed from anthers. In contrast, low humidity decreases pollen germination and results in pollen surface withering.

The wind. Dry and windy weather in the flowering period causes a quick evaporation of stigmatic extracts and the marcescence of anthers. The evaporation of nectar is enhanced by the wind. A strong wind decreases the activity of bees, especially if wind is coupled with cold temperatures. According to Smith (1970), a velocity of wind over 15 to 20 km/h prevents the flight of bees, resulting in the prevention of pollination.

In insect-pollinated (entomophilous) fruit species, wind pollination is not significant and efficient.

Further factors. Parnia et al. (1979) observed pollen shedding to be very slow or entirely prevented in some *sour cherry* cultivars, such as 'Mocanesti', 'Crisana' 'Pándy' and 'Spaniole' by genetic factors.

Nyéki (1974c) pointed out that the length of the pollination period also affected the efficiency of pollen transfer. A quick transfer of pollen is not required in the case of a long pollination period. If the pollen transfer period is short, fruit set may be poor unless pollination takes place immediately after anthesis.

Factors of flower structure. The particular structure of flowers may inhibit or significantly diminish the pollination of insect-pollinated flowers even under very favourable environmental conditions. In species, such as 'Delicious' apple (Roberts 1945), 'President' and 'Jefferson' plums, and 'Tulip' peach, the stigma is long compared to the stamens that bees often collect nectar and pollen from without having touched and pollinated the stigma. They work around the stigma and not over it.

In the 'Giovanna d'Arco' pear cultivar, petals remain partly closed during the bloom period (Schanderl 1937). Flowers of the 'Abate Fetel' pear cultivar have a lot of petals and the corolla remains partly closed during the bloom. Hence insects are hindered in pollen transfer (Branzanti 1964).

6.2. Pollen Adhesion on Stigmatic Surfaces and the Outset of Pollen Tube Growth

T. Bubán

Our knowledge on the sexuality of plants dates back to the time of Herodotus to the 5th century B.C. (Kapil and Bhatnagar 1975). Pollination, being an indispensable condition of fertilization, may also be regarded as pregamy (Bagni and Gerola 1978). From the Greek gamia means act of marrying. Therefore, pregamy is a pre-requirement for a successful union (Faust, personal communication). After arriving on stigma, pollen becomes hydrated, and free amino acids and hydrolytic enzymes emerge from its walls. They are mostly amylases, proteases, RN-ases and esterases according to the studies made on about 50 species of angiosperms (Knox and Heslop-Harrison 1970, cit. by Bagni and Gerola 1978). RNA and proteins are also synthesized during the first hours of pollen tube growth (Mascarenhas et al. 1974).

There is a great variation in pollen germination and pollen tube growth. A few conditions are listed below:

To stigma surfaces of *almond* flowers, ten times more pollen will adhere after cross-pollination than after self-pollination (Pimienta et al. 1983), but no pollen tube growth will take place within 24 hours after pollination (Godini 1981). Pollen grains do not germinate at once, i.e. at the same time on stigmas of the *sour cherry* (Bartz and Stösser 1989). In *almond*, the threshold temperature is 15.5 °C for the germination of pollen and growth of pollen tubes (Pimienta et al. 1983), while in *sour cherry* the optimum is 20 °C, for the same two processes (Kerékné 1981). In *pear*, pollen germination is the best at 15 °C, and the growth rate of pollen tubes increases with increasing temperature tendency between 5 °C to 25 °C (Vasilakis and Porlingis 1985).

In *Ribes* species, the insertion of particular flowers in an inflorescence does not affect the number of germinating pollen grains and growing pollen tubes (Al-Jaru and Stösser 1983). In *apple*, the approximate quantity of pollen grains sticking on stigma surface is between 44 and 129 in the case of pollination with diploid cultivars, while this quantity ranges from 10 to 35 in the case of triploid pollen donor cultivars with a higher rate of non-germinating pollen grains (Seilheimer and Stösser 1982). Pollen germination may start on stigma surfaces even in intersterile combinations; moreover, pollen tubes of pear cultivars can grow into styles of apple cultivars, seven to ten tubes per style on the average of three cultivars tested (Schmadlak 1965).

Pollen tube growth is always faster in styles of *apple* flowers if a large quantity of pollen has arrived on stigma surfaces, even in the case of self-pollination or in the case of pollination with the pollen of triploid cultivars (Seilheimer and Stösser 1982). The same results attained by multiple pollination in *apple* flowers, which is one of the explanations for "pseudocompatibility", when fertilization and fruit set may be influenced by pollination with a large quantity of semi-compatible pollen supposing an optimal pollen tube growth at about 20 °C (Williams and Maier 1977). The advantageous effect of a large pollen quantity on pollen tube growth is also observable in *plum* (Lee 1980) and in *apri-*

cot (Egea et al. 1991). The fertilization of *blackberry* is the best if more than 200 pollen grains are deposited on the stigma surface (Lech 1976).

Apparently, *walnut* (*Juglans regia*) flowers do not stand for deposition of high quantity of pollen. Their female flowers cease to grow a few days after hand pollination and die, while this phenomenon has not observed after free pollination or after hand pollination of other *Juglans* species (*J. nigra, J. rupestris*) (Sartorius et al. 1984). In *walnut* orchards, high quantity of pollen will also result in female flower abortion, thus a greater distance from the pollinizer trees will result in an increasing ratio of fruit set. This phenomenon is called the "autoregulation of fertility" by Szentiványi (1990), and has never been observed in other fruit species.

The prospects of receiving pollen are not always the same. It occurs that the stigma is receptive even before the anthesis, e.g., in flowers of peach (Crossa-Raynaud et al. 1984), but it becomes receptive on the first to fifth day after anthesis in sweet cherry (Guerrero-Prieto et al. 1985). The stigma is especially receptive during the second to fourth day after anthesis in apricot trees (Egea et al. 1991) and on the fourth day in pear flowers (Modlibowska 1974).

Fertilization after the arrival of pollen at the stigma is not only influenced by general physiological conditions, but also by compatibility relations. Incompatibility preventing fertilization either takes place in the stylar tissues or on the surface of the stigma and both are based on the action and interaction of the multiple alleles of an S (incompatibility) locus (Brewbaker 1957).

In the *gametophytic* incompatibility system (*GSI*), the ability of pollen to function in a given combination is genetically determined by the S alleles, present in all pollen grains. The presence of the same S allele in the pistil results in incompatibility. The inhibition is manifested in the stylar tissues during the course of pollen tube growth. Studies by Lewis and Modlibowska (1942) on diploid *pear* cultivars already led to the conclusion that the presence of compatible and incompatible tubes indicated "gametic determination of incompatibility".

In the *sporophytic* incompatibility system (SSI), the behaviour of the pollen is sporophytically determined by the maternal genotype and the inhibition is manifested on the stigma surface.

There is a single significant difference between the two systems despite their seemingly great differences. The difference is the timing of the action of the S gene during the course of microsporogenesis. In GSI, the S alleles induce the formation of inhibitory factors (or the formation of their precursors) in the anaphase of the first meiotic division or later. In SSI, the action of the S alleles will be realized before the anaphase. A further pollen cytological relation is the time of the division of the generative nucleus of the pollen, which may occur before or after pollen shedding, resulting in the formation of binucleate or trinucleate pollen, respectively. After examining many species, often, but not always, there is a correlation between binucleate pollen formation and stylar inhibition (GSI), and trinucleate pollen formation and stigmatic inhibition (SSI). It is possible that the trinucleate pollen is insufficient in certain metabolites because of the second mitotic division that has taken place almost at the end of microsporogenesis. The presence of trinucleate pollen was reported to be characteristic of *almond* by Sogomonyan et al. (1974), though the fruit species of the Rosaceae family are characterized by GSI according to our knowledge.

It appears that there is a wide range of interaction between the sexual partners before fertilization (Ende 1976).

There are proteins released from the pollen, some of which may play a role in the regulation of fertilization; i.e. they take part in cell recognition (Stanley and Linskens 1965, cit. by Ende 1976). In the case of incompatible pollination, for instance, a callose plug appears in the stigmatic papillae near the pollen, a few hours after pollination. This will not occur in compatible pollination. Such proteins of the pollen wall come from the exine. The materials of the exine originate from the tapetum during the course of pollen formation, that is why the tapetum tissue, when carried to the stigma, may also result in callose formation. Proteins synthesized in tapetum cells are stored in the cisterns of the endoplasmic reticulum and injected into the pollen wall, or settle on its surface, after the decomposition of the tapetum. Lipids and carotenoids that are also synthesized in the tapetum adhere to the surface of the pollen. The quantity of compounds elutable from the pollen surface is much greater in binucleate pollen than in the trinucleate one (Kirby and Smith 1974).

The inner layer of the pollen wall (intine) is formed by the protoplast of the haploid spore and later by the vegetative cell of the pollen. Here, also occurs a considerable protein intussusception either on the whole surface or near to the germination apertures (Heslop-Harrison 1975). The receptor location for the sporophytic (pollen wall) proteins is the pellicle of the stigma papillae. The stigma surface itself may indicate the type of compatibility. In the SSI system, papillae are well developed and there is little secretion, while in the GSI system, the stigma surface is smooth or has only small papillae associated by an intensive secretory activity. In the SSI system, if the pollen tube can overcome the rejection operating on the stigma surface, then there is no more obstacle to fertilization. Among the species discussed in this book, a sporophytic incompatibility system (SSI) exists only in *hazelnut* (Romisondo 1965, and others, cit. by Germain 1983). In this species pollen grains cannot develop tubes, or the developed tubes cannot penetrate into styles.

In the gametophytic incompatibility system (GSI), the pollen tube is inhibited in the style, and especially in its upper part. In this case, the deposition of the callose (which can also be found in normal pollen tubes inside the pectocellulose wall of the tube) becomes irregular and it may obstruct the tip of the pollen tube (Heslop-Harrison 1976). In *apple* cultivars, Kobel and Steinegger as early as in 1934 reported swollen pollen tubes that cease to grow soon and contain degenerating gametes. We can recognize these features as an indication of GSI.

Specific proteins are responsible for the incompatibility interaction of the style and the pollen tube. The composition of the protein mixtures isolated from the styles depends on the combination of the pollen and the style: if the mating is compatible, the materials of the components are originated from both the male and the female partners, but if the mating is incompatible, then only the materials coming from the female individual are incorporated in proteins (Linskens 1958, 1959, cit. by Ende 1976). Qualitative and quantitative differences in the RNA and protein synthesis occur in the self- and cross-pollinated styles within three hours after pollination. "This indicates that rejection or acceptance of pollen tubes by styles is not a straightforward reaction, but a result of sequence of activations and inactivations of a number of genes" (Donk 1974). That a number of genes are involved in incompatibility is further indicated by others.

Eleven sterility genes of ten diploid *apple* cultivars were described by Kobel et al. (1939), hence, the number of sterility factors is much greater than it would be in the same quantity of *sweet cherry* cultivars. It is just the great number of sterility genes that provides a simple explanation of the relative rarity of cross-incompatibility in the *apple*. The determination of the sterility genes and the incompatibility groups in 20 *sweet cherry* cultivars and seedlings was reported by Way (1968).

A "one way success" is often reported when self-compatible (SC) and self-incompatible plants are crossed. If the SC plant serves as the female partner, there is a possibility of success, but the reciprocal crossing is usually unsuccessful. This phenomenon is the unilateral incompatibility (UI), also mentioned as unilateral inhibition, unidirectional crossability, etc. (Abdalla and Hermsen 1972).

When chromosome number is duplicated the self-incompatibility may disappear in *pear* cultivars (Lewis and Modlibowska 1942). The diploid variant of 'Fertility' pear cultivar will bear a full crop after self-pollination, but the fruits will be seedless, while the common tetraploid type will produce seeded fruits also after self-pollination (Crane and Lewis 1942). The autotetraploid *sweet cherry* (*Prunus avium*) also proved to be self-compatible (Crane 1923, cit. by Lewis and Modlibowska 1942). Utilizing a mixture of own and some compatible pollen in one to one ratio (mentor pollen method) may help in the reduction of self-incompatibility. Another method proved to be more successful in 'Doyenne du Comice' pear cultivar. Compatible pollen was transferred to stigma 14 hours before the arrival of the own pollen (pioneer pollen method). Using this method, even the tube of the own pollen can reach the base of the style (Visser and Oost 1982).

The optimum temperature of pollen tube growth is lower in the case of the own pollen. In the 'Ne Plus Ultra' almond cultivar, which is self-incompatible, the tube growth of the own pollen is the best at 15 °C, whereas growth of the pollen tube of other cultivars in the style of 'Ne Plus Ultra' is optimal at 25 °C. In self-compatible almond hybrids (except for one hybrid), the optimum temperature for pollen tube growth was also reported to be 25 °C after self-pollination (Kester and Bradley 1976). In one of the few self-compatible almond cultivars, 'Trucito', the tube growth of the own pollen is faster than the tube growth of another cultivar at 10 to 15 °C, the same at 20 to 25 °C, and the growth of the own tube is inhibited at 30 °C (Vasilakis and Porlingis 1984). In *apple* cultivars, "an interaction is possible between environmental or physiological factors and the incompatibility reaction which allows otherwise incompatible pollination to achieve fertilization may be called *pseudo-compatibility*" (Williams et al. 1977, cit. by Deveronico et al. 1982). Further examples of the relation between compatibility conditions and temperature will be given in Section 6.6.

Incompatibility is (or may be) of great importance in plant breeding (Lewis 1956). The discovery of many important details of self-incompatibility, however, still requires further research.

6.3. Pollen Tube Growth in the Style

The growth of the pollen tube is limited to its tip. The enzymes emerging from the pollen release the materials needed for tube growth from the style; the proteins of pollen are mainly originated from the outer layer of the pollen wall (exine). That is why the pollen tube, the biologically most active part of the male gametophyte is of saprophytic nature (Ende 1976). The high concentration of physiologically effective materials in the tip of the growing pollen tube is an indication that this physiological polarity is responsible for the directed growth of the pollen tube (Kapil and Bhatnagar 1975). There are numerous enzymes, ascorbic acid and sulphydril compounds in the tip of the pollen tube. The metabolic interactions between the pollen tube and the style are illustrated in Fig. 6.1.

In Schmadlak's (1965) opinion, the *affinity coefficient*: the ratio of the number of pollen tubes penetrating into the stigma to the number of pollen tubes reaching the base of the style truly reflects the fertility-biological relationship between the two given *apple*

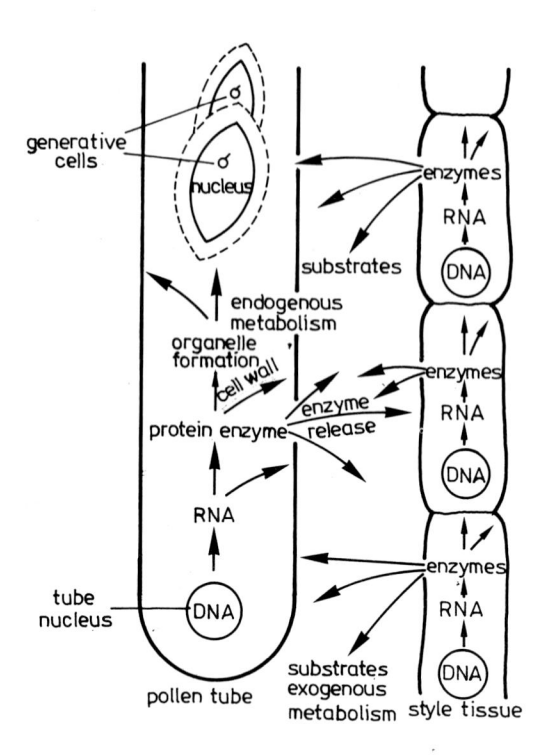

Fig. 6.1. Metabolic interactions between the pollen tube and the transmitting tissues of the style (after Linskens 1968, in Kapil and Bhatnagar 1975)

cultivars, or rather, the rate of pollen tube growth inhibition in style. As to *plum* and *apple*, excellent mathematical methods are available for the examination of the interaction of factors determining the growth of pollen tubes (Jefferies et al. 1982b, Jefferies and Brain 1984).

Pollen tubes of the *apple* reach the base of style in four to five days (Stott 1972, Anvari and Stösser 1981), but only a two-day period suffices for the fastest tubes (Schmadlak 1965). In the case of self-pollination, however, pollen tubes reach one-third ('Cox's Orange Pippin') or one-fourth ('Golden Delicious') of the full length of styles. Tubes of triploid cultivars grow slowly, only one to two tubes reach the base of the style (if any), often they soon cease to grow, the tubes become twisted and branched, and their tips become swollen (Seilheimer and Stösser 1982) as is shown in Fig. 6.2a,b. The speed of pollen tube growth rises parallel with the aging of *pear* flowers, it is fastest on the ninth day after anthesis in styles of the cultivar 'Conference' (Modlibowska 1945).

In *sweet cherry* and *sour cherry* cultivars, pollen tubes need two to three days to reach the base of styles (Stösser and Anvari 1981). This occurs similar to the own pollen tubes of self-fertile cultivars, but the tubes of self-sterile cultivars might cover the half distance during the same time (Ubatshina et al. 1976). It is worth knowing that the compatibility conditions in *sour cherry* and *plum* do not change in the case of *in vitro* pollination trials, in spite of the fact that ovule senescence is much faster. Therefore, results obtained under controlled conditions are of full value (Stösser 1980b, 1982). The growing potential of *plum* pollen tubes varies by the cultivar; the pollen of 'Czar' or 'Stanley', for instance, easily grows through the style in all studied cultivars (Lee 1980). In *almond*, the growth of pollen tubes is significantly faster in self-fertile than in self-incompatible

Fig. 6.2. The abnormal growth of pollen tubes generated by the pollen of 'Mutsu' (*a*) and 'Brettacher' (*b*) triploid apple cultivars in the style of diploid cultivars (after Seilheimer and Stösser 1982)

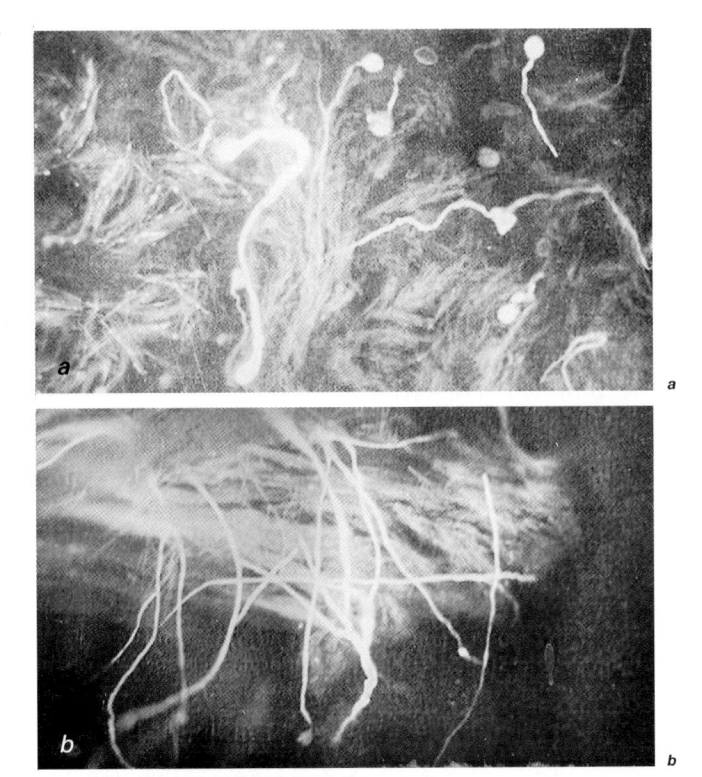

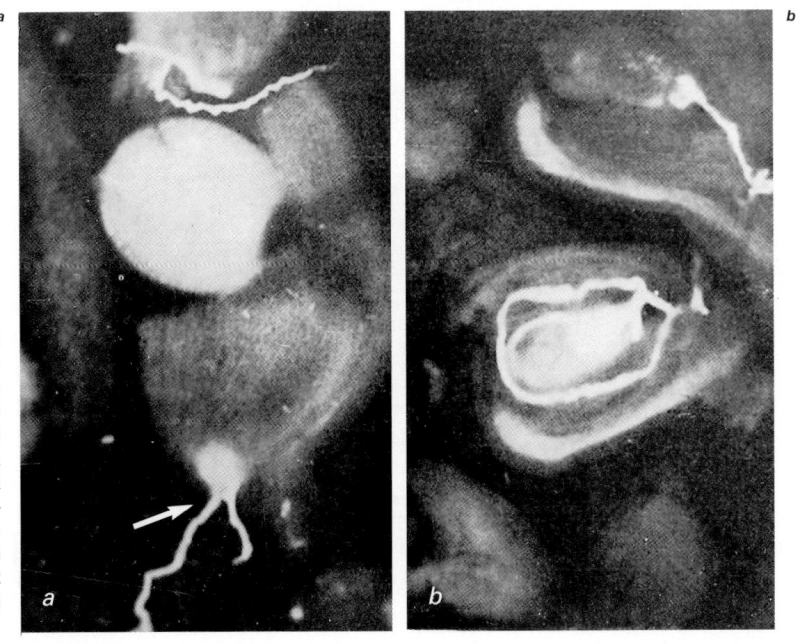

Fig. 6.3. Pollen tube growth in *Ribes* spp: (*a*) the penetration of two pollen tubes into the ovule through the micropyle in red currant; (*b*) abnormal pollen tube growth in the ovule of black currant (after Al-Jaru and Stösser 1983)

161

cultivars (Socias i Company et al. 1976), and it is also faster after cross- than self-pollination (Pimienta et al. 1983) even if it is evaluated in a self-compatible cultivar (Godini 1981).

As for *Ribes* species, the arrangement of flowers in a racemose inflorescence does not influence the pollen tube growth in *red currant*, whereas in *black currant*, which has long styles, passing through the style needs three to four days in the flowers inserted at the base of the raceme, while seven to eight days in the terminal flowers (Al-Jaru and Stösser 1983; Fig. 6.3a,b). *Walnut* pollen tubes usually need two days to grow through the style according to studies by Sartorius et al. (1984) in some *Juglans regia* cultivars and three other *Juglans* species.

6.4. Double Fertilization

The graphical summary by Lombard et al. (1971) demonstrates well that most factors determining fruit set have direct or indirect effect on fertilization (Fig. 6.4). The discussion of all factors would be the topic of another book, that is why we restrict this discussion to the immediate details of fertilization.

In the haploid, unicelled pollen grain, a vegetative and a generative nucleus comes into being during the first mitotic (and: unequal) division. The vegetative nucleus serves

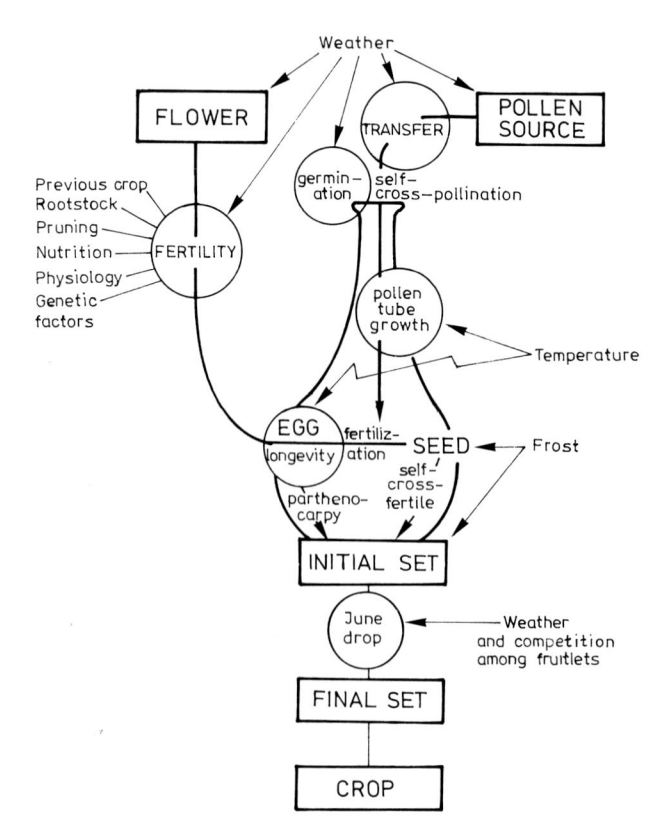

Fig. 6.4. Determinants of fruit set (modified from Lombard et al. 1971)

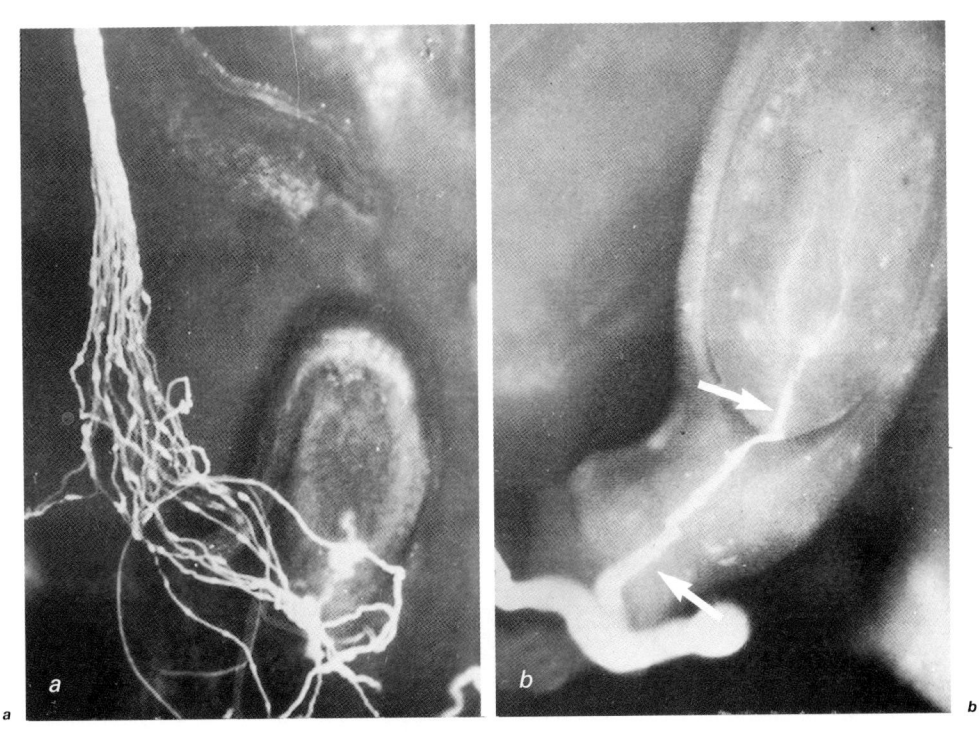

Fig. 6.5a,b. Pollen tube growing through the micropyle towards the embryo sac in apple ovule (after Anvari and Stösser 1981)

the pollen tube growth and later disappears. The sperm cells come into being during the further mitosis of the generative cell and they move to the tip of the pollen tube. The pollen tube penetrating into the embryo sac opens at its tip or at its side facing the egg cell. In the course of *double fertilization* (amphimixis), one of the sperm cells leaving the pollen tube fuses with the egg cell (zygote formation). The fusion of the other sperm cell and the secondary embryo sac nucleus forms the primary endosperm nucleus, whose repeated division results in the formation of further, at first free (free from own cell structure), endosperm nuclei. The fusions, these main points of the double fertilization, are not simultaneous. The fusion of the secondary embryo sac nucleus and one of the sperm cells often precedes the fertilization of the egg cell, and the first mitotic divisions also begin earlier than in the zygote (Kapil and Bhatnagar 1975).

Pollen tubes of the *apple*, similar to most fruit species, reach the embryo sac through the micropyle (Anvari and Stösser 1981; Fig. 6.5a,b); this type of fertilization is called porogamy. The *walnut* species (*Juglans regia, J. rupestris, J. mandschurica*) are characterized by chalazogamy. This means that the pollen tube grows to the base of the ovule between the integument and the wall of the ovary and penetrates into the nucellus through the chalaza (Sartorius et al. 1984). Then the tube grows through the nucellus (Fig. 6.6a,b) up to the embryo sac (Sartorius 1990). Pollen tubes in the nucellus of *J. nigra* flowers have never been observed in the studies mentioned above.

In contrast, the possibility of chalazogamy in *walnut* is categorically denied by Schanderl (1965). The chalazogamy of *hazelnut*, however, has been reported, e.g.

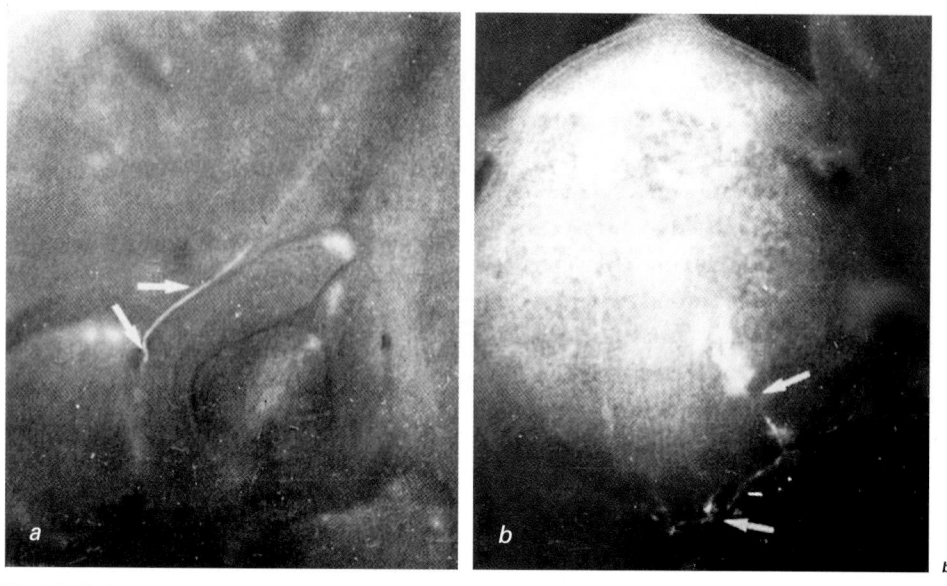

Fig. 6.6. Chalazogamy in the fertilization of walnut flowers: pollen tube growth (*a*) along the integument and (*b*) at the base of the nucellus after penetration (after Sartorius 1990)

by Geraci (1974), Thompson (1979), Benson (1983, all of them cit. by Germain 1983).

In the ovary of *apples*, there are five locules, each containing two ovules, around a central hollow. Each of the five styles of a flower belongs to a locule. This structure, however, does not imply the exclusivity of the fertilization through the only style belonging to the two ovules in question. Examinations after individual pollination of one to five styles of the flower led to the conclusion in cultivars 'Cox's Orange Pippin' and 'Golden Delicious' (Anvari and Stösser 1984) that the basic parts of carpels are not totally grown together (there are narrow gaps among them). When pollen tubes reach the central hollow, they may penetrate any locule yielding fertilization (Fig. 6.7*a,b*). Consequently, the number of pollinated stigmas has no effect on the fruit set and shape of fruits. Visser and Verhaegh (1987) argues that this phenomenon varies from cultivar to cultivar. In their similar experiment, a decrease in the number of pollinated stigmas had no negative effect on fruit set in 'James Grieve' apple and 'Bonmar' pear cultivars, but in 'Summerred' apple and 'Doyenné du Comice' and 'Conpack' pear cultivars, fruit set decreased significantly. Fruit set in 'Golden Delicious' did not decrease after pollinating only some of the stigmas but the number of seeds in fruits did decrease. It has never been observed in *apple* cultivars that all ovules had been reached by pollen tubes (Anvari and Stösser 1981). In the ovary of small fruits, there are a lot of ovules, among which the terminal ones are preferred in respect of fertilization in blackberry, for instance (Lech 1976). In the *Ribes* species, however, no such difference has been observed by Al-Jaru and Stösser (1983).

As it has been known for some time from the embryology of *apple* (Zeller 1960*c*), there also develops a haustorium in the nucellus of the ovule of stone fruits. After anthesis but before the first divisions of the primary endosperm nucleus, the embryo sac elon-

Fig. 6.7. (*a*) Pollen tubes arriving in the central hollow of the ovary (*b*) may penetrate into any of the locules through the duct between the not closely connecting basal parts of carpels (from Anvari and Stösser 1984); Z = central hollow

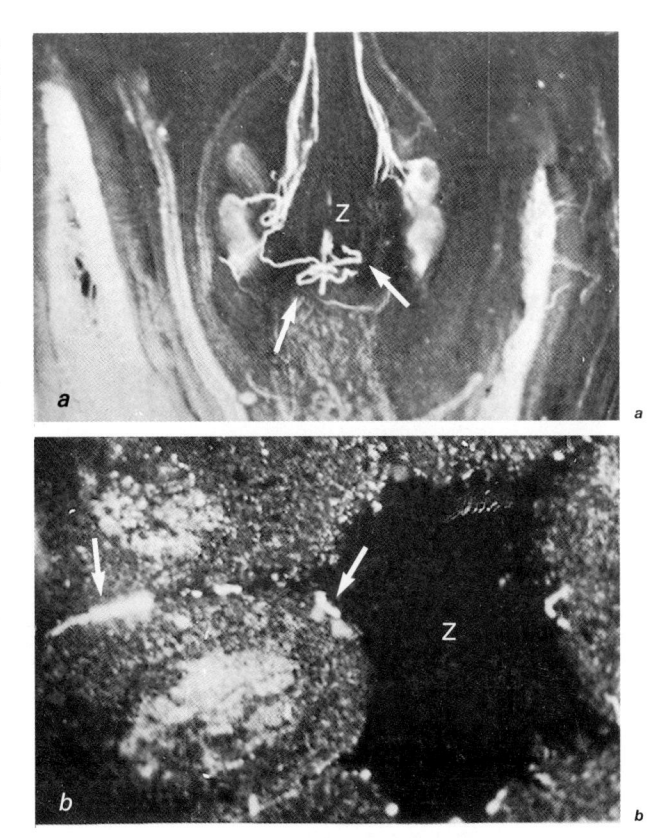

gates longer, making up a haustorial sheath reaching the chalaza, which is filled with primary endosperm, necessary for embryo development. As the vascular bundle of the funiculus ends at the chalaza, the haustorium must take part in supplying the growing endosperm (Stösser and Anvari 1978, Schauz et al. 1989). After the first one to two divisions of the fertilized egg cell in *apple*, the free (free from cell walls) nuclei of the endosperm were observed to be situated peripherally in the embryo sac, embedded in a plasmatic layer (Zeller 1960*c*).

In the nucellus of *sweet cherry*, two to three weeks after full bloom, there appears the first layer of free endosperm nuclei on the wall of the haustorium, then anticlinal cell walls begin to develop. Somewhat later but still before the formation of periclinal cell walls, the endosperm nuclei divide, forming a new layer. The entire process is repeated several times which will result in the haustorial sheath being filled with primary endosperm. The development of this cellular structure of the endosperm correlates with the end of the first period of growth of the sweet cherry fruit (Stösser 1966*a*). The embryo in *apple* ovules that is in the globular stage at this time, starts developing rapidly and uses up the primary endosperm (Zeller 1960*c*). This is followed, in both species, by the development of the secondary endosperm, which process reabsorbs the nucellar tissues. In turn, the secondary endosperm is used up in the final (fast) period of embryo development (see illustrations in Figs 6.8 to 6.11).

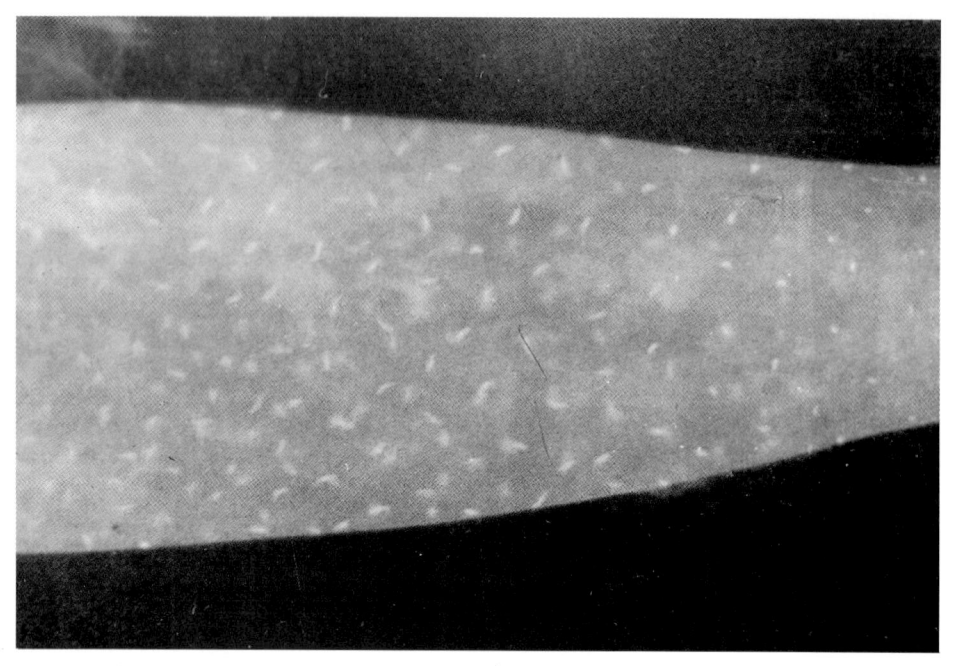

Fig. 6.8. Initiation of cell formation in the haustorium of plum (after Schauz 1989)

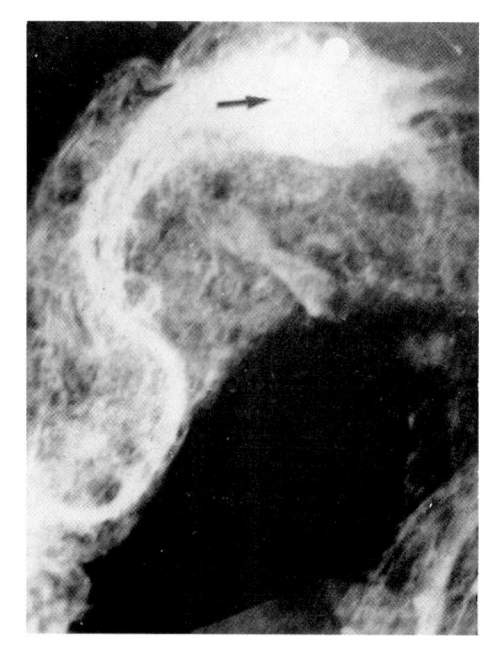

Fig. 6.9. The first divisions of primary endosperm nuclei in sour cherry (after Schauz 1989)

Fig. 6.10. Endosperm with free nuclei in sour cherry nucellus (after Schauz 1989)

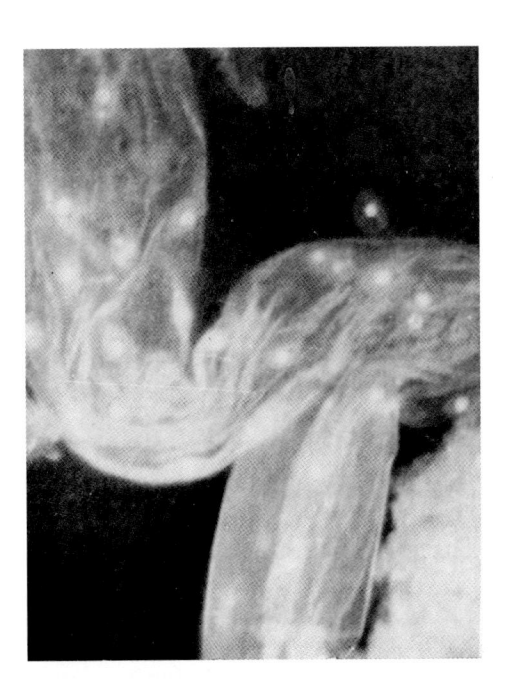

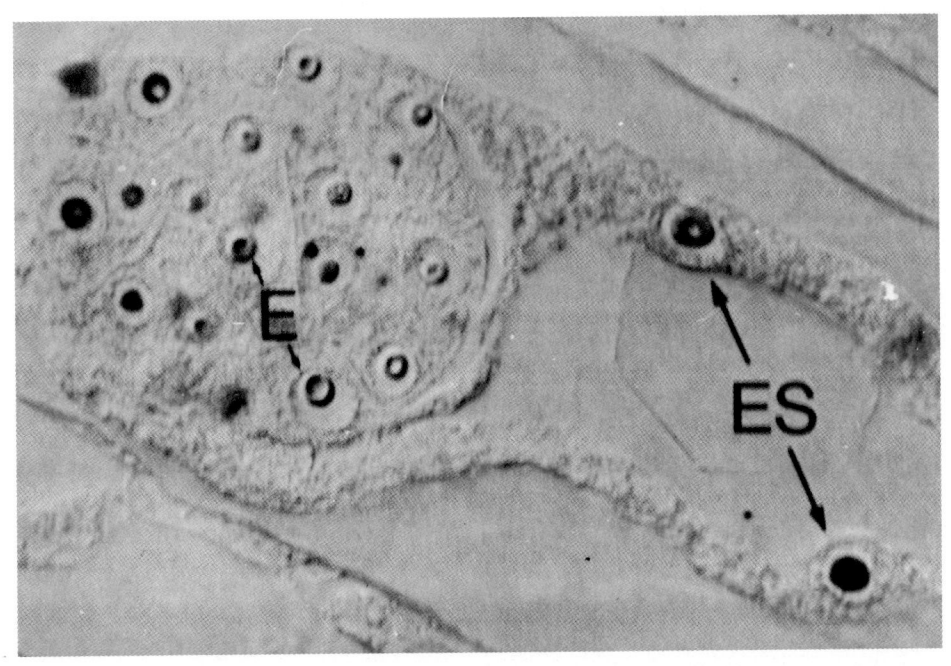

Fig. 6.11. The size difference between embryo (E) and endosperm (ES) nuclei in plum nucellus (after Schauz 1989)

Irregular phenomena may occur in the course of fertilization. Two pollen tubes were reported to have penetrated into the nucellus of some *sour cherry* ovules by Ubatshina et al. (1976). The same phenomenon was observed by Sartorius et al. (1984) in walnut (*Juglans rupestris* and *J. mandschurica*) nucelli. "Supernumerary pollen tube" (the appearance of more pollen tubes in the same embryo sac) may be concomitant with polyspermy, which means that the egg cell was fertilized by more than one sperm cells. It might also occur that synergids or antipodal cells simulate the egg cell in respect of appearance and behaviour, sometimes yielding an additional embryo of synergid origin (Kapil and Bhatnagar 1975).

The bloom and fertilization of the *hazelnut* (*Corylus avellana* L.) is a peculiarity compared to other fruit species, hence these two processes should be considered simultaneously.

Dormancy in hazelnut cultivars ends in the second part of December (Bordeianu 1967, cit. by Horn 1976). The anthesis of female flowers of certain cultivars starts as early as beginning of December in the southwestern part of France (Germain 1983), and from the end of November in Hungary (Horn 1976). Flowering of catkins lasts more than three months from the second part of December (Germain and Leglise 1973, Bergougnoux et al. 1978, cit. by Germain 1983).

The anthesis of both male and female flowers are affected by climatic conditions in winter. According to Romisondo (1978), the thresholds of temperature before the anthesis of male flowers are -16 °C, -18 °C, in the period when the catkin is elongated -8 °C, -10 °C, when anthers open, these values are -5 °C, -7 °C, before the anthesis of female flowers: -7 °C, -8 °C, during their bloom: -13 °C, -16 °C(!); these latter values are similar to those of leaf buds (-14 °C, -15 °C). Pollen shedding lasts a month or longer while the anthesis of female flowers is a two-month-long period (Germain and Leglise 1973, cit. by Germain 1983), with a critical temperature of 4 °C (Stritzke 1961, cit. by Horn 1976). A rather high percentage, i.e. 30% to 50% of pollen grains are not viable (Barbeau 1972, cit. by Germain 1983, Romisondo 1977). Hazelnut selections released as pollinizers for cv. 'Barcelona' shed their pollen with different timing, i.e. early in the season, in the mid-season or in late mid-season (Mehlenbacher and Thompson 1991).

In female flowers the area where pollen grains are received is fairly large, about four-fifth part of the surface of the style, but this fact is of little importance. What is relevant is the quantity of pollen tubes that reach the base of the style. This quantity might be little even in pistils whose receptivity seems to be good. Hence, the quality of pollen is more important than its quantity (Romisondo 1977). The optimum temperature for pollen germination falls between 15 °C and 25 °C (according to several authors); at a temperature about 0 °C the process is slow (Trotter 1947, Barbeau 1973*b*, cit. by Romisondo 1977). The temperature in the days immediately after pollination plays a crucial role: above 0 °C, it takes pollen tubes ten days to reach the base of the style even if the temperature sinks under 0 °C in a later period (Romisondo 1965*b*, cit. by Romisondo 1977).

Stigmas, from their emergence from the enclosing bud scales (red dot stage), remain receptive for at least two months. Nevertheless, there is an optimum period for the receptivity of the stigmatic surface, immediately after the beginning of anthesis, which lasts about 15 days (Geraci 1968, Germain and Leglise 1973, Bergougnoux et al. 1978, cit. by Germain 1983). Romisondo and Limongelli (1978), however, drew the conclusion on the basis of their own examinations concerning this question that there is no significant change in receptivity from the beginning of flowering up to its last phase. Furthermore, most *hazelnut* cultivars are protandrous, only a few of them are homogameous or protogynous (Germain 1983).

Schuster (1924, cit. by Germain 1983) was the first to mention that self-incompatibility of the hazelnut is of sporophytic type. Self-fertile cultivars can be found exceptionally only (Stritzke 1961, cit. by Horn 1976, Romisondo 1977).

When the pollen tube arrives at the base of the style, it ceases to grow and forms a callose cover around itself (Geraci 1974, Thompson 1979, cit. by Germain 1983) because in this period only a non-differentiated ovary tissue can be found at the style bases (Romisondo 1977). Ovary development is induced by the growth of pollen tubes; the ovaries that have not been pollinated cease to develop (Mussano et al. 1983*b*). The ovarian cavity emerges; and then in March or April, two, or sometimes four (Germain 1983) ovules appear in each ovary. At the end of April and in the first half of May the nucellus and the micropyle are already well discernible; in the upper part of the ovarian cavity the obturator comes into existence (Geraci 1974, cit. by Germain 1983, as for the obturator *see* Subsection 2.4.3).

The megaspore cells can be observed in the second half of May (De Rosa 1962, Romisondo 1965, Geraci 1974, Benson 1983, *see* in Germain 1983); the embryo sac develops from the inner megaspore cell (Geraci 1974, cit. by Germain 1983, Romisondo 1977). That time the pollen tubes, which have been inactive for about four months, begin to grow again (Romisondo 1977) and five to six days later, having reached the ovule through the chalaza, they arrive at the embryo sac (chalazogamy, *see* above in this section). Fertilization takes place at the end of May or in the first half of June, that is, four to five months after pollination. Generally, only one of the two ovules will be fertilized, and the development of embryos is so fast that it is completed not later than at the end of July (Romisondo 1977).

6.5. Period Between Pollination and Fertilization

This determining factor of fertilization depends on temperature to a great extent. In *sour cherry* flowers, pollen tubes reach the base of styles in two to three days, but it takes six to eight days to reach the majority (over 78%) of ovules (Stösser and Anvari 1981); whereas under controlled conditions (at 22 °C) or under field conditions if the weather is favourable, they can approach the micropyle as soon as after two to three days (Stösser 1980*b*, Ubatshina et al. 1976). Pollen tubes of the *plum* reach embryo sacs in four days, and in eight days, at 20–21 °C, and at 10–11 °C, respectively (Lee 1980); Stösser's (1982) similar data are three to four days at 20 °C, and seven to nine days under field conditions; whereas at 5 °C fertilization would require 16 to 20 days (Jefferies et al. 1982*b*). The growth rate of *pear* pollen tubes is 0.60, 1.75, and 2.40 mm/day at a temperature of 6 °C, 9 to 11 °C, and 15 °C, respectively; and concomitantly the time required to reach the ovule is 15, five to six, and three days (Lombard et al. 1972), and nine days in orchards (Ruiz 1977).

The time after anthesis when pollen grains are deposited on the stigma is also a determining factor in the speed of pollen tube growth. The growth of the pollen tube in the style of an *apple* flower is much better on the first to fifth day after anthesis than on the day of anthesis or on the sixth day after anthesis (Williams and Maier 1977). Styles of flowers that open very late are less suitable for pollen germination and pollen tube growth (Stott 1972). In apple cultivars, the period of time needed to reach ovule is one to two days longer at anthesis than on the second to fourth day after anthesis. When pollination occurs on the sixth to eighth day after anthesis, this period is rather shorter than longer (Braun 1984). However, pollination only results in fruit set when it takes place up to the fourth day after anthesis (Braun and Stösser 1985).

Fabbri et al. (1983) made similar observations in *sour cherry*. It takes nine, six, and three days for the pollen tubes to arrive at pollen sac if pollination has taken place on the first, fifth to sixth, and 11th day after anthesis, respectively. The best fruit set was reported when pollination occurred on the fifth to sixth day. In *sweet cherry*, pollination two to three days after anthesis proved the best in respect to fruit set (Stösser and Anvari 1983). The mean of similar data in six *apricot* cultivars fell between the second and the fourth day after anthesis. Four cultivars displayed 100% ovary penetration between two to four days after anthesis, whereas only two cultivars gave such results when pollinated on the day of anthesis (Egea et al. 1991).

The gradual ripening of the pistil, examined in biological test plants, may play a role in the phenomena discussed above: the stigma is active three to five days before anthesis while the style and the ovule are active only two to three days after anthesis (Ende 1976). This was corroborated by data concerning the *almond* (Pimienta et al. 1983). An observation concerning the *Japanese (oriental) pear* indicated that self-pollination at anthesis resulted in a total self-sterility, but self-pollination in the balloon stage resulted in a fruit set of 10%, and cross-pollination in the same period provided a result of 20 to 90% fruit set (Hiratsuka et al. 1985).

It is a general observation that pollen tube growth will decelerate in the ovary. In *apple* cultivars, pollen tubes reach the base of styles in four to five days, but the rate of pollen tubes that reach embryo sacs in 'Golden Delicious' is 22 to 24% after seven days, and 44 to 46% after ten to twelve days. In 'Cox's Orange Pippin' a rate of 72 to 76% was obtained after an eight-to-ten-day period (Anvari and Stösser 1981). *Sour cherry* pollen tubes grow through the style in two to three days but on the fourth day the embryo sac is reached only in 42 to 52% of flowers (Anvari and Stösser 1978*b*). In *almond* flowers, it also takes more time to cover the shorter distance from the style base to the ovule (Pimienta et al. 1983). In *blackberry* flowers, the pollen tubes spend four to five days in the ovary; fertilization takes place seven to nine days after pollination (Lech 1976). The *Japanese (oriental) pear* is an exception in that it requires the same time, two days each, to grow through the style and the ovary (Hiratsuka and Tezuka 1980). The *walnut (Juglans regia)* ovule is reached by pollen tubes in five to six days, while in other *Juglans* species, the pollen tubes get in the nucellus as soon as in three to four days (Sartorius 1990, Sartorius et al. 1984).

6.6. The Effect of Temperature and Environmental Pollution on Fertilization

Temperature exerts a determining influence from the time pollen grains begin to germinate until the fertilization is completed. It is known from Jona's (1967) *in vitro* examinations that *apple* pollen begins to grow an hour after setting on culture medium at 20 °C. This time the vegetative nuclei are already indiscernible. A large-scale pollen germination takes place three to four hours later. The generative nucleus starts to move in the pollen tube four to five hours later and reaches its tip after the 13th hour. Between the 17th and 21st hours it divides and completes the mitosis in two hours. This process is not influenced by a temperature between 20 and 25 °C, but a temperature of 15 °C or 10 °C decelerates it: there is no mitosis in the first 32 to 34 hours at 15 °C and within 96 hours at 10 °C, and hardly any mitosis occurs even later. This is to say, that cold weather at the beginning of pollen germination results in an irreversible damage. Contrary to this, after a 13-hour-long initial period of 25 °C, temporary short periods of 4 °C or 0 °C later have no influence on the chronology of mitosis. The rate of the growth of apple pollen in the style at 5, 14, and 24 °C is 1.0, 1.33, and 8 mm/day, respectively; the growth in the basic style zone (where the styles are fused) is somewhat faster (Child 1967).

In the previous discussion we dealt with the effect of temperature on the processes of fertilization. From now on the effects on fruit set will be focused on. At 15 °C, which is the optimum temperature for the fertilization of *pear* flowers, pollen tubes reach the ovule in three to four days (Lalatta et al. 1978*a*). They reach the base of styles in 24, 72, and 120 hours at 21, 15.5, and 10 °C, respectively. The optimum temperature for fruit set, however, is between 17.3 and 18.2 °C (Mellenthin et al. 1972). Other reports somewhat differ with this, e.g. Vasilakis and Porlingis (1985) report that only 70% of style bases have been reached by pollen tubes after four and three days at 20 and 25 °C, respectively.

In *apricot*, low temperature (under 10 °C) stimulates the growth of pollen tubes in the styles (Gülcan and Askin 1989). This corresponds to the observations of Egea et al. (1991) in southeastern Spain where high temperature during flowering drastically shortens the stigma receptivity period and the effective pollination period. More exactly, the difference between the best percentage of fruit set, established by pollination on the second day after flowering, and the lesser result received when pollination was carried out on the fourth day, was greater under warmer climatic conditions (Burgos et al. 1989). For instance, in a region where the mean temperature is 13.8 °C during flowering the two rates of fruit set, mentioned above, were 50 to 54% and 32 to 41%, whereas in a region associated with a mean temperature of 15.7 °C the two rates were 50% and 17%. In even warmer regions with 17.5 to 18.5 °C mean temperatures 28 to 44% and only 0.8 to 3.9% were reported for fruit set.

In *plum*, the tendencies concerning fruit set are contrary to the experiences with apricot. Thompson and Liu (1973) compared two years in which embryo sacs were normal at anthesis and there was no frost damage in those years, but they differed in the following: in year one the weekly mean temperatures in the first three weeks after full bloom were 5.7 to 9.1 °C lower and in 74% of time temperature was under 10 °C. The corresponding temperature in the other year under 10 °C was 49%. In the cooler year set was 1 to 13% and in the warmer year the primary fruit set was 36 to 64%.

In *sweet cherry*, under high temperature the proportion of embryo sacs containing functional egg cell decreased rapidly after anthesis. Another year of low temperature, the maximum proportion of functional embryo sacs at any time was much lower than in that year with high temperature, and their percentage began to decrease within the first two days after anthesis (Eaton 1959*b*). As for fruit set in *blackberry*, the optimum temperature is 14 to 16 °C. At 6 °C pollen tube growth is blocked (Lech 1976). The percentage of the germination of *hazelnut* pollen *in vitro* shows a decreasing tendency parallel with an increasing temperature between 15 and 30 °C (Kim et al. 1985).

The temperature optimum for growth is not the same in the case of compatible and incompatible pollen tubes. Examinations concerning *apple* pollen demonstrate that both pollen tube types are sensitive to temperature to the same degree during the first ten hours; then the incompatible pollen tube will cease to grow at 25 to 30 °C, while it keeps on growing, although, at 15 °C. This corresponds to the fact that at a higher temperature incompatibility reaction and the expression of inhibition are faster (Modlibowska 1945). Temperature optimum for self pollen tubes in apple is 20.5 °C, whereas for pollen tubes of cross-pollinating cultivars is 26.6 °C (Williams and Maier 1977).

The same holds for self-incompatible *almond* cultivars: the optimum temperature for pollen tube growth is lower (15 °C) after self-pollination than after cross-pollination (25 °C), like in self-fertile hybrids of almond (Socias i Company et al. 1976). In styles of self-fertile selections, pollen tube growth at 12 °C and 22 °C is similar both after self-pollination and after cross-pollination (Socias i Company and Felipe 1987). The rate of pollen tube growth in self-compatible *sour cherry* cultivars with increasing temperature increased faster than in self-incompatible cultivars (Fabbri et al. 1983). The 'Bartlett' ('Williams Bon Chrétien') pear cultivar, however, was totally self-incompatible at 15 °C, while at a higher temperature, some self pollen tubes managed to reach the base of styles (Lombard et al. 1972).

The other environmental factor that has an influence on fertilization is *environmental pollution*. If the air is polluted with hydrogen fluoride and hydrogen chloride, pollen tube growth in *apricot* styles will be reduced. Higher concentration during short exposure will result in a greater pollen tube growth reduction than smaller doses during a longer period (Facteau and Rowe 1977). Sulphur dioxide hinders pollen tube growth in both, *apricot* and *sweet cherry* (Facteau and Rowe 1981). Pyrene and fluoranthene exert a negative effect on two sweet cherry cultivars, but not on apricot (Facteau and Chestnut 1983). Chemical plant protectants applied for reducing diseases may also have an effect on pollen germination (*see* Subsection 2.4.1, too).

Williams et al. (1987) had shocking experiences with the triadimefon–bitertanol fungicide mixture applied biweekly from April to August as an inhibitor of ergosterol biosynthesis during a four-year period. The yield of 'Cox's Orange Pippin' trees halved as a consequence of the poor fruit set, which was a result of a decrease in the longevity of ovules, and not the inhibition of pollen tube growth. After the low fruit load in the year under examination, a greater extent of bloom was expected in the succeeding year, but it did not take place. However, the application of triadimefon alone before bloom or at petal fall had a stimulating effect on fruit set in 'Cox's Orange Pippin' trees (Church et al. 1984). The insecticide chlorpyriphos and the growth regulator paclobutrazol decreased fruit set. The application of the majority of twelve different fungicides before bloom but with various timings was reported to have no influence on fruit set; dodine is an exception. Its application sometimes resulted in a slight reduction in fruit set. The application of sulphur at the pink bud or green cluster stage was sometimes observed to increase fruit set, perhaps as an indirect result through shoot growth reducing effect of sulphur (Church and Williams 1983).

6.7. Effective Pollination Period

The concept "effective pollination period" (EPP) has been proposed by Williams (1966d). Its meaning is the difference between the longevity of the embryo sac and the time needed for the pollen tube to reach the embryo sac. The EPP is expressed in days. As we discussed previously, the degradation of style papillae and transmitting tissues during the early days immediately after anthesis will not decelerate pollen tube growth. Obviously, the longevity of the ovule and/or the egg cells play the decisive role in governing fruit set. If the viability of the inadequately developed ovule is only a few days and pollen tube growth requires as long as a week (because of a low temperature, for instance), then there is no possibility for fertilization.

Depending on cultivar and temperature (*see also* Fig. 6.4), the EPP is one to ten days in *pear* and two to ten days in *apple* (Williams 1966d), but according to Child (1967), the

EPP is a maximum of six days at a mean temperature of 11 °C. In cultivars such as 'Golden Delicious', 'Imperator' and 'Yellowspur', EPP is eight days, while cultivars 'Stayman', 'Richared Delicious' and 'Wellspur' are characterized by an EPP shorter than six to eight days. Because there is a four-day difference in anthesis between the terminal and lateral flowers in a given inflorescence, the EPP concerning the entire inflorescences in the above-mentioned two cultivar groups is ten to twelve days, and less than ten days, respectively (Marro and Lalatta 1982).

In *pear* cultivars the mean EPP is three to four days (Ruiz 1977, Lalatta et al. 1978a) but, depending on cultivar, it may be one to two days ('Comice'), six to seven days ('Bartlett'), or nine to ten days ('Bosc') (Lombard et al. 1971). According to a detailed study (Vasilakis and Porlingis 1985), if the daily number of hours with temperatures over 10 °C is 6.0, then the EPP is 11 days, but if these hours are 13.6 or 9.4, then the EPP is nine days. However, if there is a standard temperature of 8 or 20 °C, then the EPP is 13 and seven days, respectively.

Eaton (1959b) found that *sweet cherry* cultivars were characterized by a very short (sometimes two-day-long) EPP, but Stösser and Anvari (1983) reported four to five days. The EPP of *sour cherry* cultivars is five to six days according to Fabri et al. (1983), while Furukawa and Bukovac (1989) observed three to five days. In *apricot* there is a two-to-four-day-long EPP (Burgos et al. 1989). The *plum* is characterized by an EPP of three to six days depending on the year (Lee 1980). Contrary to our knowledge on other species, there is a theoretical possibility for a nine-day-long EPP in *gooseberry*, but practically its EPP is shorter because of the early degeneration of the stigma and/or the style (Jefferies et al. 1982a). The *hazelnut* is characterized by the longest, but not exactly known, EPP (Marro 1982).

6.8. Embryo Development

Embryo development in *apple* consists of two phases: a slow initial phase and a very fast final phase (Zeller 1960c, 1964a). The proembryo, which consists of not more than four cells two weeks after bloom (Fig. 6.12a,b), will increase only concerning size and cell number in the following three to four weeks, and there is still a spherical proembryo in the ovule in the fifth to sixth week after bloom, called the globular stage. Six weeks after bloom, when the polarity manifests itself in the proembryo, that is, cotyledons are initiated and radicules become discernible, the former proembryo is to be regarded as a real embryo (Fig. 6.13a). Then the development of the embryo will accelerate and the development is completed during the following four weeks (Fig. 6.13b,c,d).

Zeller claims that the most important phases of embryo development coincide with periods of fruit drop of apple. The first drop (in May) is associated with the four-cell proembryo stage, and the so-called June fall occurs when differentiation of the embryo starts. According to histochemical examinations of Weinbaum and Simons (1974a), 18 days after bloom and two days after the NAA-treatment for fruit thinning, the proembryos are smaller in fruitlets that will drop than in those that will persist, but the quantity of nucleic acids, protein and starch deposition in the embryo or in the endosperm has not decreased.

The development of *sweet cherry* embryos are also characterized by a slow initial phase. At the end of the first (fast) period of fruit growth the length of the proembryo, which is in the globular stage, is 0.1 to 0.2 mm (Stösser 1966a). The radial symmetry of the proembryo will turn to bilateral when cotyledon primordia appear. This is similar to

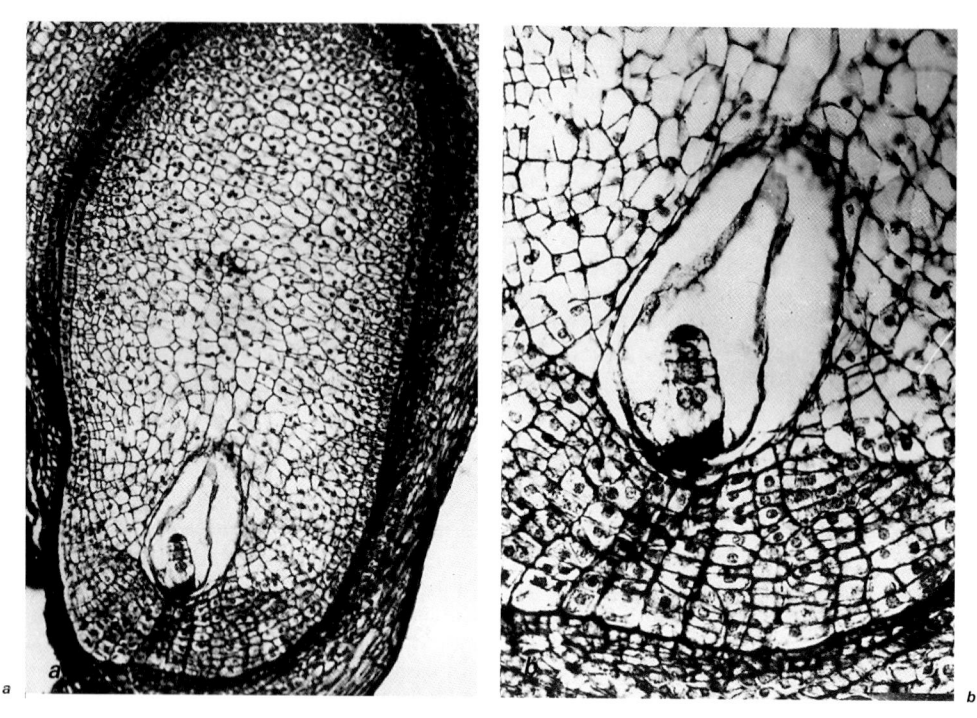

Fig. 6.12a,b. Four-celled proembryo of apple two weeks after bloom (after Zeller 1964a)

apple, see Fig. 6.13a,b. The second (slow) period of fruit growth coincides with the morphological differentiation of the embryo and its development. Isolated during the second period of fruit growth, and then *in vitro* cultured *sour cherry* embryos reach their complete size in about two to three weeks (Poulsen 1983).

In *peach*, five weeks after bloom, in the haustorial sheath around the proembryos (already in the globular stage), the cellular structure of the endosperm begins to evolve, which process is called cytokinesis (Bubán 1992; Fig. 6.14a,b). The initiation and rapid development of the cotyledon primordia start from end of the second month after bloom (Fig. 6.15a,b). The leaf primordia of the plumula appear in the 11th to 13th week after bloom.

In *almond*, in the early flowering 'Ne Plus Ultra' cultivar the diameter of the globular proembryo is 200 microns, two times as wide as that of the proembryo in the late flowering 'Texas' cultivar, and the former has a longer suspensor consisting of more cell layers (Grigorian 1974). After further 50 days the number of leaf primordia of the plumula is almost the same in the two cultivars (seven and five, respectively), and the great difference in the initial period will disappear during final period of embryo development.

In *tangerine* (*Citrus reticulata*, Blanco cv. 'Clementine') embryo sac and endosperm formation begins soon after fertilization but the first division of the zygote only falls between the eighth and the tenth week after anthesis (Eti and Stösser 1987). The embryo is in the globular stage between the tenth and the twelfth week, and it takes the embryo (further) four weeks to reach its final size. In the *walnut* zygote the first cell divisions fall between the second and the fourth week after full bloom, immediately after this begins

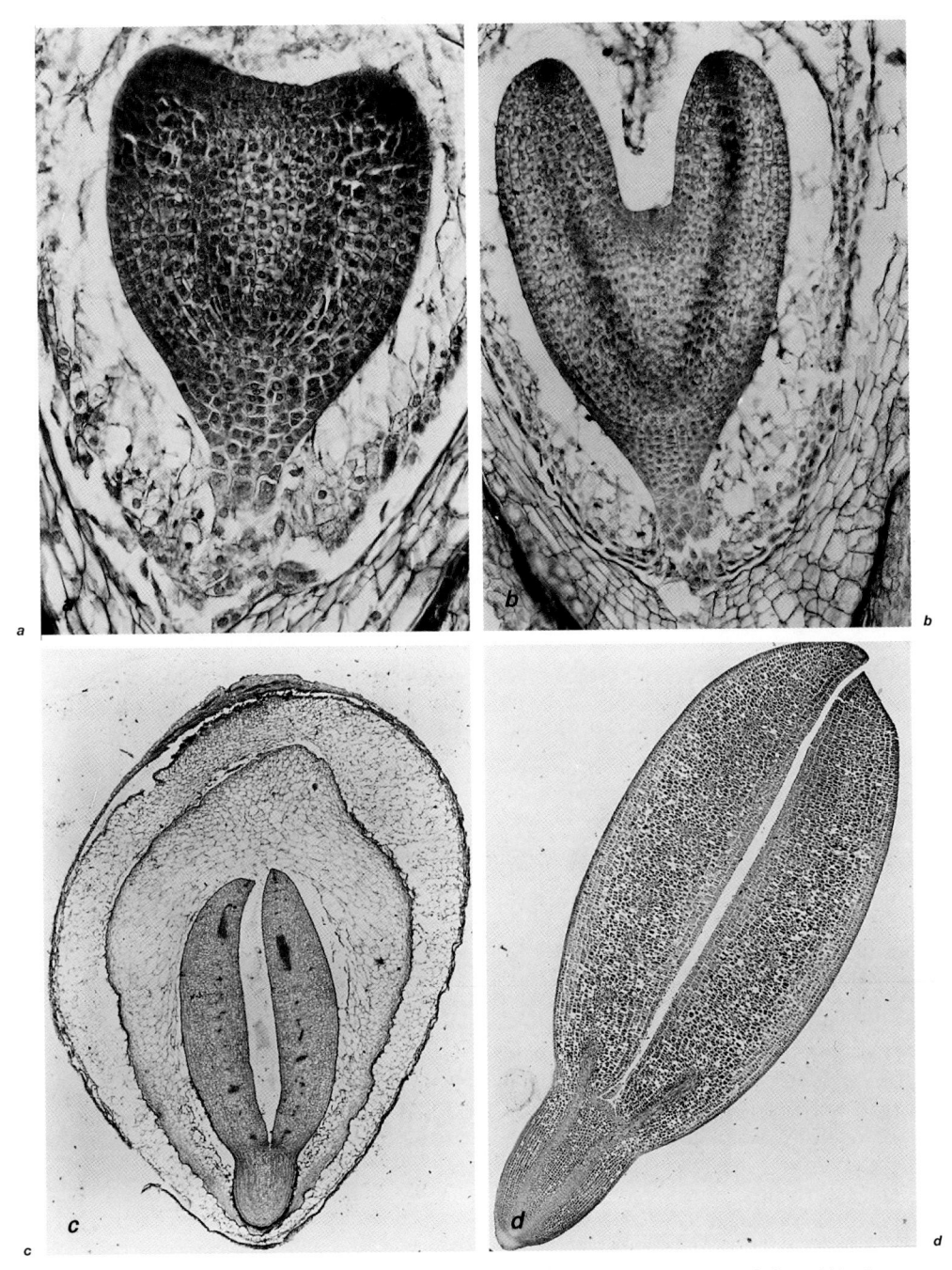

Fig. 6.13a-d. Apple embryo (a) six weeks after bloom and (b-d) its development to completion within the next month (after Zeller 1964a)

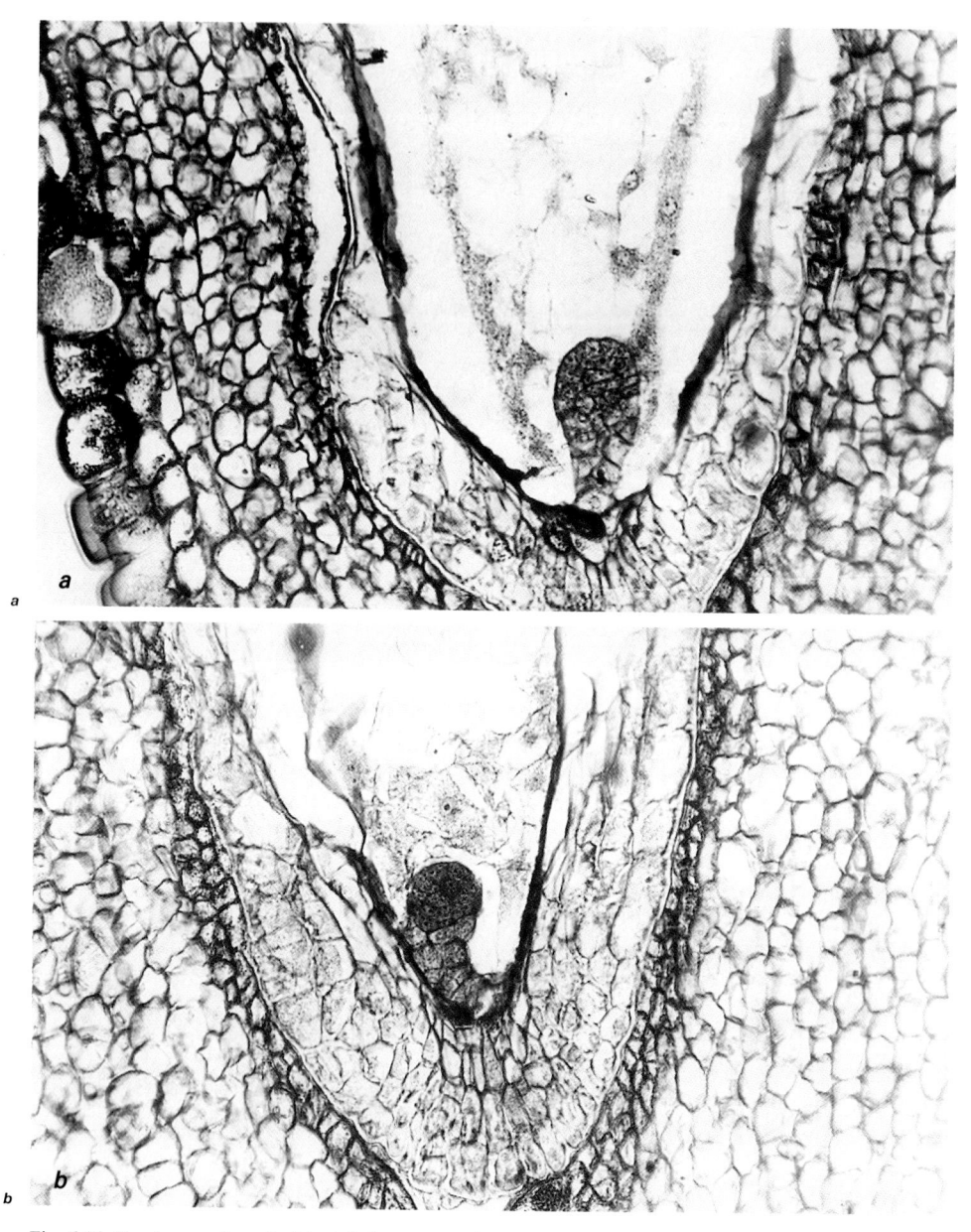

Fig. 6.14. Peach proembryo in (*a*) globular stage and (*b*) at the outset of cytokinesis five weeks after bloom (after Bubán 1992)

the formation of cell walls in the endosperm, and it is in the sixth week after bloom that the embryo is in the globular stage (Sartorius 1990).

There are several ways to supply nutrients to the developing embryo (Stösser 1966a). The two ovules are supplied by two funicular vascular bundles branching from the fruit

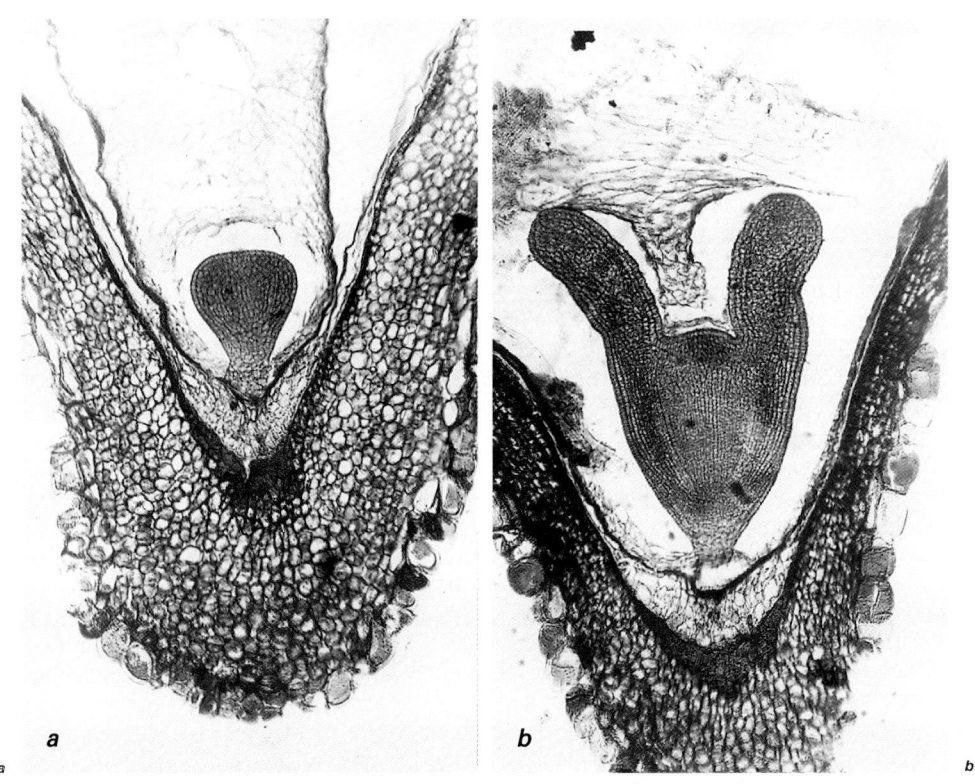

Fig. 6.15*a,b*. Initiation of cotyledon in peach embryo (*a*) two months after bloom and (*b*) its fast further development (within two weeks) (after Bubán 1992)

pedicle. The histological features of the chalaza and that of the basic part of the nucellus as well as the development of the haustorium indicate that there is a transport of substances from the chalaza through the haustorium to the embryo. There is a third, indirect, way to supply the embryo: through a special tissue, that is, the endosperm (*see also* Section 6.4). However, the details of this process are less known.

Embryo development has a special case, called apomixis, when viable embryos and seeds develop without fertilization. The first definition of *apomixis* is that "the substitution of sexual reproduction for another, asexual, reproduction process, including no fusion of cells and nuclei" (Winkler 1908, cit. by Sartorius 1990) and in this sense apomixis may be considered as somatic reproduction (Faust, personal communication). Winkler's statement, however, also involves the reproduction by runners and viviparous proliferation. Recently, in fact, the pseudo-viviparous proliferation of some species has been considered as vegetative apomixis (Borhidi 1993).

Gustavsson (1946, cit. by Sartorius 1990) was the first to describe the apomictic seed development. He regarded the way of reproduction with apomictic seeds as agamospermy.

Besides vegetative apomixis (*see* above), the other basic form of the apomixis is the agamospermy, which is a pseudo-sexual process resulting in seeds consisting of maternal tissues only (Borhidi 1993). There are four types of agamospermy.

a. Parthenogenesis: an embryo forms from an intact egg cell, it can either be haploid (generative parthenogenesis, Zatykó 1962) or diploid (if the meiosis has not taken place previously).

b. Apogamy: the embryo originates from diploid synergids or antipodal cells of the embryo sac. It occurs often in ferns (e.g. *Dryopteris*) but rarely in angiosperms. Embryo development beginning with haploid synergids or antipodals is the generative apogamy (Zatykó 1962).

c. Apospory: there is no embryo sac formation, the role of the embryo sac will be played by one integument cell developing into an embryo. According to Sartorius (1990), however, there are two types of apospory. Somatic apospory (i.e. apospory itself) is when a non-reduced embryo sac forms by the mitosis of a somatic cell of nucellus. In the case of generative apospory (or diplospory) a diploid embryo sac develops from an embryo sac mother cell without previous (or incomplete) meiosis.

d. Pseudogamy: one cell of the nucellus—due to the stimulation exerted by the pollination—forms an embryo. Later on, this embryo grows into the embryo sac (it is called: adventive embryony). The embryo—now within the embryo sac—gets in touch with the endosperm arising from the fusion of the secondary embryo sac nucleus and one of the sperm cells. More than only one embryo can also develop in the embryo sac, this is polyembryony. Contrasting Borhidi's description, Sartorius (1990) postulated that formation of somatic embryos in this way should be called induced adventive embryony. In his opinion, pseudogamy sets out from the egg cell stimulated by pollination, however, without fertilization. Finally, nucellar embryony will be realized when the embryo does not arise from a cell of the embryo sac but directly from a somatic cell of the nucellus (Sartorius 1990). Regarding the terminology of apomixis, vegetative apospory and diplospory refer to the gametophyte development while parthenogenesis, pseudogamy and adventive embryony belong to the development of sporophyte (Sartorius 1990).

The progenies resulting from somatic parthenogenesis, apospory, or adventive embryony are of clone value. Szentiványi (1976*a*) argues that this statement does not hold for *walnut* because the plants developing from apomictic seeds are heterogeneous.

It should also be noted that apomixis can either be obligatory, meaning that all progenies are identical in respect of genetic features (Schmidt 1967, cit. by Sartorius 1990), or optional, which means that, among the apomictic progenies, there are progenies coming from the fertilization of the egg cell.

Apospory is characteristic of some *Malus* subspecies (Hjelmquist 1957, 1959, cit. by Zatykó 1962) and the American *Rubus* species (Pratt and Einsett 1955).

Among the *Malus* species, some are apomictic, for instance, *M. sikkimensis* and *M. toringoides*, they are characterized by apospory and parthenogenesis (Sax 1931, Dermen 1940, Campbell and Wilson 1961, Schmidt 1964, cit. by Koloteva and Zhukov 1984). Koloteva and Zhukov (1984) found parthenogenesis also in a cultivated apple. They also mention that Grevcova (1978) was able to observe adventive (nucellar) embryo development, which is very rare in apple cultivars, in the cv. 'Antonovka'. Apomictic progenies of the cv. 'Anna' were described by Sherman et al. (1978). Sherman et al. also stated that mainly the triploid and tetraploid variants are subject to apomixis among the *Malus* species.

The self-fertile *sour cherry* cultivars, but not the self-sterile ones, show a tendency for apomixis but the occurrence of this phenomenon varies year to year (Ostapenko and Zhukov 1985).

In *walnut*, apomictic fruit set was reported first by Zarubin (1949), but the details of such fruit set are known from the work of Sartorius (1990). Seedlet growth does not differ considerably in young fruits developing from the isolated and non-isolated flowers of the apomictic 'Eszterházy II' and 'No. 26' cultivars when isolation was carried out to prevent pollination. However, it is notable, that the free nuclei were observable in the endosperm of the isolated flowers of 'Eszterházy II' only 22 to 27 days after anthesis. This is eight to 14 days later than the appearance of nuclei in the non-isolated flowers. The beginning of the apomictic embryo development occurred 20 to 27 days after full bloom in cultivar 'No. 26', and somewhat later, 27 to 32 days after full bloom in 'Eszterházy II'. The apomictic embryo developed in every case within the embryo sac more precisely in its micropylar area. This location of the embryo excludes the possibility that the embryo may develop from a somatic cell of the nucellus. The differentiation of the apomictic embryos and their later development were similar to those embryos developed in non-isolated flowers. The formation of cell walls in the endosperm starts at the same time in seedlets of fruits of both the isolated and the non-isolated flowers. At the time when the cell wall begins to form the embryo consists of eight or 16 cells. It has recently been reported (Sartorius and Stösser 1992) that apomictic cultivar 'No. 26' only sets fruit when open pollinated.

It is to be noted, that Schanderl (1965) observed that apomictic embryo development in *walnut* always comes from a cell of the nucellus. Moreover, he observed that the nucellus of the apomictic walnut types often (40 to 79%) lacks the embryo sac. Walnut cultivars in northern countries, e.g. Germany, are inclined to apomixis to a greater extent.

The occurrence of types of apomixis were *summarized* in a *review* by Golubinsky et al. (1979); the following data and references, resp. come from this study.

Malus domestica: apogamy (Kobel 1957).

M. sieboldii: apomixis, mainly pseudogamy (Olden 1953),

M. sieboldii and *M. hupehensis*: apospory (Poddubnaya and Arnoldi 1976).

M. toringoides and *M. sikkimensis*: apomictic inclination (Lipaeva 1965).

M. silvestris, *M. prunifolia*: apospory (Smargon 1964),

M. prunifolia, *M. niedwetzkyana*, *M. coronaria*: no parthenogenetic seed develops (Kitshina 1970).

Semi-cultivated types of apples: apospory (Lizner 1968).

Diploid cultivars, e.g. 'Antonovka', 'Melba', 'Papirovka': apospory (Rodionenko and Smitshenko 1972), but nucellar embryony is also suspected.

Triploid apple cultivars: apospory (Gorczynski 1934); but Rodionenko (1972) argues that there is no correlation between the frequency of triploidy and that of apomixis. A great extent of apomixis is characteristic of summer cultivars and takes place as a result of unfavourable weather at bloom.

Pear: experimentally induced apomixis (Tschermak von Seysenegg 1951), nucellar embryony (Grevcova 1971, 1974), and most often apospory and polyembryony, were recorded in cultivars 'Favourite of Clapp' and 'Saint-Germain' (Kaimakan and Kovsova 1976).

Quince: subject to apomixis to a certain extent (Shtsherbenev 1972), apogamy from synergids (Samushia 1978).

Peach: reduced parthenogenesis (Pratasenia 1938, 1939), apomixis suspected (Ryabov 1979).

Apricot: reduced parthenogenesis (Samushia 1972).

Sour cherry: haploid embryo, reduced parthenogenesis (Haritonova 1968), in cultivated cultivars apomictic embryo (Zhukov and Havrilina 1977); *Prunus cerasus fruticosa*: nucellar embryony (Shoferistov 1974).

Plum: apomixis (Olden 1960, Semtshenko 1974), nucellar embryony (Kobel 1957).

Raspberry: optional apomixis possible (Rozanova 1934), pseudogamy (Petrov 1939, 1958), optional induced apomixis (Kantor 1976).

Blackberry: apomixis possible (Liddforss 1914).

Black currant: apogamy from synergids (Kobasnidze 1971*a,b*). Black currant cultivars are inclined to apomixis in direct ratio to their capacity of self-fertilization but apomixis may occur in all cultivars (Smirnov 1972, Bavtuto 1976, 1977).

Strawberry: apomixis may occur in *Fragaria ananassa* but it is not usual (Tsherevatenko 1974). Spontaneous pseudogamy might be typical (Suhareva 1978). *F. orientalis*: haploid pseudogamy (Petrov et al. 1968), *F. vesca*: reduced parthenogenesis (Shangin and Berezovsky 1962).

Grossularia reclinata Mill.: inclined to apomixis (Rudloff and Schanderl 1950, Bavtuto 1977).

Crossing *gooseberry* and *black currant*: apomictic seed development (Sergeeva 1970).

Apomixis can be induced, for instance in *black currant* with synthetic auxins (Zatykó 1962) or dimethyl-sulphoxide. The latter results in somatic parthenogenesis (Vermel and Sololova 1973*a,b*, cit. by Golubinsky et al. 1979). In *pear* trees, auxin and gibberellin, and sometimes vitamins, were used to induce apomictic seed development and fruit set (Golubinsky et al. 1977). A detailed description on the utilization of haploid apomixis and ways of haploid embryo creation in stone fruits is available in Soldatov (1982).

Anomalies in embryo development may occur as early as the formation of embryo sac. In the nucellus of both young and senescencing *apple* or *pear* flowers, beside the megaspore mother cell, a few cells with big nucleus may remain. The staining of the cytoplasm in these cells is very intensive. In other cases a secondary, anomalous proliferative activity can be observed in the two nucleated stages of macrosporogenesis, or haploid pseudoembryos may develop in an unfertilized embryo sac, or a secondary embryo sac may develop in the degenerated primary embryo sac (Deveronico et al. 1982). The frequency of these anomalies is greater in cultivars with a long bloom period than in those in which bloom is short.

Embryo sac degeneration may occur to a certain percentage in 'Richard' apple flowers as early as full bloom but it is significant at petal fall. Degeneration occurs in trees grafted on seedlings more often than in trees on rootstock M 9 regardless of time (Marro 1976). Consequently, in trees with seedling rootstock, fruit set is lower at any arbitrary point of time after pollination. In flowers of seven apple scion-rootstock combinations, the frequency of anomalous ovules was reported not to be consistent (Marro and Deveronico 1978). This included the triploid 'Stayman Winesap' cultivar also. Different quantities of proliferating nuclei were observed in flowers that have not been pollinated.

In flowers of 'Granny Smith' trees the number of ovules with intact egg apparatus was greater in the balloon stage up to the full bloom stage, and decreased during the period after petal fall (Cobianchi et al. 1978). The trees with rootstock M 9 were reported to be characterized by a prolonged life of embryo sac in comparison with the trees with seedling rootstock. The authors reported a casual occurrence of ovules with two nuclei and nucellus with two embryo sacs. In nucellus of cultivated apple cultivars such as 'Papirovka' and 'Jonathan', Krilova (1967, cit. by Golubinsky et al. 1979) found more than one normal embryo sacs. The duplication of various organs or other growths, such as style, nucellus, embryo sac, etc., generally has a negative effect on fertility.

According to Simons' (1965) histological examinations, in the potential drop fruits an abscission zone may evolve at the base of the funiculus of the ovule through the funicular bundle. There are anastomosing (connecting) vascular bundles at the base of the ovule.

At the aborted ovule, senescing placental tissue can be found. Simons believes, however, that "abscission zone formation in the suspensor may be phase 1 of many phases that eventually cause the fruit to abscise". Weinbaum and Simons (1974b) based on electron microscope comparisons between persisting and potential drop fruits of apple concluded that "fruit commitment to abscise preceded by ultrastructural evidence of degenerative or functional changes in embryo or endosperm tissue". In other words, "embryo degeneration, which has been associated with fruit abortion, occurs subsequent to the commitment of fruit to abscise". This statement does not support the theory that embryo abortion is the primary factor of fruit abscission in the post-bloom period (Weinbaum and Simons 1974a).

As much as 26% of the ovules of the 'Constant' apricot cultivar the integuments were short and the nucellus was in direct contact with the wall of the ovary (Eaton and Jamont 1965). Embryo sac degeneration occurred in every stage of development. Between anthesis and petal fall only 22% of the ovules had a functional egg cell.

In warm years, when bloom takes place early, 4 to 13% of embryo sacs were incomplete or degenerated in *sour cherry* flowers as early as anthesis (Furukawa and Bukovac 1989), but 75% of embryo sacs had eight nuclei which is considered normal at this time. It was seven days after anthesis that the synergids, which play an important role in fertilization, began to degenerate, although the egg cell and the polar nuclei were still characterized by structural integrity.

The abscission of *walnut* female flowers is also derivable from the processes that begin in the ovule (Catlin and Polito 1989). In one case, the embryo sac does not develop

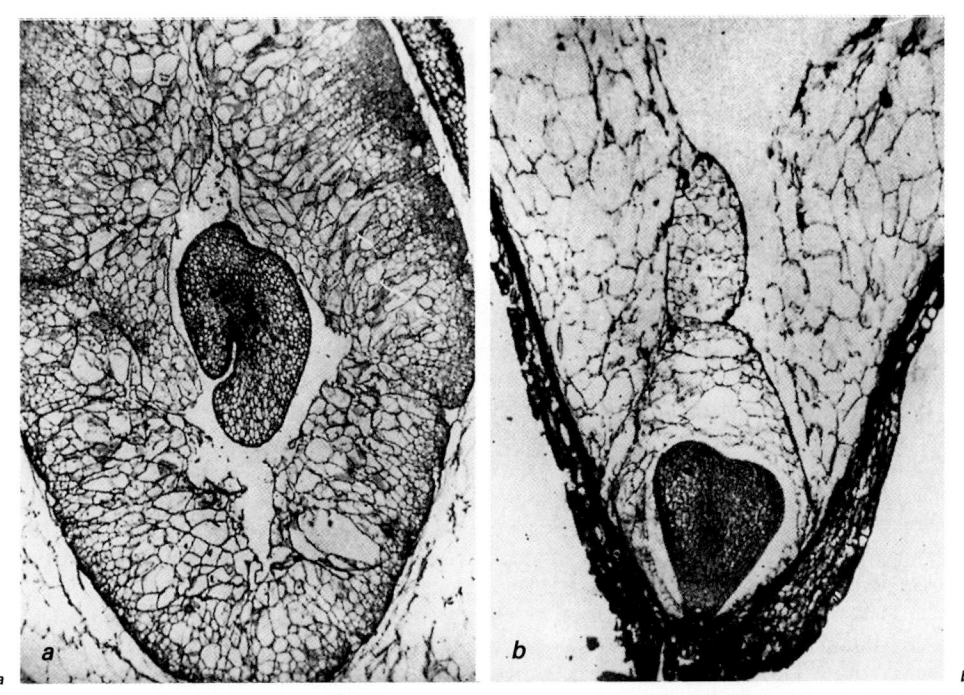

Fig. 6.16. Anomalies in the development of sweet cherry embryo: (*a*) embryo without suspensor; (*b*) the incomplete development of secondary endosperm (after Stösser and Anvari 1978)

and cell degeneration is limited to the nucellus. In another (more usual) case, cell necrosis begins at the top of stigmas or in the apical part of the integument or the apical part of the evagination, but never before the two or four-nucleate stage of macrosporogenesis (for evagination *see* Subsection 2.4.3). The impairment of tissues spreads basipetally, to all the cells sooner or later. There is no cell necrosis in the nucellus. It seems that the development of the embryo sac is normal and the early embryogeny also seems to be normal.

As was discussed in Section 6.4, *hazelnut* female flowers still lack a developed ovary at bloom. The development and the growth of the female gametophyte requires the existence of a compatible pollen tube in the style (Mussano et al. 1983*b*). Female flowers that have not been pollinated or have been pollinated with incompatible pollen will fall in the first week of June.

A large scale of developmental anomalies in *sweet cherry* embryos already during their differentiation was demonstrated by Stösser (1966*a*). It occurs rather frequently that two ovules fused together are linked to one and the same vascular bundle. One of them will die as it happens also to one of only two ovules existing usually in the ovary of stone fruits. One of these two ovules is generally (or becoming rapidly) rudimentary and in it no normal haustorial sheath, endosperm or embryo will develop. It is not rare that there is no real connection between the suspensor and the embryo, causing the embryo to be situated entirely isolated in the endosperm. An extreme case is the deformation of the spatial orientation of the embryo, which might range to 180° (Fig. 6.16*a,b*).

6.9. Research Methodology

The traditional methods of histological examinations concerning the phytoanatomical characterization of *flower bud differentiation* are well known. However, the methods of cytochemistry are suitable for detecting the first signs of flower induction. Based on preliminary methodological examinations (Hesemann and Bubán 1973), microspectrophotometric measurements were carried out in fruit trees, i.e., in sections made from apices of buds induced to bloom and those destined not to form flowers. We measured the level of DNA, DNA + RNA, and nucleohistons in nuclei, stained with Schiff's reagent, gallocyanine-chrome alum, and Fast Green FCF at 550, 570, and 630 nanometer wave-length, respectively. Further details on informative value of these methods are available in Bubán and Hesemann (1979) and Bubán and Faust (1982).

There are different methods for the examination of *pollen viability*. Parfitt and Ganeshan (1989) provide a comparison between the usefulness of seven assays (hanging drop slide and agar-plate germination, acetocarmine, three tetrazolium-based stains, and Alexander's staining procedures). The two *in vitro* germination tests proved the most reliable. Staining procedures are not reliable; e.g., the acetocarmine and Alexander's staining stained also aborted pollen. However, there is a reagent, a special composition of components, such as malachite green, acid fuchsin and orange G stain among others, that proved suitable for distinguishing aborted pollen grains from non-aborted ones in the case of both angiosperm and gymnosperm microgametes (Alexander 1969).

Furthermore, pollen viability can be characterized on the basis of fluorescence and starch content in the same pollen grain (using fluorescein diacetate and potassium iodide) (Jefferies 1977). The use of fluorescein diacetate as a vital stain is successful in determining pollen viability in more than 30 species (Peterson and Taber 1987). The small size of chromosomes and the density of plasm often cause some difficulty in such examinations.

Fig. 6.17. Pollen tubes in the histological sections of transmitting tissues of sweet cherry: (*a*) longitudinal section; (*b*) intercellularly growing pollen tubes (cross-section) (after Stösser 1980*a*)

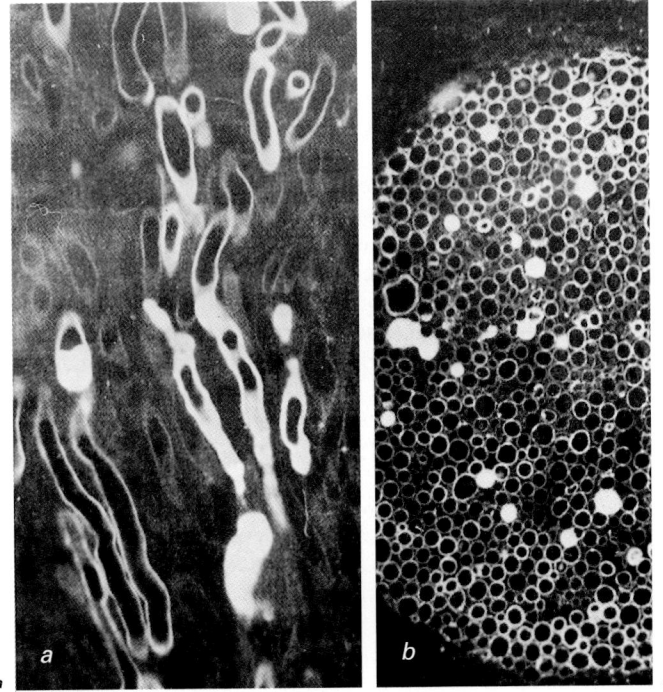

Lespinasse and Salesses (1973) offered the following procedure to stain chromosomes in meiosis: double fixation on Carnoy's solution at room temperature, then high temperature exposure at 65 °C, then an extended period of staining with acetocarmine by the method of Belling.

The traditional method of the investigation of *pollen tube growth* is the following: fluorescent microscopic observations after staining with aniline blue solution of 0.1% (Linskens and Esser 1957, Martin 1959, cit. by Egea et al. 1991). This method scarcely works, or does not work at all, in the case of species with pollen tubes containing very little callose deposition (e.g. *plum*), but a simultaneous use of 0.1% aniline blue and fluorescent brightener "Calcofluor White M2R New" makes a perfect solution (Jefferies and Belcher 1974). Staining with aniline blue is a reliable stain not only for traditional squash preparations but also for microtome sections. The localization of pollen tube will be much more precise if microtome sectioning is used (Stösser 1980*a*; Fig. 6.17*a,b*).

The most obvious way of characterizing the *fertility and compatibility relations* is to execute appropriately designed experiments concerning pollination. Certain factors of field conditions (e.g. temperature), however, often make it difficult to compare results. Satisfactory result can be obtained when detached flowers are kept on a simple medium of 1% agar and 12% sucrose to observe pollen tube growth *in vivo*. The result of this semi *in vitro* method did not differ significantly from those of the usual *in vivo* method. Using this method, there is no consequential change in compatibility relations under controlled conditions, at least in *sweet cherry* and *plum* (Stösser 1980*b*, 1982).

In *hazelnut*, where all cultivars are self-incompatible, and often cross-incompatible as well, laboratory tests are useful. It takes only a week to determine the percentage of styles with an actual pollen tube growth. A reliable correlation has been established between laboratory tests and fertilization under field conditions (Thompson 1971).

In detailed observations paraffin sections can be used. They are double stained with haematoxylin and aniline blue so that pollen tube growth and embryo sac development can be observed in the same ovule of the flower (Jefferies 1979).

In order to estimate the prospects for *embryo development*, the ovary (or eventually the ovule extracted from the ovary) is to be stained with aniline blue. Ovules that display a significant fluorescence are not viable, no pollen tube will grow into them (Anvari and Stösser 1978a). It is of great importance that this method is suitable for indicating the effect of frost, even for lack of any external signs, in *sweet cherry* and *sour cherry* (Stösser and Anvari 1982). In *apple*, the histochemical examinations concerning the comparison of embryos in potential abscissing young fruits and those in persisting young fruits have indicated no convincing difference (Weinbaum and Simons 1974a).

7. Fertilization Conditions
J. Nyéki

Fruit cultivars can be classified on the basis of their fertilization type as follows:

A cultivar can be (1) self-sterile, (2) partially self-fertile, (3) self-fertile, (4) inclined, or not, to parthenocarpy or (5) apomixis.

Two cultivars either (1) fertilize each other, or (2) one fertilizes the other, or (3) neither fertilizes the other.

7.1. Self-Fertility and Self-Sterility

A cultivar is regarded as self-fertile if pollinated with its own pollen yields fertilization and fruit set, and development of viable seeds in fruits.

Cultivars are usually divided into three groups according to the degree of self-fertility:

1. Self-sterile (autosterile) cultivars: they are always fertilized by a different cultivar (or different cultivars); hence they are to be planted in company with pollinizers.
2. Self-fertile (autofertile) cultivars:

a) there is no need for cross-pollination, hence they can be planted as single cultivars,
b) pollinizers increase fruit set.

3. Cultivars self-fertile to an insufficient degree: a sufficient degree of fertilization and a high rate of fruit set require pollinizers with cultivars belonging to this group.

A cultivar is practically self-sterile if its pollination with pollen of the same cultivar results a low percentage of fruit set (e.g. in *apple, sour cherry, plum, apricot*), therefore it cannot be regarded as self-fertile for economical fruit growing. In other words, self-fertilization might take place in certain cultivars, to a varying degree year by year, but this extent of self-fertilization is insufficient for the production of a satisfactory yield.

In fruit growing, a cultivar is to be regarded as self-fertile if it can be planted in single cultivar orchards without risking a poor crop, whereas we consider a cultivar self-sterile if it brings no fruit, or a poor or unstable crop without having been pollinated by a different cultivar that serves as a pollinizer.

Pseudo-fertility refers to the phenomenon when self-incompatible pollen tubes manage to reach the ovary and carry out fertilization; but very few egg cells are expected to be fertilized this way. In certain self-sterile cultivars partial self-fertilization is observable under exceptional physiological or environmental conditions [e.g. the pollination of very "young" or very "old" flowers with a great amount of pollen, or very high temperatures (20 to 25 °C) occur at bloom]. Pseudo-compatibility is of no practical importance. The percentage of fruit set is low (1 to 5%), and it rarely occurs. The percentage of fruit set in self-sterile cultivars is as high as if the cultivars were partially self-fertile.

In partially self-fertile cultivars, the degree of self-fertilization displays a great deviation year by year. Under conditions optimal for pollination and fertilization, partially self-fertile cultivars are characterized by a high degree of fruit set: they behave as if they were self-fertile. Under unfavourable weather conditions, however, they are self-sterile, or fertile to a very low degree.

Factors affecting the extent of self-fertilization. The capability of self-fertilization is a genetically determined cultivar-specific feature. There are great differences in the percentage of self-fertilization between different cultivars. The genetic character of a cultivar (self-sterile or self-fertile) is a constant property but the extent of fertilization in the case of a cultivar may vary under different ecological conditions. If a cultivar is self-fertile in every growing area, we can say that its self-fertility is genetically determined. If the rate of self-fertilization in a cultivar displays a significant variation year by year, this cultivar must not be planted alone but only in company with pollinizer cultivars.

The following factors affect the extent of self-fertilization:

1. Ecological conditions and fruit growing methods: (*a*) self-fertilization takes place earlier at sites that are warm during bloom (in comparison to cooler sites); (*b*) a high rate of nutrients, as well as the lack of nutrients, may result in sterility.
2. The age, physical and physiological state influencing the vigour and health of trees (infection by virus).
3. The efficiency of self-pollination.

Self-fertile cultivars bear several advantageous properties compared to self-sterile ones:

a) A constant and great crop can only be expected of self-fertile cultivars. Fruit set and yield in self-sterile cultivars are usually fairly poor, or unsteady and these depend on climatic conditions and cross-pollination during and after bloom.

b) Self-fertile cultivars can be planted in solid cultivar blocks, without pollinizers, which makes it possible to apply cultivar-specific phyto- and agrotechnology.

c) Self-fertile cultivars are especially advantageous in early blooming fruit types (e.g. stone fruits), whose bloom usually takes place under unfavourable conditions (when the temperature varies quickly and ranges from low to high values, in rainy and windy weather) when bee pollination is not efficient.

d) The placement location and time of bloom in the case of self-fertile cultivars are irrelevant to their combination in an orchard. The bloom period of self-fertile cultivars and their simultaneous bloom with other cultivars are to be considered only in the case of their application as pollinizers.

In examinations concerning self-fertilization, we should face the question whether the data in the literature can be taken for granted.

Often, contradictory data are published on self-fertilization. The deviation of data is partly explainable by the use of distinct methodology. Natural (spontaneous) self-pollination usually results in a low extent of self-fertilization. Inside an isolator usually only a part of flowers are pollinated. Artificial self-pollination usually provides a higher rate of fertilization than does natural self-pollination.

There are also additional factors that may play a role in the yearly fluctuation of self-fertilization efficiency.

a) When individual trees are examined in different years results may differ because each tree has its own physiological state.

b) The cultivar is genetically heterogeneous, that is, it consists of distinct types (clones) (e.g. *sour cherry* cultivars such as 'Pándy meggy', 'Cigánymeggy', 'Montmorency', 'Schattenmorelle'; the plum cultivar 'Besztercei', and the apricot cultivar 'Magyar kajszi').

c) There is no entirely uniform nomenclature. Data concerning the fertilization of one and the same cultivar are published as if they were two distinct cultivars, such as the *sour cherry* cultivars: the 'English Morello' and 'Schattenmorelle' pair, or the 'Fanal' and 'Heimanns Konserwen' pair.

To sum up, the data in the literature concerning self-fertilization of given cultivars show great differences perhaps because data were taken under distinctly different ecological conditions or circumstances and methods of examination were not uniform. Therefore, such data can be applied under new conditions only after several years of checking.

Apple. Self-sterility in *apple* cultivars was pointed out by Waite (1898). Afify (1933) argued that complete self-incompatibility was rare in apple, and the extent of self-fertilization was fairly variable. According to Kobel (1954), every Swiss apple cultivar is to be regarded as self-sterile. Further examples concerning the self-fertilization of certain apple cultivars are available in Crane (1927/28), and later in Knight (1963) and in Williams and Wilson (1970).

Vondracek (1962) studied self-fertilization in 33 apple cultivars in the years 1952 and 1961. He concluded that the majority of apple cultivars had only a weak inclination to self-fertilization. However, there were differences in the rate of self-fertilization between the cultivars he examined. According to the work of Soltész (1982*a*), 58% of the cultivars was entirely self-sterile; and no cultivar was able to provide a considerable amount of fruit in every year under self-pollinating conditions. An autogamic pollination of 55,457 flowers in 141 apple cultivars resulted in a fruit set of 0.2%.

Table 7.1 is a summary of data concerning the self-fertilization and fruit set of self-fertilized 'Jonathan' flowers available in the literature. This cultivar is to be regarded as partially self-fertile.

Self-fertilization based on fruit set in apple cultivars, such as 'Golden Delicious' and their possible replacement cultivars was examined by Branzanti et al. (1978) (Table 7.2).

Table 7.1. Data Concerning Self-Pollination of 'Jonathan' Apple Cultivar

Fruit set (%)	Author (year)
0–1.6	Maliga (1953*a*)
2–8.3	Kandaurova et al. (1973)
0.7–1.2	Griuner and Medvezhova (1977)
1.5–7.1	Karatsharova (1977)
0.5	Soltész (1982*a*)
4.1	Kandaurova (1985)

Table 7.2. Open and Self-Pollination and the Number of Viable Seeds in the Cultivars of Golden Delicious Group (Modified after Branzanti et al. 1978)

Cultivar	Fruit set (%) open	Fruit set (%) self-pollination	No. of viable seeds per fruit open	No. of viable seeds per fruit self-pollination
C.D.A. Summerland 9 E 13-47	56.2	7.1	8.5	4.0
C.D.A. Summerland 10 C 18-33	25.6	7.5	10.2	8.0
Ed Gould Golden	65.8	1.7	8.8	0.6
Missouri A 2071	11.9	0.2	7.2	1.7
Ozark Gold	26.7	7.4	6.8	2.5
Virginia Gold	35.6	2.9	7.9	3.0
Golden Delicious	61.3	3.6	8.5	2.4
Stark Spur Golden Delicious	20.9	1.3	8.1	2.5
Mutsu	50.8	0.0	3.1	0.0
Prime Gold	46.9	6.4	7.0	5.2

The following cultivars are also practically self-sterile: 'Missouri A 2071', 'Stark Spur Golden Delicious', 'Ed Gould Golden', 'Virginia Gold', 'Golden Delicious' (fruit set between 0.2 and 3.6%). Partially self-fertile (7.1 to 7.5%) are the Canadian selections 'C.D.A. Summerland 9E 13-47' and 'C.D.A. Summerland 10C 18-33', and the 'Ozark Gold'. Kovács (1968) claimed that 'Egri piros' was a self-fertile apple cultivar: having self-pollinated 570 flowers, he recorded a fruit set of 8.6%. The number of viable and non-viable seeds, however, was not published.

Lobanov and Basina (1969) observed the following cultivars to be self-fertile regularly: 'Zorenka' (a three years' mean of fruit set was 31.5%, with 2.2 viable seeds per fruit) and 'Narodnoe' (12.8% fruit set and 3.7 viable seeds per fruit). Spiegel-Roy and Alston (1982) published data concerning cultivars cultivated in England. 'Malling Kent' (11%) and 'Malling Greensleaves' (8.2%) are partially self-sterile, while 'Cox's Orange Pippin' and the triploid 'Malling Suntan' (0 to 1.4%) are self-sterile.

Kandaurova (1985) examined self-sterility in apple cultivars under Ukrainian ecological conditions. 'Jonathan' displayed the greatest inclination to self-fertilization (4.1%). After free fertilization 47.1% fruit set. Self-fertilization in most apple cultivars was not greater than 0.5 to 3.2%. Her data also corroborate the statement that in apple cultivars, self-pollination will never provide a great amount of marketable yield.

Vondracek (1967) classified apple cultivars on the basis of the extent of their self-fertilization into two groups:

1. the majority of the apple cultivars under examination displayed a negligible inclination to self-fertilization;
2. those cultivars that inclined to self-fertilization can be divided into the following groups:

a) cultivars such as 'Ontario', 'Wagener', 'Ribston' and 'Laxton's Superb' are self-fertile to a greater degree;

b) cultivars, such as 'Jonathan' and 'Cox's Orange Pippin' were inclined to self-fertilization in certain years, but not in other years;

c) certain cultivars are not inclined to self-fertilization at all.

Karamysheva and Blinova (1976) divided the apple cultivars into three groups on the basis of the extent of self-fertilization: (1) where the amount of fruits exceeds 10%, (2) falls between 6% and 10%, (3) is under 5%.

Kandaurova et al. (1973) proposed three groups, also:

1. the cultivars are inclined to self-fertilization in a percentage of 3 to 10% (e.g. 'Jonathan');
2. self-fertilization occurs rarely;
3. there is no self-fertilization at all (e.g. 'Delicious').

A fruit set of 3 to 10% from self-fertilization is generally regarded as sufficient for a good yield.

Having pollinated with its own pollen, 'Starking' was not fertilized (Maliga 1953*b*). Lalatta (1982*a*) argued for the existence of apple cultivars (e.g. 'Delicious') that would bring no yield self-pollinated, that is, display an entire self-sterility.

The apple is not a typical self-sterile fruit species, even though several cultivars are characterized by self-sterility (e.g. 'Starking', 'Húsvéti rozmaring', 'Staymared'). Soltész (1982a) observed a low level (1 to 2%) of self-fertilization in 42% of the cultivars, such as 'Jonathan', 'Egri piros', 'Golden Delicious', 'Rome Beauty', 'Idared', 'Summerred', 'Jerseymac'.

Lobanov and Basina (1969) considered self-fertility a cultivar-specific property but they also argued that the environment would exert a great influence. According to the examinations executed by Kandaurova et al. (1973), the possibility for fertilization after self-pollination varies from cultivar to cultivar, mainly depending on the activity of generative cells and the length of the period when embryo sacs are viable.

Self-sterility is not so frequent and its extent is not so high in triploid apple cultivars than in diploid ones. Hence, despite the high level of genetic (gametic) sterility, a greater percentage of fruit set is characteristic of self-fertilized triploids than self-fertilized diploids (Modlibowska 1945). As is argued by Vondracek (1962), triploid apple cultivars are more often inclined to self-fertilization than do diploids. According to Knight (1963), in the triploid 'Baldwin' and 'Bramley's Seedling' cultivars a sufficient fruit set may take place even without pollinizers. However, Bidabe's (1955) data (Table 7.3) do not support Modlibowska's and Vondracek's conclusions.

The evaluation of self-fertility in triploids is more difficult because they have a certain extent of parthenocarpy, low seed content, sterility of pollen, and partial sterility of embryo sacs.

Fruits resulting from self-fertilization contain generally fewer viable seeds than those produced by cross-pollination. In partially self-fertile apple cultivars, three to five seeds will develop at most for each fruit as a result of self-pollination, while five to eight, or even ten, seeds can be found after cross-pollination (Teskey and Shoemaker 1972).

Table 7.3. Fruit Set of Open and Self-Pollinated Diploid and Triploid Apple Cultivars (Between 1948 and 1952) (after Bidabe 1955)

Cultivars	Self-pollination final set (%)	No. of seeds (pce/fruit)	Cross-pollination final set (%)	No. of seeds (pce/fruit)
Mean of diploid cultivars	0.06	5	14.8	7.5
Mean of triploid cultivars	0.2	2.7	13.3	2.3

The extent of self-fertilization in apple shows an increasing tendency parallel with the age of the tree according to Crane's (1923) observations. Maliga (1953*b*) claims that if the soil is rich in nutrients and conditions are optimal for growth, concomitantly an increased extent of self-fertilization may also be expected in various apple cultivars. Vondracek (1962) emphasizes the strong effect of climatic circumstances on inclination to self-pollination. Karamysheva and Blinova (1976) demonstrated that under different ecological conditions the extent of self-fertilization of the same apple cultivar is different. Under the conditions typical of Krasnodar, for instance, 'Golden Delicious' was always self-fertile (to a low degree), and 'Jonathan' was partially self-fertile. In Karamysheva and Blinova's (1976) opinion, if a cultivar is self-fertile in every ecological condition, then self-fertility is to be regarded as genetically determined. They regard self-fertility in apple as a cultivar-specific property which may appear to different degrees depending on environmental conditions and whose percentage may vary year by year.

'Golden Delicious' flowers are self-sterile. Near the Mediterranean Sea in the southern part of France, however, this cultivar is self-fertile and brings a good yield without pollinizers if the weather is dry and warm during bloom (Hugard 1978). A great flower density is characteristic of 'Golden Delicious' so that the fertilization of a fairly small percentage (3 to 4%) of flowers may result in a good yield (50 to 60 tons/hectare). Whereas in years when the weather is unfavourable, cool and wet during bloom, flowers will be fertilized poorly without cross-pollination, which entails a poor crop or the complete fruitlessness of the trees. Lalatta (1982*a*) observed in Milan that 'Golden Delicious' was partially self-fertile in certain years; the fruits contained few seeds but they were not parthenocarpic.

These examples suggest that data on self-fertility obtained under different ecological conditions must not be taken for granted. The extent of self-fertility of a cultivar is to be determined in each growing area, and the pollinizers best to the given cultivar are also to be chosen individually.

All in all, the data on self-fertilization available in the literature concerning the apple can be summarized as follows:

1. Some apple cultivars are partially self-fertile;
2. some apple cultivars, under certain ecological conditions, are sometimes self-fertile and display an acceptable extent of fruit set without cross-pollination;
3. in apple an inclination to self-fertilization is more usual than natural parthenocarpy;
4. pollinated with its own pollen, an apple cultivar will not set so many fruits than cross-pollinated with an appropriate pollinizer;
5. to assure a great extent of fruit set, diploid pollinizers are required for every apple cultivar;
6. the apple is essentially a cross-fertilization-requiring species.

Pear. It is regarded as a self-sterile species since 1894 (Waite 1894). The data available in the literature concerning self-sterility in different pear cultivars, however, are contradictory. Some cultivars had been reported to be self-fertile (e.g. Tufts and Philp 1923, Wellington et al. 1929, Reinecke 1930). Kamlah's (1928*b*) observations supported the claim that the pear cultivars were practically self-sterile. Kobel and Steinegger (1934) also reported self-sterility in the cultivars they examined. Out of 81 cultivars, Schanderl (1934*a,b*) found only one which brought viable seeds in the case of self-pollination. In the other cultivars he observed a parthenocarpic fruit set. Hence, he concluded that the pear was self-sterile. In 1937 Schanderl published that out of 122 pear cultivars, two displayed a good extent of fruit set after self-pollination but the fruits were parthenocarpic.

A genetic explanation of the self-sterility of the pear was given by Lewis and Modlibowska (1942). Crane and Lewis (1942) examined self-sterility in 19 diploid, eight triploid, and two tetraploid pear cultivars. Pollinated with its own pollen, nine of the diploid cultivars and four of the triploid ones did not bring any fruit at all. The tetraploid pear cultivars, however, were self-fertile (Crane and Lewis 1942): they produced a fairly high percentage of viable seeds (14%) with a mean of 4.1 seeds per fruit. The diploid pear cultivars were characterized by a very low percentage (0.3%) of self-fertilization and a mean viable seed content of 1.5. The triploid cultivars might display a certain percentage of self-fertilization similar to the diploid cultivars, but the mean number of viable seeds in the triploid cultivars was only 0.1. This decrease in seed number in the case of the triploid cultivars is to be attributed to sterility caused by the unbalanced number of chromosomes.

In Manzo's (1956) examinations only two cultivars were self-fertile to a low degree out of 54 cultivars. The others were entirely self-sterile. Esaulova (1960) reported the cultivars examined to be self-sterile. According to Botez et al. (1960), certain pear cultivars were self-sterile, while others were self-fertile. Nagy (1960) studied self-fertilization in eight cultivars. One was entirely self-sterile (0% mature fruits), two were practically self-sterile (0.1 to 0.9% mature fruits), while four were self-fertile to a low degree (1.0 to 6.0% mature fruits) and one was self-fertile to a high degree (50% of mature fruits). Maliga (1961) reported 13 cultivars to be entirely self-sterile out of those examined by him and 12 cultivars displayed different degrees of self-fertility. Nyéki (1972a) studied self-fertilization by natural self-pollination (autogamy) in 52 pear cultivars, and geitonogamy in 27 cultivars. In his studies self-fertilization varied from cultivar to cultivar year by year. Most pear cultivars were entirely self-sterile, or self-fertile to a low degree, which depended on the year (e.g. 'Madame Du Puis', 'Olivier de Serres'). Nevertheless, even these cultivars are to be regarded as practically self-sterile and to be planted accompanied by pollinizers. Only the 'Pringall' butter-pear set fruit regularly year after year by artificial self-pollination (in 1969: 9.4% and one viable seed, in 1970: 4.5% and two viable seeds).

Nyéki (1972a) observed the following results of self-pollination:

1. Either no fruit had been set (entirely self-sterile cultivar),
2. or the fruits developed

a) were either parthenocarpic,
b) or contained only non-viable seeds,
c) or contained both viable (perfect) and non-viable seeds at the same time.

Few self-fertile pear cultivars are cultivated in the world. One of them, 'Blanca de Aranjuez', which is fairly widespread and cultivated everywhere in Spain, can be planted without pollinizers (Ruiz 1977). Self-pollination in flowers of this cultivar will result in a fruit set of 5.2 to 6.7%. Nevertheless, cross-pollination can significantly increase the percentage of fruit set (8 to 22.4%). This cultivar is not inclined to parthenocarpic fruit development.

The tetraploid pear cultivars ('Fertility Improved', 'Double William') and the diploid 'Spalding' are self-fertile (Bellini 1987). 'Spartlet' is partially self-fertile. In 'Packham's Triumph', 1.4% of 7,000 flowers set fruit (Dwyer and Bowman 1933).

Misic et al. (1979) published the following data on self-fertilization in different pear cultivars: 'Curé': 0.5% mature fruits, 'Bosc Flaschenbirne': 1.2%, 'Bartlett' ('Williams' Bon Chretien'): 5.2%, 'Passe Crassane': 9.5%. Bellini (1987) attributes a varying extent of self-fertility to the following pear cultivars: 'Andy', 'Monchallard', 'Leopardo', 'Redspire', 'Santa Maria Morettini', 'Spalding', 'Flemish Beauty'. All the other pear cultivars are to be regarded as practically self-sterile.

Reviewing the relevant data in the literature, one can conclude that the reported percentages concerning self-fertilization in particular pear cultivars are fairly different and contradictory. Most researchers do not publish the number of viable (germinative) seeds in fruits set after self-fertilization. In addition to fruit set, the number of viable seeds contained by a fruit can serve to characterize self-fertilization.

Comparisons concerning self-incompatibility (Moffet 1931, Crane and Lewis 1942, Modlibowska 1945) clearly show that self-sterility is much more typical of the pear than of the apple, characterized by a gradual self-incompatibility, but less typical of it than of the sweet cherry. The diploid pear cultivars are entirely, or almost entirely, self-sterile.

Factors affecting the extent of self-fertilization. Several factors affect the extent of self-fertilization in pear. According to Reinecke (1930), capability for self-fertilization in different pear cultivars strongly depends on the site, the age and health of trees, and the soil. Bordeianu et al. (1955) argued that favourable weather might make certain pear cultivars (e.g 'Bartlett', 'Bosc Flaschenbirne', 'Passe Crassane') self-fertile. Nagy (1960) concluded that self-fertility in pear varied from year to year, as being a function of the state of trees, the weather, and other ecological circumstances, hence the cultivars could be divided into groups according to self-fertility. Hoffman (1965) observed that the variation of the extent of self-fertilization was greater in pear than in apple due to nutritional and climatic conditions. Certain cultivars, such as 'Flemish Beauty' and 'Conference', are known as sufficiently self-fertile pear cultivars in the U.S.A. Batjer et al. (1967) pointed out that the degree of self-fertility in some pear cultivars was a function of the year and the growth condition of the trees.

Table 7.4 summarizes data concerning self-fertilization in 'Bartlett' under different ecological conditions. Griggs and Iwakiri's (1954) detailed examinations of the self-fertilization of 'Bartlett' in different habitats in the U.S.A, however, showed that the extent of fruit set in the case of self-fertilized flowers in solid cultivar 'Bartlett' orchards was almost the same as in the case of cross-fertilized flowers in orchards where the former cultivar had been planted mixed with pollinizers. 'Bartlett' thus is a weakly self-fertile cultivar, as is supported by the number of viable seeds per fruit. Hooper (1912) and Florin (1924) considered it self-sterile, whereas Griggs and Iwakiri (1954) partially self-fertile.

In the 'Comice' pear cultivar, also, a significant geographical dependence can be observed in respect of fruit set. Jaumien (1968), Bini and Bellini (1971), and Lombard et al. (1971) found it self-sterile. In the southern part of Oregon, however, this pear cultivar brings a normal yield in orchards where there are no pollinizers (Callan and Lombard 1978). Stephen (1958), also in Oregon, reported a very low extent of fruit set, and no parthenocarpy. Tufts and Philp (1925) held the cultivar acceptably self-fertile in the inner valleys of California, but self-sterile in the submountain region of the Sierra.

The two pear cultivars discussed above serve as evidence for the dependence of inclination to self-fertilization on ecological conditions.

Quince. Statements on the self-fertility of the *quince* in the literature are contradictory. Quince cultivars are often held to be self-fertile but self-sterile and partially self-

Table 7.4. Data Concerning Self-Pollination of 'Bartlett' Pear Cultivar

Fruit set (%)	No. of complete seeds (pce/fruit)	Year examined	Author (year)
7.0	–	1894	Waite (1894)
4.0	–	1923	Tufts and Philp (1923)
2.2	0.2 (mean of	1926	Kamlah
8.8	two years)	1927	(1928a,b)
0	–	1931	Schanderl (1932)
3.0	4.0	1933	Branscheidt (1933)
2.6	0.2	1934	Kobel and Steinegger (1934)
2.8	–	mean of 12 years	Cummings et al. (1936)
6.4–9.0*	0–1.0	1949	
0–35.9	0–0.08	1950	Griggs and Ivakiri (1954)
6.0–12.8	0–0.3	1952	
0.8	0.1	1953	Breviglieri and
0	–	1954	Baldassari (1956)
0	–	1954–56	Maliga (1961)
1.5	–	1962	Blaya (1962)
0	–	1968	
0	–	1969	Nyéki (1972a)
0–1.1	–	1970	

* End values observed in different parts of U.S.A.

fertile cultivars are also mentioned. There is no consensus among specialists either. Singular quince trees in little private gardens will bring a good yield, whereas in homogeneous orchards the trees are practically fruitless (e.g. 'Bereczki birs').

Examinations of self-fertilization are required to decide whether the extent of self-fertility in quince cultivars suffices for the receipt of good yield by their planting in solid cultivar orchards, or mutual cross-pollination is more useful, or required, to obtain a sufficient yield (25 to 30 tons/hectare).

Schanderl (1965) argues that the majority of quince cultivars are self-sterile. According to Angelov (1966), cross-pollinated, partially self-fertile quince cultivars would bring a yield two to three times as much as self-pollinated. Ershov and Hrolikova (1977) concluded that "practically the quince is to be regarded as a self-sterile fruit cultivar". Duganova and Hrolikova (1977) also states that the quince requires cross-fertilization. The extent of self-fertility in quince is a function of ecological conditions (mainly temperature) and varies year by year.

On the basis of the relevant literature and our own experience we have concluded that the flower density of quince is low in comparison with that of other fruit species, such as apple and pear. Therefore, a fruit set of at least 20 to 25% is required to obtain a good yield. The quince cultivars are regarded to be self-sterile even if they have a fairly high fruit set of 7 to 12% (Stancevic 1963, Ershov and Hrolikova 1970, Roman and Blaya 1984). The contradictory data available in the literature concerning self-fertility in quince cultivar 'Bereczki birs' serves as an illustration (Table 7.5).

Table 7.5. Data Concerning Self-Pollination of 'Bereczki' Quince Cultivar

Author	Final fruit set after self-pollination (%)
Kordon (1934)	8
Stancevic (1963)	12
Brózik (unpublished)	19.3
Maliga (1966b)	0–1.1
Ershov and Hrolikova (1970)	10–24
Angelov (1975)	0
Aeppli (1984)	1.1–15.9

Stancevic (1963), Angelov (1975) and Duganova and Hrolikova (1977) divided the quince cultivars into three groups (self-sterile, partially self-fertile, self-fertile); whereas Ershov and Hrolikova (1977) have made four groups:

1. Self-fertile cultivars: fruit set of 7 to 10% in single cultivar plantations;
2. partially self-fertile cultivars: fruit set of 3 to 8% when planted in the company of pollinizers. The extent of self-fertility in these cultivars varies from year to year;
3. practically self-sterile cultivars. They set fruit in a very low percentage (1 to 2%), hence, they require cross-pollination;
4. entirely self-sterile cultivars (0% fruit set). They are to be cross-pollinated.

Sweet cherry. The fertilization biology of *sweet cherry* attracted the attention of researchers as early as in the first decades of the 20th century (Gardner 1913, Schuster 1922, 1925, Crane 1925, 1927/28, Tufts et al. 1926, Wellington 1926, Kamlah 1928b, Crane and Lawrence 1929, 1931, Kobel and Sachov 1929, Branscheidt 1931, 1933, Kobel 1931, Einset 1932, Krümmel 1932, 1933, Schanderl 1932). They pointed out that the sweet cherry cultivars were self-sterile, and it was rare that a certain (very low) percentage of fruits would come from self-pollinated flowers. We would like to highlight two groups of researchers. Kobel and Steinegger (1933) managed to verify the complete self-sterility in 75 sweet cherry cultivars. Crane and Brown (1937) examined self-fertilization in 66 cultivars. Fruit set was only 0.06% in 49,160 self-pollinated flowers. It has been shown also in the case of new hybrids and selections that self-sterility is an inherent property of the sweet cherry (Baldini 1950, Vahl 1956, 1959, De Vries 1968, Way 1968, 1974, Mattusch 1970, Ryabov and Ryabova 1970, Krapf 1971, Fogle et al. 1973, Nyéki 1989). All sweet cherry cultivars grown in Hungary are self-sterile (Brózik 1962, 1971). Auto-sterility in sweet cherry is of gametophytical nature: it is manifested when the pollen tube is growing through the style (Bargioni 1978).

Nevertheless, there are data on self-fertilization published in the literature which contradict the above data. Such self-fertilization of data is rare and often doubtful. Pashkevic (1930), for instance, claimed some sweet cherry cultivars to be self-fertile at a high rate ('Zholtaya Denisena', 'Bigarreau Napoleon': 44%, 'Tshornaya Daybera Napoleon': 57%, 'Tshornaya Daybera': 64%). Oratovsky (1935, 1940) reported the 'French Black' cultivar to be self-fertile (12.5%), and 'Tshornaya Daybera' (12%). Maliga (1952) observed a fruit set of 1 to 5% after self-pollination in seven sweet cherry cultivars, while 16 cultivars proved to be entirely self-sterile. Koleshnikov (1953, 1959) also published data in support of the appearance of self-fertility in certain sweet cherry cultivars ('Red

Late Byutner': 28%, 'Yellow Dragan': 26%, 'French Black': 25.7%). Ryabova (1961) examined 33 cultivars: 31 out of them were entirely self-sterile while 'Orlica' and 'Prekrasnaya iz Toskany' set fruit after self-pollination in 3% and 6%, respectively. Stösser (1966a) considered 20 sweet cherry cultivars self-sterile but 'Dritte Schwarze' was self-fertile in 2.2 to 3.1%. Stösser regarded the occurrence of self-fertilization in sweet cherry as a result of pseudo-fertility or experimental mistakes. Stancevic (1971) published data on the self-fertilization of 16 sweet cherry cultivars. The fruit set coming from self-fertilized flowers fell between 0 and 15.1%. Only the cultivars 'Primavera', 'Sodus' and 'Lambert' were self-sterile (0%). 'Jaboulay', 'Hedelfinger Riesenkirsche' and 'Schneiders Späthe Knorpelkirsche' were self-fertile in 0.3, 6.5, and 2.0%, respectively. Two cultivars displayed a very high level of self-fertility ('Black Tartarian': 12.5%, 'Emperor French': 15.1%). Data on the self-fertilization of Moldavian sweet cherry cultivars were provided by Bespetshalnaya (1981). 'Valery Tshkalov' and 'Melitopolskaya plodnaya' were reported to be self-sterile, whereas 'Krupnoplodnaya' was self-fertile in 22.1%.

Proven self-fertile sweet cherry cultivars. Within each fruit species there are self-sterile and self-fertile cultivars. The majority of sweet cherry cultivars are self-sterile. Nevertheless, Hugard (1978) reported a self-fertile sweet cherry cultivar named 'Cristobalina', well known in Spain for a long time, which has come into existence by spontaneous mutation. Its use is not widespread. According to Caillavet (1973), this cultivar blooms very early and it is often damaged by spring frosts. There has also been found a spontaneously developed cultivar in Tunisia, named 'Bou Argoub'. Furthermore, in sweet cherry, self-fertile S-gene spontaneous mutations may occur (Lewis 1948), and such mutations can also be produced by X-ray (Lewis and Crowe 1953, 1954). Lewis and Crowe (1954) managed to isolate three self-fertile mutations.

The first self-fertile sweet cherry cultivar with good quality fruits, named 'Stella', was created by Lapins (1971) in 1956 by crossing 'Lambert' with seedling 'John Innes No. 2420'. This cultivar blooms very early or early. Table 7.6 demonstrates Lange's (1979) data concerning self-fertilization in 'Stella'. The self-fertilization of emasculated 'Stella' flowers and their pollination by the pollen of 'Van' provided almost the same fruit set (31 to 34%). Lange (1979) considers 'Stella' a steady self-fertile cultivar. Pollinizer 'Van' caused no increase in fruit set. Bee pollination was advantageous as it increased the extent of fruit set.

The cultivar 'Compact Stella' is a semi-dwarf tree. It is also self-fertile. Lapins created it in 1966 by the mutation of 'Stella' with X-rays. It fertilizes some sweet cherry

Table 7.6. Fruit Set and Fruit Weight of 'Stella' Sweet Cherry Cultivar After Different Methods of Pollination (after Lange 1979)

Pollination type	Final set (%)	Fruit weight (g/fruit)
Control (insect- and wind-pollinated)	51.3	7.87
Flowering branches of Van tied to Stella branches (insect- and wind-pollinated)	45.6	8.61
Flowering branches isolated with textile bags (wind-pollinated)	22.9	9.23
Stella ×Stella (hand-pollinated, flowers were emasculated)	30.9	8.44
Stella ×Van (hand-pollinated, flowers were emasculated)	33.7	7.72
Wind pollination of emasculated flowers	3.8	8.62

Table 7.7. Self-Fertile Sweet Cherry Cultivars

Cultivar	Origin	Author
Stella	cross (Lambert ×JI. 2420)	Lapins (1970)
Temprana de Sot	local cultivar	Orero (1974)
Bou Arkoub	local cultivar	
Compact Stella	induced mutation	Lapins (1975)
Cristobalina	spontaneous mutation	Hugard (1978)
Starkrimson	cross (Stella ×Garden Bing)	Frecon (1980)
New Star	cross	Lapins (unpublished)
Sunburst, Lapins	cross (Van ×Stella)	Sansavini and Lane (1983)
Kronio	local selection	Calabrese et al. (1984)

cultivars ('Chinook', 'Lambert' and 'Van') well (Lapins and Schmid 1976). A detailed description of two additional self-fertile sweet cherry cultivars, 'Sunburst' and 'Lapins' is available by Sansavini and Lane (1983). 'Sunburst' is a medium-late-blooming cultivar. Its pollen is compatible with every other sweet cherry cultivar, and its fertility will reach a fairly high extent even in years unfavourable for pollination and fertilization.

'Lapins' is a mid-period-blooming cultivar. It is very fruitful and serves as a valuable universal pollinizer because its pollen can fertilize every sweet cherry cultivar.

Table 7.7 lists the most important self-fertile sweet cherry cultivars.

Presently, in orchard setting the self-fertile cultivars are still the exceptions. In the future, however, the number of self-fertile sweet cherry cultivars with fruits of better quality is likely to increase because self-fertility is inherited and can be transmitted to new cultivars through hybridization (Matthews 1970, Theiler 1985).

Sour cherry. Fertility in *sour cherry* cultivars has been studied since the beginning of the 20th century. Middlebrook (1915–16) described nine self-fertile sour cherry cultivars. Kostyna (1926–27, 1928) reported nine cultivars to be self-fertile out of 29 cultivars. Crane and Lawrence (1929) argued that among the sour cherry and the sour cherry × sweet cherry ('Duke') cultivars there were completely self-sterile as well as highly self-fertile cultivars. Other researchers (e.g. Roberts 1922, Schuster 1922, 1925, Crane 1923, Kamlah 1928b) also led to similar conclusions in connection with the self-fertility of the sour cherry. Ryabov (1930) divided the sour cherry cultivars into three groups on the basis of the extent of their fertility: self-sterile, self-fertile, and partially self-fertile. He established that most sweet cherry ×sour cherry ('Duke') hybrids were self-sterile. The earliest data on self-fertility in Hungarian sour cherry cultivars are available in writings of Magyar (1935), Maliga (1942) and Husz (1943).

Groups of self-fertility. Sour cherry cultivars are usually divided into three or four groups in the literature according to the extent of their fertility (Table 7.8). As can be seen in Table 7.8, different percentages of self-fertility are chosen by various authors to separate the groups. The percentage above which a certain fertility group is assigned strongly differs from author to author. Thus, there is a great need for the standardization of the levels of fertility in sour cherry as well as in other stone fruits.

Among the cherry ×sour cherry ('Duke') cultivars, several intermediate fertility types can be found ranging from self-fertility to complete self-sterility. In partially self-fertile cultivars fruit set is strongly affected by ecological conditions and shows a significant variation from year to year. When cultivar combinations are designed, these cultivars must be regarded as self-sterile and pollinizers need to be assigned with them. Ryabov and Ryabova (1970) came to the conclusion that in sour cherry and the sweet cherry

Table 7.8. Fertility Groups of Sour Cherry

Author	Grouping on the basis of self-pollination	Fruit set (%)
Brózik (1969)	1. very good self-fertility	15–30
	2. weak and medium self-fertility	3–5
	3. self-sterility	0–3
Enikeev (1973)	1. very good yield	above 30
	2. good yield	21–30
	3. medium yield	11–20
	4. poor yield	0–10
Blazek et al. (1974)	1. reliable self-fertility	above 30
	2. good self-fertility	10–30
	3. not reliable self-fertility	2–5
	4. self-sterility	under 2
Stancevic (1974)	1. self-fertility	above 20
	2. partial self-fertility	5–10
	3. self-sterility	under 5
Stancevic (1975)	1. good self-fertility	above 15
	2. partial self-fertility	5.1–15
	3. self-sterility	under 5
	3a full self-sterility	0
Tamássy et al. (1975)	1. self-fertility	above 5
	2. partial self-fertility	1.1–5.0
	3. self-sterility	0–1.0
Misic et al. (1975)	1. definitely self-fertile	above 30
	2. self-fertile	20.1–30
	3. partially self-fertile	10.1–20
	4. practically self-sterile	under 10
Albertini (1982)	1. self-fertility	above 20
	2. partial self-fertility	4–20
	3. self-sterility	under 4
Redalen (1984a)	1. self-fertility	above 15
	2. partial self-fertility	1.5–15
	3. self-sterility	under 1.5
Ostapenko and Zhukov (1985)	1. self-fertility	above 10
	2. partial self-fertility	0.1–10
	3. self-sterility	0

×sour cherry ('Duke') cultivars the capability for fruit set after self-pollination is a stable, cultivar-specific property. The fertility level remains from year to year the same at different sites, at least within a given climatic zone. The above property changes only in partially self-fertile cultivars.

Table 7.9 is a summary of the proportions of the self-fertility in sour cherry cultivars. In general fertility of sour cherry cultivars is characterized depending on the geographic location where the characterization occurred as:

Table 7.9. Sour Cherry Cultivar Fertility Levels

Author	No. of cultivars examined	Self-sterile %	Partially self-fertile %	Self-fertile %
Cociu and Gozob (1960–61)	13	69	23	8
Blazek et al. (1974)	33	40	12	48
Stancevic (1975)	36	53	14	33
Cociu (1981)	33	52	30	18
Gozob et al. (1981)	19	58	16	26
Albertini (1982)	15	47	40	13
Fabbri et al. (1983)	13	39	15	46
Montalti and Selli (1984)	21	33	43	24
Redalen (1984a)	35	46	26	28
Ryabov and Ryabova (1970)				
a) sour cherry	13	46	39	15
b) sweet ×sour cherry (Duke)	16	75	25	0
Average	22	50	26	24

1. the proportion of self-sterile cultivars is high (33 to 75%);
2. 50 to 100% of cultivars require pollinizers;
3. most (75%) sweet cherry ×sour cherry ('Duke') cultivars are self-sterile.

For example: out of the 91 sour cherry cultivars grown in Russia, 11 are self-fertile (12.2%), 24 are partially self-fertile (26.4%), 28 are self-sterile (30.7%), and the extent of self-fertility of 28 cultivars (30.7%) has not been unambiguously determined yet (Ostapenko and Zhukov).

Self-fertilization in sour cherry cultivars and the factors affecting it. The following two basic questions should be answered here:

1. Is a cultivar self-sterile?
2. How stable is the extent of self-fertility at different sites and years?

Self-fertility in sour cherry cultivars is a genetically determined property. There are great differences in its extent across the cultivars. The self-sterile 'Pándy meggy' and the self-fertile 'Schattenmorelle' serve as excellent illustrations. In sour cherry, the maximum fruit set resulting from self-fertilization may exceed 50%. Blazek et al. (1974) reported 64,2% fruit set, Wociór (1976a) 55.2%, Misic et al. (1977) 50.0%.

To obtain a good yield in sour cherry, 25 to 30% of flowers are to be set and produce mature fruits (Gozob et al. 1981). A high percentage of fruit set in self-fertile sour cherry cultivars may be facilitated by cross-pollination (Krapf 1976). The extent of self-fertilization in the same cultivar, may vary year by year. The cultivar may behave in one year as if it were self-sterile, and in another year as if it were self-fertile. Ryabov and Ryabova (1970) argue that the extent of fruit set in highly self-fertile sour cherry cultivars in the case of self-pollination is stable and less dependent on year and growing area. Self-fertilization in the partially self-fertile sour cherry and sweet cherry ×sour cherry ('Duke') cultivars is a function of climatic factors. Their fruit set is similar to that of the self-fertile and the self-sterile cultivars under favourable and unfavourable conditions, respectively. A rainy and cool weather during bloom has a negative effect on the yield of self-sterile sour cherry cultivars, that is, their yield will decrease to a great extent (Enikeev 1973). Enikeev (1973) observed an immense year by year variation of yield in self-sterile sour cherry cultivars in comparison with the self-fertile ones. In other words, the latter were characterized by a capability for greater and more stable yield. Wociór (1976a) came to the conclusion that from the viewpoint of pollination and fruit set in self-sterile and partially self-fertile cultivars, the favourable years were those when the temperature during bloom was not very high. Whereas in self-fertile cultivars high temperature facilitates an increase in fruit set.

Plum. a) European plums and gages (Prunus domestica). The self-sterile nature of plum cultivars was pointed out as early as in the first decades of the 20th century (Hendrikson 1918, Dorsey 1919). Self-sterility in plums is widespread, especially in the oriental plum cultivars. In the European plum cultivars, similar to sour cherry, several intermediate fertility types can be found ranging from completely self-sterile to highly self-fertile cultivars. Stösser (1984) reported that the self-sterile nature of plum and gage cultivars are to be regarded as "absolute", similar to that of sweet cherry fertility. This means that no fruit set will result from self-pollination in a self-sterile cultivar.

In plum self-fertility is a cultivar-specific property, whose extent does not depend on ecological and climatic factors (Cociu and Bumbac 1968). Hoffman (1965) thought that nutritional conditions and the yield in the preceding year affect the extent of self-fertilization in plum the most.

Tóth (1969) studied the relation between self- and open pollination in 120 plum cultivars. The mean percentage of open fertilization was 18.1% in self-fertile, 10.9% in partially self-fertile, and 7.0% in self-sterile cultivars. These data clearly show that self-fertile plum cultivars set fruit to a much higher degree (%) than partially self-fertile, or especially self-sterile, cultivars. In the case of insufficient insect visitation, which may be a result of unfavourable weather, the extent of open pollination is much greater in self-fertile plum cultivars than in self-sterile ones. Kobel (1954) did not see the necessity of cross-pollination in self-fertile plum cultivars. He thought that bee pollination sufficed for a good fertilization. Krapf (1976) observed that in years when the weather during and just after bloom was unfavourable cross-pollination would exert a positive effect even in self-fertile plum cultivars (e.g. 'Fellenberg'). Belmans and Keulemans (1985) observed a definitely increasing effect of cross-pollination also in the case of self-fertile cultivars. They concluded that even self-compatible cultivars require compatible cross-pollination. Iliev (1985), however, came to the opposite conclusion on the basis of his observations on some self-fertile plum cultivars ('Paczelt szilvája', 'Besztercei szilva', 'Sopernica', 'Renklod Hramovyh'). In his opinion, self-pollination cannot be excluded even when insects are not present.

As the plum flower is hermaphrodite, that is, bisexual, it is often assumed that the self-fertile cultivars need no external pollen transmitters but pollination may take place within a single flower (autogamy). This assumption is correct only to a certain extent.

During the course of Hungarian experiments it occasionally occurred that researchers had no time to pollinate isolated but not emasculated flowers of some plum cultivars. When they checked the results later, they observed a certain extent of fruit set also in the unpollinated flowers in the case of the self-fertile cultivars. The following questions arose: First, are the values, concerning fruit set, to be regarded as equivalent to those values obtained with flowers that they could pollinate? Second, is it necessary at all to carry out an artificial pollination in such experiments, or may it be sufficient to isolate the flowers simply to exclude foreign pollen? The omission of the manual pollination would be very advantageous because this is a labour requiring procedure during the fairly

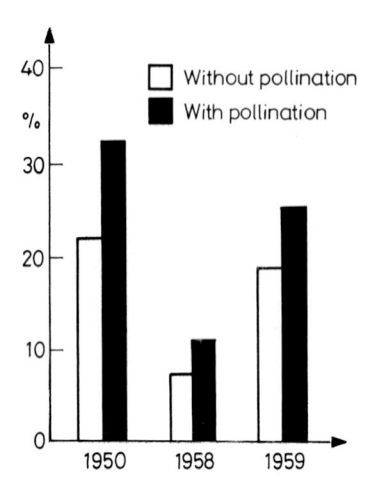

Fig. 7.1. Self-fertilization of plum cultivars with or without artificial pollination (after Tóth 1980, in Nyéki 1980)

short bloom. Tóth (in Nyéki 1980) designed a three-year-long experiment to investigate this problem with self-fertile plum cultivars. He compared two methods. Figure 7.1 illustrates the results. The columns clearly show the favourable effect of artificial self-pollination on the extent of fruit set. In all three years a higher rate of fruit set was achieved with the aid of artificial self-pollination than without it. These increases in fruit set were 45.7%, 58.7%, and 31.1% in the years 1950, 1958, and 1959, respectively. Thus, the mean increase of the three years examined was 45.2%. These data suggest that stigmas in isolated plum flowers will receive an insufficient quantity of pollen by gravitation and limb movement due to the wind.

Groups of self-fertility. Plum cultivars are divided into groups depending on the extent of self-fertility (Table 7.10). Generally three groups are distinguished in plum as well as in the fruit species discussed so far, but sometimes four or seven groups are described. Various authors assigned various differing limits for various groups of fertility. The Italian researchers (Faccioli and Regazzi 1972, Bellini and Bini 1978, Faccioli and Marangoni 1978, Bellini 1980) consider a cultivar to be self-fertile in the case of a fruit set over 5%, while Tóth (1967, 1969) argues that only a fruit set over 10% is to be regarded as self-fertile. Paunovic (1971) qualifies a cultivar as highly self-fertile in the case of a fruit set over 20%. Levitzkaya and Kotoman (1980) regard a plum cultivar as self-fertile if its fruit set is over 15%.

Tóth (1969) carried out self-fertilization experiments upon about 124,700 flowers of 119 plum cultivars from 1950 to 1969 in Hungary. Each experiment lasted at least two years. He divided the cultivars under examination into the following four groups according to fertility:

1. self-sterile: 0.0% mature fruit;
2. practically self-sterile: 0.1 to 1.9% mature fruits;
3. weakly self-fertile: 2.0 to 9.9% mature fruits;
4. highly self-fertile: the proportion of mature fruits is over 10.0%.

The cultivars were included into each group on the basis of the means of data collected during several years. The following distribution of cultivars presented itself:

Table 7.10. Fertility Groups in Plum

Author	Grouping on the basis of fertility	Fruit set (%)
Tóth (1967)	1. self-fertile 2. partially self-fertile 3. self-sterile	10 0.1–9.9 0
Tóth (1969)	1. highly self-fertile 2. weakly self-fertile 3. practically self-sterile 4. self-sterile	10 2.9–9.9 0.1–2.8 0
Paunovic (1971)	1. very good self-fertility 2. good self-fertility 3. medium self-fertility 4. weak self-fertility 5. partial self-fertility 6. partial self-sterility 7. self-sterility	30.1 20.1–30 15.1–20 10.1–15 5.1–10 2.1–5 0–2
Faccioli and Regazzi (1972), Bellini and Bini (1978), Faccioli and Marangoni (1978)	1. self-fertile 2. partially self-fertile 3. self-sterile (fully or practically)	above 5 1–5 under 1

Bellini 1980, in Baldini and Scaramuzzi 1980	1. self-fertile
	2. partially self-fertile
	3. self-sterile
	  a) practically self-sterile
	  b) fully self-sterile
		above 5
		1–5
		
		0.1–1
		0

Nyéki et al. (1985)	1. highly self-fertile 2. partially self-fertile 3. self-sterile	above 10 2–9.9 0–1.9

1. self-sterile: 37 cultivars (30.84%),
2. practically self-sterile: 34 cultivars (28.33%),
3. weakly self-fertile: 16 cultivars (13.33%),
4. highly self-fertile: 33 cultivars (27.50%).

These data show that a sufficient extent of fruit set would result from self-pollination in quite few plum cultivars, about one quarter of the cultivars examined by Tóth (1969), though, two-thirds of cultivars did set fruit after self-pollination. Finally, almost one-third of the plum cultivars were completely self-sterile. The extent of self-fertilization in particular cultivars ranged widely, for example in the case of 'Penyigei szilva', from 0.0% to 66.3%. The 66.3% fruit set means so great fruit load that the tree cannot develop normally, hence, fruit thinning should be carried out. Table 7.11 summarizes the proportions of the plum cultivar numbers according to fertility.

Data in the literature concerning fertility in plum cultivars are sometimes contradictory, or at least differing. Some authors attribute these differences to the different ecological conditions of the experiments. Others regard capability for self-fertilization as an intrinsic, inherited, cultivar-specific property, affected by external environmental factors to a negligible extent.

Table 7.11. Rates of Plum Cultivars According to Self-Pollination

Author	No. of cultivars examined	Self-sterile	Partially self-fertile	Self-fertile
Tóth (1966)	119	37	37	26
Paunovic (1971)	24	21	25	54
Faccioli and Marangoni (1978)	62	48	12	40
Iliev (1985)	14	36	14	50

In Hungary Tóth (1966, 1969) scrutinized the problem in the case of plum cultivars. He carried out experiments concerning self-fertility with certain plum cultivars and compared his data to those available in the literature concerning the same cultivars. His data coincided 95% of the time with the 875 references he collected, mainly from foreign sources. He came to the conclusion that self-fertility of plum cultivars could not be affected by external environmental factors. Otherwise, there could not have been so obvious coincidence between the data concerning self-fertility. His experience indicates that the capability of plum cultivars for self-fertilization is a genetically determined, very stable property. Consequently, the data in the literature concerning self-fertility of various plum cultivars should be accepted.

Nyéki (1989) has studied the question and reviewed the relevant literature (Table 7.12). The cultivars 'Besztercei szilva' and 'Stanley' are uniformly held to be self-fertile in the literature. The extent of fruit set in both cultivars shows a great variation (0 to 30% in 'Stanley', 5.8 to 37.5% in 'Besztercei szilva') depending on localities and years. 'President' is almost always described as a self-sterile cultivar, only Paunovic (1971) has indicated that it is partially self-fertile (3.9%). 'Reine-Claude d'Althan' is unanimously

Table 7.12. Fertility of Some Important Plum Cultivars Reported by Various Authors

Cultivar	Self-sterile	Partially self-fertile	Self-fertile
Reine-Claude d'Althan	4, 7, 9, 10, 15, 16, 18, 19, 20	2	
Besztercei szilva			1, 2, 3, 4, 7, 9, 10, 14, 15, 17, 19, 20
Stanley		5, 17	1, 3, 4, 7, 8, 9, 10, 13, 14, 15, 18, 19
French prune		2, 3, 4, 7, 9, 10, 13	2 (French plum 1)
Montfort	2, 7, 10		
Ruth Gerstetter	4, 5, 7, 9, 10, 12, 13, 14, 15, 19		
Tuleu gras	4, 5, 7, 9, 10, 12, 14, 15		
Green gage	10, 13, 14, 15, 17	2	3, 8
Italian prune		5, 15	2, 3, 4, 7, 9, 10
Bluefre		5, 8	4, 7, 9, 10, 13, 17
President	2, 4, 5, 7, 8, 10, 11, 13, 14, 15, 18, 19	3	

Authors: 1 = Hoffman (1965); 2 = Tóth (1967); 3 = Paunovic (1971); 4 = Faccioli and Regazzi (1972); 5 = Bellini (1973); 6 = Krapf (1976); 7 = Bellini and Bini (1978); 8 = Bradt et al. (1978); 9 = Faccioli and Marangoni (1978); 10 = Bellini (1980), in Baldini and Scaramuzzi (1980); 11 = Levitskaya and Kotoman (1980); 12 = Roman (1981); 13 = Bellini et al. (1982); 14 = Stösser (1983); 15 = Stösser (1984); 16 = Belmans and Keulemans (1985); 17 = Iliev (1985); 18 = Belmans (1986); 19 = Kellerhals (1986); 20 = Roman and Radulescu (1986)

held to be self-sterile with the exception of Tóth (1967). 'Stanley', a self-fertile cultivar, is qualified as partially self-fertile in two reports (Bellini 1973, Iliev 1985). 'French prune' is partially self-fertile. 'Montfort' is self-sterile. 'Ruth Gerstetter' is usually considered self-sterile, only Tóth (1967) and Paunovic (1971) included it in the group of the partially self-fertile cultivars. 'Tuleu gras' is male sterile. Data concerning the 'Green gage' are fairly contradictory: this cultivar has been included in all the three groups of fertility. 'Italian prune' plum cultivar is self-fertile, only Bellini (1975) and Stösser (1984) hold it as partially self-fertile. 'Bluefre is described as self-fertile in six publications, and as partially self-fertile in two reports (Bellini 1973, Bradt et al. 1978).

Fertilization of self-sterile plum cultivars with self-fertile and self-sterile cultivars. Tóth's (1968) experimental results shed light on advantages of self-fertile plum cultivars. He sought appropriate pollinizers for five partially or completely self-sterile plum cultivars. He pollinated them with their own pollen and the pollen of nine other cultivars during a several-year-long period. Half of the pollinizers were highly or partially self-fertile ('French prune 1' and '2', 'Besztercei szilva', 'Italian prune', 'Paczelt szilvája'), and the further pollinizers were self-sterile. The results can be seen in Table 7.13. It appears that the pollen of the self-fertile pollinizers resulted in a significantly higher fruit set than the self-sterile pollinizers. Hence, the compatibility of self-fertile plum cultivars to other plum cultivars is generally better.

b) Oriental plums. Cobianchi et al. (1978b) and Costa and Grandi (1982) investigated the fertility in oriental plum cultivars. Among the cultivars they examined, Cobianchi et al. (1978b) observed a great fruit set (16%) after self-fertilization only in 'Premier'. The 'Early Golden' cultivar was partially self-fertile (1.4%) in 1976. The majority of oriental plum cultivars, however, were self-sterile. According to Costa and Grandi (1982), all oriental plum cultivars were self-sterile with the exception of 'Santa Rosa', which was observed to be partially (1.3%) self-fertile. 'Methley', another oriental plum cultivar, is described as partially self-fertile by Bellini and Bini (1978), whereas Bradt et al. (1978) described it to be self-fertile. 'Burbank' is unanimously qualified as self-sterile by both Bellini and Bini (1978) and Bradt et al. (1978). In an excellent summary of the topic, Sansavini et al. (1981) pointed out that the majority of oriental plum cultivars are self-sterile; consequently, they must not be planted in an orchard without appropriate pollinizers. Myrobalan is an appropriate pollinizer for oriental plums. Myrobalan serves as pollinizer, only its fruits are not harvested.

Nyéki (1989) examined the self-fertilization of the oriental plum cultivars in two Hungarian growing regions (Balatonújhely and Helvécia). In Balatonújhely 0.5% of 7031 isolated flowers set fruit in six cultivars. Four cultivars, 'Burbank', 'Duarte', 'Elephant Heart', and 'Shiro' were completely self-sterile (0%) and in two cultivars, 'Methley' and 'Santa Rosa', a low percentage of fruit set (1.3 to 1.6%) could be observed. The data obtained in Helvécia corroborated the previous results: 2087 flowers had been isolated, and a mean fruit set of 0.5% was observed.

Apricot. Rudloff and Schanderl (1935) published data on fertility of 31 *apricot* cultivars, 19 of which belonged to the self-fertile group. Morettini (1943) examined 23 apricot cultivars and reported five of them to be self-sterile. Duhan (1944) found 20 self-fertile cultivars among 27 apricot cultivars; he observed two cultivars ('Large Krems', 'Schöllschitz') to be self-sterile and could not include five cultivars in any group. Schultz (1948) examined fertility in six apricot cultivars and reported two to be self-sterile ('Riland', 'Perfection'). He also made a comparison between the fertilization of flowers self-pollinated *in vivo* and with those pollinated by hand. He concluded that hand pollination was better. It increased fruit set compared to the *in vivo* pollinated flowers.

Table 7.13. Fertilization of Self-Sterile Plum Cultivars After Being Pollinated by Self-Fertile and Self-Sterile Cultivars (after Tóth 1968)

Cultivars		Self-fertile pollinizer cultivars				
		French 1	French 2	Besztercei	Italian	Paczelt
Reine-Claude d'Althan	a	3520	1679	5377	6099	1519
	b	16.2	12.2	12.7	11.1	13.3
Debreceni muskotály	a	1103	1091	1139	1107	1055
	b	11.3	13.8	5.5	11.5	16.7
Ruth Gerstetter	a	843	944	798	1065	1058
	b	5.9	3.2	4.8	6.2	6.1
Sermina	a	1239	1257	1353	1046	1264
	b	1.8	1.1	0.6	2.1	0.6
Green gage	a	3454	1172	4905	5937	1155
	b	9.7	11.6	9.2	3.1	10.9
Average	b	9.0	8.4	6.6	6.8	9.5

Cultivars		Self-sterile pollinizer cultivars				
		R-C d'A.	Debr.m.	Ruth G.	Sermina	Green g.
Reine-Claude d'Althan	a	7597	1073	947	1102	4742
	b	0.4	2.0	3.2	3.5	6.6
Debreceni muskotály	a	1360	1229	1098	917	1222
	b	7.4	0.2	5.1	4.8	8.5
Ruth Gerstetter	a	940	896	702	954	899
	b	11.0	2.0	0.9	1.7	7.5
Sermina	a	1052	953	1264	3345	1235
	b	1.7	1.0	0.3	0.1	0.2
Green gage	a	5097	954	1190	1289	7627
	b	3.7	6.4	1.3	4.5	0.5
Average	b	6.0	2.9	2.5	3.6	5.7

Note: fruit set resulting from self-pollination is not included in the average; a = number of flowers examined, b = average fruit set (%)

Löschnig and Passecker (1954) carried out an experiment with 16 apricot cultivars. Two of them were self-sterile, and the other 14 cultivars were self-fertile. Welkerling (1954) reported that only one cultivar ('Damasca') was self-sterile out of 18 examined. Crossa-Raynaud (1957) examined the fertility of 27 European and American apricot cultivars in different areas of Tunisia for five years. With the exception of two self-sterile cultivars, they were reported to be partially or completely self-fertile. Out of eight local (North African) cultivars, six were self-sterile, and two were partially self-fertile. In the experiments designed by Got (1958), 11 apricot cultivars were self-fertile and three cultivars, including 'Paviot', were self-sterile. Kostyna (1970) divided the apricot cultivars into three groups on the basis of the extent of their self-fertilization (self-fertile, partially self-fertile, self-sterile). A cultivar had been qualified as self-fertile if fruit set over 10%. Kostyna (1970) and Smykov (1974) examined the fertility in apricot cultivars of different geographical origin. The data summarized in Table 7.14 demonstrate that most European apricot cultivars are self-fertile, whereas the majority of Middle Asian and Iranian–

Table 7.14. Fertility of Apricot Cultivars With Different Geographical Origin

Geographical origin	Kostyna (1970)				Smykov (1974)		
	No. of culti-vars	self-fert. (%)	partially self-fert. (%)	self-ster. (%)	No. of cultivars	self-fert. (%)	self-ster. (%)
European	42	88.1	9.5	2.4	25	80	20
Middle-Asian	33	18.1	9.2	72.7	14	7.1	92.9
Iranian-Caucasian	17	5.9	0	94.1	6	0	100

Caucasian ones are self-sterile. Sanagian (1968) published data on 13 apricot cultivars. According to his observations, eight cultivars were self-sterile, two cultivars were partially self-fertile, and three were self-fertile. Out of the cultivars under examination only 'Krasnoshtshoky', which is identical to 'Magyar kajszi', was self-fertile in a high percentage (10%). Limongelli and Cappellini (1978) and Cappellini and Limongelli (1981) found some cultivars self-fertile to a very low extent among those cultivated near Vesuvius (Italy). Lamb and Stiles (1983) divided the apricot cultivars that they examined into three groups according to fertility as follows: five cultivars were included in the group of self-sterile cultivars, one cultivar was qualified as partially self-fertile, and six were classified as self-fertile. Pugliano and Forlani (1985) studied ten apricot cultivars, all of which were self-fertile. Moreover, all the cultivars studied by Pugliano and Forlani (1985) were characterized by a fruit set over 20% when flowers were isolated and self-pollinated. This percentage of fruit set indicates a great extent of self-fertilization in apricot.

In Hungary, Maliga (1966a) was the first to study the fertilization of apricot cultivars. 'Auvergnei', 'Blanchet' and 'Nancy' were self-sterile, and 'Montgament' was practically self-sterile. Further apricot cultivars ('Magyar kajszi', 'Késői rózsa', 'Borsi rózsa', 'Rakovszky') were self-fertile. Maliga concluded that the extent of self-fertilization in cultivars 'Magyar kajszi' and 'Késői rózsa' was great enough to plant them in solid cultivar orchards without the risk of an insufficient fruit set and yield. Nyujtó et al. (1982, 1983, 1985) formed three groups according to fertility: (1) self-sterile (0% fruit set), (2) partially self-fertile (fruit set of 0.1 to 9.9%), (3) self-fertile (fruit set over 10%).

Nyéki (1989) examined self-fertilization in 19 cultivars, all foreign in origin. The 'Bergeron' was regularly characterized by a great (24%) or very great (33%) percentage of self-fertilization, whereas the following cultivars were self-sterile: 'Early Divinity', 'Stella', 'Sun Glo', 'Perfection'. The self-sterility of the last cultivar, 'Perfection', had already been known since 1948 (Schultz 1948).

The extent of self-fertility in the cultivars with large (giant) fruits, selected in Hungary, had not been established, which made further examinations necessary. Nyéki (1989) studied them ('Ceglédi óriás', 'Ligeti óriás', 'Szegedi mammut', 'Nagykőrösi óriás') in orchards in Pomáz, and in Kecskemét in the years 1985–86, and 1985, 1988 to 89, respectively. All cultivars mentioned above were completely self-sterile (0%) (Nyéki 1989, Szabó and Nyéki 1989).

Peach. Self-fertility in *peach* was studied by a large number of researchers (e.g. Branscheidt 1933, Detjen 1945, Maliga 1961, Ryabov and Kantsherova 1970, Fogle 1977, Tudor 1981, Perfileva 1982, Nyéki 1989). They advocated the self-fertile nature in the majority of cultivars. Only some cultivars were claimed to be male sterile. Ryabov and Kantsherova (1970) divided peach cultivars into four groups according to their fertility: (1) self-sterile cultivars (0% fruit set), (2) male (pollen) sterile cultivars, (3) partially self-fertile cultivars (fruit set up to 10%), (4) self-fertile cultivars (fruit set over 10%). The authors mentioned above argued that there was no correlation between self-fertility and flower type (rose/bell shape). Detjen (1945) evaluated self-fertilization in 37 peach cultivars for four years. His conclusion was that manual self-pollination would result in a higher extent of fruit set than self-fertilization of isolated flowers but a sufficient extent of fruit set could be obtained also in the latter case. Maliga (1961) pointed out that in self-fertile peach cultivars cross-pollination was unfavourable because the high extent of fruit set would result in a decrease in fruit size.

Nyéki (1989) studied the self-fertility of the peach cultivar groups. His classification is listed below:

a) Fresh market types. (1) fruit set in the majority of cultivars fell between 10% and 25%; (2) the greatest extent of fruit set was observed in 'Early Redhaven' (32.9%) and 'Elberta' (28.8%); the least extent of fruit set was in the male sterile 'J. H. Hale' cultivar (9.6%).

The yearly variation in self-fertilization ranged from 0% to 56.8%. A self-fertilization over 50% was observed in the following cultivars: 'Sunhaven': 56.8% (1978), 'Merrill Sundance': 55.7% (1983), 'Summerset': 54.2% (1978), 'Michelini': 53.3% (1981), 'Early Redhaven': 52.1% (1982).

b) Processing, clingstone types. In these cultivars self-fertilization has always been above 10%. Mean fruit set was greatest in 'Sudanell' (34.8%). In two clingstone cultivars fruit set was over 50% in 1983 ('Babygold 7': 62.4% and 'Sudanell': 69.9%). These high percentages of self-fertilization resulted in an unfavourable "over-set" of fruits.

c) Nectarine cultivars. The typical mean fruit set values fell between 10 and 20%. The nectarine cultivar with the least percentage of self-fertilization (9.0%) was 'Fusador'. The greatest percentage of self-fertilization (21.7%) was observed in 'Silver Lode'. Self-fertilization over 50% was not observed. There were only two cases when fruit set, after natural self-pollination, exceeded 40% (in 'Morton': 49.4% in 1983, in 'Silver Lode': 45.8% in 1978).

Nyéki (1989) concluded on the basis of his experiments that:

1. the least mean fruit set (15.7%) could be observed in nectarine cultivars;
2. the extent of fruit set in fresh market peach cultivars (20.9%) was almost the same as that in the processing clingstone group (21.7%);
3. peach cultivars are generally characterized by self-fertilization values between 15 and 20%.

Summary of the data concerning self-fertility in peach cultivar groups (after Nyéki 1989)

Cultivar group	No. of cultivars examined	No. of isolated flowers altogether	Mean self-fertilization (%)
Fresh market cultivars	29	10,466	20.9
Clingstone cultivars	11	4,259	21.7
Nectarine cultivars	14	6,849	15.7

Distribution of stone fruit cultivars according to the extent of self-fertilization. Nyéki (1989) based on his experiments summarized fertility of stone fruit cultivars, and formed the following five groups:

1. completely self-sterile: fruit set of 0%,
2. self-sterile: fruit set between 0.1 and 1%,
3. partially self-fertile: fruit set of 1.1 to 10%,
4. self-fertile: fruit set of 10.1 to 20%,
5. highly self-fertile: fruit set over 20%.

Figure 7.2 provides the proportion of the number of cultivars that belong to particular groups.

In *sweet cherry*, 73.7% of cultivars are completely self-sterile, and 26.3% are partially self-fertile. In *sour cherry* cultivars and types, 62.8% are completely self-sterile, 20.9% are self-fertile, and only 7.0% are highly self-fertile. In *oriental plum*, 60% of cultivars are completely self-sterile, 20% are self-sterile, and 20% are partially self-fertile. Hence, all ten cultivars examined by Nyéki (1989) require appropriate pollinizers. In *European plum* cultivars, the majority also need some pollinizer (60.4%). Out of the 27 *apricot* cultivars that Nyéki studied, 70.3% require pollinizers. In *peach*, however, only 7.4% of the cultivars were partially self-fertile. The other cultivars were self-fertile (46.3%) or highly self-fertile (46.3%) so 92.6% of peach cultivars can be planted without pollinizers, they can expected to produce a good yield in solid cultivar orchards.

A comparison of self-pollination types in stone fruits. The following two types of self-pollination are to be compared in stone fruits:

1. natural autogamic pollination, when pollination can take place within flower without pollen carrier;
2. clonal geitonogamic pollination, when pollen is carried from flowers of different trees of the same cultivar either by hand or by bees.

In comparison with natural autogamy, clonal geitonogamic pollination increased fruit set percentages in self-fertile *sour cherry* cultivars (Tamássy et al. 1975), but, the effect of a clonal geitonogamic pollination on fruit set was less effective in partially self-fertile cultivars.

Nyéki (1989) pointed out in sour cherry that:

1. the extent of autogamic pollination was enough (20.9%) to achieve a good yield (10 to 15 tons/hectare) only in 1972 one out of the three years examined;
2. geitonogamic pollination by hand increased fruit set in all the three years. Therefore, self-fertile cultivars also require efficient pollen transmission, that is, the application of bee pollination;
3. a mean fruit set could be increased by 9% in sour cherry by geitonogamic pollination;
4. there is a regression in fruit set resulting from autogamic and geitonogamic pollination (Fig. 7.3). The correlation coefficient is: $r^2 = 0.73$. At low autogamic fruit set geitonogamic pollination may double fruit set. However, at very high autogamic fruit set the effect of geitonogamic pollination on fruit set is negligible.

It was demonstrated in *apricot* (Nyujtó et al. 1982, 1983, 1985) that geitonogamic pollination had an increasing effect on the extent of fruit set in comparison with auto-

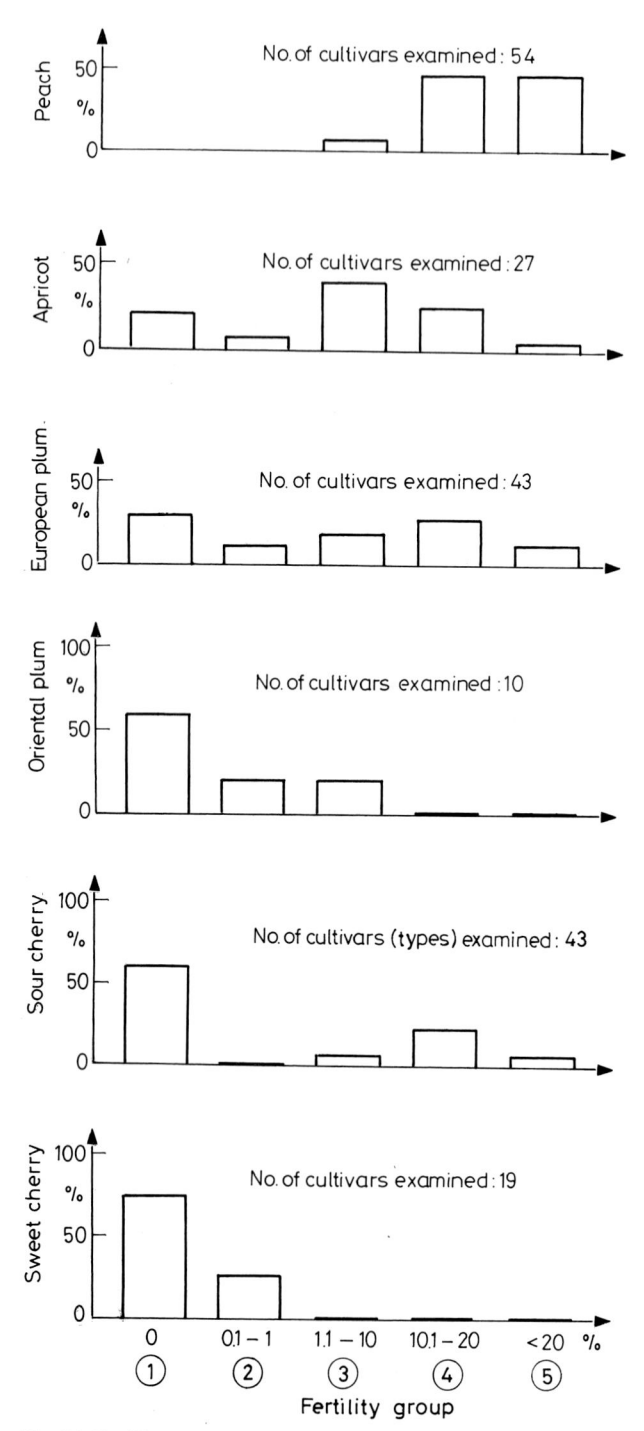

Fig. 7.2. Fertility distribution of stone fruit cultivars (after Nyéki 1989)

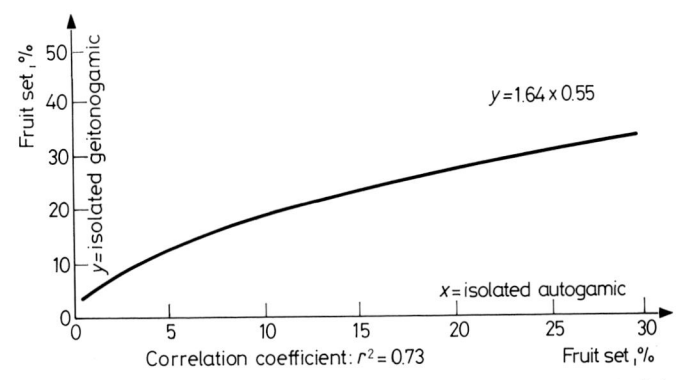

Fig. 7.3. Correlation between autogamic and geitonogamic pollination in fruit set in sour cherry cultivars (after Nyéki 1989)

gamic pollination. The same is true for the *peach*: geitonogamic pollination would increase fruit set to a great extent compared to autogamic pollination (Nyéki et al. 1980). These studies verified the necessity of bee pollination also in self-fertile peach cultivars.

Almond. Taylor (1918), Tufts (1919) and Tufts and Philp (1922) were the first to point out the self-sterile nature of the *almond*, which later has been corroborated by several researchers (Lutri 1935, Wood 1947, Gagnard 1954, Kester and Griggs 1959a, Brózik 1967, and Pejovics 1968). Only a few self-fertile cultivars are known in almond cultivated around the world (Kester and Bradley 1976). To achieve a good yield, the extent of self-fertilization should exceed 25 to 30% (Kester and Griggs 1959a).

The discovery of self-fertile almond cultivars occurred between 1935 and 1960 (Reino et al. 1986). Lutri (1935) was the first to publish data on self-fertile almond cultivars. Reino et al. (1986) assumed that the self-fertility was transmitted to the almond (*Amygdalus communis* L.) by spontaneous hybridization with the bitter almond (*Amygdalus webbii* Spach) a long time ago.

Grasselly and Olivier (1976) studied self-fertilization in 35 almond cultivars. Five of them were self-fertile through three years. The percentage of self-fertilization was less than that of open pollination.

Herrero et al. (1977) examined self-fertility in 16 almond cultivars. One of them, 'Tuono', was self-fertile. After self-pollination mature fruits resulted from 48% of flowers, and pollen tubes reached the ovary in 65% of the pollinated flowers.

Muhturi and Stylianidu (1979) reported that 'Trucito' almond cultivar was self-fertile, and the extent of fruit set after self-pollination was the same as that after cross-pollination.

Vasilakakis and Porlingis (1984) achieved a fruit set of 34.4% with autogamic, and 56.8% with geitonogamic pollination by hand. 'Le Grand', an additional almond cultivar, is partially self-fertile (Winbaum 1985).

Reino et al. (1986) examined 25 "types" of almond in a population in Puglia in respect of self-fertility. He formed three groups on the basis of the extent of self-fertilization:

Group 1, completely self-sterile: these cultivars are absolutely auto-incompatible, that is, no fruit has been set as a result of self-pollination (this group consisted of five "types").

Table 7.15. Self-Fertile Almond Cultivars

Cultivar	Author
José Diaz	Almeida and Marques (1945)
Duro Italiano	Almeida (1949)
Francesco Nocciolara	Milella (1959)
Arrubia	Milella (1959)
Mollisona	Crescimanno (1960)
Romana	Palazon (1965)
Mazzetto	Jaovani (1973)
Filippo Ceo	Grasselly and Olivier (1976)
Genco	Grasselly and Olivier (1976)
Occhiorosso	Grasselly and Olivier (1976)
Tuona	Grasselly and Olivier (1976)
Exinograd	Grasselly and Olivier (1976)
5 hybrids:	Kester and Bradley (1976)
54–39E	
54–42E	
55–13W	
56–88	
56–98	
Tuono	Herrero et al. (1977)
Bari Rachele	Monastra and Fideghelli (1978)
Trucito	Vasilakis and Porlingis (1984)

Group 2, partially self-fertile: fruit set always remained at a low level (less than 5%) so these cultivars are to be regarded as self-sterile (four "types" belong to this group).

Group 3, self-fertile: these cultivars were characterized by a consistently good fruit set so they can be cultivated without pollinizers.

The self-fertile almond cultivars have not been suitable for commercial growing so far because their quality is not satisfactory.

Nowadays an increasing number of self-fertile almond cultivars have become known. Table 7.15 is a summary of these cultivars.

Walnut. Walnut cultivars (*Juglans regia* L.) are self-fertile (Krapf 1971). Germain et al. (1975) pointed out little differences in the extent of self-fertility. Walnut cultivars 'Parisienne' and 'Grandjean' are less self-fertile than 'Corne', 'Morhat', 'Franquette', and 'Mayette'. The walnut cultivars examined so far can be fertilized well with their own pollen as well as the pollen of other trees reciprocally (Szentiványi 1980a, in Nyéki 1980).

All walnut cultivars cultivated are self-fertile in a high percentage. They are not to be planted in homogeneous orchards without pollinizers, however, because the bloom periods of the flowers of different sex, staminate and pistillate flowers, differ in every year, hence a tree can pollinate only a small portion of its own flowers (Szentiványi in Nyéki 1980).

Chestnut. There is no consensus concerning self-fertility in *chestnut* in the literature. In Breviglieri's (1951) experiments concerning self-fertility, both the ordinary chestnuts and the marron types provided many seedless burs and few well-developed nuts indicating partial self-fertilization. Breviglieri concluded that the chestnut was to be regarded as self-sterile in practice and to be planted in company of pollinizers.

Solignat (1958, 1973) examined self-pollinated trees of 15 chestnut cultivars during an eight-year-long period. Every cultivar produced burs but the extent of set was very little, so they were practically self-sterile. The mean of the 15 self-pollinated cultivars was 10.7 nuts per 100 burs, whereas in the case of open pollination this value was 171. The number of fruits resulting from self-pollinated flowers in individual cultivars ranged from 0.25 to 45. These experimental results indicate the self-sterile nature of the chestnut. Hence, cross-pollination is required. McKay (1969) also corroborated that all chestnut species and cultivars were self-sterile and added that isolated trees would bring only a limited crop.

To sum up, in chestnut, pistillate flowers require cross-pollination. Furthermore, the flowers of different sex in a tree do not bloom at the same time. The inclination to self-pollination is so limited that even in years when the relevant blooming periods coincide we cannot expect great enough self-fertilization for a sufficient yield.

Hazelnut. Romisondo (1963*a,b*) pointed out that the *hazelnut* cultivar 'Tonda Gentile' was practically self-sterile. Its pollen was sterile. Zielinski and Thompson (1967) reported that self-pollination in 'Barcelona', another hazelnut cultivar, resulted in a fruit set of 6.2%, which meant a weak self-fertility. Thompson (1971) pointed out that all hazelnut cultivars were to be regarded as self-sterile. Reviewing the literature concerning self-fertility in hazelnut, Modic (1974) established that in hazelnut

1. there are completely self-sterile cultivars;
2. only a few cultivars are characterized by a sufficient extent of self-fertility;
3. the majority of hazelnut cultivars are weakly self-fertile but are to be regarded as self-sterile for practical purposes.

Modic (1974) also concluded that the hazelnut is not worth planting in homogeneous orchards. Romisondo (1977, 1978) stated that self-sterility was fairly widespread in hazelnut. Self-fertile cultivars are rare, and the extent of their self-fertility varies. Examples for self-fertility are: 'Carello', 'Santa Maria del Gesù', 'Tonda' and 'Romana'. Szentiványi (in Nyéki 1980) pointed out that the blooming periods of its own male and female flowers in a hazelnut cultivar did not coincide; hence, an essential condition of self-pollination is not met.

Strawberry. As was demonstrated by Sheliahudin's (1960) examinations, *strawberry* cultivars are self-fertile to various extents. The author has divided the cultivars into three groups as follows:

1. self-fertile to a slight extent: 0 to 30%,
2. self-fertile to a medium extent: 30 to 60%,
3. self-fertile to a great extent: over 60%.

The cultivars of an early ripening ('Laxton's Royal Souvereigne', 'Captain Cook', 'Rügen') were entirely self-sterile (0%).

Only about 10% self-fertilization was observed in male sterile strawberry cultivars (Sheliahudin and Brózik 1966). These cultivars need cross-pollination.

The extent of fruit set ranged from 70% up to 100% in highly self-fertile cultivars.

Strawberry cultivars differ from each other in the extent of self-fertilization.

Raspberry and blackberry. The cultivated *raspberry* cultivars are partially or completely self-fertile. Several authors reported that individual cultivars show various extents of fruit set. The authors and the percentages they reported are as follows: Logintsheva (1958): 72 to 90%, Sheliahudin and Brózik (1967): 40 to 90%, Daubeny (1969): 11 to 75%, Kollányi (1976): 23 to 79%. Hence, the cultivated raspberry cultivars can become

Table 7.16. Fertility of Raspberry Cultivars (after Redalen 1977)

Cultivar	Artificial self-pollination fruit set (%)	Artificial self-pollination seed set (%)	Artificial pollination by pollen of II.-2 LGP and own pollen fruit set (%)	Artificial pollination by pollen of II.-2 LGP and own pollen seed set (%)	Open pollination fruit set (%)	Open pollination seed set (%)
Asker	82	70	79	61	93	75
Camenzind	30	38	29	37	81	67
Glen Clova	82	59	84	62	82	78
Lloyd George	75	53	73	42	85	53
Malling Jewel	89	65	86	51	96	61
Norna	66	38	71	43	92	42
Prussen	88	20	85	40	93	60
Veten	98	37	92	47	95	73
II.-2 LGP	77	46	73	44	89	56
13/182	76	31	83	37	88	58
Average	76	46	76	46	89	62

fertile and bring fruit to a sufficient extent also in homogeneous plantations. Nevertheless, cross-pollination always results in a greater extent of fruit set and a better yield (Kollányi 1976). Contrary to this, Redalen (1977) concluded on the basis of his experiments from ten raspberry cultivars (Table 7.16) that there was no difference in fruit and seed load between self-pollination and cross-pollination.

Different cultivars of the *blackberry* are characterized by various extents of self-fertility. Most blackberry cultivars, as well as most raspberry cultivars, are self-fertile but there is also partial or complete self-sterility among them (Kollányi in Nyéki 1980).

Black and red currant. Different cultivars of the *black currant* are characterized by various extent of self-fertility (Tamás 1963). There are completely self-sterile cultivars (e.g. 'Coronet') as well as highly self-fertile ones (e.g. 'Brödtorp'). Sheliahudin and Brózik (1967) classified the black currant cultivars into three categories according to their self-fertility:

1. practically self-sterile cultivars,
2. cultivars self-fertile to a sufficient extent (20 to 40%),
3. cultivars self-fertile to a high extent (40 to 80%).

Smirnov (1974) formed five categories for the same purpose:

1. self-sterile,
2. self-fertile to a slight extent,
3. self-fertile to a medium extent,
4. self-fertile to a high extent,
5. self-fertile to a very high extent.

A third categorization is due to Porpáczy (1974):

1. cultivars self-sterile to a high extent (0 to 5% of self-pollinated flowers become a fruit),

2. cultivars characterized by partial self-fertility (5 to 45% of self-pollinated flowers become a fruit),
3. cultivars are self-fertile to a high extent (more than 45% of self-pollinated flowers become a fruit).

Few cultivars belong to the third category ("highly self-fertile"). Out of those examined by Porpáczy (1974), only 'Brödtorp' was such a cultivar (79.5% fruit set).

In *red currant* three categories have been formed (Sheliahudin and Brózik 1967):

1. self-fertile to a slight extent: 3 to 20% of self-pollinated flowers produced fruit,
2. self-fertile to a medium extent: 20 to 40% of self-pollinated flowers produced fruit,
3. self-fertile to a great extent: 40 to 60% of self-pollinated flowers produced fruit.

In *gooseberry*, Kirtvaya (1965) observed the following effects of self-pollination upon the *gooseberry*: the berries had become smaller (4 to 5 g) and their shape had become irregular. As a result of cross-pollination, large (7 to 10 g) and uniform berries had been produced. Sheliahudin and Brózik (1966) classified the gooseberry cultivars into five categories according to their fertility:

1. practically self-sterile: fruit set of 1 to 5% after self-pollination,
2. self-fertile to a slight extent: fruit set of 5 to 20% after self-pollination,
3. self-fertile to a medium extent: fruit set of 20-35% after self-pollination,
4. self-fertile to a high extent: fruit set of 35-50% after self-pollination,
5. self-fertile to a very high extent: fruit set of over 50% after self-pollination.

In cultivar 'Zöld óriás', a self-fertilization of 70% was observed. Zenina (1967) made a comparison between the self-fertilization of eight gooseberry cultivars and compared autogamy with geitonogamy (Table 7.17). The extent of self-fertilization of the individual cultivars differed in the two cases. In 'Record', a male sterile gooseberry cultivar with functional female flowers, self-fertilization was 2%, whereas open fertilization was 80%. In general, as a result of cross-pollination, Zenina (1967) observed an increase in yield compared with self-pollination.

Table 7.17. Self-Pollination of Gooseberry Cultivars (Modified from Zenina 1967)*

Cultivar	Self-pollination by autogamy		Geitonogamy	
	No. of flowers isolated	Fruit set (%)	No. of flowers isolated	Fruit set (%)
Smena	893	47.6	824	34.1
Russki	876	31.9	847	32.8
Severni Vinograd	928	26.8	878	30.3
Jubileyni	887	25	905	28.6
Finik	909	24.8	908	28.5
Moskovski Krasni	934	54.4	845	55.8
Kolhozni	820	57.8	825	52.7
Record	886	2.0	698	2.7

* Data were obtained at Rossoshan from 1964 to 1966

Table 7.18. Self-Fertility of Gooseberry Cultivars (Újfehértó 1977) (Modified from Bubán et al. 1980)

Cultivar	Fruit set (%)
Zöld óriás	91
Szentendrei fehér	86
Grüne Kugel	92
54/23. G.	90
Delamere	65
Pallagi óriás	91

Bubán et al. (in Nyéki 1980) stated that self-pollinated gooseberry cultivars showed a fruit set over 80%, which was sufficient for planting them in homogeneous plantings, that is, without pollinizers (Table 7.18).

Classification of fruit species according to fertility. The following generalization can be made across all temperate zone fruit species. Temperate zone fruit species can be divided into four basic classes according to their behaviour in respect of fertility:

1. self-fertile species;
2. species whose cultivars occupy every kind of intermediate positions from completely self-sterile to highly self-fertile;
3. species whose cultivated cultivars are usually self-sterile but there are already bred and selected self-fertile cultivars in the group;
4. species that require cross-pollination.

Table 7.19 demonstrates the inclusion of fruit species in the classes above.

Apple cultivars are typically partially self-fertile. In *pear*, the majority of cultivars are completely self-sterile and only a few cultivars are self-fertile to some (low) extent. Most *quince* cultivars are self-sterile. In the current choice of commercial *sweet cherry* cultivars, the self-fertile ones are only exceptions. In *sour cherry* and *European plum*, the whole range of fertility occurs from completely self-sterile to highly self-fertile. The

Table 7.19. Grouping of Fruit Species According to Their Self-pollination (Nyéki, Original)

Self-fertile	Full scale from self-sterile to self-fertile (3)	Majority of cultivars is self-sterile	Cross-pollination is needed
tetraploid pears	sour cherry	sweet cherry	apple
peach (1)	European plums	almond	pear
raspberry	apricot		quince
walnut (2)	hazelnut		oriental plums
	strawberry (1)		walnut
	blackberry		chestnut
	currants		hazelnut
	gooseberry		

(1) = some cultivars are male sterile; (2) = dichogamy hinders self-pollination in protandrous cultivars; (3) = pollinizers are needed for a) self-sterile and b) partially self-fertile cultivars

majority of *oriental plums*, however, are self-sterile, only a few cultivars are partially self-fertile. A large percentage of *apricot* cultivars are self-fertile but completely self-sterile cultivars are also cultivated. *Peaches* are self-fertile, they often "over-set". Cultivated *almond* is self-sterile, only a few self-fertile cultivars are known. Despite the fact that cultivars of the *walnut* are self-fertile, they require pollinizers because of dichogamy. The *chestnut* is to be regarded as self-sterile for practical purposes. In *hazelnut*, self-sterility is fairly widespread. Few partially self-fertile cultivars are known. The *strawberry* and the *raspberry* have cultivars with various extents of self-fertility; the cultivated cultivars are self-fertile. The majority of *blackberry* cultivars are self-fertile. The *black currant* has cultivars with various extents of self-fertility but there are also few highly self-fertile cultivars. Among the *red currant* cultivars, the self-sterile ones are rare. *Gooseberry* cultivars show a wide diversity of self-fertility values but those cultivated nowadays are self-fertile.

7.2 Cross-Pollination

There are numerous ways to increase yield or to improve fruit quality. For instance, we should find the best site, cultivate modern fruitful cultivars, apply intensive methods of cultivation, permanently improve the level of agrotechnics (pruning, nutrition, plant protection, irrigation), and select the appropriate pollinizers.

In order to select the appropriate pollinizer(s) for a cultivar, we should carry out experiments with isolated flowers and hand pollination to determine the best partners in cross-pollination. While the cross-pollination of self-sterile or partially self-fertile cultivars are being examined, the pollen of the parent plant also takes part in pollination. The same process takes place in the orchard when the own pollen falls on the stigma. If we intend to exclude this kind of self-pollination, then the flowers should be emasculated.

In examinations concerning fertilization, the concept of generative compatibility (interfertility) is to characterize the possibility for a generative connection between two cultivars, that is, the intact unification of the gametes and the development of a viable progeny from the fertilized egg cell.

The following six combinations are available if two cultivars self-fertile to different extent are cross-pollinated (the two cultivars pollinate each other reciprocally):

a) self-sterile cultivar × self-sterile cultivar;
b) self-sterile cultivar × partially self-fertile cultivar;
c) self-sterile cultivar × self-fertile cultivar;
d) partially self-fertile cultivar × partially self-fertile cultivar;
e) partially self-fertile cultivar × self-fertile cultivar;
f) self-fertile cultivar × self-fertile cultivar.

The six combinations of pollination listed above only concern the case of one pollinizer. In an orchard, usually two or more pollinizers take part in pollination, which is called a pollination with pollen mixture.

If the cross-fertilization of two cultivars is to be determined, the following cases can arise:

1. the two cultivars reciprocally fertile toward each other to some extent,
2. the first cultivar in one direction:

a) has not been fertilized, or
b) has not fertilized the other cultivar,

3. the two cultivars reciprocally fertile toward each other (0%), that is, the combination is inter-incompatible.

The following factors affect fruit set resulting from cross-pollination:

a) Concerning the cultivar to be pollinated: (1) flower density, (2) the time and the duration of bloom, (3) pistil viability, (4) the time of pollination, (5) the amount, cultivar-composition, and germinativity of pollen grains that have come to the stigma, including the number of pollen tubes, (6) the length of the effective pollination period, (7) the state of flowers when pollinated and the length of the pollination process, (8) capability for fertilization, (9) degree of relationship between the given cultivar and the pollinizer.
b) Concerning the cultivar to serve as a pollinizer: (1) flower density, (2) the coincidence of its blooming period with that of the cultivar to be pollinated, (3) pollen shedding from its anthers, the amount of its pollen, (4) the capability of its pollen for tube growth, (5) the capability of its pollen for fertilization, (6) the number, proportion and arrangement of pollinizers in an orchard.

According to its capability for fertilization, a cultivar that provides pollen can be qualified as a (1) good or (2) poor pollinizer. The capability above can also be qualified by the comparison of the extent of fruit set resulting from pollination with the pollen of the cultivar in question to that resulting from open pollination. At the evaluation of particular combinations of pollination, a cultivar is to be regarded as a good pollinizer if it has resulted in a greater extent of fruit set than open pollination has. The ability of pollinizers to fertilize varying from year to year, also depends on cultivar combinations to a high extent. On the basis of our 20-year experience (Nyéki 1989) and our experimental results, we concluded that weather had a greater influence upon interfertility differences than the cultivars. Roh (1929a–c) also observed a large variance in cross-pollination in a given cultivar combination. The extent of fruit set varied among trees of the same cultivar as well as within a single tree that it sometimes reached the level of sterility even in highly interfertile combinations.

In the case of fixed cultivar combinations, the extent of fruit set is affected by: (1) ecological conditions, (2) site, (3) rootstock, (4) canopy structure, (5) age of trees, (6) the state and nutrition of trees, (7) the agrotechnical level of the orchard, (8) meteorological conditions in the period of pollination and fertilization, (9) the way of isolation and pollination was carried out by hand, (10) the direction of pollination, etc.

There are two ways to study cross-fertilization:

1. in open pollinated flowers,
2. in isolated and emasculated flowers that have been pollinated with the pollen of one or more cultivars artificially.

In practice, if one aims at selecting pollinizers, it is worth examining reciprocal fertilization only in the case of cultivars that bloom at the same time.

Several factors should be considered when evaluating cultivars for cultivar combinations. It should be decided which year's (good/bad/average) fruit set value is to be considered so as to select pollinizers and to establish their capability for fertilization. Soltész

(1982a) and Nyéki (1989) stated that years favourable for pollination and fertilization should be used, lots of flowers should be pollinated, and these data are to be taken into account at the evaluation. If few flowers are examined for many years, the mean of the data resulting from this method will be of a lesser value.

Are the data available in the literature concerning cross-fertilization of different cultivars to be taken for granted? Reviewing the international as well as the Hungarian literature, Nyéki (1989) came to the conclusion that fruit set values concerning interfertility in different cultivar combinations were not to be accepted and not to be applied for cultivars to be cultivated under Hungarian ecological conditions without checking. This is different from self-fertility or self-sterility discussed in the previous sections which are less affected by environment. Numerous researchers verified (examples will be mentioned at the discussion of particular fruit species) that the extent of fruit set after cross-pollination would not be constant but strongly dependent on geographical and ecological circumstances of the growing area. Data resulting from experiments carried out in different countries are often contradictory. Most local cultivars cultivated in certain countries have not been examined in other localities. Different authors evaluate the results of cross-pollination in different ways (e.g. before or after first drop, or before harvest), which means difficulty in comparing of fruit set values. A further problem is that the number of pollinated flowers are often unpublished, or few flowers (less than 50) are pollinated which results in data that bear a slight informative value.

Apple. Apple cultivars are practically self-sterile so their fertilization and fruit set require cross-pollination. In examinations concerning the cross-pollination of a particular apple cultivar, emasculation is usually not necessary because of the essentially self-sterile nature of the apple (Maliga 1956a). Some experimental results, however, seem to refer to the fact that the presence of the pollen of the mother cultivar may modify the extent of fertilization. In heterogeneous orchards, where two or more cultivars have been planted together, the effect of self-pollination and cross-pollination upon fruit set is simultaneous. Hence, the effect of the pollen of the mother cultivar in cross-pollination also needs to be considered. Nesterov (1956) studied the role of the pollen of the mother cultivar in the reciprocal crosses of apple cultivars (Table 7.20). In most cultivar combinations, the pollen of the mother cultivar had an increasing effect on fruit set, with the exception of apple cultivars 'Melba' and 'Lavfram'. In the orchard, flowers are always pollinated with a pollen mixture because their own pollen grains always fall on their stigmas.

In apple and pear, the four most widespread cross-combinations are the following:

$♀ × ♂$
1. diploid ×diploid
2. triploid ×triploid
3. diploid ×triploid
4. triploid ×diploid

$2× =$ diploid cultivars ($2n = 34$);
$3× =$ triploid cultivars ($3n = 51$)

In the cross-combinations listed above, fruit set depends on the number of chromosomes in the pollinizer.

Brittain and Eidt (1933) pointed out in apple that triploid ×triploid combinations were very sterile. Afify (1933) stated that the fact, that different cultivars have different capability for fertilization, is due to the secondary polyploid nature of apple cultivars,

Table 7.20. Effect of Mother Cultivar's Pollen on Reciprocal Cross of Apple Cultivars (after Nesterov 1956)

Mother cultivar (ripening time)	Self-pollination		Pollinizer cultivars
	No. of flowers self-pollinated	Final fruit set (%)	
Summer			
Sedli	350	1.4	Prevoshodnoe
Borovinka	338	1.2	Melba
Melba	323	0.9	Borovinka
Autumn			
Pepin Litovski	316	2.2	Melba
Sneznoe	343	1.2	Belfler Kitayka
Sneznoe			Lavfram
Winter			
Lavfram	349	0	Sneznoe
Prevoshodnoe	329	2.1	Sedli
Sum	2348		
Average		1.3	

Pollinizer cultivars	Without pollen of mother cultivar		With pollen of mother cultivar	
	No. of flowers pollinated	Final fruit set (%)	No. of flowers pollinated	Final fruit set (%)
Prevoshodnoe	303	10.9	323	13.0
Melba	523	8.0	498	10.0
Borovinka	549	5.6	539	5.6
Melba	813	9.7	759	12.6
Belfler Kitayka	111	6.3	106	6.8
Lavfram	412	3.2	347	6.9
Sneznoe	226	6.2	252	5.2
Sedli	207	2.9	234	6.8
Sum	3144		3058	
Average		7.1		9.1

which is a result of two sterility factors in every gamete of a diploid apple cultivar and four sterility factors in their somatic cells. Kobel and Steinegger (1934), and Kobel et al. (1939) observed the same situation in other diploid fruit species (e.g. in sweet cherry) as in diploid apple cultivars.

Crane and Lawrence (1931) observed that in triploid apple cultivars the fertilization of flowers was followed by the abortion of the majority of embryos as a result of the unbalanced combination of the chromosomes.

Modlibowska (1945) studied the cross-pollination of apple and pear cultivars in different pollination types. She has found that most diploid ×diploid combinations are compatible but there are a few cases of incompatible combinations.

Table 7.21. Intercompatibility of Diploid and Triploid Apple Cultivars (after Bidabe 1955)

Cultivar	Final fruit set of controlled crosses	
	with diploid	with triploid
	pollinizer cultivars	
	%	%
Mean of 12 diploid cultivars	6.6	2
Mean of 4 triploid cultivars	2.5	0

Note: mean of open pollination of 16 cultivars: 14.7%

In the triploid ×diploid combinations, there are compatible as well as incompatible cultivar combinations. The frequency of incompatible combinations of the triploid by triploid combinations, however, was higher than the frequency of incompatible combinations in the diploid ×diploid combinations.

In the diploid ×triploid combinations, there was no completely incompatible cultivar combination. The low level of fruit set in these crosses can be attributed to the high level of pollen sterility. The results in the triploid ×triploid type (slight fruit set) were similar to those in the diploid ×triploid type.

Bidabé (1955) pointed out that triploid apple cultivars were weak pollinizers because of the bad quality of their pollen (Table 7.21). Consequently, appropriate pollinizers are to be selected from the diploid cultivars.

Soltész (1982a) did not observe a significant difference between the fertilization of different diploid apple cultivars in a given year, but the differences between the years were significant. In triploid cultivars, capability for becoming fertile was greater than capability for making fertile. According to Schmadlak (1962), the fertility of cross-combinations shows the differentiated sexual affinity of the sexual partners where the mother cultivar is the dominant partner.

Evaluation of cross-combinations. The capability of pollinizers for making fertile can be qualified on the basis of percentages of fruit set as follows (Maliga 1953b):

1. cultivars that can fertilize another one to a negligible extent: fruit set between 0 and 2%;
2. cultivars that can fertilize another one to a slight extent: fruit set between 2 and 10%;
3. cultivars that can fertilize another one to a medium extent: fruit set between 10 and 20%;
4. cultivars that can fertilize another one to a high extent: fruit set between 20 and 30%;
5. cultivars that can fertilize another one to a large extent: fruit set over 30%.

Soltész (1982a) formed the following three groups to characterize cross-fertilization in apple cultivars:

1. incompatible: fruit set of 0%,
2. partially compatible: fruit set of 0.1 to 10%,
3. compatible: fruit set over 10%.

Table 7.22. Fruit Set and Number of Viable Seeds Per Fruit in Pear (after Crane and Lewis 1942)

Crosses (mother ×father)	Final fruit set (%)	No. of viable seeds per fruit
2× self-fertilization	0.3	1.5
3× self-fertilization	0.2	0.1
4× self-fertilization	14.5	4.1
2× *2x	5.6	6.6
2× *3x	3.6	3.2
3× *2x	5.6	1.3
3× *3x	0.8	0.5
2× *4x	7.8	5.9
4× *2x	19.0	5.8

Note: 2x = diploid cultivar (2n = 34); 3x = triploid cultivar (3n = 51); 4x = tetraploid cultivar (4n = 68)

Vondracek (1962) also studied cross-pollination in apple. He thought it was impossible to characterize pollinizers according to their capability for making another cultivar fertile as acceptable, less acceptable, or unacceptable. A cultivar would be either able to fertilize another cultivar, or it is never able to do so.

Pear. Crane and Lewis (1942) reported fruit set percentages and the numbers of viable seeds in *pear* cross-combinations of different ploidities (Table 7.22). Diploid and triploid pear cultivars were practically self-sterile but tetraploid cultivars were self-fertile. The triploid × triploid combinations were fruitless since triploid cultivars were weak pollinizers. In diploid× triploid combinations, in spite of the low extent of fruit set, the mean seed number was more than twice of that in reciprocal crosses. The authors (Crane and Lewis 1942) attributed the low seed content of fruits of triploid× diploid crosses to the fact that there was no possibility for selection from female gametes. The triploid egg cell could be fertilized by the diploid pollen but the developed zygote would often be non-viable because of the unbalanced number of chromosomes in the egg cell. Hence, lots of fruits would be set but they would contain few viable seeds.

Chollet (1965) crossed pear cultivars and reported 13%, and 3% fruit development as a result of fertilization, and parthenocarpy, respectively. The yield resulting from crossing triploid cultivars did not differ from the yield that had resulted from vegetative parthenocarpy. Crossing diploid cultivars, however, resulted in fruit set values similar to those that open pollination resulted in. The author noticed no significant differences between the diploid cultivars as pollinizers.

Nyéki (1973a) points out that in certain pear cultivars capability for becoming fertile is better than capability for making fertile, and in other cultivars vice versa. The variance of pear cultivars' capability for becoming and making fertile may be quite large even in the case of cultivars that serve as good pollinizers to each other. It may happen in certain years that the extent of fruit set is very low or reaches 0%.

Evaluation of cross-combinations. Capability for becoming and making fertile in pear cultivars is evaluated by Nyéki (1973a) according to the following five categories:

1. very weak: mature fruit is below 1%,
2. weak: mature fruit is between 1.1 to 3.0%,

3. medium: mature fruit is between 3.1 to 5.0%,
4. good: mature fruit is between 5.1 to 10.0%,
5. very good: mature fruit is over 10%.

Misic et al. (1979) divided pear cultivars into three groups according to their capability for pollination:

1. good pollinizers: more than 10% of the flowers pollinated with their pollen have resulted in mature fruit;
2. sufficient pollinizers: produced 5 to 10% fruits;
3. weak pollinizers: the amount of resulting fruits was below 5%.

The number of seeds developed in apple and pear fruits. In the course of the evaluation of interfertility of different cultivars one should consider not only percentages of fruit set but the seed content of fruits.

The mean number of viable (completely developed germinative) seeds in a fruit is a value relevant to the quality of fertilization, that is, the degree of compatibility in cross-combinations of certain cultivars.

A classification of seeds:

1. completely developed, viable seeds,
2. aborted, non-viable seeds that are small, flat, "empty", degenerated, or incomplete "seed tube".

The small, incomplete, indolent seeds result from stenospermocarpic fruit development (Kozma 1963). The egg cell aborts before or after fertilization.

Modlibowska (1945) emphasizes that in examinations concerning the cross-pollination of *apple* and *pear* cultivars, seed content should be established early, before June drop, and the number of seeds should be compared to the number of embryo sacs.

Murneek (1954a,b) observed in *apple* that in the case of self-fertilization the mean seed content per fruit would be at three to five, whereas in the case of a good cross-pollination the same value would fall between five and eight but there would also be found fruits with ten seeds. When a tree has a large crop, fruits with less than three seeds usually abscise.

Kobel et al. (1939) called attention to the necessity that seed content is always to be specified in the description of crossing compatibility in diploid *pear* cultivars. The triploid cultivars produce seedless fruits in a higher percentage than the diploid cultivars.

According to Nagy (1960), the physiologically more active pollen results in more viable seeds in *pear*. The number of viable seeds is greater in cultivars easier to make fertile. The fruits, that contain more viable seeds due to better fertilization, are more resistant to nutritional disorders that may occur in the course of their growth than those with less viable seeds. The quantity and quality (viable or non-viable) of seeds in a fruit vary depending on pollinizers.

Nyéki (1977) demonstrated in *pear* that a weak fertilization (the rate of mature fruits below 3%) might be accompanied with many viable seeds (eight to ten per fruit), and a very good fertilization (the rate of mature fruits over 10%) may be accompanied with few viable seeds (0.5 to 1.3).

The number of seeds in *pear* refers to the combined sexual capability or sexual affinity of the pollen donor cultivar and the mother cultivar (Nyéki 1977).

Table 7.23. Good Pollinizer Cultivars of Quince

Cultivar	Recommended pollinizer	Year examined	Fruit set (%)	Author
Bereczki	Portugal	1971	20.3	Angelov (1975)
		1972	12.5	
	Vranja	1981	36.0	Aeppli (1984)
Champion	Vranja	1983	11.0	Aeppli (1984)
	Bereczki	1983	17.7	
Portugal	Bereczki	1971	31.7	Angelov (1975)
		1972	18.2	

Maliga (1966b) and Nyéki (1977) regarded a cultivar to be a better pollinizer if the pollen of this cultivar has resulted in higher yield, compared to other cultivars tested. If capability of two cultivars in producing yield is equal we regard the one whose pollen has resulted in fruits with a greater mean viable seed content as the better pollinizer.

Quince. In *quince*, only few experiments have been carried out to determine the appropriate pollinizers on the basis of values of cross-fertilization. Table 7.23 provides a summary of the cultivar combinations to be taken into account in the course of the arrangement of cultivars.

Maliga (1966b) states that cross-pollination in quince cultivars have increased fruit set percentages compared to self-pollination.

Out of the quince cultivars, 'Bereczki' (Duganova and Hrolikova 1977) and 'Moldovenesti' (Roman and Blaya 1984) are excellent pollinizers.

Sweet cherry. Fruit set groups have been formed in different cross-combinations in order to evaluate cross-fertilization. Table 7.24 summarizes the classification for fertility grades of six authors. Scrutinizing the table, one may conclude that different classifications are available in the literature. A comparison of the work of De Vries (1968) to that of Keulemans (1984) can serve as an illustration that the opinion of various authors on degree of satisfactory fruit set widely differ.

There has been no uniform method of evaluation to select the suitable sweet cherry cultivars to be pollinizers so far. Kamlah (1928a) thinks that a fruit set of 20 to 30% means a good yield. The same author reports the usual occurrence of fruit set about 50 to 60% as a result of artificial pollination, which has never been observed after open pollination. Kobel and Sachov (1929), and Krümmel (1933) came to the conclusion that there were only two kinds of cultivar combinations in sweet cherry to be distinguished: certain combinations were fertile (compatible) while other combinations were sterile (incompatible). Stösser (1966a,b) also thinks that there is no need for making a distinction between "good" and "poor" pollinizers in sweet cherry. Every cultivar that the author examined could serve as an acceptable pollinizer, that is, they could all result in a sufficient fruit set (12 to 15%). According to a summary by Brózik and Nyéki (in Nyéki 1980), those sweet cherry cultivars are to be regarded as good pollinizers that could result in a fruit set over 25%. Cociu et al. (1981) stated that a fruit set of 25 to 30% was required to achieve a good yield in sweet cherry. Finally, we mention Keulemans' (1984) observations, who also intended to select sweet cherry cultivars to be pollinizers. The extent of fruit set in different cultivar combinations depends on the pollinizers but large differences could be observed in different years. For instance, out of 18 cultivars under exami-

Table 7.24. Fruit Set Groups in Sweet Cherry After Cross-Pollination

No.	Fruit set (%)	Grouping	Author
1	0–1	no fertilization	
2	1–5	wrong fertilization	
3	5–10	weak fertilization	Maliga (1952)
4	10–20	medium fertilization	
5	20–30	good fertilization	
6	above 30	very good fertilization	
1	0	no yield	
2	1–2	poor yield	
3	3–10	weak yield	Brózik (1962)
4	11–15	medium yield	
5	16–25	good yield	
6	above 25	very good yield	
1	below 2	incompatible	
2	2–4	possibly incompatible	De Vries (1968)
3	4–6	possibly compatible	
4	above 6	compatible	
1	0	no yield	
2	0.1–1.9	poor yield	
3	2–4.9	weak yield	Brózik (1971)
4	5–9.9	medium yield	
5	above 10	good yield	
1	1	sterile combination	
2	20	wrong fertilization	Mihatsch and Schuman (1971)
3	30	good fertilization	
4	above 30	very good fertilization	
1	0–5	no fertilization	
2	6–25	wrong fertilization	
3	26–50	medium fertilization	Keulemans (1984)
4	51–75	good fertilization	
5	above 76	very good fertilization	

nation, the 'Blauwe Bigarreau' was the worst pollinizer in 1980 while two years later this cultivar was the best.

Sour cherry. Nyéki (1976a) and Brózik and Nyéki (1979) classified *sour cherry* pollinizers (Table 7.25). They suggested pollinizers that can yield a fruit set of 30% for self-sterile cultivars.

Now we mention a few essential observations available in the literature concerning cross-pollination in sour cherry.

1. The values of cross-fertilization of a given cultivar combination may vary depending on the growing area and the year (Brózik 1969, Nyéki 1989).

2. The fertilizing effect of a pollinizer changes from year to year. Nagy (1965) pointed out that when temperature changes the best pollinizer to a given cultivar also changes. Milovankic (1974) corroborates that a cultivar's capability for fertilizing depends on meteorological conditions, and this capability may change from year to year. As

Table 7.25. Fruit Set Groups in Sour Cherry After Cross-Pollination (after Nyéki 1976 and Brózik and Nyéki 1979)

No.	Fruit set (%)	Grouping
1	0	incompatibility
2	0.1–10	weak compatibility
3	10.1–20	medium compatibility
4	20.1–30	good compatibility
5	above 30	very good compatibility

was observed by Wociór (1976a-c), the self-sterile cultivars (e.g. 'Pándy meggy') would become fertile to a high extent in years when there is relatively low temperature during bloom.

3. The sour cherry is sensitive to pollinizers. Contrary to the sweet cherry, in the case of the sour cherry it is very important to know which cultivar is compatible with which one (Gozob et al. 1979). For increasing fruit set and improving fruit quality, cross-pollination is required even in self-fertile sour cherry cultivars.

Plum. a) European plum cultivars. The majority of *European plum* cultivars are self-sterile or partially self-fertile. The cultivars self-fertile to an insufficient extent require the selection of pollinizers. There are few data in the literature concerning the question as to which plum cultivars are "good" pollinizers. Roman and Radulescu (1986), for instance, stated that a fruit set between 10 and 20% could be achieved with the aid of "good" pollinizers (Table 7.26). Tóth (1966) observed that the best pollinizers of self-sterile and partially self-fertile plum cultivars usually belonged to the group of self-fertile cultivars.

According to the data available in the literature (Tóth 1967, Paunovic 1971, Faccioli and Regazzi 1972, Bellini and Bini 1978, Faccioli and Marangoni 1978, Keulemans 1980, Roman 1981, Stösser 1984, Belmans and Keulemans 1985), fertilization in European plum cultivars varies depending on cultivar combinations. 'Stanley' is usually a good pollinizer (Paunovic 1971, Faccioli and Regazzi 1972, Faccioli and Marangoni 1978). Even in the case of a fixed cultivar combination, however, fruit set percentages may vary across the years. Faccioli and Regazzi (1972) provides a good illustration: in the 'Tuleu gras' ×'Italian prune' cross-combination. The extent of fruit set in this combination was 7.8% in 1969, and 53.7% in 1970. Another excellent example of so great a variance (Nyéki 1989) is the case of the cross between 'Tsatsanska naibolia' and 'Tsatsanska lepotitsa' (6% in 1986 and 45.8% in 1985).

Nyéki (1989) summarized the crossing possibilities among European plum cultivars.

Table 7.26. Fruit Set Groups in European Plum after Cross-Pollination (after Roman and Radulescu 1986)

No.	Fruit set (%)	Grouping
1	under 10	weak pollinizer
2	10–20	good pollinizer
3	above 20	very good pollinizer

1. The European plum cultivars are to be divided into three clusters according to their demands for pollination:

   i) the following cultivars are to be planted in company with one or more pollinizers: 'Tsatsanska naibolia', 'President', 'Ruth Gerstetter';
   ii) the pollination and fertilization of the following cultivars are more certain in company with pollinizers: 'Tsatsanska lepotitsa', 'Bluefre', 'Stanley';
   iii) the following cultivars can be planted without pollinizers: 'Tsatsanska rodna', 'Besztercei'.

2. In the highly self-fertile European plum cultivars cross-pollination has a significantly increasing effect on fruit set percentages. If one aims to achieve a great yield (20 to 25 tons/hectare) then, one should plant the self-fertile plum cultivars also in the company of other cultivars within an orchard.

b) Oriental plums. Cross-fertilization in *oriental plum* cultivars has scarcely been studied so far. In Europe only Italian researchers have dealt with the topic (Bellini and Bini 1978, Cobianchi et al. 1978, Bellini et al. 1982, and Costa and Grandi 1982). In addition to widely planted cultivars bred in the U.S.A., they have studied also Italian cultivars ('Morettini 355', 'Sorriso di Primavera', 'Sangue di Drago', etc.).

In oriental plum cultivars the usual classification according to fruit set percentages consists of four classes (Table 7.27). It is somewhat surprising that certain cultivars have been qualified as good pollinizers at a fruit set below 10% (Cobianchi et al. 1978, Costa and Grandi 1982).

Nyéki (1989) concluded that in oriental plum cultivars:

1. there are large differences in fruit set in different cross-combinations;
2. a fixed cross-combination shows a large fluctuation of fruit set across years;
3. oriental plums are characterized by lower fruit set rates than European plums;
4. the Italian cultivar 'Sorriso di Primavera' is undoubtedly an excellent pollinizer.

Apricot. Few studies concerning the selection of pollinizers for self-sterile *apricot* cultivars are available in the literature. Schultz (1948) published data on crossing. Fruit set percentages show a fairly wide variation ranging from 0 to 61%. The self-sterile apricot cultivars 'Riland' and 'Perfection' are good pollinizers for other cultivars. The following apricot cultivars can serve as good pollinizers for 'Perfection': 'Blenheim',

Table 7.27. Fruit Set Groups in Oriental Plums after Cross-Pollination

No.	Fruit set (%)	Grouping	Author
1	under 1	no fertilization	
2	1–4	weak fertilization	Cobianchi et al. (1978)
3	4–10	good fertilization	
4	above 10	very good fertilization	
1	under 1	no fertilization	
2	1–5	weak fertilization	Costa and Grandi (1982)
3	5–10	good fertilization	
4	above 10	very good fertilization	

'Riland', 'Royal', 'Tilton', 'Wenatchee Moorpark' while 'Blenheim', 'Perfection', 'Royal', and 'Tilton' are good pollinizers for 'Riland'.

Nyéki (1989) have sought pollinizers for the self-sterile apricot cultivars with "óriás" (giant) fruit. Flowers of 'Ceglédi óriás' can be fertilized well by the pollen of several apricot cultivars ('Bergeron', 'Borsi-féle kései rózsa', 'Gönci magyar kajszi', 'Magyar kajszi C. 235') but neither of them can be used as an ideal pollinizer because the bloom period of 'Ceglédi óriás' precedes that of the cultivars tested. Thus, Nyéki (1989) could not offer a certain pollinizer for 'Ceglédi óriás'. For 'Szegedi mammut', 'Ligeti óriás', and 'Nagykőrösi óriás', 'Magyar kajszi C. 235' is an appropriate pollinizer. Crossing 'Szegedi mammut' with 'Rakovszky' is also successful.

Peach. Fruit set percentages in different *peach* cultivar combinations show a fairly wide variety (Branscheidt 1933). Cross-pollination increases these percentages compared to self-pollination (Maliga 1961).

In peach, as well as in apricot, few experiments concerning cross-pollination have been carried out and described in the literature. Generally, pollinizers are sought for male sterile peach cultivars. The self-sterile and male sterile 'J. H. Hale' peach cultivar, for instance, was examined by Tsukanova (1974). All the five cultivars tested were good pollinizers for 'J. H. Hale'.

Nyéki (1980) also studied the 'J. H. Hale' cultivar. All the pollinizers tested ('Sunbeam', 'Elberta', 'Champion', 'Sunhaven' and 'Redhaven') resulted in a fruit set over 10%. The best were 'Sunbeam' (26.1%) and 'Elberta' (25.9%)

Capability for becoming and making fertile in different cultivars. The capability of a cultivar for becoming fertile means the female receptivity, whereas the capability of a cultivar for making fertile means the capability of a cultivar for providing pollen. It is important to know the capability of a given cultivar for becoming and making fertile when choosing cultivars for arrangement in the orchard.

The perfect selection of pollinizers and the determination of their proportions and placing require the knowledge of female and male fertility.

The 'Müncheberger Frühernte' *sweet cherry* cultivar with its early blooming period is a poor male partner (pollinizer) and also a poor female partner (pollinated) in a cultivar combination (Mihatsch and Schumann 1971). The pollen of the 'Harlem' sweet cherry cultivar is characterized by a poor capacity for fertilizing in the majority of cross-combinations. 'Early Rivers' and 'Márki korai', two sweet cherry cultivars with an early bloom, however, are excellent pollinizers.

Brózik (in Brózik and Nyéki 1975) made a comparison between the mean fertility of sweet cherry cultivars. The results are summarized in Table 7.28. The conclusion was that certain cultivars were more fertile as male partners, whereas other cultivars were better as female partners. The following sweet cherry cultivars were characterized by a poor capability (fertilization rate below 5%) for becoming fertile: 'Szegedi óriás', 'Badacsonyi óriás', 'János cseresznye', 'Schneiders Späthe Knorpelkirsche'. A high rate (over 15%) of capacity for becoming fertile was characteristic of 'Lyana', 'Van', 'Bigarreau Burlat'. The following cultivars were excellent pollinizers: 'Gyöngyösi szívcseresznye', 'Bigarreau Moreau', 'Badacsonyi óriás', 'Pomázi hosszúszárú'. A similar experiment upon sweet cherry cultivars cultivated in Germany was carried out by Stösser (1979a,b). He observed greater differences than Brózik (1975) did between the capability for becoming fertile and the capability for making fertile in individual cultivars. He reported 'Hedelfinger Riesenkirsche' to be an excellent pollinizer; its mean capability for making fertile was 52.7% calculated on the basis of different cross-combinations with this cultivar as a male partner. Its capability for becoming fertile,

Table 7.28. Capability of Sweet Cherry Cultivars for Making and Becoming Fertile (1958 to 64) (adapted from Brózik 1975)

	Fruit set (%)	
Cultivar	becoming fertile (used as female)	making fertile (used as male)
Márki korai	10.2	12.6
Müncheberger Frühernte	14.2	8.3
Pomázi hosszúszárú	7.1	15.0
Jaboulay	9.8	9.8
Bigarreau Moreau	9.5	16.0
Bigarreau Burlat	20.3	8.9
Szomolyai fekete	13.5	10.5
Gyöngyösi szívcseresznye	11.7	18.9
Szegedi óriás	2.6	8.8
János cseresznye	4.6	6.9
Solymári gömbölyű	10.0	9.7
Lyana	15.9	9.2
Van	18.3	10.2
Schneiders Späthe Knorpelkirsche	5.0	12.7
Badacsonyi óriás	4.4	15.4
Hedelfinger Riesenkirsche	9.6	9.2

however, was also great (36.1%). 'Van', another sweet cherry cultivar, was also characterized by a great capability for becoming as well as making fertile.

Sour cherry cultivars have also been compared with each other both as male and as female partners in different cross-combinations. According to Hruby's (1963) examinations, 'Pándy' is a poor pollinizer whereas the self-fertile sour cherry cultivar, 'Schattenmorelle', is good as a male and as a female partner.

Data concerning *plum* and *gage* cultivars are published by Stösser (1983*a*) (Table 7.29). Among the plum cultivars tested, 'Opal', 'Ortenauer' and 'Zimmers' can be fertilized well. The following cultivars are good at making fertile, that is, they are appropriate pollinizers: 'Stanley', 'Ersinger', 'Ruth Gerstetter', 'President'. Stösser (1984) also compared the variation of capability for making fertile across years in different plum cultivars. He acknowledged 'Stanley' as the best pollinizer among the few sufficient polliniz-

Table 7.29. Capability of Plum Cultivars for Making and Becoming Fertile (after Stösser 1983*a*)

	Fruit set (%)	
Cultivar	becoming fertile (used as female)	making fertile (used as male)
Ersinger	19.0	37.8
Zimmers	43.2	12.4
Ortenauer	50.5	18.9
Stanley	27.4	50.6
President	14.8	32.0
Opal	51.2	34.9
Tiroler Zuckerzwetsche	30.9	36.8
Ruth Gerstetter	5.4	37.7

Table 7.30. Changes in Capability of Plum Cultivars for Making Fertile in Different Years (after Stösser 1984)

As pollinizer	1980 Fruit set (%)	1980 No. of crosses	1981 Fruit set (%)	1981 No. of crosses	1982 Fruit set (%)	1982 No. of crosses
Ersinger	32.8	6	31.6	4	37.8	9
Zimmers	20.9	5	8.4	2	12.4	6
Ortenauer	28.7	8	25.5	7	18.9	4
Stanley	51.3	8	31.2	8	50.6	9
President	39.3	6	19.5	6	32.0	9
Opal	38.2	4	29.8	5	34.9	10

ers. This cultivar resulted in a fruit set (31 to 51%) above the overall mean in all the three years of the experiment (Table 7.30). Capability for fertilizing in a given cultivar may vary from year to year. It may be high in a year and low in another year. As early as in 1929 Roh (1929a–c) observed that the fruit set of the pollinated (female) cultivar in a fixed cross-combination showed a wide variation from year to year. In certain years fruit set was so low that it bordered on the level of sterility. Roh (1929a-c) also pointed out than the capability of a cultivar for becoming fertile showed a greater variance than the capability of the same cultivar for making fertile.

Walnut. Szentiványi (1992) developed a method to regulate fertilization in *walnut*.

The self-regulation of reproduction in walnut includes the female flower abortion when a large number of pollen grains land on the stigma. In this case the stigma turns brown early, and the fruitlet falls. This stigma browning takes place in any walnut orchard, and the rate of flower abortion shows an increasing tendency parallel with an increase in the number of trees in the orchard.

This autoregulation of reproduction can be prevented by regulating the production and shedding of pollen during the period of female bloom.

If one intends to create a new walnut orchard, one should use cultivars with distinct periods of male and female bloom. To achieve the required low pollen density required in the period of female bloom, the proportion of pollinizer trees mixed with the trees of the mother cultivar in the orchard should not be higher than 2 to 3%.

Chestnut. According to (Solignat 1973), a cross-combination has been very successful if it produces more than 75 *chestnut* per 100 burs on the average. We can summarize Solignat's (1973) observations as follows:

1. with a given mother cultivar, the amount of yield can show a wide variation depending on pollinizers;
2. intercompatibility in chestnut cultivar combinations can be evaluated according to the following schedule (the number of fruits per 100 burs):

a) good: more than 100 chestnuts,
b) medium: 50 to 75 chestnuts,
c) weak: less than 20 chestnuts.

Hazelnut. The greatest percentage of fertilization in *hazelnut* cultivar combinations was reported to be 62.4% by Zielinski and Thompson (1967). According to Thompson (1971), fruit set rates in various cross-combinations fell between 49 and 59%. Modic

Table 7.31. Cross-Combinations of Black Currant Cultivars of Various Degrees of Self-fertility (after Porpáczy 1974)

Pollinated cultivars	Pollinizer cultivars			
	Parents of different fertility	Highly self-sterile	Weakly self-fertile	Highly self-fertile
Highly self-sterile (0 to 5 %)		varying, often wrong		good or very good
Weakly self-fertile (5 to 45 %)		varying, often wrong	varying, without extremities	varying, sometimes very good
Highly self-fertile (above 45 %)	good result, lower diversity			

(1974) offered five classes to evaluate the data concerning effectiveness of hazelnut crosses. The author stated that interfertility was sufficient for commercial purposes if in every year, or in the majority of years, 25 hazelnut fruits developed from 100 female flowers.

Strawberry. The suitability of *strawberry* cultivars to serve as pollinizers in cross-combinations was evaluated by Sheliahudin and Brózik (1966) in five categories. Cultivars that have made fertile another to a:

1. negligible extent: 1 to 20%,
2. slight extent: 20 to 40%,
3. medium extent: 40 to 60%,
4. good extent: 60 to 80%,
5. eminent extent: 80 to 100%.

Black currant. Smirnov (1974) distinguished three groups in order to evaluate *black currant* pollinizers:

1. good male partner (pollinizer): fruit set is over 45%,
2. sufficient pollinizer: fruit set is between 30.1 to 45%,
3. poor pollinizer: fruit set below 30%.

Porpáczy (1974) formed three groups to evaluate self-fertility in black currant (highly self-sterile, partially self-fertile, highly self-fertile), and crossed these types with each other. The results are in Table 7.31. The effect of cross-pollination in the different self-fertility groups was different.

7.3. Incompatibility and Inter-Incompatibility

Incompatibility is defined as a genetic mechanism that prevents angiosperms from self-fertilization and from fertilization between closely related plants.

Inter-incompatibility (reciprocal sterility, mutual incompatibility) means that the pollen of a cultivar is not able to germinate on the stigma of another cultivar, or cannot develop a pollen tube, or cannot penetrate into the style, and vice versa, nor can the pollen of the latter cultivar fertilize the egg cell of the former cultivar. Inter-incompatibility thus requires (at least) two cultivars.

Nowadays the expressions incompatibility and inter-incompatibility are used replacing the former self- and cross-sterility.

In the case of incompatibility, either the pollen does not germinate on the stigma or the pollen tube cannot penetrate into the style. If the pollen tube develops, it does not reach the embryo sac, excluding fertilization and seed development. In the case of incompatibility, the reproductive organs (pistil and pollen) are functional, they are not sterile, neither in morphological nor cytological aspects.

Partial incompatibility (semi-sterility), a further concept to be introduced here, occurs when the crossing partners bear a uniform sterility (S) allele, and the pollen that contains the common allele does not develop a pollen tube of full value on the diploid stigma.

Incompatibility groups. Two, or more than two, cultivars that do not fertilize each other make up an incompatibility group. The members do not fertilize each other reciprocally, that is, fruit set will take place regardless of whether the cultivar serves as the male or female partner.

Inter-incompatibility usually occurs in diploid fruit species, such as *sweet cherry, almond, oriental plum*. Whereas in polyploid species, such as *apple, pear* and *plums* of the *Prunus domestica* group, the inter-incompatibility is more varied, due to an interaction between S alleles. The more complex the polyploidy is, the less feasible is the occurrence of a complete incompatibility. As polyploidy increases, the possibility of meeting two cultivars with the same genetic makeup (which could constitute an incompatibility group) decreases.

During the course of examining incompatibility, the degree of relationship between the partners should be determined. The parents of a cultivar are usually unknown. The widespread use of a cultivar in breeding for developing new cultivars will result in the number of incompatible cultivar combinations because they are closely related.

Unidirectional incompatibility. Many researchers have reported cases when a crossing between a self-fertile and a self-sterile cultivar had been successful in one direction. If the self-fertile cultivar played the female role, crossing was successful, whereas the reciprocal crossing had usually failed. This phenomenon has called unidirectional incompatibility (or unidirectional crossability). Unidirectional incompatibility can occur also between two self-fertile cultivars or between two self-sterile cultivars (Abdalla and Hermsen 1972). Unidirectional incompatibility is usual in polyploid cultivars but rare in diploid ones (Crane and Lewis 1942). The occurrence of unidirectional or reciprocal inter-incompatibility is difficult to determine in triploid ×diploid apple or pear cultivar combinations because of the high extent of pollen sterility in triploid cultivars.

The occurrence of inter-incompatibility in temperate zone fruit species. Fruit cultivars can be classified according to their origin as follows:

1. derived from a controlled cross (both parents are known);
2. the mother, but not the male parent, is known;
3. both parents are unknown (random seedlings);
4. mutations (the spur types also belong to this class).

If one aims to study inter-incompatibility (in order to select pollinizers), one should determine the genetic and taxonomic relations between the cultivars under examination. The following questions should be answered:

1. Are the mutations of a cultivar compatible with each other?
2. Can the spur types and the standard initial cultivars fertilize each other reciprocally?
3. Can the hybrids fertilize their own parent cultivars?
4. Can the hybrids fertilize each other if (one of) their parents are the same?

In temperate zone fruit species, it is to be assumed that the cultivars within a taxonomic unit of a species are closely related. So are the hybrid cultivars and their parents closely related.
The following are to be regarded as close relations:

a) offsprings of a definite parent,
b) offsprings of two known parents,
c) mutation derived from a known cultivar.

The relationship between hybrid offsprings and parents is important and therefore one must consider relationships such as:

a) female parent × hybrid offspring,
b) male parent × hybrid offspring,
c) female parent × male parent,
d) crossing between sister offsprings.

The related cultivars can be classified according to their origin as follows:

a) one of the parents is shared,
b) both parents are shared (they come from the same family).

Some stone fruit cultivars will serve as an illustration of these classifications:

a) the following cultivar groups are considered closely related: 'Pándy meggy' types; apricots with "óriás" (giant) in their name; sour cherry cultivars 'Kántorjánosi', 'Újfehértói fürtös' and 'Debreceni bőtermő';
b) the cultivars that belong to the 'Besztercei' group are distinct in origin;
c) the group of 'Cigánymeggy' cultivars are fairly heterogeneous.

Taxonomically, these groups are to be regarded as cultivar groups (Terpó 1968).
Apple. Inter-incompatibility in *apple* cultivars occurs in few cultivar combinations. Examples of inter-incompatible crosses are as follows:

'Reine des Reinettes' × 'Peasgood's Nonsuch'
'Cox Pomona' × 'Peasgood's Nonsuch'
'Cox's Orange Pippin' × 'Kidd's Orange Red'
'Golden Delicious' × 'Maigold'.

There are gametophytically determined S alleles in apple (Speigel-Roy and Alston 1982). The place of incompatibility is in the stigma (Modlibowska 1945).
In ten apple cultivars 11 S genes were identified by Kobel et al. (1939), on the basis of which the authors concluded that there were 55 incompatibility groups in apple. Krapf

Table 7.32. Fertility of Cultivars of 'Golden Delicious' Group Different Cross-Combinations (Modified from Branzanti et al. 1978)

Female	Male									
	Golden Delicious		Stark Spur Golden Delicious		Stark Delicious		Open pollination		Mean of cultivars	
	Fruit set %	No. of seeds	Fruit set %	No. of seeds	Fruit set %	No. of seeds	Fruit set %	No. of seeds	Fruit set %	No. of seeds
C.D.A. Summerland 9 E 13–47	29.8	2.1	47.0	6.5	59.1	8.5	56.2	8.5	38.0	5.9
C.D.A. Summerland 10 C 18–33	21.7	10.5	18.0	10.7	20.3	10.0	25.6	10.2	17.7	9.9
Ed Gould Golden	9.9	1.7	64.8	7.5	68.6	8.7	65.8	8.8	44.8	5.5
Missouri A 2071	49.0	7.8	36.5	8.0	33.7	7.3	11.9	7.2	28.8	6.4
Ozark Gold	50.7	6.8	44.3	7.3	40.7	6.8	26.7	6.8	34.3	6.0
Virginia Gold	35.5	8.7	19.9	8.3	30.4	8.4	35.6	7.9	24.4	7.3
Golden Delicious	–	–	45.4	7.0	50.7	8.3	61.3	8.5	40.3	6.6
Stark Spur Golden Delicious	3.9	8.7	–	–	13.9	7.9	20.9	8.1	10.0	5.3
Mutsu	9.8	2.5	24.7	2.9	36.3	3.1	50.8	3.1	28.6	2.3
Prime Gold	40.1	8.3	50.0	7.7	32.9	6.3	46.9	7.0	35.5	6.9
Mean of pollinizers	27.8	5.7	38.9	7.3	38.1	7.5	39.3	7.6		

(1966), however, argued that there were no real sterility "groups" in apple. Kobel et al. (1939) observed more inter-incompatible cross-combinations between triploid apple cultivars and diploid pollinizers than between diploid cultivars.

Einset (1934) was the first to call attention to inter-incompatibility between a given cultivar and its bud mutations. Bud mutations do not fertilize each other (e.g. 'Starkrimson Delicious' × 'Redspur Delicious'). The mutation has not concerned the genes of incompatibility.

Usually mutations cannot be fertilized by the parent cultivar, with the exception of certain 'Golden Delicious' spurs, which can be fertilized by the parent plants (Sansavini and Bassi 1977).

Branzanti et al. (1978) studied cross-fertilization in 'Golden Delicious'-like apple cultivars; the results of crosses can be seen in Table 7.32. The 'Stark Spur Golden Delicious' could make certain 'Golden Delicious'-like cultivars fertile to such a high extent that it is hard to believe that they are closely related. The spur clone of 'Golden Delicious' is a better pollinizer than the standard 'Golden Delicious'. In the 'Ed Gould Golden' × 'Golden Delicious' cultivar combination the close relation manifested itself on the basis of the low percentage of fruit set. In the case of the triploid 'Mutsu' cultivar a high level of fruit set and a low level of seed number were observed, which may be attributed to meiotic abnormalities or disorders in the course of gamete development.

We concluded that the cultivars derived by bud mutation are inter-incompatible with the cultivars they are derived from, with a few exceptions. In respect of self- and cross-fertilization, their bahaviour is the same as that of the standard cultivars.

Incompatibility between closely related apple cultivars can appear (1) between parents and hybrids derived by crossing and (2) between hybrids coming from the same parents.

Nevertheless, inter-incompatibility may also appear between apple cultivars that are not relatives. 'Melrose', a hybrid of 'Jonathan' × 'Red Delicious', is inter-incompatible with both 'Jonathan' and 'Red Delicious' (Krapf 1966) as well as the cultivars that belong to the cultivar groups of the two parents (Hugard 1978).

Table 7.33. Compatibitity of Hybrid Cultivars with Their Parent Cultivars (after Deveronico and Marro 1980) [1,2]

Pollinated cultivar		Year	Jonathan	Golden Del.	Red Del.	Rome Beauty	Open poll.
				Parent			
				Fruit set %			
Maigold	Franc.Roseu × Golden Del.	1978 1979 1980		34 1 47	30 23 57		19 16
Jonagold	Jonathan × Golden Del.	1978 1979	19 27	22 24			10 25
Idared	Jonathan × Wagener	1978 1979	18 23				18 8
Jonnee	Blackjon mutation	1978 1979	7 3				8 18
Rubra precoce	Imperatore × Jonathan	1978 1979	38 15			43 11	10 12
Jonadel	Jonathan × Delicious	1978 1979	8 24		9 6		6 6
Monroe	Jonathan × Rome Beauty	1978 1979	41 38.5			48 29	29
Stark Redgold	Golden Del. × Richard	1978 1979		33 6	24 5		17 16
Chieftain	Jonathan × Red Delicious	1978 1979	10 19		10 3		8 7
Melrose	Jonathan × Red Delicious	1978 1979	17 13		13 19		5
Jerseymac	New Jersey × Julyred	1978 1979	64 29	24 21	27 19		17 12
Rome Beauty Spur		1978 1979				10 4	
Summerland	Kendal × Golden Del.	1978 1979		34			98 25
Edgould	Delicious seedling	1978			41		17

[1] Fruit set was evaluated 25 days after full bloom; [2] Number of viable seeds was not published.

Examples of inter-incompatible combinations with common parents are:

'Maigold' × 'Golden Delicious' (Krapf 1969, Lalatta et al. 1978),
'Golden Delicious' × 'Charden' (triploid) (Lalatta et al. 1978),
'Mutsu' (triploid) × 'Golden Delicious' (Lalatta et al. 1978).

'Golden Delicious' is closely related to 'Jonagold', the latter being a progeny of the former (the parents are 'Golden Delicious'× 'Jonathan'). 'Jonagold' is a triploid cultivar. Pollinated with the pollen of 'Golden Delicious', it provided a fruit set of 6%, whereas in the cultivar combination 'Jonagold' × 'Gloster 69', a fruit set of 49% was reported (Blasse 1986b). Examinations concerning fertility corroborate that there is an incompatibility between the parent cultivars. Cultivars having 'Golden Delicious' as one parent, such as 'Elstar', are not to be expected to serve as good pollinizers for 'Golden Delicious'.

It is not always true that the hybrid cultivar is inter-incompatible with its parents. 'Goro', for instance, which resulted from a cross between 'Golden Delicious' and 'Swiss Orange' ('Ontario' × 'Cox's Orange Pippin'), is compatible with both 'Golden Delicious' and 'Cox's Orange Pippin'. 'Idared' is compatible with its parent 'Jonathan' (Hugard 1978). The hybrid cultivar 'Gloster 69' ('Glockenapfel' × 'Richard Delicious') is reciprocally compatible with the parent cultivars (Saure 1975).

Now let us summarize Deveronico and Marro's (1980) observations on compatibility between hybrids and parents (Table 7.33). They could not find any incompatibility between the pollen of parents and the female gametophyte of the derived daughter cultivars. They used the case of 'Maigold' for example. The pollen of 'Golden Delicious' resulted in a good fruit set (34 and 47%) in this cultivar in two out of three years. However, results differ with this cultivar. Krapf (1969) reported that 'Maigold' and 'Golden Delicious' had not fertilized each other (reciprocally). Deveronico and Marro (1980) concluded, based on their experience with 14 combinations, that the combination of the apple cultivars with their parents would not cause problems in incompatibility.

According to Soltész's (1982a) experiments, certain closely related cultivars are able to fertilize each other to a certain extent. This includes 'Jonathan' × 'Éva', 'Jonathan' × 'Idared', 'Jonathan' × 'Fertődi téli' and 'Starking' × 'Chieftain'.

Pear. Osterwalder's (1910) experiments and observations are classic in this area. Flowers of the *pear* cultivar 'Bonne Louise d'Avranches' were pollinated with 'Bartlett' pollen. The author observed that the pollen tubes in the style stopped growing and their end became swollen, as in self-pollination. Fertilization had not taken place. Unfortunately, the author had drawn no conclusion from this phenomenon. Otherwise, he could have been the discoverer of inter-incompatibility groups in fruit species because 'Bonne Louise d'Avranches' and 'Bartlett' are inter-incompatible cultivars.

Table 7.34 is a summary of the inter-incompatible cultivar combinations available in the literature.

Table 7.34. Inter-Incompatible Cultivar Combinations in Pear

Cultivar combination	Author
Bergamotte d'Avranches × Bartlett	Osterwalder (1910)
Bartlett × Seckel	Marschal et al. (1929)
Bergamotte d'Avranches × Seckel	Johannson and Callmar (1936)
Belle Lucrative × Seckel	Johannson and Callmar (1936)
Fertility (4×) × Fertility (2×)	Crane and Lewis (1942)
Beurré d'Amanlis × Conference	Crane and Lewis (1942)
Alexandrine Douillard × Duc de Bordeaux	Chollet (1965)
Alexandrine Douillard × Epine du Mas	Gauthier (1974)
Olivier de Serres × Beurré Hardy	Nyéki (1977)

Table 7.35. Inter-Incompatible Cultivar Combinations in Quince

Cultivar combination	Author
Smirne × Vranja	Scaramuzzi (1951)
Smirne × Mostruosa di Bazine	
Czech × Portugal	Angelov (1966)
Anzherskaya × Izobilnaya Krimskaya	Ershov and Hrolikova (1970)
Azerbaidzhanskaya Grushevidnaya × Bereczki	
Nikitskaya Rannaya × Azerbaidzhanskaya Okruglaya	

The cultivars derived from bud mutation are reciprocally incompatible with the initial cultivars and the former do not differ from the latter in respect of compatibility:

'Bartlett' × 'Max Red Bartlett',
'Bartlett' × 'Dougs Red Bartlett'
'Bartlett' × 'Rosey',
'Clapp's favourite' × 'Starkrimson',
'Beurré Hardy' × 'Red Hardy'.

Quince. Fruitlessness because of inter-incompatibility is not characteristic of the *quince*. This phenomenon occurs only in a few cultivar combinations (Table 7.35).

Sweet cherry. Among fruit species, the incompatibility system of the *sweet cherry* is the best known. Inter-incompatibility is widespread in sweet cherry (Crane and Brown 1937). The S genes that determine incompatibility have a lot of alternative alleles: $S_1,...,S_x$ (Crane and Brown 1937, 1955). S genes are assumed to consist of two parts, one of which controls the activity of the pollen while the other controls that of the style (Lewis and Crowe 1954). The cultivars that bear the same S genes are incompatible.

Nine S alleles (S_1, S_2, ..., S_9) have been revealed in 66 sweet cherry cultivars (Crane and Lawrence 1929, 1931, Crane and Brown 1937, Lewis and Crowe 1954). A complementary fertility allele (S_f) has also been identified, which inactivates the preventive effect of the style (Lewis 1948, 1949). Incompatibility is caused by the sterility genes in sweet cherry. Six S alleles have been known so far (Brown 1955, Knight 1969). Fourteen incompatibility groups have been derived from different combinations of alleles S_1 to S_6. Matthews (1966), however, argued for the existence of 18 incompatibility groups in sweet cherry. Moreover, there is a Group 0, which consists of the sweet cherry cultivars compatible with every member of the above-mentioned 14 incompatibility groups. It is called the group of universal pollinizers. The incompatibility groups are denoted by the Roman numbers I to XIV. The S genes have been established in groups I to IX and XIII to XIV (Matthews and Dow 1969, Way 1968, 1974, Bradt et al. 1978, Trefois 1986). Table 7.36 serves as a summary of the sweet cherry incompatibility groups determined by different researchers. Their results show differences. The 'Vic' cultivar, for instance, has been included in Group XIII (Matthews and Dow 1969) as well as Group VII (Bradt et al. 1978, Trefois 1986).

Within an incompatibility group the cultivars are not able to fertilize each other (with the exception of Group 0). In contrast, two cultivars coming from different incompatibility groups can fertilize each other (reciprocally) to a sufficient degree.

In different countries with an intensive sweet cherry growing the inter-incompatible cultivar combinations have been determined, and the cultivars have been included in

Table 7.36. Incompatibility Groups in Sweet Cherry

Group	S alleles	Cultivar
I	S_1S_2	Black Tartarian (3), Early Rivers (3,4)
II	S_1S_3	Merton Bigarreau (2), Sodus (3), Van (1,2,3, 4), Venus (1,2,3,4), Windsor (2,3,4)
III	S_3S_4	Bigarreau Marmotte (1), Bigarreau Reverchon (1), Bing (1,2,3,4), Schneiders Späthe Knorpelkirsche (1), Napoleon (1,2,3,4), Lambert (1,2,3,4), Star (1,3,4), Vernon (1,3)
IV	S_2S_3	Sue (1,3,4), Victor (1,3,4), Velvet (1,3), Viva (2,3,4), Amber (3,4)
V	S_3S_5	N.Y. 1625 (4)
VI	S_3S_6	Gold (2,3)
VII	S_4S_5	Bigarreau Burlat (1), Bigarreau Moreau(1,4), Hedelfinger Riesenkirsche (1,2,3,4), Vic (3,4)
VIII	S_2S_5	Schmidt (2,3,4)
IX	S_1S_4	Chinook (1,2,4), Giant (3), Hudson (2,3,4), Rainer (2,3,4), Ursula (3)
X	S genes not known	Jaboulay (1)
XI	S genes not known	
XII	S genes not known	
XIII	S_2S_4	Ulster (1,2,3,4), Vic (1)?
XIV	S_1S_5	Valera (3,4)
O		Corum (2), Sam (2), Seneca (3), Stella (4), Vista (1,2,3,4), Vega (3,4)

Authors: (1) Matthews and Dow (1969); (2) Way (1974); (3) Bradt et al. (1978); (4) Trefois (1986)

incompatibility groups. Data are available in the following studies: England: Crane and Brown (1937, 1955), Matthews (1966), Matthews and Dow (1969); Switzerland: Kobel et al. (1938), Kobel (1954), Krapf (1976); Germany: Vahl (1965), Stösser (1966, 1979), Mattush (1968, 1970), Mihatsch and Schuman (1971), Koch (1979), Stösser and Neidhart (1975); the Netherlands: De Vries (1968); the former Soviet Union: Ryabov and Ryabova (1970); in the U.S.A: Way (1968, 1974), Fogle et al. (1973); in the former Yugoslavia: Stancevic (1971); France: Sanfourche (1972), Trefois (1986); and in Italy: Bargioni (1982). Kobel (1954) provided an evaluation of the relevant literature and a detailed summary especially on the examinations concerning fertility in sweet cherry carried out in the 1920s. The incompatibility groups published have a decreased value from the viewpoint of an international comparison because of the non-uniform cultivar names and the great number of lesser-known (mainly local) cultivars. In addition to the "old" sweet cherry cultivars, whose origin is not known, Ryabov and Ryabova (1970) examined a number of new sweet cherry cultivars, whose origin is known. They concluded on the basis of an analysis of the inter-incompatible cultivar combinations that inter-incompatibility was usually characteristic of closely related cultivar pairs, that is, of those which either shared a parent or came from the same family. Vahl (1965) had also studied fertility in 526 hybrid seedlings derived by crossing sweet cherry cultivars. Every offspring of cultivar 'Rube' was compatible with the mother cultivar while crossing between the sister seedlings of the progeny had always resulted in inter-incompatibility. Stösser and Neidhart (1975) studied sweet cherry cultivars used in the distilling industry. In these cultivars, which constitute a fairly heterogeneous group regarding origin, inter-incompatibility had been observed only exceptionally. Reviewing the literature, we have also found some contradictory data. Stancevic (1971), for instance, reported the 'Schneiders Späthe Knorpelkirsche' ×'Hedelfinger Riesenkirsche' cross-combination to result in a fruit set of 3.8% and the reciprocal combination to result in 0.5%. The author, however, qualified this cultivar combination as inter-incompatible.

In most sweet cherry cultivars in Hungary, it is not known which incompatibility group they belong to, that is, what their genetic structures are like. Maliga (1952) reported the following two cultivar combinations to be inter-incompatible: 'Schneiders Späthe Knorpelkirsche' ×'Olivet', and 'Badacsonyi óriás' ×'Olivet'. To ascertain the identity of cultivars in a combination is important. Brózik (personal information) stated that among the sweet cherry cultivars of Hungary, only the cross between 'Solymári gömbölyű' and 'János cseresznye', selected near Solymár, was inter-incompatible. Detailed examination of the origin of these cultivars and comparison of their characteristics revealed that they were the same cultivar under different names. Brózik and Nyéki (in Nyéki 1980) verified the following incompatibility groups in Hungary:

1. 'Márki korai' × 'Kelebia korai' × 'Hulin korai',
2. 'Schneiders Späthe Knorpelkirsche' × 'Magyar porc' × 'Olivet' × 'Badacsonyi óriás',
3. 'Jaboulay' × 'Ramon Oliva' × 'Nagypáli cseresznye'.

Observations of Nyéki (1989) are relevant, who had implemented a series of crossing experiments with sweet cherry cultivars in Hungary:

1. In cultivar combinations with one common allele and one distinctly different allele, very low fruit set percentages were observed (0 to 4.8%). The following cross-combinations were reported to be incompatible: 'Van' ×'Chinook', 'Venus' ×'Schneiders Späthe Knorpelkirsche', 'Bigarreau Reverchon' ×'Van', and 'Victor' ×'Van'.
2. In the cross-combination 'Schneiders Späthe Knorpelkirsche' ×'Sam', the pollinizer is thought (Way 1974) to belong to Group 0 according to incompatibility, that is, this cultivar is assumed to be compatible with any sweet cherry cultivar. In other words, 'Sam' is "officially" included in the group of the universal pollinizers (in the literature). Contrary to this, Way (1974) found a low fruit set in two years (3.5% in 1986, 0.4% in 1987) in this cross-combination.
3. In the 'Schneiders Späthe Knorpelkirsche' ×'Tardif de Vignola' cross-combination incompatibility could be observed. The author assumed that both alleles were common in these two cultivars. Hence, they were to be included in Group III according to incompatibility.

Some factors make it difficult to determine the incompatibility groups (Nyéki 1989). Cultivar names are often imprecise and non-uniform, there are a lot of synonyms especially in the case of old cultivars. For example, the cultivar 'Schneiders Späthe Knorpelkirsche', is the same as 'Magyar porc', 'Olivet', 'Badacsonyi óriás', 'Karmazsin' and 'Germersdorfi óriás'. Hence, it is difficult to identify a cultivar and distinguish it from others. The name of a cultivar may vary from country to country, from site to site. Often the same cultivar is cultivated under different names, making it impossible to compare the experimental results from different sites to each other. Furthermore, it requires much time and labour to identify S alleles. Schmidt (1982) pointed out that the only question that could be answered on the basis of an experiment with sweet cherry was whether a cultivar combination was sterile or fertile, and it could not be decided how many alleles the compatible partners shared, one or zero.

Sour cherry. In *sour cherry*, as a result of tetraploidy, the incompatibility mechanism rarely becomes apparent because of the complicated nuclear divisions. The S genes of sour cherry cultivars are unknown. Maliga (1944, 1953a) mentions an incompatibility group in sour cherry, which consists of 'Pándy' and 'Podbielszky', which is supposedly another 'Pándy'-like cultivar. Hruby (1963) reports reciprocal incompatibility in the cross-combination of 'De Oliver' ×'Brune de Bruxelles'. De Vries (1968) found an incompatibility cultivar group among three sour cherry cultivars, 'Rode Waalse', 'Montmorency' and 'Bruine Waalse'. Milovankic (1972) reported that 'Pándy' was inter-incompatible with sour cherry cultivars such as 'Spanyol üvegmeggy', 'Szentesi meggy', and 'Ostheimi meggy'. Furthermore, the 'Spanyol üvegmeggy' is incompatible with the 'Ostheimi meggy' (Milovankic 1974). Ubatshina et al. (1976) observed inter-incompatibility in cultivar combination 'Krassa severa' ×'Vladimirskaya'.

Plum. As was observed by Afify (1933), 'President' and 'Late Orange', two *plum* cultivars, were self-sterile and their crossing proved inter-incompatible. Flowers of a third cultivar, the 'Cambridge' gage, could be fertilized by the former two cultivars only to a slight extent (3%) but the reciprocal pollination had resulted in a sufficient yield in both cultivars.

Tóth (1975) described the following cultivar combinations as inter-incompatible:

'Lützelsachser' × 'Zimmer',
'Zöld ringló' × 'Italian Blue',
'Jefferson' × 'Ruth Gerstetter',
'Jefferson' × 'Tragedi'.

Bellini and Bini (1978) enumerate 11 inter-incompatible cultivar combinations in plum. We mention the 'President' ×'Italian prune' combination out of them. They underline that there is no inter-incompatibility between self-fertile plum cultivars, whereas this phenomenon often occurs in partially self-fertile and self-sterile cultivars. Finally, we mention two cultivar combinations whose inter-incompatibility has been demonstrated by Roman and Radulescu (1986): 'Pescarus' × 'Vision' and 'Ialomița' × 'Vision'.

Apricot. Inter-incompatibility is a rare phenomenon in *apricot* (Duhan 1944, Schultz 1948, Kostyna 1970).

According to the fertilization, values coming from an experiment with apricot cultivars with "óriás" (giant) in their name scarcely fertilized, or did not fertilize, each other producing 0 to 5.2% fruit set (Nyujtó et al. 1989).

Nyéki (1989) also pointed out that the self-sterile apricot cultivars with "óriás" in their name did not fertilize each other, hence they were to be included in the same in-

Table 7.37. Inter-Incompatible Cultivar Combinations in Almond

Cultivar combination	Author
Nonpareil × I.X.L.	Tufts (1919)
Nonpareil × Texas x Languedoc	Tufts and Philp (1922)
Harpareil × Jordanolo	Wood and Tufts (1938)
Coco Grado × Coco Mindo	Almeida (1949)
Nonpareil × Tardy Nonpareil	Kester (1966)
Cressey × Ballico	Kester (1966)
Texas × Ballico	Kester (1966)
Burbank seedling × Budatétényi 70	Brózik (1967)
I.X.L. × Profuse	Kester and Asay (1975)
Languedoc × Texas × Ballico	Kester and Asay (1975)
Reams × Jubilee	Kester and Asay (1975)
Rivers Nonpareil × Kutsch × Sultana × Bigelow	Kester and Asay (1975)
Smith XL. × Drake	Kester and Asay (1975)
Primorski × Yaltinski	Herrero et al. (1977)

compatibility group. The author observed also in orchards planted with combinations of the cultivars with "óriás" name that they did not increase each other's fruit set. He conjectured that these cultivars with "óriás" name and also with large fruit ('Ceglédi óriás', 'Nagykőrösi óriás', 'Szegedi mammut') had been derived from the same family, that is, their genetic makeup was the same.

Peach. According to Branscheidt (1933), reciprocal incompatibility has not been observed in *peach*. This is somewhat unexpected because the genetic base of peach is very narrow.

Reviewing the special literature on inter-incompatibility concerning stone fruits and carrying out some experiments, Nyéki (1989) concluded that in stone fruits (sweet cherry, sour cherry, plum, apricot) the self-sterile initial cultivars and the cultivars derived from them by bud mutation were inter-incompatible, because during the course of mutation several fruit- or tree-related features, such as way of growth, fruit colour, shape and maturation time changed but the S genes remained unchanged. Thus, in stone fruits as well as in apple and pear, making experiments concerning inter-incompatibility in such cross-combinations is superfluous. Some typical examples of cultivars derived from self-sterile cultivars by bud mutation are 'Santa Rosa' ×'Late Santa Rosa', 'President' ×'Late President', etc.

It is likely that the bud mutations of self-fertile cultivars (e.g. 'Stella' ×'Compact Stella', 'Redhaven' ×'Early Redhaven') can fertilize the initial cultivar.

Almond. Several reciprocally sterile cultivar groups are known. Table 7.37 serves as a summary of the inter-incompatible cultivar combinations.

Hazelnut. Thompson (1971) reviewed the potential pollinizers compatible or incompatible with 'Barcelona' *hazelnut* cultivar. Out of the 74 cultivars under examination, 47 had resulted in a sufficient fruit set in 'Barcelona', and with 24 male partners practically no fruit had been set. The fact that almost one-third of pollinizers have been incompatible with a distinguished hazelnut cultivar suggests that inter-incompatibility must be fairly widespread between hazelnut cultivars.

Romisondo (1978) also summarized the inter-incompatible cross-combinations between hazelnut cultivars. His results are given in Table 7.38.

Table 7.38. Inter-Incompatible Cultivar Combinations in Hazelnut (after Romisondo 1978)

Cultivar	Pollinizers
A Frutto Grosso	Cosford, Sodlinger
Barcelona	Avelline de Piemont, Avelline de Provance, Avelline Rouge Ronde, Blanche Ronde, Brixnut, Camponica, Clackmas, Comun Type A, Culplá, D'Alger, Emperor, Ennis, Fitzgerald 30, Gironell, Goodpasture, Grossal, Grosse Blanche, Haleski Oriaski, Kruse, Late Barcelona, Montebello, Rian, Ross, Tonda Romana, Tonda di Giffoni, White Barcelona
Braunschweiger	Lambert Filbert, White Lambert
Cosford	Bergeri, Daviana, Impératrice Eugenie
Daviana	Cosford, Longue d'Espagne, Impératrice Eugenie
Fertile de Coutard	Ennis, Grossal, Tonda di Giffoni
Gentile di Viterbo	Sodlinger
Grossal	Fertile de Coutard, Ross
Hallesche Rieselnuss	Cosford
Hempels Zellernuss	Langliche Rieselnuss, Meraviglia di Bollwiller
Impératrice Eugenie	Bergeri, Cosford, Daviana, Longue d'Espagne
Istrski Debelopodni Leski	Cosford, Sodlinger
Lambert Filbert	Braunschweiger, White Lambert
Lange Zellernuss	Sodlinger
Langliche Rieselnuss	Hempels Zellernuss, Meraviglia di Bollwiller
Lansing	Fertile de Coutard
Meraviglia di Bollwiler	Gunslebert, Hempels Zellernuss, Langliche Rieselnuss
Negret	Culplá, Grifoll, Negret primerenc
Pauetet	Culplá
Romische Zellernuss	Bandnuss, Sodlinger
Tonda Gentile Delle	Barcelona × T.G.L. (C 3/4), Locale di Romagna, Tonda di Giffoni
Trenet	Culplá, Ross

7.4 Parthenocarpy

In 1898, Müller-Thurgau observed in *pear* cultivar 'Joséphine de Malines' that fruits had developed from flowers whose pistil had partially frozen. Noll (1902) applied the expression "parthenocarpy" to seedless fruits that had developed without pollination or other external stimuli. Ewert (1907) was the first to observe parthenocarpic fruits and to report that this phenomenon was usual mainly in pear. The authors observed the development of long, thin and empty seed shells, from emasculated, isolated and unpollinated flowers which had turned brown by ripening.

In parthenocarpic fruit development the completely seedless fruit becomes mature without the ovule's having been fertilized.

Figure 7.4 illustrates the ways of possible parthenocarpic fruit set.

a) Natural parthenocarpy. Vegetative parthenocarpy means that the own stamens of flowers are taken out and both self- and cross-pollination are excluded.

In the case of induced parthenocarpy, some parthenocarpic fruit set takes place as a result of the stimulatory effect of the pollen deposited on the stigma. Fruit development thus is due to pollination but it is not due to fertilization. Induced parthenocarpy can be triggered by the own pollen of a flower or that of another flower of the same cultivar, some kind of pollen extract, aborted or sterile pollen, or the pollen of another species.

Two sorts of parthenocarpy, (1) obligatory or (2) optional, may occur in cultivars inclined to parthenocarpy.

In the case of obligatory (inherited) parthenocarpy, a cultivar always produces seedless fruits, whereas if a cultivar is inclined to optional parthenocarpy, it develops both seedy and parthenocarpic (seedless) fruits.

b) Artificial parthenocarpy. Artificial parthenocarpy means that fruit set has taken place due to the stimulating effect of external or internal factors, or chemicals. Some examples of external factors are high or low temperature during bloom, parasite injury or secretion. Other stimulatory factors include: bark ringing and wounding, removal of certain parts of stigma and style from emasculated flowers, a high hormone content of ovary, etc.

c) Stenospermocarpic fruit development. Stenosperm seed development is the development of a seed that was fertilized and began to grow but aborted as early as at the

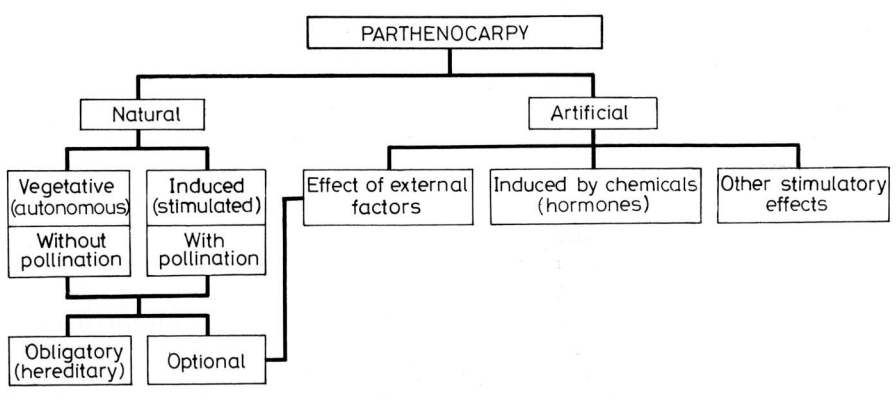

Fig. 7.4. Types of parthenocarpy (after Vazart 1955, modified)

beginning of its growth; a fruit that has developed this way is to be called a stenospermocarpic fruit (Kozma 1963). It is common in raisins and seedless grapes. In stenospermy, as a result of fertilization, or at least the penetration of the pollen tube into the ovule, fruit develops with non-viable, degenerated, small, and incomplete seeds. They are not viable (Kozma 1963). There are no germs in the seeds, and they contain only endosperms. The egg cell or the definitive nucleus aborts before or after fertilization. Thus, stenospermocarpy must not be confused with parthenocarpy.

Economic importance of parthenocarpy. The cultivars highly and regularly inclined to parthenocarpy are valuable for practical purposes and worth cultivating because they may bring good yields even in years when frost damages the flowers during spring, resulting in stigma abortion, or the weather is unfavourable during bloom for bee activity and pollination and/or fertilization do not take place. Parthenocarpic fruit set is a factor that helps to increase the expectable amount and certainty of yield.

Factors that induce or affect parthenocarpy. Gustafson (1942) observed that the ovary of cultivars inclined to parthenocarpic fruit set contained more auxin than that of fruits that had come into existence by regular fertilization. Gorter and Visser (1958) assumed that ovules in parthenocarpic fruits would abort later than in regularly fertilized fruitlets. Nevertheless, the co-ordinated activity of hormones in the course of fruit set has not been sufficiently revealed so far (Nyéki 1989). The problem is that techniques measuring hormones changed immensely during the last ten years and old data are questionable on the basis of incomplete methodology.

The extent of parthenocarpy is affected by several factors: (1) the nutritional state of the tree and its flowers, (2) the age of the tree, (3) differences between trees of the same cultivar, (4) the amount of yield per tree (rivalry with fruits set as a result of cross-pollination), (5) ecological features of the site (temperature), (6) the amount of flowers damaged by frost, etc.

The effect of temperature on parthenocarpic fruit set. Frost may induce parthenocarpy in *pear* (Lewis 1942, Rudloff and Schanderl 1950). Mittempergher and Roselli (1966) observed in pear cultivars that a large extent of fruit set had resulted from flowers damaged by frost without fertilization. Definitely high rates of parthenocarpic, mature fruits were reported in 'Sakesbirne': 85%, 'Beurré d'Amanlis': 61%, 'Marguerite Marillat': 52%, 'Goodale': 51%. In the majority of pear cultivars, however, parthenocarpy could not be observed. The following cultivars brought parthenocarpic fruits in lower percentages: 'dr. Jules Guyot' (22%), 'Bartlett' (16%), etc.

We can conclude on the basis of these data that it is not true for every pear cultivar that parthenocarpic fruits develop from flowers damaged by frost. The extent of parthenocarpy is mainly a cultivar-specific property in pear.

Natural parthenocarpy is not usual in *apple*. Lalatta (1982a), however, reported that there are apple cultivars, such as 'Reinette du Canada' that show an inclination to parthenocarpy, but this phenomenon is not of genetic nature, rather, it has something to do with stress, such as the damage of ovary caused by late frosts. Gorter and Visser (1958), and Karnatz (1960) failed to verify the statement of the authors mentioned above that spring frosts would induce parthenocarpy. Karnatz (1962) studied the fruit set of flowers damaged by frost in 'Précoce de Trévoux' pear cultivar. She has found a fruit set of 66% in healthy flowers, contrary to the damaged flowers, which had a fruit set of only 18%. Her experiments with *apple* and *pear* cultivars have questioned the positive effect of frost on parthenocarpy. Instead, she thought frost hindered parthenocarpy.

High values of temperature during bloom were reported to exert a positive effect on parthenocarpic fruit set (Breviglieri and Baldassari 1956, Schmadlak 1962, Nyéki 1974a). Nyéki (1974a), for instance, observed that temperature in the range of 20 to 25 °C during bloom had induced parthenocarpic fruit set in certain pear cultivars. In 1968, when the temperature was 10 to 15 °C, a negligible extent of parthenocarpic fruit set was observed (0.5%), whereas temperature values of 20 to 25 °C during the 1969 bloom had resulted in a parthenocarpic fruit set of 5.1%.

The survival of fruits set by parthenocarpy greatly depends on the meteorological conditions during the periods of May and June drops. Warm and wet weather in a period of fruit drop in California or South-Africa will result in regular and high parthenocarpic fruit set in 'Bartlett'.

'Bartlett' would set fruit in high percentages in solid cultivar orchards in California from year to year through parthenocarpy (Griggs and Iwakiri 1954). In the eastern states of U.S., however, this cultivar must be cross-pollinated for fruit set. Chollet (1965) verified that a high solar radiation during bloom stimulated vegetative parthenocarpy.

There is a strong nutritional and hormonal rivalry between parthenocarpic fruits and fruits set as a result of cross-pollination. In other words, the fruits of the same tree with different numbers of seeds compete with each other for water and nutrients. The fruits with a complete seed content are preferred in this competition to the seedless ones. The majority of parthenocarpic fruitlets fall at the time of June drop, or somewhat later but before ripening. Karnatz (1963) pointed out that rivalry was much more between parthenocarpic *apples* and apples with seeds, than between parthenocarpic and seeded *pear* fruits.

Inclination to vegetative parthenocarpy is held to be a cultivar-specific feature (Ewert 1907, Gorter and Visser 1958, Karnatz 1960, Maliga 1961, Chollet 1965, Nyéki 1974a, Smirnov 1974, etc.). It is an inherited capability of certain cultivars while other cultivars do not bear this capability. Some cultivars show the phenomenon of parthenocarpy regularly in every year, whereas others will never set fruit this way. A third group of cultivars is inclined to parthenocarpy only in certain years. In cultivars, inclined to parthenocarpy, the extent of fruit set shows a wide variability across the years (Gorter and Visser 1958). The 'Précoce de Trévoux' pear cultivar is an excellent example for this with a fruit set rate variability from 7 to 69%.

The inclination of fruit trees to natural parthenocarpy in various cultivars is determined by hereditary factors and a large year-dependent variation. Environmental factors, the nutritional state and the health of the trees, the type of rootstocks all facilitate the prevalence of inclination to parthenocarpy. The influences are cultivar- and year-dependent.

In pear the inclination to parthenocarpic fruit set is rather a genetic trait than a consequence of environmental factors (Stancevic 1970).

In the course of an examination concerning natural parthenocarpy, the following data are to be evaluated: (1) In what percentage does this phenomenon appear? (2) In which cultivars? (3) How frequently? (4) What is the fruit set rate variability with parthenocarpy?

Now let us discuss the correlation of inclination to parthenocarpy with self-fertility in different fruit cultivars. 'Klarapfel' *apple* cultivar was reported to be completely self-sterile and inclined to parthenocarpy at the same time (Karnatz 1963). 'Bath', another apple cultivar, was inclined to self-fertilization but not to parthenocarpic fruit set. Maliga (1961) pointed out that self-fertility was usually characterized by *pear* cultivars inclined to parthenocarpy. Soltész (1982a) has shown a correlation between parthenocarpy and self-fertility. These two characteristics are often concomitants in an apple cultivar. According to Smirnov (1974), the self-fertile *black currant* cultivars, and not the self-sterile ones, are inclined to parthenocarpy. Kobel (1954) stated that triploid cultivars were inclined to parthenocarpy to a higher extent than diploid cultivars. Karnatz (1960), however, observed that triploid apple and pear cultivars showed neither a stronger inclination to parthenocarpy nor a stronger inclination to self-fertilization than diploids did.

The market value of parthenocarpic fruits as goods. The data concerning the weight and size of fruits available in the literature are contradictory. Wellington et al. (1929) reported that parthenocarpic 'Bartlett' fruits were smaller than cross-pollinated ones. Stancevic (1970) concluded on the basis of experiments with pear cultivars that the parthenocarpic fruits were smaller in weight by 30% compared to fruits resulting from open pollination. The size of parthenocarpic fruits is often not uniform. Other authors observed no differences in weight between seeded and seedless fruits (Griggs et al. 1957, Chollet 1965). In trees where parthenocarpic pear fruits had been prevented competing with fruits coming from open pollination, their weight exceeded the commercial fruit weight characteristic of the given cultivar (Karnatz 1963).

In addition to fruit size, the number and distribution of seeds in apple or pear fruit show a correlation with the genetically determined fruit shape. The seeds will determine the shape of the fruit in an indirect way, due to their effect on the ovary (Marcucci and Visser 1983).

Seedless (parthenocarpic) 'Bartlett' fruits are often claimed to be cylindrical, lengthened and longish compared to seeded fruits (Wellington et al. 1929, Griggs et al. 1957, Mittempergher and Roselli 1966, etc.). Schanderl (1955) observed similar elongation in 'Conference' pear cultivar. In certain pear cultivars (e.g. 'Arabitka', 'Beurré Hardenpont', 'Olivier de Serres'), however, this kind of elongation could not be observed (Nyéki 1974b). The effect of parthenocarpy on fruit shape is cultivar-specific in pear (Chollet 1965). In certain cultivars seedless fruits did not differ from seeded ones while in other cultivars fruits had lengthened. Karnatz (1963) observed in seedless apples no change in shape, whereas among seedless pears along with those characteristic of the cultivar, she has also found cylindrical types.

The core of seedless pear fruits is smaller in diameter than that of those with seeds (Marcucci and Visser 1983). The diameter of the fruit had decreased by the diameter decrease of the core. Hence, parthenocarpic pear fruits were much longer than fruits coming from fertilization, containing seeds.

Acid content was observed to be higher in parthenocarpic fruits than in those set by open pollination (Karnatz 1960). According to Gardner et al. (1952), seedless fruits are characterized by a slower growth than seedy ones. The parthenocarpic fruits of pear cultivars with a harvest period during summer or fall mature 1 to 1.5 weeks later than the cross-pollinated seeded fruits (Griggs et al. 1957, Nyéki 1974b). Parthenocarpic apples and pears bear a commercial value only if their quality is similar to that of fruits coming from cross-pollination.

Inclination to natural parthenocarpy in different fruit species. Apple. Soltész (1982a) stated that parthenocarpy played a negligible role in *apple* fruit set. The author carried out experiments upon 41,000 flowers to study natural parthenocarpy in 53 apple cultivars. In 72% of apple cultivars parthenocarpy could not be observed. In the remaining 28% of cultivars the extent and occurrence of parthenocarpy were low and they were restricted to certain years.

The following cultivars have been reported to yield parthenocarpic fruits to a higher extent: 'Ontario' and 'Herrnhut' (Schmadlak 1962, Karnatz 1963), 'Reinette du Canada' (Lalatta et al. 1978a).

Pear. Natural parthenocarpy is usual in *pear*. It is easy to confuse this phenomenon with self-fertilization. Kobel (1939) observed parthenocarpy in 23 pear cultivars out of 41 cultivars he studied.

Pear cultivars are inclined to natural parthenocarpy to a greater extent and more often than apple cultivars are (Gorter and Visser 1958). Parthenocarpy in apple occurs rarely and in low percentages (Karnatz 1963), but Chollet (1965) estimated that parthenocarpic fruits develop from 6 to 8% of pear flowers.

Only a few pear cultivars are inclined to genetically determined natural parthenocarpy to a greater extent regularly in every year. Such cultivars are the 'Précoce de Trévoux' (Table 7.39) and 'Arabitka', a local cultivar occurring in Hungary (Table 7.40). These two cultivars are inclined to natural parthenocarpy to so high an extent and so regularly that they can also bring large yield without pollinizers in homogeneous or-

Table 7.39. Data Concerning Natural Parthenocarpic Fruit Set of 'Précoce de Trevoux' Pear Cultivar

Year	Parthenocarpic final set (%)	Author
1948	38.0	
1949	17.0	
1950	69.0	
1951	45.0	Gorter and Visser (1958)
1952	34.0	
1953	44.0	
1954	7.0	
1955	37.0	
.		Milovankic (1968)
.		
1962	5.5	
.		
1964	9.1	
1965	6.6	
1966	4.1	
1967	16.0	Stancevic (1970)
1969	24.0	
1970	13.3	

Table 7.40. Parthenocarpic Fruit Set of 'Arabitka' Local Pear Cultivar in Hungary

Year examined	Parthenocarpic final set (%)	Author
1954–56	21.0	Maliga (1961)
1955	51.0	Nagy (1960)
1969	34.6	Nyéki (1974*b*)
1970	21.9	Nyéki (1974*b*)

chards. The other pear cultivars are to be planted and cultivated in company with other pear cultivars in spite of their slight inclination to parthenocarpy.

There are pear cultivars in which a zero percentage of parthenocarpic fruit set has been observed. These are: 'Triomphe de Vienne' (in four years; Milovankic 1968), 'Le Lectier' (in three years; Stancevic 1970), 'Beurré Hardy' (in three years; Nyéki 1974). Data in literature concerning several pear cultivars are contradictory. 'Doyenné du Comice', a pear cultivar grown all over the world, is an excellent example. Stephen (1958), Callan and Lombard (1978), Mittempergher and Roselli (1966), and Bellini (1987) all took positions against its parthenocarpic nature, whereas others (Bini 1972, Drescher and Engel 1978) provided evidence for its low inclination to parthenocarpy. Crane and Lewis (1942) observed that 'Doyenné du Comice' would produce parthenocarpic fruits pollinated by the pollen of 'Beurré Bedfort'.

There are different classifications of pear cultivars in the literature according to the extent of their natural parthenocarpy. Stancevic (1970) has included the cultivars in six groups: (1) cultivars that are not inclined to parthenocarpy (0%), (2) inclined to it to a negligible extent (1%), (3) to a slight extent (1 to 1.5%), (4) definitely inclined to parthenocarpy (5.1 to 10%), (5) inclined to it to a high extent (10.1 to 15%) and (6) to a very high extent (over 15%). Bellini (1987) has formed three groups in order to evaluate pear cultivars: (1) parthenocarpy is fairly usual, (2) parthenocarpy is partial, (3) parthenocarpy is rare, or there is no parthenocarpy. Table 7.41 demonstrates Nyéki's (1974b) six groups.

Nyéki (1973a) has also analysed the genetic determination and year-dependent variation of inclination to natural parthenocarpy in various pear cultivars. He has distinguished four categories: (1) cultivars with no genetic inclination to parthenocarpy, (2) cultivars that abscise parthenocarpic fruitlets during the course of the drop periods, (3) those that are inclined to parthenocarpy only in certain years, and (4) those with regular parthenocarpic tendency in every year (Table 7.42).

Quince. Natural parthenocarpy has never been observed in *quince* so far. Ewert (1929) and Maliga (1966b) pointed out that quince is not inclined to parthenocarpy at all. No fruit has ever developed from emasculated, isolated, unpollinated flowers.

Sweet cherry. Natural parthenocarpy has never been observed in *sweet cherry* so far (Kamlah 1928a,b, Zwitzscher 1962). Integuments are often empty in seeds of early ripening cultivars because of the early degeneration of embryos.

Sour cherry. Nyéki (1973a) emasculated and isolated 9917 flowers in ten 'Pándy' types. No fruit had resulted from the emasculated flowers. He concluded that there was no fruit set by natural parthenocarpy in 'Pándy'.

Table 7.41. Grouping of Pear Cultivars on the Basis of Natural Parthenocarpy (Modified from Nyéki 1974b)

None	Very weak	Weak	Medium	Strong	Very strong
(0)	(0.1-1.0)	(1.1-5)	(5.1-10)	(10.1-20)	(above 20)
Clairgeau, Decaisne Henrik, Beurré Hardy, Curé	Beurré Diel, Bartlett	Bosc Flaschenbirne, Madame Du Puis, Beurré Hardenpont, Olivier de Serres, Doyenné d'hiver	Clapp's favourite, Pringall	Passe Crassane	Arabitka

Parthenocarpic groups — Final fruit set (%)

Table 7.42. Characteristics of Parthenocarpy in Pear Cultivars (Modified from Nyéki 1973)

Genetically not parthenocarpic	Parthenocarpic fruitlets abscise during fruit drop periods	Yearly irregular tendency to parthenocarpy	Yearly regular parthenocarpic fruit set	Strong genetic tendency to yearly regular parthenocarpic fruit set
Curé, Decaisne Henrik	Clairgeau, Beurré Hardy	Bosc Flaschenbirne, Clapp's favourite, Beurré Diel, Madame Du Puis, Olivier de Serres, Bartlett	Beurré Hardenpont, Passe Crassane, Pringall, Doyenné d'hiver	Arabitka

Plum. Natural parthenocarpy has never been observed in *plum* so far (Bellini and Bini 1978, Keulemans 1980).

Apricot. Natural parthenocarpy has never been described in the relevant literature so far (Nyéki 1989).

Peach. Natural parthenocarpic fruit set between 22 and 42% was observed in 'Shen chon' cultivar (Shing and Feng 1936). A bit lower (14 to 21%) extent of parthenocarpic fruit set was reported in another peach cultivar 'J. H. Hale' (Bulatovic 1961). Parthenocarpic fruits could grow mature only when well supplied with water and nutrients (nitrogen especially). Nyéki (1989) could not find natural parthenocarpic fruit set in peach and nectarine cultivars.

To sum up, stone fruits are not characterized by natural parthenocarpy. Occurrence of parthenocarpy is regarded to be as exceptional in stone fruit species and if it occurs its rate is very low. In early ripening sweet cherry and peach cultivars the early abortion of embryos, or the abortion of embryos in young fruitlets at a low temperature.

Walnut. In certain years, when meteorological conditions are unfavourable during bloom, a small amount of fruits can be found that are free from germ and cotyledons (Szentiványi 1980*a*). Parthenocarpic walnuts are often small so they are easy to separate from normally fertilized walnuts. Parthenocarpic set in walnut is of no practical importance because we grow walnut for its seed, for the walnut cotyledons. Therefore, parthenocarpic set produces no marketable product.

Chestnut. The removal of catkins and the isolation of stigmas before their receptive period often result in small burs containing small and aborted chestnuts (Breviglieri 1951). When fruits are harvested, these small aborted burs are deformed, soft, empty, and they soon dry.

Parthenocarpy is unusual in chestnut.

Almond. Pejovics (1966) observed parthenocarpy induced by pollen in *almond* cultivar 'No. 61/6'. Brózik (1980*a*) reported parthenocarpic fruit set in seedlings 'No. 2/4' and '2/11'. Parthenocarpic (seedless) fruit development in almond growing is disadvantageous. Similar to walnut, in almond the seed is the marketable product. Occurring rarely, parthenocarpy plays a negligible role in almond growing.

Strawberry. Sheliahudin and Brózik (1965) observed parthenocarpic fruit set in some *strawberry* cultivars. Receptacles may develop when strawberry achenes are parthenocarpic but the fruits developed this way are small and have no commercial value (Thompson 1963).

Raspberry. Sheliahudin and Brózik (1967) reported a slight extent (1 to 2%) of parthenocarpic fruit set in some *raspberry* cultivars. Fruit development without fertilization, however, plays no significant role in raspberry cultivars.

Table 7.43. Characterization of Fruit Species Based on Natural Parthenocarpic Fruit Set

	Natural parthenocarpic fruit set		
Usual	Not regular, low level	Rare	No occurrence
pear[1]	apple	peach walnut[2] chestnut[2] almond[2] strawberry raspberry gooseberry black currant[3]	quince sweet cherry sour cherry plum nectarines apricot

[1] Some cultivars are genetically inclined to parthenocarpy and develop parthenocarpic fruits in high percentage; [2] Parthenocarpy is harmful, seedless nuts have no market value; [3] Natural parthenocarpy was only observed in self-fertile cultivars.

Gooseberry. Parthenocarpic fruit development was observed in 'Piros győztes' gooseberry cultivar (Tóth and Tóthné 1959).

Black currant. This species is inclined to parthenocarpic fruit development. In cultivars that are highly self-fertile, parthenocarpic fruit development could often be observed. The percentage of parthenocarpic fruit development in black currant showed an extremely wide variability from year to year. 'Golubka' ranged from 0 to 16%, 'Pamiat Mitshurina' from 0 to 49%).

Classification of temperate zone fruit species according to their inclination to natural parthenocarpy. Inclination to natural parthenocarpy is species-dependent. Table 7.43 provides a classification of fruit species according to their inclination to natural parthenocarpy. In *pear*, parthenocarpic fruit development is usual and its extent is definitely high in certain cultivars. Hence, this phenomenon plays a role also in the production of pears. In *apple*, parthenocarpy occurs rarely and in low percentages. In *stone fruits* there is no natural parthenocarpy. In *nuts* parthenocarpy is detrimental, seedless nuts are worthless as marketable product. In *small fruits* the parthenocarpic way of fruit development occurs rarely and in low percentages.

7.5. Research Methodology

Fletcher (1908) was the first to work out methods of pollination and crossing and thoroughly described them in the U.S.A. In Europe, Ewert (1906, 1907, 1909) dealt with phytobiological methodological research in Germany, and Osterwalder (1909, 1910) in Switzerland. Methods of pollen collection, emasculation and pollination were reviewed by Barrett and Ariaumi (1952). A detailed description of examinations concerning fertility with an evaluation of the methods is available in a study by Manaresi (1953). Recently, Layne (1983) has provided a summary on novel methods with a special regard to breeding (hybridization).

Reviewing the literature, one may conclude that methods have been worked out for two purposes: (1) they intended to elucidate conditions of fertilization in different cultivars, or (2) to produce new hybrids.

Stages of pollination (crossing). Pollen collection and storage. The time of pollen collection is determined by the degree of maturity of pollen. Anthers collected too early ('immature') will not dehisce, or provide little pollen. They usually parch without having dehisced. The quantity and quality of pollen depend on fruit species and cultivars, weather and part of the day (Stanley and Linskens 1974). Pollen is to be collected from buds about to bloom and from closed anthers. It can also be collected from isolated or opened flowers. Mature pollen can be obtained only from flowers that have already opened. The forced "young" pollen comes from the forced flowers on cut branches. At bud swelling some 1 to 1.5 m long branches are cut, put in water, and forced at 20 to 25 °C. The branches are to be collected from several trees of the same cultivar.

Barrett and Ariaumi (1952) worked out a fast method of pollen collection. The anthers are to be rubbed against a mesh with tiny holes (1.5 to 2 mm.). Anthers fall through the mesh tissue onto clear paper trays. Anthers can be extracted from flowers or buds also with tweezers. The paper trays must be dry, clean and white or can be substituted for Petri dishes. It is important to record the time of collection and the name of the cultivar. Then the pollen gathered is dried in a dry and dark or shaded place at room temperature (20 to 25 °C) spread thinly for 24 to 28 hours. During this period the anthers dehisce and shed their pollen from the pollen sacs. Then the pollen is to be cleared from tehe remains of anthers. Finally, pollen is stored in glass ampoules loosely closed with cotton or screw cap.

Pollen storage methods were reviewed by Visser (1955), King (1965), Stanley and Linskens (1974). In a desiccator the pollen in closed glass ampoules is placed above calcium chloride or $CaSO_4$ or other desiccators. The pollen of most fruit species remains viable when relative vapour content is between 0 and 30% (Galletta 1983). The pollen is stored in refrigerators at 0 to 2 °C for a few weeks.

Washing the hand and the tools, 70% ethyl alcohol can be used to clean pollen of staining and to avoid contamination or mixing on the tools.

Flower isolation. The purpose of isolation is to prevent unintended and uncontrollable cross-pollination by insects, or the wind. Flowers are to be enclosed with isolators before flower bud opening and anther dehiscence. Flowers of completely self-sterile cultivars are isolated without emasculation in the balloon stage. Flowers about to bloom or opened are removed when isolating the branch. In non-self-sterile cultivars unopened flowers are emasculated in order to exclude an unintended self-pollination. After pollination flowers are isolated again.

The ways of isolation: (1) the whole tree is isolated with an isolator cage, (2) bags are used to exclude cross-pollination (parchment, cheese cloth, cellophane), (3) vaseline is applied. The size of isolator bags falls between 15 to 25 cm × 15 to 40 cm. In apple, pear, and stone fruits, isolators 25 × 35 cm in size are widespread. In chestnut one to three staminate and three to 15 pistillate flowers are closed in a bag (Breviglieri 1951). Too large bags are disadvantageous because they easily get damaged. Rains accompanied by strong wind, for instance, may result in their breaking. The isolators are to be placed on the tree as follows: at a medium height in the canopy, uniformly distributed towards the four cardinal points on the same branch. The ends of limbs with flowers are to be isolated. The bags are to be fastened to limbs with fine wires or strings. Every isolator is associated with a number and a label, which contains also the number of pollinated flowers. After the bloom of the tree, when the stigmas in the isolators have turned brown and dried, the isolators are to be removed carefully but small "collars" are to be left for the purpose of better perceivability.

Isolation with parchment bag. The isolator is a white, water-resistant, translucent bag. It is quite resistant to unfavourable weather. The bag protects flowers from external

Table 7.44. Effect of Isolation Type on Fruit Set of Sour Cherry Cultivars in the Case of Natural Self-Pollination (Autogamy) (after Nyéki 1975)

Cultivar	Year examined	Parchment bag No. of flowers isolated	Parchment bag Fruit set (%)	Cheese cloth bag No. of flowers isolated	Cheese cloth bag Fruit set (%)
Pándy clone 48	1972	958	0.0	631	0.2
	1973	771	0.0	769	3.3
	1974	861	0.0	627	2.6
Pándy clone 10–1	1972	1514	0.0	371	0.0
	1973	972	0.0	562	2.6
	1974	804	0.0	421	1.4
Pándy clone 279	1972	1279	0.0	393	2.0
	1973	1456	0.0	432	0.2
	1974	975	0.0	929	1.9
Schattenmorelle	1972	738	29.0	211	40.2
	1973	453	12.8	404	18.9
	1974	94	13.8	327	21.1
Röhrings Weischel	1972	576	15.8	652	26.0
	1973	359	30.4	531	35.5
	1974	540	40.4	766	43.5

severe meteorological factors (rain, fog, wind). The bags may break only if they have been soaked with rainwater as a result of long rainy periods and shaken by stormy winds. When the wind shakes the bags, the pollen scatters within the bag. The bags are to be placed on limbs of the canopy far enough from each other to exclude the risk of the breaking of bags as a consequence of their collision.

Manaresi (1953) argues that an industrious and patient worker is able to isolate 600 flowers daily.

Isolation with cheese cloth. Tóth and Tóthné (1959) demonstrated in *gooseberry* that isolation with cheese cloth made a low extent of cross-pollination possible, hence it was not suitable for isolation. Cheese cloth (gauze) isolation was used in the course of examinations concerning fertility in *apricot* by Kostyna (1970), in sweet cherry by Ryabov and Ryabova (1970), and in sour cherry by Ryabov and Ryabova (1970), Enikeev (1973), and Wociór (1976a). Table 7.44 is a summary of Nyéki's (1975) observations on the effect of different ways of isolation on fruit set in *sour cherry* cultivars. Natural self-fertilization (autogamy) in the completely self-sterile 'Pándy' types ('Pándy 48', 'Pándy 10-1', 'Pándy 279') fell between 0 and 3.3% in cheese cloth bags, which clearly shows that the cheese cloth isolator could not prevent cross-pollination from taking place. Thus, its use is not recommended. Also in self-fertile sour cherry cultivars ('Schattenmorelle', 'Röhrings Weischel'), greater fruit set percentages were observed in the case of the use of cheese cloth than in the case of parchment bags. Examinations in *pear* resulted in similar conclusions (Nyéki 1974c). The pollen of insect-pollinated fruit species, transmitted by the wind, could penetrate the cheese cloth (gauze) bags with holes, 2 mm in diameter, to a very low extent, or alien pollen could fall to the flowers from other parts of the tree due to gravitation. Bees may also visit the cheese cloth bags and shed alien pollen on stigmas (Nyéki 1975).

Isolation with cellophane pouches. Tóth and Tóthné (1959) observed in *gooseberry* that the air temperature may increase in cellophane pouches to a high value that damages flowers. In cool and wet weather, the flowers wounded by emasculation will abort in a high percentage.

Cellophane was not suited for isolation in *stone fruits*. As a result of the glare of the sun and high temperature in the bag, flowers would "burn up" and dry out (Nyéki 1974*b*). Nyéki (1974*b*) demonstrated in *sour cherry* that stigmas were characterized by the shortest viability (24 to 48 hours) in cellophane pouches, the emasculated flowers dried out in a day.

Isolator effect. Isolation may affect pollination and fruit set. In certain cases it is difficult to find out whether a change in fruit set rate is due to isolation itself or pollination. Isolation may exert a negative effect upon flowers compared to the state of free flowers. As early as in 1906, Ewert observed that flowers closed in bags received lower light, and the air inside was warmer and wetter. Microclimatic conditions under isolators are different from the state of free flowers. Isolators affect stigma receptivity, pollen emission, the chance of pollen grains for falling on stigma, flower development, pollen germination, pollen tube growth, etc. In warm and sunny weather, for instance, very high temperature values (30 to 40 °C) were measured in parchment isolators (Nyéki 1974*b*), and the flowers had been "cooked" under isolators. Tamássy and Nyéki (1976) observed in sour cherry that during the night and at dawn the temperature was 3.5 to 5 °C lower in parchment isolators than around free flowers. Sour cherry flowers in parchment isolators had frozen at a higher rate, approximately by 20 to 25%. Isolated flowers received less light than free flowers. For lack of air motion, it is warmer by day and cooler by night than in the open air, and vapour content is also higher because of the transpiration of leaves than in other parts of the canopy. Nyéki (1989) pointed out that secretion extraction began sooner on stigmas of flowers under warmer circumstances in isolators, and this process was faster and more intensive than in the case of free flowers. Stigma senescence and anther dehiscence also began sooner and took place faster. Redalen (1981) carried out detailed examinations concerning the effect of different methods of isolation on fruit set in *raspberry* (Table 7.45). Higher fruit set rates were observed when paper bags had been used. The temperature and relative humidity in paper bags were higher than in gauze cages and in the open air. Flowers in gauze cages were under circumstances similar to those of non-isolated flowers. In paperbags the daily variation of relative vapour content decreased so the surroundings of flowers were wetter. In the upper part of bags it was warmer by day and cooler by night. Light rays were absorbed by brown paper to a higher extent than by white paper. Consequently, white paper bags were recommended by Redalen (1981). Unfortunately, paper bags might easily break, especially in wet, windy weather.

Emasculation. Emasculation is a procedure in the course of which bisexual flowers are deprived of stamens before anther dehiscence to prevent self-pollination. Flowers of

Table 7.45. Effect of Isolation Type on Fruit Set and Conditions Around the Flowers in Raspberry Cultivars (after Redalen 1981)

Isolation type	Fruit set (%)	Set of single fruits (%)	Temperature (average of 12 days) (°C)	Relative humidity (average of 6 days) (%)
Without isolation	67	21	15.6	58
Gauze	53	27	16.3	60
White paper bag	82	33	18.4	74
Brown paper bag	86	61	17.3	70

cultivars not completely self-sterile are to be emasculated at the full balloon stage before isolation. In warm (20 to 25 °C), dry and windy weather anther dehiscence may precede petal opening. Such case was observed in certain years in *apricot* by emission, the chance of pollen grains for falling on stigma, flower development, pollen germination, pollen tube growth, etc. In warm and sunny weather, for instance, very high temperature values (30 to 40 °C) were measured in parchment isolators (Nyéki 1974*b*), and the flowers had been "cooked" under isolators. Tamássy and Nyéki (1976) observed in sour cherry that during the night and at dawn the temperature was 3.5 to 5 °C lower in parchment isolators than around free flowers. Sour cherry flowers in parchment isolators had frozen at a higher rate, approximately by 20 to 25 %. Isolated flowers received less light than free flowers. For lack of air motion, it is warmer by day and cooler by night than in the open air, and vapour content is also higher because of the transpiration of leaves than in other parts of the canopy. Nyéki (1989) pointed out that secretion extraction began sooner on stigmas of flowers under warmer circumstances in isolators, and this process was faster and more intensive than in the case of free flowers. Stigma senescence and anther dehiscence also began sooner and took place faster. Redalen (1981) carried out detailed examinations concerning the effect of different Brózik et al. (1978). Before emasculation every flower that is about to bloom or has opened must be removed. Tools of emasculation are tweezers, scalpels, or small scissors (cut depth can be controlled). Emasculation requires much care and manual labour.

Ways of emasculation by hand.

1. *Anther emasculation*: anthers (and often stamens) are removed, but petals are left. In *apple*, *pear*, and *quince* anther emasculation is usually executed in Europe but not in the U.S.

2. *Radical emasculation*: anthers, as well as corollas, petals, and sepals, are removed. *Plum* is susceptible to this method because of the serious wounds caused by it. Nevertheless, radical emasculation is a simpler, faster, and more reliable method than anther emasculation. The former method has become widespread in stone fruits, for instance. *Sour cherry* flowers, however, are very sensitive to radical emasculation according to Cociu (1961) and Way (1968). *Apricot* flowers are more difficult to emasculate than

Table 7.46. Effect of Emasculation on Fruit Set of Gooseberry Cultivars (Flowers Were Pollinated with Their Own Pollen) (after Kirtvaya 1965)

Cultivar	Fruit set (%) without emasculation	with emasculation
Finik	73	38.8
Zeleni butilotshni	86.5	35
Varshavski	34	12
Jubileyni	82	35
Russki	35	23
Avenarius	55	50

those of other *Prunus* species (Layne 1983). Many flowers break in the course of emasculation.

According to an observation of Visser (1951), radically emasculated flowers are rarely visited by bees. Kirtvaya (1965) reported in *gooseberry* that the rate of fruit set had become significantly lower as a result of emasculation (Table 7.46). This decrease could be attributed to physical injury. The perianth being removed, the stigmas remain overt, evaporate much water, and dry out. Kamlah (1928a,b) pollinated emasculated flowers two to three days later, whereas Layne (1983) pollinated emasculated flowers on the same day.

Field experimentations concerning fertility.

Experiment design. Trees of the same cultivar and same age that are in good condition are to be chosen. The mean number of trees to be examined that belong to a cultivar is between five and ten. The treatments within a tree are to be arranged randomly and to be repeated at least four times. Very bendable limbs are to be avoided as the wind may move them so strongly that the isolators on them get damaged.

Pollination. At pollination both stigmas and pollen must be mature. According to the age of flowers, there are different sorts of pollination: flowers can be pollinated (1) in balloon stage, (2) when they are about to bloom, (3) on the day when they just open, (4) two to three days after their opening. In *stone fruits* a pollination in the early phase of flowering proved to be the most successful. This means that pollination should follow immediately the emasculation in the bloom stage. In *sour cherry* the highest rate of fruit set resulted in a pollination of flowers just about bloom. Therefore, pollination should follow emasculation in the balloon stage by one to two days.

Preparation of flowers for pollination. Flowers that are undeveloped or have opened are to be removed. A whole tree can be pollinated with the pollen of the same cultivar but different kinds of pollen can also be used, but each kind of pollen should be associated with a distinct limb. Before pollination the viability of pollen is to be checked. Pollen can be placed on stigmas with the aid of a toothpick, camel brush, glass stick, rubber, or leaf, and each flower should be touched at least twice. After five to ten touches the pollinizing tool should be dipped into the pollen again.

Pollen rubbing is the simplest way of pollination. The flowers to be pollinated are to be isolated before opening. Immediately after picking the pollinizing flowers with dehisced anthers are to be tied together in brush-like bunches and to be touched to the stigmas to be pollinated. Waite (1894, 1898), Kamlah (1928), Stösser (1966) and others used the pollen rubbing method. Nyéki (1975) pointed out in sour cherry that the way pollen was placed on the stigma has influenced fruit set rates. Pollen rubbing with flowers tied together in brush-like bunches proved to be the most efficient method of pollination. The author also demonstrated that the capability of pollen of different ages for making fertile was different. The pollen coming from flowers that have opened resulted in the highest fruit set rates. The "immature" pollen coming from "forced flowers" had a worse capability for fertilizing.

Tools are to be sterilized with alcohol of 70 to 95% after use, or between treatments with different kinds of pollen, in order to clean them from unintended pollen.

Each pollination is to be registered separately. The data about the isolator, the cross-combination, the number of pollinated flowers, and the time of emasculation and pollination are to be written on a label as well as on a work sheet.

Self-pollination. The examinations serve the purpose of determining which cultivars are self-fertile, which cultivars are not self-fertile to a sufficient extent, and which cultivars are self-sterile.
Methods for determining the rate of self-pollination.

a) Natural (spontaneous) self-pollination: buds are to be counted and isolated, and then to be retained isolated to the end of petal fall. This procedure ensures that pollination may take place inside the isolators. Usually stigmas of isolated flowers are pollinated with their own pollen, that is, stamens touch stigmas. Natural self-pollination is facilitated by the motion of limbs due to the wind. The motion of isolator bags results in the pollen falling on stigmas spontaneously.

b) Artificial self-pollination, by hand may take place in several ways:

1. Pollination of a flower with its own pollen (autogamy).
2. Pollination of a flower with the pollen of another flower of the same inflorescence or the same tree (geitonogamy).
3. Pollination of emasculated flowers with the pollen of other trees of the same cultivar (clonal geitonogamy).

Experimental design. The isolators should be placed at a medium height in the canopy, uniformly distributed towards all four directions. In the course of an examination concerning self-pollination, five to ten isolated flowers of each cultivar are to be pollinated (Nyéki 1974*a*). Rudloff and Schanderl (1950) argue that the pollination of 100 to 200 flowers of each cultivar ensures a reliable result.

Determining natural parthenocarpy (fruit set without fertilization).

Methods for determining rate of parthenocarpy:

1. the rate of seedless fruits after uninfluenced flowering,
2. the rate of seedless fruits developed from isolated flowers,
3. the number of seedless fruits developed from emasculated and isolated flowers (this method is the most usual),
4. the number of seedless fruits developed from flowers damaged by frost when stamens and pistils have been frozen.

Checked (or controlled) crossing (seeking pollinizer partners). The capability of a cultivar for fertilizing another cultivar can only be checked if the pollen donor is known. Therefore, in the case of cultivars self-fertile to any extent, emasculated flowers are to be pollinated with the pollen of the selected cultivar(s). Emasculation is not required in self-sterile cultivars.

Methods used for determining cross-pollination. Determining the rate of cross-pollination can be carried out in several ways:

a) a cultivar is pollinated with its own pollen and the pollen of another cultivar simultaneously;
b) emasculated flowers are pollinated with the pollen of another cultivar;
c) emasculated flowers are pollinated with a mixture of the pollen of other cultivars.

In the case of controlled crossing the partners (mother and male cultivars) are known. There may be two directions of pollination:

a) unidirectional pollination: A ♀ × B ♂,
b) reciprocal pollination: A ♀ × B ♂, and B ♀ × A ♂

The pollinizing partner (♂) is referred to with several names in the literature: father cultivar, male partner, pollinizer partner, pollen-supplying cultivar, fertilizing cultivar.

The examinational method of cross-pollination is the same as that of self-pollination (discussed above). Kamlah (1928*a,b*) pollinated 30 flowers under isolator in each cultivar combination. Stösser (1966) pollinated 200 to 300 flowers in *sweet cherry*. Nyéki (1974*a*) argues that eight to ten isolated flowers of each cultivar combination (at least 100 flowers altogether) should be pollinated in order to ensure the certainty of the experimental results concerning cross-fertility.

Additional pollination. Usually pollination is executed once (in flowers that are about to open or have opened); it is called a single pollination. In order to achieve higher fruit set rates, however, it is useful to pollinate at least twice (double or repeated pollination). In double pollination, we should repeat pollination one to two days later with the pollen of the same cultivars. Manaresi (1953) carried out the procedure described above in *apple* and observed that the mean fruit set rates had increased by 14% due to the second pollination.

Double pollination is preferred over single pollination because it better approximates the natural conditions, where every flower is pollinated repeatedly. Layne (1983) provides a summary of the restrictive factors imposing some effect on the success of crossing that includes meteorological, biotic and pedological factors. Nyéki (1989) pointed out that the cultivar-specific properties are constant, genetically determined, and usually prevail also under various environmental conditions. This includes the hereditary inclination of the mother cultivar for fertilization.

Open pollination. Pollination takes place in natural circumstances, spontaneously, without any kind of control. In addition to its own pollen, a cultivar is pollinated with the pollen mixture of other cultivars. In this pollen mixture the cultivars and their proportions are not known. In open pollination flowers are pollinated by wind and/or by insects. Affecting factors are the flower density of the cultivars, the number, proportion, arrangement and distance of cultivars in the surroundings, the weather during bloom, the strength and direction of the wind, and the number of insects (bees) visiting the flowers. The cultivars used in such experiment are usually located in cultivar collections, so conditions of open pollination are to be regarded as "ideal". Flowers under field conditions may be pollinated once or several times. Thus, recurrent or additional pollination also frequently occurs.

Experimental design. Four branch segments, each containing 100 to 250 flowers, are to be selected at the medium height in the canopy on each side of the tree to evaluate fruit set.

Evaluation of fruit set. There are three periods of drop in most fruit species.

1. In the first period flowers that have not been fertilized or have degenerated in the course of their growth fall. In *stone fruits* this period immediately follows petal fall (Nyéki 1978).

2. The second period when fruits fall is the June drop. Fruit remaining after June drop is called "initial" fruit set. The initial fruit set can be evaluated four to six weeks after pollination (Kamlah 1928*a,b*, Stösser 1966).

3. In the third period, mature fruits fall. In sweet cherry and sour cherry there is a period, called "red drop", when those fruits fall which have coloured early (Blasse and Barthold 1970, Nyéki 1978). Fruit set rates are worth evaluating just before harvest. This is the value that most truly characterizes self- and cross-pollination in fruit cultivars. Generally, the fruit set values before harvest are those which are published.

In addition to genetic factors, the extent of fruit set is affected by storms, pests and pathogens, bird and ice damage, and it may also happen that machines damage a part of yield during the course of cultivation. Fruit set is evaluated by various researchers at different points of times. Nyéki (1978) provided a detailed review on examination of fruit set. The estimation of yield is sometimes carried out a few weeks after pollination, whereas in other cases this task is done only before harvest. It may also occur that high fruit set rates were established at an early evaluation in certain cultivar combinations, but then many fruits have fallen by harvest greatly changing fruit set rates, which is the case in apricot cultivars with 'óriás' name. Hence, comparing fruit set rates available in the literature is difficult.

In evaluating fruit set percentages many researchers use the following formula in examinations concerning fertility:

fruit set (%) = number of mature fruits × 100/number of flowers

8. Requirements for Successful Fruit Set in Orchards
M. Soltész

8.1. How to Determine Cultivar Combinations?

Cultivars are selected for production according to market demands and planted in separate blocks, or mixed within a given block. The main purpose of mixing is to ensure appropriate cross-fertilization in order to achieve the maximum extent of fertility in cultivars. The cultivars (their number and proportion) are selected for satisfying ecological, technological, growing site, and pollination conditions of cultivation. Instead of entering into general aspects of cultivar selection, the reader is referred to Tomcsányi's (1979) and Gyuró's (1977) summaries.

An important criterion of cultivar selection is concerned with ensuring conditions for pollination in the orchard. Concerns include: bloom time, fertilization conditions, and demand for pollination in cultivars. The more information is available as to bloom and fertilization, the better the decision will be. In order of importance the reliability of information concerning blooming can be listed as follows (Soltész 1992):

a) results coming from local observations and examinations of sufficiently long periods;

b) observations coming from areas with the same characteristics as those of the growing site of the orchard to be planted (the same latitude, height above sea level, temperature, precipitation, configurations of the soil, etc.);

c) results coming from sites with warmer climates where springs are earlier;

d) information coming from sites with cooler climates where springs are later;

e) data on bloom of an unknown origin from an ecological point of view.

Especially in the two last cases, pieces of information can be used only with restrictions; in such sites adaptative observations are required. Information listed under *d*) is more unreliable in cultivars with an early blooming period (e.g. *almond, apricot, currant*).

In addition to mean values resulting from multi-year-examinations, partial results and extremes also need to be considered. Regardless where the information is coming from, there is a need to know the growing site, the rootstock and the age, shape, and health of trees, etc.

The ranking of the above criteria may also help in the evaluation of information on fertilization. The extent of self-fertility, parthenocarpy, and fruit set may widely vary according to the growing sites.

Another kind of difficulty, arising during the course of interpreting results available in the literature, may result from the fact that investigators execute experiments using different trees from year to year. It often occurs that the cultivar under examination is not uniform genetically, that is, it consists of distinct clones (e.g. 'Besztercei' plum, 'Magyar kajszi', sour cherry cultivars 'Montmorency' and 'Schattenmorelle'), or the identity of the cultivar is not reliable. Furthermore, cultivar names are often interchanged; different

Table 8.1. Distribution of Recommended Pollinizer Varieties on the Basis of Time of Bloom (after Data by Krapf 1976)

Species	No. of combinations	Bloom time of recommended pollinizers compared to cultivars to be pollinated			
		same	earlier neighbouring bloom group	later	significantly different
Apple	482	169	64	57	192
Pear	570	169	132	149	120
Sweet cherry	451	179	136	125	11
Plum	132	72	24	28	8
Sum	1635	589	356	359	331
	100%	36%	22%	22%	20%

researchers publish their experimental data concerning one and the same cultivar under different cultivar names (this phenomenon is typical of species such as *sweet cherry, sour cherry* and *pear*).

Data on bloom periods are also often contradictory. The time of blooming in pollinizers is often unpublished, or the classification of such data is inconsistent. A work of Krapf (1976) can serve as example. We examined the pollinizer cultivars recommended by Krapf according to the distribution of bloom. Results are listed in Table 8.1. Only 36% of combinations bloomed as described by Krapf (1976).

Fruit species and/or cultivars can be grouped in three categories according to the extent for necessity of cross-fertilization:

A) Species and/or cultivars require cross-fertilization in any case.

B) Species and/or cultivars that produce sufficient yield with self-pollination but cross-fertilization will ensure higher yield with less risk and better fruit quality.

C) Species and/or cultivars in which self-fertilization is common and its extent is sufficient for good yield. Cross-fertilization in certain years may result in an exaggerated extent of fruit set which decreases fruit quality, or requires high rate of thinning.

When one intends to determine cultivar combinations, a cultivar is to be regarded as requiring cross-fertilization if it provides a better yield with cross-fertilization.

Group *C)* contains the self-fertile *apricot, peach,* and *plum* cultivars cultivated at favourable growing sites. As pointed out by Maliga (1961), cross-fertilization might be disadvantageous in *peach* and *apricot*, because it increases fruit set in species which are difficult to thin thus decreasing fruit size and quality.

In self-fertile *plum, sour cherry, apricot* and *almond* cultivars, however, cross-fertilization is advantageous (Rubin 1968). Other researchers came to the same conclusion in self-fertile cultivars of the following species: *apricot* (Pugliano and Forlani 1985), *sour cherry* (Vasiliev and Rodeva 1978) and *plum* (Tóth 1967, Faccioli and Regazzi 1972, Kellerhals 1986, Mesnil 1987, Szabó 1989).

Kirtvaya (1965) and Zenina (1967) stated that cross-fertilization exerted an increasing effect on yield in *gooseberry*. Harmat (1987) claims that importance of cross-fertilization in gooseberry lies in the fact that it increases the stability of yield.

Cross-pollination is also advantageous in species such as *raspberry* and *strawberry*, which are regarded as sufficiently self-fertile. More drupelets are set in fruits, which increases the amount of yield and the market value of fruits (Logintsheva 1960, Daubeny 1969, Redalen 1977, Kollányi 1980). Cross-pollination helps to achieve the maximum fertility mainly in cultivars that are able to bring a great amount of drupelets and large fruits. Cross-fertilization plays a crucial role in flowers of early cultivars of *strawberry* and especially in the first flowers to open (Lemaitre 1978). If the proportion of pistils that have ceased to grow, reaches 30% within the inflorescence, the amount of yield will decrease and the fruits will be deformed. The deformed parts will usually appear at the top of the larger fruits first to ripen.

In *red* and *black currant*, cross-fertilization is also held to be advantageous (Lucka and Lech 1974, Bergfeldt 1980), especially in the case of black currant. The fairly long bloom period of black currant requires an extraordinary pollen supply (Porpáczy 1987). Cross-fertilization has a positive effect also in *black sorbus, cornel-berry, blueberry*, and *black elder* (Porpáczy 1987).

As illustrated by the brief summary above, cross-fertilization is to be discussed in almost every fruit species because they belong to either Group A) or Group B). Obviously, the determination of cultivar arrangement in Group A) (*apple, pear, sweet cherry, walnut, chestnut, hazelnut* and *almond*) requires an extremely thorough examination. There is no applicable general method available.

Before entering into the species-specific details of the determination of cultivar arrangement, we are going to summarize the aspects to be studied in every species in the course of pollinizer selection. There is no universal pollinizer which is to be regarded as perfect in every respect. The satisfiability of the criteria to be listed below depends (among others) on the pollination system where the pollinizer is used. It is rare that the role of the pollinizer is limited to pollen supply only. In most cases we need the yield of the pollinizer as well.

Aspects of pollinizer selection considering the period of bloom.

* The bloom periods are to coincide with each other, or at least a sufficiently long overlap in bloom is required;
* the pollinizer is to belong to the same bloom time group as the cultivar to be pollinated or at least to the next group;
* the period of pollen shedding is to coincide with the receptive period of stigmas of the cultivar to be pollinated;
* a long bloom time of the pollinizer is advantageous. Cultivars with very short bloom times are not worth selecting to be pollinizers;
* a regular bloom is required in every year;
* the bloom time of the cultivars is to be stable.

The quantity and quality of pollen.

* A great amount of sufficiently viable pollen is to be provided by the pollinizer;
* anthers are to contain a sufficient amount of pollen, which is to be shed well: 70-100% of pollen grains are to be shed at the right time;
* pollen is to be uniform and of a good quality; pollen grains are to develop tubes quickly and in a high percentage;
* pollen shedding is to last long, be uniform and resistant to unfavourable weather;
* pollen must not be sensitive to (chemical) plant protectives.

Requirements concerning the capability of pollinizers for making and becoming fertile.

* The pollinizer is to be characterized by a regularly great fertilization ability;
* the pollinizer is to fertilize well and ensure a sufficient extent of fruit set and seed content;
* the pollinizer must not have a disadvantageous effect (xenia, metaxenia);
* the pollinizer is to be self-fertile (as much as possible) and easily fertilized unless its role is limited to supplying pollen only;
* the pollinizer is not to be closely related to the cultivar to be pollinated;
* the pollinizer and the cultivar to be pollinated are to fertilize each other reciprocally, as much as possible;
* the combination of the pollinizer with any other cultivar in the orchard must not be inter-incompatible.

The cultural value of pollinizers.

* Their demand for growing site and their ecological tolerance should be similar to the cultivar to be pollinated;
* the pollinizer must produce a great flower density;
* the pollinizer cultivars are to be suitable for insect pollination, and their attractivity to pollinating insects must be similar to the cultivar to be pollinated;
* pollinizers must bloom and produce fruit regularly in every year;
* the pollinizer and the cultivar to be pollinated should begin to bear at the same time;
* in traditional growers the cultivars are to be characterized by similar vigour;
* the cultivars should require similar management technology and protection against pests as much as possible;
* the pollinizer should not show an unfavourable reaction in response to growing methods applied to the other cultivars in the orchard.
* the maturity time of the pollinizer should be well adjusted to the harvest season of the varieties of the plantation, according to necessity, its immediate and distant maturity time should be equivalent to that of the cultivars to be pollinated;
* the fruits of the cultivars are of either the same goods value and maturity time, therefore can be harvested at the same time, or the fruits can be well differentiated.

In apple modern growers use crab apples, "Manchurian crab" that is a narrow, columnar tree with small, useless fruit. However, they require small space and can be planted in the row.

Factors affecting the proportion of pollinizers.

* The level of simultaneous blooming to be achieved,
* the number of cultivars in the orchard,
* the capability of the pollinizer for making fertile,
* the capability of the cultivar to be pollinated for becoming fertile,
* the period of harvest and the growing and market value of the pollinizer.

The pollination system of an orchard (see below) *is to be designed on the basis of the above mentioned criteria.*

* Mixed plantation with a great proportion of the cultivar to be pollinated;
* mixed plantation of some cultivars of the same value in the same proportion;
* the application of special pollinizers (e.g. *Malus* sp., *Pyrus* sp., trees of a cultivar combination) in an orchard where there is only one cultivar of market value;
* artificial pollination in addition to the versions mentioned so far,
* only artificial pollination.

When one designs the arrangement of cultivars, one should strive for a pollination as reliable as possible. However, no kind of pollination system can help to avoid certain problems associated with the growing site or growing method. As stated by Hugard (1978), the cultivar to be pollinated is the more stable factor of a cultivar arrangement in an orchard, the fertility and profitability of an orchard can be controlled by means of pollinizers.

As for the *apple*, it is worth knowing that in addition to the triploid cultivars, some diploid cultivars are also poor pollinizers (e.g. 'Melba', 'McIntosh', 'Staymared'). A very high proportion of triploid cultivars in an orchard is disadvantageous as they attract bees to a great extent due to their high nectar production, which may prevent cross-pollination (Soltész 1988). Fifty per cent overlap in flowering is usually enough but certain cultivars, such as 'Delicious' require 70%.

It is to be regarded as a useful general rule that at least three cultivars are to be planted in an apple orchard. However, this rule is absolutely necessary (Soltész 1982a) if one of the cultivars is

* a poor pollinizer,
* or the egg cells of the cultivar to be pollinated are inclined to partial sterility (e.g. 'Cox's Orange Pippin', 'Delicious'),
* the effective pollination period is short, such in 'Delicious',
* if any of the cultivars is inclined to a high rate of irregular flower bud differentiation (e.g. 'Melba', 'Húsvéti Rozmaring', 'Reine des Reinette'),
* any cultivar that requires a level of simultaneous blooming over 50%,
* if the bloom period is short (e.g. 'Red Astrachan', 'Delicious'),
* if the bloom time group is unstable (e.g. 'Early Red Bird', 'James Grieve'),
* the duration of its full bloom and its bloom time varies with the age of trees (e.g. in cultivars in which flowers develop from the lateral mixed buds of long shoots),
* if any of the cultivars in the combination is characterized by a great extent of ecological dependence (e.g. 'Delicious', 'Granny Smith', 'Elstar', 'Jonagold').

In order to continuously provide pollen during the entire blooming period for the cultivar to be pollinated, we should use more than one pollinizer. This allows multiple pollination, whose favourable effect on fruit set has been pointed out (Visser 1984).

In *pear*, there are also diploid cultivars that are not suited to serve as pollinizers (e.g. 'Passe Crassane', 'Giffard') (Gautier 1983). Cultivars inclined to second bloom (e.g. 'General Leclerc') should not be planted as pollinizers (Weber 1984). In pear two cultivar combinations are disadvantageous even if they belong to the same blooming group (Soltész 1975, Soltész and Nyéki, in Nyéki 1980), because pears produce less pollen and are visited by bees to a lesser extent. Hence, in pear a greater proportion of pollinizers is necessary than in apple.

In *sweet cherry* and *sour cherry*, the sufficient level of necessary simultaneous blooming is over 70% but reaches 80% in 'Pándy' sour cherry (Nyéki 1974c). In order to ensure a favourable level of pollen in *sweet cherry*, Götz (1970) suggests to plant five to eight cultivars in an orchard, while Ryabov and Ryabova (1970) suggest three to four cultivars, and Koch (1981) three cultivars.

In the case of *sour cherry* and *sweet cherry*, Brózik (1971) suggests three cultivar mixes in 1:1:1 proportion, and Sanfourche (1972) suggests a 1:10 proportion. Maliga (1980) and Nyéki (1989) recommend that in sour cherry cultivars that are regarded as regularly self-fertile (over 20 to 30% fruit set by self-pollination) it is sufficient to plant solid cultivar orchards. Maliga (1980) suggests that two reciprocally fertile sour cherry cultivars should be planted in a 1 to 1 or 1 to 2 proportion.

According to Brózik and Nyéki (in Nyéki 1980), a high proportion of pollinizers is to be avoided in cherry because of the risk of a high rate of fruit set in cultivars inclined to a high level of fruit set. Krapf et al. (1972) argue that in sweet cherry orchards the choice of pollinizer trees needs much care because the bloom periods may be different. According to Bargioni's (1982) opinion the proportion of cultivars depends on whether there are any incompatible combinations in the orchard. Kellerhals (1986) thinks that a pollinizer should not belong to any of the incompatibility groups.

Way (1974) calls attention to the fact that a more careful cultivar arrangement would be required in *sweet cherry* orchards in the western U.S. because the periods of bloom in given cultivars separate to a greater extent here than in other sites, that is, the extent of simultaneous blooming is less than elsewhere. Hugard (1978) demonstrated a similar effect for various growing areas in France.

Egg cell degeneration in certain sweet cherry cultivars is faster than in others. In these cultivars pollen supply is to be enhanced by ensuring simultaneous blooming (Eaton 1959a, Fogle et al. 1973). Hugard (1978) recommends that an effective plant protection requires cultivars with the same time of maturity to be planted in sweet cherry orchards. Bargioni (1979) suggests continuously maturing cultivars in the case of hand harvesting, and cultivars with the same time of maturing in the case of a machine harvest. In the machine harvest, however, different cultivars are not worth planting mixed within rows.

An intensive investigation of cultivar arrangement in *sour cherry* orchards is characteristic of countries where the cultivation of the self-sterile 'Pándy' sour cherry has been widespread (Montalti and Selli 1984). Brózik et al. (1980) thoroughly worked out different versions of cultivar arrangement in *sweet cherry* and *sour cherry* orchards. They argue against the (formerly customary) mixed plantation of 'Pándy' sour cherry and some *sweet cherry* cultivars with simultaneous bloom periods on the basis of the following evidence:

* the ecological and growing demands of sweet and sour cherry are different,
* the sensitivity of sweet and sour cherry to plant protectants are also different;
* the vigour of the two species and the time when they begin to produce fruit also differ;
* bees prefer sweet cherries to the sour cherry;
* the 'Pándy' sour cherry is not suited to serve as a pollinizer, hence at least two sweet cherry cultivars are required to be planted.

Paarmann (1980) recommends against planting of 'Pándy' *sour cherry* in blocks because if mixed with other cultivars, it is discriminated by pollinating insects to a lesser extent.

The stability of yield in self-fertile *plum* cultivars can be increased by a mixed plantation of two or three cultivars (Tóth in Nyéki 1980, Bellini et al. 1982). Chiriac et al. (1981) recommend for a mixed plantation of at least three to four cultivars in an orchard. Cobianchi and Sansavini (1984) state that among the plum cultivars, one may expect a regular, sufficient, yield without pollinizers only in the case of 'Stanley', the other self-fertile cultivars require a mixed plantation. Roman and Radulescu (1986) support this opinion. They say that self-fertile cultivars also need cross-pollination because of the possibility of an unfavourable weather during the bloom period.

In *oriental plum* cultivars, *Myrobalan* pollinizer trees are worth planting within rows in order to increase the stability of yield (Bellini et al. 1982). The majority of oriental plum cultivars are not really suitable for serving as pollinizers but there are some exceptions, such as 'Burbank', 'Elephant Heart', 'Santa Rosa' and 'Wickson'. In oriental plums there are more incompatible cultivar combinations than in European plums.

In *apricot*, Maliga (1966a) suggests that the blooming periods of two cultivars that reciprocally fertilize each other should not coincide perfectly. This makes it possible to avoid an overset as well as to decrease the risk of frost damage in the bloom period. Gautier (1977) recommended for a mixed plantation of apricot the use of late blooming and long endodormancy cultivars. Lomakin (1974) reports that the pollen of apricot cultivars in lowland sites is less viable. Hence, a greater proportion of pollinizers is to be planted in such growing sites. In apricot, self-sterile cultivars are not worth planting alone, without pollinizers. The self-fertile cultivars serve as the best pollinizers (Nyujtó 1980) in apricot.

As for the *peach*, self-fertile cultivars do not require a mixed planting. According to Nyéki's (1989) observations at Hungarian growing sites, in certain years, an exaggerated fruit set will result in mixed plantations that decreases fruit size or requires a high rate of thinning. Consequently, a mixed plantation of cultivars is only required with self-sterile, or male-sterile cultivars. In such cases the combination of two to three pollinizers is suggested (Ryabov and Kantsherova 1970).

In *black currant*, only self-fertile cultivars are held to be good pollinizers (Porpáczy in Nyéki 1980). In racemes of self-fertile cultivars, berry size is balanced and every flower is fertilized. In self-sterile cultivars, such combinations are to be used that ensure a very high level of simultaneous blooming (at least 80 to 90%). Usually, three to four cultivars are planted together in an orchard. When one intends to work out a good cultivar arrangement, one should determine the effect of pollination on fruit size in each cultivar combination, in addition to the question as to which cultivars have simultaneous blooming periods. Thus, one should pay attention to these factors in order to receive more and larger berries at the same time, in the racemes.

In *gooseberry*, as little proportion of pollinizers as 1% is claimed to be sufficient by Harmat (1987).

Strawberry cultivars can usually fertilize each other reciprocally very well. Only partially male-sterile cultivars (e.g. 'Senga Sengana') require more care in the course of cultivar arrangement (Szilágyi in Nyéki 1980). 'Pandora', a male-sterile cultivar, should also be planted in company of pollinizers (Simpson and Blanke 1989).

Blackberry cultivars are polyploid. Pollen sterility is typical of cultivars with an odd chromosome apparatus but may appear in cultivars with even chromosome numbers if they are of hybrid origin (Kollányi 1970). Sterility in *purple raspberry* cannot be eliminated even by means of pollinizers because some egg cells are characterized by a decreased viability.

In *raspberry* cultivars, arrangement according to the species the cultivar belongs to, was demonstrated by Zych (1965):

Fertile combinations:

female	male
Rubus occidentalis	×*R. occidentalis*
R. idaeus	×*R. idaeus*
R. occidentalis	×*R. idaeus*.

Infertile combinations:

R. idaeus   ×*R. occidentalis*.

In *highbush blueberry*, it is advantageous to plant more cultivars mixed in planting because cross-fertilization is favourable.

In *Aronia melanocarpa*, cross-fertilization has an increasing effect on fruit set but this species cannot be crossed with other sorbus species (Porpáczy 1987). The *black elder*, characterized by a fairly long period of bloom, requires a mixed plantation of cultivars that flower simultaneously to an extent of 80 to 90% to result in a similarly high percentage of fruit set. The unisexual, dioecious *sea buckthorn* can produce fruit only by cross-fertilization. A 10% proportion of pollinizers provides sufficient amount of pollen in this species.

In *walnut*, the extent of dichogamy varies from year to year but cultivar-specific features remain constant. Some cultivars are rather protogynous while others are protandrous. It is useful to mix protogynous and protandrous cultivars in an orchard (Yedrov et al. 1982). Rudloff and Schanderl (1941) recommend a 20 to 1 proportion between cultivars to be pollinated and pollinizers. Other researchers suggest higher percentages (5% by Stebbins 1971, 5 to 6% by Germain et al. 1975 and 10% by Forde 1970). According to Szentiványi (in Nyéki 1980), in warmer and wet sites several pollinizers are required, and their proportion is to reach, or rather exceed, 10%. During the whole period, when pistillate flowers of the cultivar to be pollinated are open, a sufficient extent of pollen supply is required, which is to be achieved by planting carefully selected pollinizers. It is useful to plant cultivars as pollinizers that are characterized by a high level of homogamy; in this way the fruit set of the pollinizer will reach a sufficient level without further pollinizers.

Nevertheless, a rich pollen supply, usually held to be as advantageous in most fruit species may have a negative effect in the case of the *walnut* (Kavetzkaya and Tokar 1963, Szentiványi 1992, Pór and Pórné 1990). Though other researchers, Nedev and Stefanova (1979), for instance, argue against the negative effect of a high pollen concentration, cultivar arrangement in walnut orchards needs to be carefully evaluated.

In *hazelnut*, Horn (1976) suggests that one should plant at least three pollinizers with gradual bloom periods in company with the cultivar to be pollinated in order to achieve the optimum simultaneous blooming. This method is especially favourable in years when catkins shed pollen early due to the warm spring. The recommended proportion of pollinizers in hazelnut is 6 to 20%.

Cultivar arrangement in various fruit species is summarized in Table 8.2. This table also contains information on the greatest possible distances from pollinizers beyond which pollinizers are largely ineffective.

Table 8.2. Cultivar Combination Requirements of Various Fruit Species (after Soltész 1989a, Modified)

Species	Fertilization group*	Minimum level of simultaneous blooming (%)	No. of cultivars in an orchard	distance from pollinizers (m)
Apple	A	50	2	25
	triploid, A	50	3	10
Pear	A	50	2	20
	triploid, A	50	3	8
Quince	A	70	3	7–10
Medlar	C	50	2	7–10
Sweet cherry	A	70	3	6–8
	C	70	3	12–16
Sour cherry	A	70	3	6–8
	B	50	2	20–30
Plum	A	70	3	15–20
	B	50	2	30–40
Apricot	A	70	3	20–25
	B	50	2	30–40
Peach	A	70	3	20–25
	B	50	2	30–40
Almond	A	80	4	6–8
Walnut (seedling)	A	80	2	100
Walnut (grafted)	A	80	4	50
Chestnut (seedling)	A	80	2	100
Chestnut (grafted)	A	80	4	50
Hazelnut	A	80	4	50
Gooseberry	B	50	2	30
	C	70	2	20
Red currant	C	50	2	30
Black currant	A	90	4	3–4
	C	80	3	6–8
Black currant x gooseberry	C	70	2	20
Raspberry	C	50	2	30
Purple raspberry	A	70	2	10
Blackberry	A	80	2	6–10
	C	80	2	12–20
Strawberry	A	80	3	15–20
	C	60	2	25–30
Black elder	C	80	2	10
Black sorbus	C	80	2	15–20
Sea buckthorn	A	90	2	8–10

*A = obligatory cross-fertilization; B = self-fertile, cross-pollination may be disadvantageous; C = self-fertile, cross-pollination is advantageous.

8.2. Xenia and Metaxenia

Focke (1881) was the first to use the term "xenia", which meant the effect of the pollen of a cultivar on the fruit of another cultivar as well as the seeds growing in it. This author formed three groups to classify the effects observed. The three classes: the zygote, the endosperm, or the tissue of the mother plant can be affected. The effect on the tissue of the mother plant was described in apple by Bradley much sooner. It was noted in 1739 (cit. by Turner 1820, cit. by Nyéki 1989), but was not given any term that time. Later it was called "carpoxenia" and then "metaxenia". The term "metaxenia" had become widespread in the special literature due to Swingle (1928).

The use of the two terms "xenia" and "metaxenia" and the concepts behind them were fairly contradictory during the last decades; the two effects were usually examined separately, and even nowadays the relation between the two phenomena often remains vague. Xenia in our terminology refers to the effect on the seed, metaxenia refers to the effect on the maternal tissues of the fruit which, according to the species, may differ in origin. In some fruit species (*walnut, chestnut, hazelnut*), it suffices to consider only xenia but in most species the direct and indirect effects of the pollen of the pollen donor are also to be considered.

Xenia means the appearance of the pollinizer parent's properties in embryonic mother tissues of the hybrid seed. As a result of double fertilization, both the embryo and the endosperm contain the successional material of the pollen donor plant, that is, the dominant features of the pollinizer exert an influence upon the growth of the embryo as well as the endosperm. It is fairly difficult to trace this process because the endosperm is used up in the course of embryo development.

In *walnut*, Porpáczy (1951) investigated the effect of pollinizers. Features, such as the size and taste of kernels and the grooving and width of the nuts of the pollinizer cultivar of *Juglans nigra*, appeared in the pollinated *Juglans regia* fruits. Intermediary types appeared often when crossing *J. regia* cultivars.

In *walnut, chestnut* and *hazelnut*, the determination of cultivar arrangement, should consider not only simultaneous bloom periods and fertilization ability but the pollinizer should also produce nuts of an excellent market value. It is important that xenia effect causing inferior nut quality is prevented by choosing the right pollinizers, but metaxenia is important in the green walnut marketed. Szentiványi (in Nyéki 1980) demonstrated the usefulness of this method in walnut as follows. He selected some Hungarian local clones and compared their nut quality at their original growing site with nut quality produced using pollen donor cultivars of better nut quality. Nut quality was much better with the better donors.

In *chestnut*, a change in fruit size and maturity time was mentioned as a xenia effect (Jaynes 1964). According to Breviglieri (1951), it is very difficult to separate xenia from other affecting factors. Variability of the pollinizer may be so great even within the same tree, because of the long bloom period, that a short period of time does not suffice to reach general conclusions.

Later experiments (Solignat 1966) more or less proved the tendency that pollination with pollen of a cultivar with larger fruits would result in larger fruits in the cultivar pollinated. The year-dependent variation, however, is dominant also in such cases. The xenia effect of pollinizers is definitely cultivar-dependent so this phenomenon is also to be examined in each possible cultivar combinations. The xenia effect of the pollinizer can be pointed out only in appropriate sites because otherwise unfavourable weather conditions (especially a cool and foggy weather before maturity) may exert a more significant effect on the size of nuts (early pericarp dehiscence). Solignat's (1973) experimental results concerning nut set in a chestnut cultivar named 'Marigoule 15' serve as an illustration (Table 8.3). The pollinizer and the cultivar to be pollinated both must produce good fruits, otherwise advantages of xenia would not be utilized.

Among the insect-pollinated fruit species, the most observations about xenia have concerned the *apple*. These observations, however, have been limited to morphological changes of seed in almost every case (Goryaczkowski 1926, Zaderbauer 1926, Roh

Table 8.3. Effect of Pollinizer on Fruit Size of 'Marigoule 15' Clone (after Solignat 1973)

Pollinizers	Mean weight of 100 fruits* (g)		
	1969	1970	average
M. 75 (CA-75)	100**	100	100
Précoce Migoule (CA-48)	107	130	118.5
Couderc 51 (CA-112)	114	121	117.5
L 43 bis (CA-43)	115	134	124.5
MARAVAL 74 (CA-74)	120	133	126.5
Belle Epine (CA-114)	120	134	127
AW. 07 (CA-07)	126	143	134.5

*Mean weight of 100 fruits expressed in % of data given by 'CA-75' (*Castanea mollissima*) pollinizer; **no statistical data are available.

1929a,b, Nebel 1930, Nebel and Trump 1932, Abramov 1955, Plock 1966, Saure 1967, Stösser 1967). Bryant (1935) reported that endosperm development had been affected by the pollinizer. Krumbholz (1932), and Tufts and Hansen (1933) observed no change in seed shape because of xenia.

Stoll and Krapf (1973) report that certain pollinizers made seeds of 'Golden Delicious' turn brown sooner without exerting any influence upon maturity of the fruit. Another consequence of this phenomenon is that browning of seeds cannot be used as a sign of maturity because of the possibility of xenia.

In addition to browning, Mackowiak (1974) observed in *apple* xenia effect on seed size and shape and fullness of seed (Table 8.4). As a result of the two first pollinizers in the table, seeds turned brown 15 days sooner compared when 'Starking' served as a pollinizer. Furthermore, as a result of 'Yellow Transparent', a decrease was observed in seed respiration intensity. Pollination coefficient is an outdated term. It is adapted by early plant physiologists, but it is meaningless in plants. It is useful only in mammalian systems. Another important observation is that pollinizers affect seed content, but there is no correlation between this change and their own seed content. The influence of the pollen donor definitely affects seed size because the seed bears the properties of both parents. The author, however, mentions Dzieciol's (1967) observation that seed weight in 'McIntosh' apple is affected by two factors: seed number and the effect of the pollinizer.

It is worth mentioning here that Williams et al. (1979) apply the phenomenon of xenia to prove that pollen dispensers to be fixed on hives are effective. On the basis of a method developed by Lewis and Crane (1938), Williams et al. (1979) could point out the participation of indicator pollen in fertilization thus they were able to evaluate the effectiveness of bee-dispensed pollen.

Table 8.4. Xenia in 'Wealthy' Apple Cultivar (after Mackowiak 1974)

Pollinizer	Weight of 100 seeds (g)	Length	Width of seeds (mm)	Fullness
King of the Pippins	6.29	8.71	4.39	1.87
Yellow Transparent	5.86	8.53	4.35	1.91
Starking	6.00	8.79	4.33	1.93

Table 8.5. Length, Width and Weight of 'Bartlett' Seeds (200 Seeds Measured Except Where Noted) (after Tufts and Hansen 1933)

Pollinizer	weight (mg)	Orchard A Average width (mm)	length (mm)	weight (mg)	Orchard B Average width (mm)	length (mm)
Winter Nellis	29.8	3.97	7.88	29.4	4.06	8.10
Easter Beurré	29.9	4.02	7.88	28.7	4.07	8.13
Comice	30.4	4.10	7.92	28.2	4.10	7.91
Anjou	29.8	4.02	8.22	28.4	4.08	8.03
Beurré Hardy	31.5	4.00	8.09	–	–	–
P. Barry	–	–	–	28.4	4.05	7.98
Bosc	–	–	–	26.5	4.04	7.77
Bartlett	34.7	4.13*	8.09*	29.1	4.18**	8.17**

*= 63 seeds; **= 100 seeds

Tufts and Hansen (1933) could demonstrate a clear appearance of xenia in 'Bartlett' pear (Table 8.5). Stösser (1966) reported xenia in sweet cherry. Kobel (1954) described as a kind of xenia the case when the effect of pollinizer influenced the whole chromosome apparatus. For instance, the *sweet cherry* egg cell of eight chromosomes was fertilized by *sour cherry* pollen of 16 chromosomes so the embryo, now containing 24 chromosomes, was larger. In contrast, the mean size of sour cherry seeds will decrease if they come from fertilization with sweet cherry pollen instead of sour cherry pollen. The resulting sour cherry (16) ×sweet cherry (8) seeds will have only 24 chromosomes instead of sour cherry (16) ×sour cherry (16) giving 32 chromosomes and a larger seed.

Chaparro and Sherman (1988) report a similar heterosis effect: pollinated with 'Nonpareil' almond, the *peach* embryo became larger.

In the last decades metaxenia was most often observed in *apple* and *pear*. A brief summary of the numerous observations available in the literature is summarized in Tables 8.6 to 8.9 for apple, and in Tables 8.10 to 8.12 for pear. It is to be noted, however, that there have been arguments against the occurrence of metaxenia in apple as well as pear since the beginnings (Kobel 1954, Sedov 1958, Way 1971, Hugard 1978). Several researchers have regarded this phenomenon as irrelevant (Höstermann 1924, Muth and Voigt 1928, Schanderl 1932, Tufts and Hansen 1933, Degman and Auchter 1935, Gerin 1972, Childers 1975, Pheasant 1985).

Childers (1975), for instance, argued that the metaxenia effect of the pollen donor was due to nothing else but the distinct seed content. Wickes (1918, cit. by Soltész 1992) also emphasized that pollen exerted some influence only upon the development of seeds, which stimulate neighbouring tissues through the auxin content. It is pointed out in several later studies that metaxenia cannot be examined independently of xenia, that is, effects on seeds (Gardner et al. 1952, Nesterov 1952, Ollivier 1967, Thibault 1971, Gushtshin 1972, Gautier 1983). According to Kobel (1954), earlier data on metaxenia in *apple* and *pear* were not really convincing because usually variability within trees had been ignored or very few fruits had been examined.

There are also numerous observations concerning fruit development in apple and pear that do not serve as arguments against metaxenia but clearly demonstrate that metaxenia cannot be studied separately from additional affecting factors. The effect of metaxenia, mainly on the basis of data on apple and pear, can be summarized as follows:

Table 8.6. Occurrence of Metaxenia in Apple

Character of change caused by metaxenia	Author (year)
Fruit shape	Lewis and Vincent (1909), Zaderbauer (1926), Horn (1927), Nebel (1930), Kovács (1977), Soltész (1982b)
Ribbing	Bach (1928a,b), Husz (1942), Maliga (1953b, 1956a), Dániel (1962), Kovács (1966, 1977), Soltész (1982b), Gautier (1983)
Fruit size, weight	Krumbholz (1932), Hibbard (1933), Kravtshenko (1955), Dzheneyev (1958), Gushtshin (1972), Karatsharova (1977), Soltész (1982b), Soltész et al. (1984)
Overcolour, russeted skin	Lewis and Vincent (1909), Husz (1942), Kravtshenko (1955), Dzheneyev (1958), Dániel (1962), Kovács (1966), Ollivier (1967), Kovács (1977), Soltész (1982b)
Fruit flesh colour	Nebel and Kertész (1934)
Strength of fruit skin	Kravtshenko (1955), Kovács (1977), Soltész (1982a)
Time of maturity	Dzheneyev (1958), Palocsay (1961)
Fruit characteristics (sugar, acid, vitamin)	Nebel and Kertész (1934), Nebel (1936a,b), Kravtshenko (1955), Dzheneyev (1958), Gushtshin (1972), Kovács (1977)
Storability (time, weight loss, physiological disorders)	Nebel and Kertész (1934), Nebel (1936a,b), Kravtshenko (1955), Ollivier (1967), Misotáné (1972), Mackowiak (1974), Soltész (1982b), Soltész et al. (1984)

Table 8.7. Effect of Pollinizer Cultivars on Fruit Set and Fruit Weight of 'Jonathan' Apple (after Karatsharova 1970)

Pollinizer	Fruit set (%)	Mean fruit weight (g)
Golden Delicious	28.2	109
Starking	26.9	121
Starkrimson Delicious	22.9	118
Royal Red Delicious	22.7	114
Stark Delicious	20.8	122
Jonathan (self-pollination)	4.4	105
Open pollination	7.1	114

Table 8.8. Metaxenia Effect of Pollinizer Cultivars on Composition of 'Severny sinap' Apple Cultivar (after Gushtshin 1972)

Pollinizer	soluble solid	Index* (%) sugar	acid	vitamin C
(1) Open pollination	100.0	100.0	100.0	100.0
(2) Antonovka obiknovennaya	105.0	100.0	75.0	116.7
(3) Bogatyr	98.6	99.0	88.8	97.2
(4) Suvorovets	101.0	103.0	80.6	100.0
(5) Pobeditel	101.0	103.0	72.7	116.7
(6) Pepin Shafranny	105.5	104.0	108.4	119.4
(7) Antonovka novaya	103.5	96.0	130.4	129.2
(8) Pamyat Mitshurina	105.5	109.1	108.2	119.4
(9) Oranzhevoe	110.4	108.2	102.8	123.6
2 + 3	103.5	99.0	122.2	116.7
2 + 4	119.4	104.0	166.6	108.3

* 100% equals in soluble solids: 14.4 g/100 g; sugar: 9.8 g/100 g; acid: 0.54 g/100 g; vitamin C: 7.2 mg/100 g

Table 8.9. Effect of Pollinizer Cultivars on Fruit Characteristics and Storability of 'Jonathan' Apple Cultivar (1981 to 83) (after Soltész et al. 1984)

Pollinizer	Fruit set (%)	No. of viable seeds	Fruit length (mm)	Fruit diameter (mm)	Shape index (l/d)
Éva	11	5.6	56	65	0.86
Idared	8	5.4	54	63	0.85
Mollie's Delicious	8	5.6	55	65	0.85
Starking	16	5.9	53	64	0.83
Staymared	7	4.9	55	66	0.84
Reine des Reinettes	16	6.4	56	66	0.86
Húsvéti rozmaring	8	4.9	56	66	0.85
Golden Delicious	10	4.5	55	65	0.84
Granny Smith	15	5.8	56	65	0.86
Jonathan (self-pollination)	2	3.8	55	65	0.85
Significant at 5%	+	+	NS	NS	NS

Pollinizer	Fruit weight (g)	Weight loss (%)	Rate of fruits with Jonathan or lenticel spots (%)	Spots (pce/fruit)
Éva	139	10.6	58	12
Idared	122	11.0	75	13
Mollie's Delicious	133	9.5	57	9
Starking	128	8.7	39	7
Staymared	131	7.5	59	9
Reine des Reinettes	132	7.3	54	9
Húsvéti rozmaring	132	9.6	63	7
Golden Delicious	126	10.7	58	12
Granny Smith	130	9.7	57	8
Jonathan (self-pollination)	140	7.5	54	9
Significant at 5%	+	+	+	+

	1981	1982	1983
Date of pollination	24th April	12th May	20th April
Days until harvest	138	127	139
Storage (day) in same condition	189	181	200

Table 8.10. Occurrence of Metaxenia in Pear

Character of change caused by metaxenia	Author (year)
Fruit shape	Horn (1927), Nagy (1957), Chollet (1965), Nyéki (1970, 1971, 1972a,b, 1976)
Fruit size, weight	Kim (1946), Nagy (1957), Nyéki (1970, 1972a,b)
Overcolour, russeted skin, russet stripe	Chollet (1965), Nyéki (1970, 1971, 1972)
Maturity time	Nyéki (1971, 1972a,b)

Table 8.11. Occurrence of Metaxenia in Pear (after Chollet 1965)

Year	Pollinizer exerts effect on fruit skin colour		Pollinizer exerts effect on fruit shape	
	female (♀)	male (♂)	female (♀)	male (♂)
1955, 1956	Beurré Hardy,	Williams	Alexandrine Douillard,	Comice, Passe Crassane,
	Williams	Comice, Passe Crassane, Beurré Hardy,	Passe Crassane,	A. Douillard, Beurré Hardy, Comice
	Comice, Alexandrine Douillard,	Jeanne d'Arc, Comice	Williams	A. Douillard, Comice,
	Jeanne d'Arc	Comice	Comice	A. Douillard, Duc de Bordeaux, Passe Crassane, Jeanne d'Arc,
1957	Williams	Beurré Hardy	Beurré d'Amanlis, Duc de Bordeaux, Comice,	Alexandrine Douillard, Passe Crassane, Duc de Bordeaux,
1958			Beurré Hardy, Jeanne d'Arc, Passe Crassane	Passe Crassane, Comice, A. Douillard

Table 8.12. Changes Caused by Metaxenia in Pear (after Nyéki 1971, 1972a,b, 1976)

Female (♀)	Male (♂)	Fruit characteristics (♀)
Clapp's favourite	Bartlett	elongated, expressed neck
	Virgouleuse	elongated
	Curé	russet stripe
Madame Du Puis	Pisztráng	bright red overcolour
Beurré Hardenpont	Pringall	stubby, earlier maturity
	Madame Du Puis	stubby
	Clapp's favourite	stubby, fruit skin colour is similar to pollinizer's
	Bartlett	elongated
Beurré Hardy	Bartlett	fruit skin is similar to ♂
Olivier de Serres	Bartlett	elongated
Passe Crassane	Virgouleuse	elongated
	Curé	russet stripe
Bartlett (Williams)	Pringall	stubby
	Madame Du Puis	stubby
	Olivier de Serres	stubby
	Curé	expressed neck

a) hormones produced by seeds (Swingle 1928, Gardner et al. 1952, Nesterov 1952, Childers 1975) may affect the fruit;

b) differences in embryo, endosperm, seed structure and growth (Crane and Brown 1942) may account for the metaxenia effects;

c) hormonal effect of pollen tube growth (Husz 1942) may cause the appearance of metaxenia;

271

d) the appearance of metaxenia depends on the competition between fruits (Krumbholtz 1962, Nyéki 1975), but fruit size is essentially determined by the locality in the canopy, and not the position of individual fruiting woods and flowers (Müller 1976);

e) the metaxenia effect depends also on weather conditions (Husz 1942, Maliga in Okályi and Maliga 1956, Nagy 1957, Dzheneyev 1958, Nyéki 1975);

f) metaxenia is more apparent in young trees of the cultivar to be pollinated (Palocsay 1961).

In addition to the factors listed above, there are several methodical problems in connection with metaxenia. Here we list some of them on the basis of experimental results with apple trees.

The tree has a major influence upon fruit variability (Gushtshin 1972). Care must be taken that the flowers pollinated are to be of the same morphological and physiological values when examining metaxenia (Müller 1976). Only the flowers pollinated with the pollen of the pollen donor cultivar are to be retained on the sample branch of the mother tree (Kravtshenko 1955). Experiments should be carried out simultaneously at more growing sites (Kravtshenko 1955), and on trees grafted on different rootstocks or with different canopy shapes (Palocsay 1961). Flowers of partially self-fertile apple cultivars are to be emasculated in order to make differences in metaxenia clear-cut. Terminal flowers are to be removed (Gerin 1972).

Only the first 50 days during fruit development is the period when the endosperm and the embryo determine fruit size. Later, they play a diminished role (Abbott 1959) and the crucial role is played by the amount of assimilated material per fruit (Magnes et al. 1931, Murneek 1954*a*, Zatykó in Nyéki 1980).

In addition to seed distribution, fruit shape is influenced by illumination, forces of gravity upon the fruit (Schanderl 1955*b*) and the amount of assimilating materials per fruit (Murneek 1954*a*).

As a result of cool weather in the bloom period, irregular fruit shapes may appear (Way 1971, Bubán 1980*b*). In fruits of 'Delicious' apple, lengthened and conal shape and the bulges near the calyx cavity are not typical cultivar characters but they are influenced by climate. Metaxenia does not play any role in the shape of fruit. A three to four week-long cool period after bloom governs this phenomenon (Way 1971).

The time of bloom may have a strong influence upon fruit shape (Shaw 1911, Sullivan 1965). How fruits become ribbed may also depend on the time of bloom, and the temperature existing at bloom time (Shaw 1911).

Lengthened fruits will develop in warmer growing sites (Roemer 1966), or at a higher (Dozier et al. 1980) or very low (Simons and Lott 1963) temperature after petal fall, but the temperature of soil may also be a determining factor (McKenzie et al. 1971). Warmer weather during fertilization enhances seed content, and more seeds have an influence on fruit diameter rather than on fruit length (Tufts and Hansen 1933), so fruits will seem to be flattened.

In addition to ecological factors, fruit shape is influenced by chemical growth regulators (Bubán 1980*b*). The major factor determining the final fruit size is the temperature after bloom (Tukey 1956).

Hanging fruiting woods produce less fruits, that are smaller and greener, than those on horizontal or on acute angle branches. Fruit weight and dry weight are strongly affected by the position occupied in the canopy (Tustin et al. 1988). According to Porpáczy (1964), much care is required when deciding whether a change is due to metaxenia or either to a deficient fertilization or somatic mutations.

Luckwill (1959) stated that in fruits with few seeds there was a strong correlation between seed content and fruit size. It is more difficult to point out the correlation between seed content and fruit size in cultivars inclined to parthenocarpy. Shape index changes are relevant during the course of fruit growth, when metaxenia is investigated. It is also difficult to choose the appropriate control. Because of self-sterility, the effect of pollinizers must be compared to each other. One cannot compare artificially pollinated and open pollinated fruits in evaluating metaxenia because fruit set conditions in open pollinated flowers are usually different from those in artificially pollinated flowers (Soltész et al. in Nyéki 1980). Furthermore, it is difficult to repeat experiments concerning metaxenia. Some of the features that play an important role in metaxenia (e.g. cover colour, ribbedness, stripiness, fruit taste) are subjective. References to certain aspects connected with metaxenia in the literature are usually based upon specific morphological features based on a limited number of fruits instead of being based on results which are verified by quantitative measurements and statistical methods (Soltész 1982b).

There are lesser and later data on metaxenia in stone fruits (Table 8.13). According to Apostol et al. (1977), metaxenia in *sweet cherry* is practically negligible and is not a factor when determining the arrangement of cultivars. Terpó et al. (1978a,b) stated, how-

Table 8.13. Occurrence of Metaxenia in Stone Fruits

Character of change caused by metaxenia	Species	Author (year)
Fruit shape	plum	Tóth (in Nyéki 1980)
Fruit size	plum	Constantinescu (1939), Crane and Brown (1942), Tóth (in Nyéki 1980)
	sour cherry	Cociu and Gozob (1962), Brózik and Nyéki (in Nyéki 1980)
	sweet cherry	Terpó et al. (1978)
Overcolour	sweet cherry	Cociu and Gozob (1962)
Maturity time	plum	Tóth (1962)
	sweet cherry	Stancevic (1971)
	sour cherry	Cociu and Gozob (1962)
Fruit constituents	plum	Tóth (in Nyéki 1980)
(sugar, acid, vitamin)	sour cherry	Pandele (1962–63)

Table 8.14. Berry Size of Black Currant Cultivar 'Coronet' After Pollination by Different Pollinizers (after Porpáczy 1987)

Pollinizer	min.	Berry size max.	diameter	Mean fruit weight (%) compared to that of berries resulting from open pollination
Open pollination	0.1	2.1	0.76	100
Ri 270	0.4	3.1	1.35	176
Ri 428	0.1	1.8	1.08	142
Silvergieter	0.2	1.3	0.54	71
Ri 1166	0.3	0.5	0.40	52
Stahanovka	0.1	0.6	0.22	48
Ri 330	0.2	1.5	0.74	125

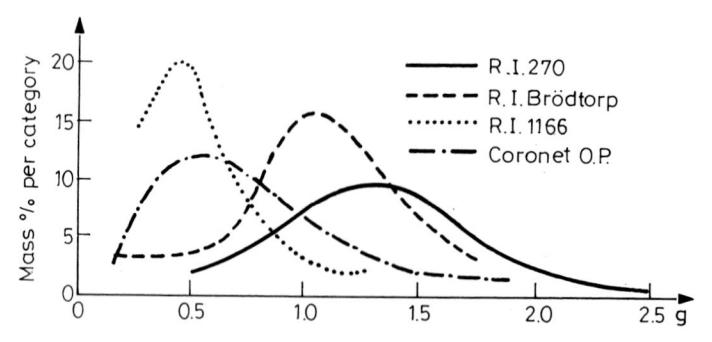

Fig. 8.1. Berry size variation in 'Coronet' black currant cultivar pollinated by different pollinizers and open pollination (after Porpáczy 1987)

ever, that the metaxenia effect required an investigation in the case of each possible cultivar combination. Bernhard et al. (1951) did not observe metaxenia in *plum*.

Pavlov (1973) reports that in *raspberry*, fruits with large, globular drupelets have grown as a result of the pollen from a cultivar named 'King'. Bauer (cit. by Brózik and Nyéki 1975) also reported drupelet change due to metaxenia. In tetraploid *blackberry* cultivars the number of drupelets was not influenced by pollinizers (Perry and Moore 1985), but fruit weight had increased.

Porpáczy (1987) crossed a *red currant* cultivar, 'Jonkheer van Tets', with a white currant cultivar, 'Jüterborg', and received berries with a red half and a white half, with some striped berries. In the berry size of 'Coronet', a *black currant* cultivar, metaxenia can be demonstrated (Table 8.14); and it plays an important role in berry size variation (Fig. 8.1).

It would be a mistake to place exaggerated importance on metaxenia or totally ignore its existence. In the future, more attention should be focused on the relationship between xenia and metaxenia, and the factors affecting this relationship should be thoroughly investigated. What is to be strived for is to demonstrate the effect of pollinizers.

8.3. Locating Cultivars in an Orchard

Fruit growers had already been aware of the importance of cross-fertilization when they noticed that the proportion and placing of the cultivars in the orchard and distance from pollinizers were also of a great importance (Table 8.15). Early researchers also recognized the importance of the distance from the pollinizer (in pear: Fletcher 1907, in sweet cherry and sour cherry: Tukey 1925 cit. by Soltész 1982*a*). Examinations begun in the forties of this century when investigations started on the problem why there was no yield or poor yield in huge orchards in certain years. However, widespread observations were made only in the last two to three decades on the importance of pollinizer placement in the orchard.

The figures and tables to be mentioned here serve as illustrations of the effect of the distance of pollinizers relative to that of the cultivars to be pollinated upon fruit set and yield: Figures 8.2 to 8.3: *apple*, Table 8.16 and Fig. 8.4: *pear*, Tables 8.17 to 8.18 and Fig. 8.5: *sour cherry*, Table 8.19: *plum*, and Fig. 8.6: *apricot*.

Table 8.15. Listing of Research on Placement of Pollinizer Cultivars on Fruit Set and Yield in the Literature

Species	Author (year)
Apple	Shitt and Metlitsky (1940), Canisius and Voundenberg (1959), Maliga (1961), Free and Spencer–Booth (1964), Liwerant (1966), Baillood and Mottier (1966), van Lier (1968), Krapf (1967), Wertheim (1968), Maggs et al. (1971), Reichel (1972), Frick (1974), Gyuró et al. (1976), Milovankic (1976), Kurennoy (1977), Soltész (1977), Nyéki and Soltész (1978a), Soltész (1980, 1982a), Lalatta (1982a), Kurennoy (1984), Wertheim (1986)
Pear	Fletcher (1907), Williams (1955), Free (1960a, 1962), Free and Spencer–Booth (1964a), Chiusoli (1966), Williams (1966b), Soltész (1975), Nyéki and Soltész (1976), Gyuró et al. (1976), Nyéki and Soltész (1977a, 1978b), Gautier (1983), Wertheim (1984)
Quince	Angelov (1981)
Sweet and sour cherry	Tukey (1925, cit. by Soltész 1982a), Laczy (1943), Lewis and Crowe (1954), Free and Spencer–Booth (1964b), Pejkic (1966), Nyujtó (1967), Bugarcic (1968), Brózik et al. (in Brózik and Nyéki 1975), Nyéki et al. (1976), Nyéki and Soltész (1977b)
Apricot	Szabó (1987, 1988), McLaren et al. (1992)
Plum	Free (1962), Tóth (1967), Keulemans (1980), Mesnil (1987), Szabó (1988, 1989)
Almond	Griggs et al. (1983)
Chestnut	Watanabe et al. (1964)
Walnut	Pór and Pórné (1990)
Black currant	Smirnov (1974)

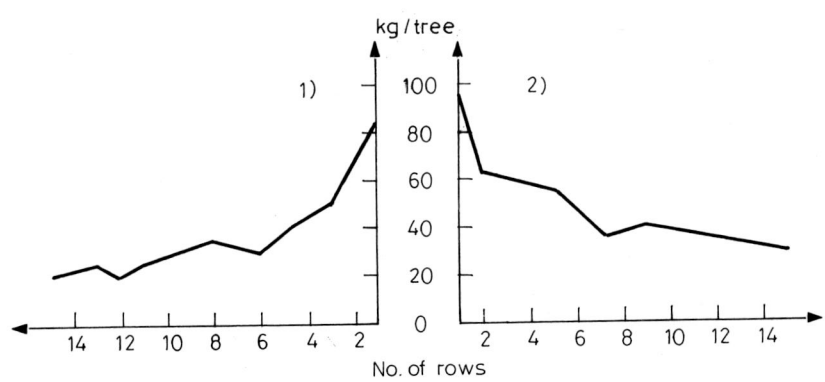

Fig. 8.2. Effect of width of cultivar blocks on yield of 'Koritshnoe polosatoye' (1) and 'Antonovka' (2) apple cultivars planted together (after Shitt and Metlitsky 1940)

The effect of the placing of cultivars can be pointed out only in the case of a good cultivar arrangement. Even the best cultivar ratios can be counteracted by poor cultivar arrangement. In order to continuously and reliably ensure a sufficient amount of pollen, pollinizers are to be planted at an appropriate distance from the cultivars to be pollinated in an appropriate proportion. Both these requirements are to be satisfied, they cannot substitute each other in a pollination system. There is no universal method because there

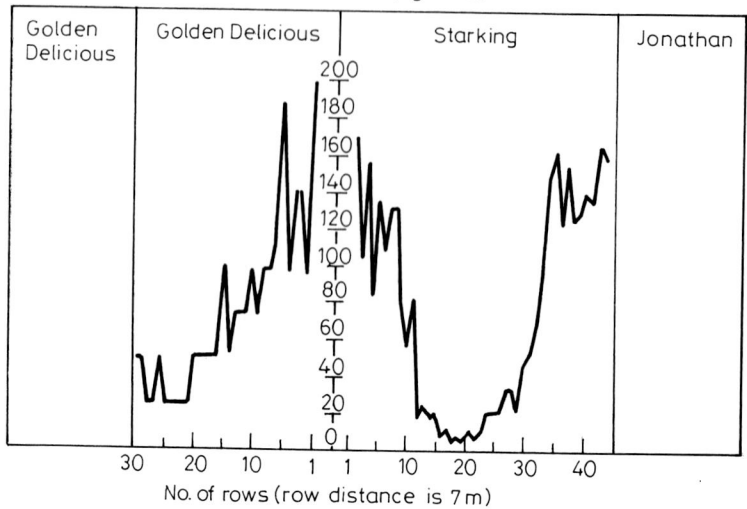

Fig. 8.3. Effect of cultivar arrangement on yield of 'Starking' and 'Golden Delicious' apple cultivars in the spindle orchard of Debrecen State Farm in 1976 (after Soltész 1982a)

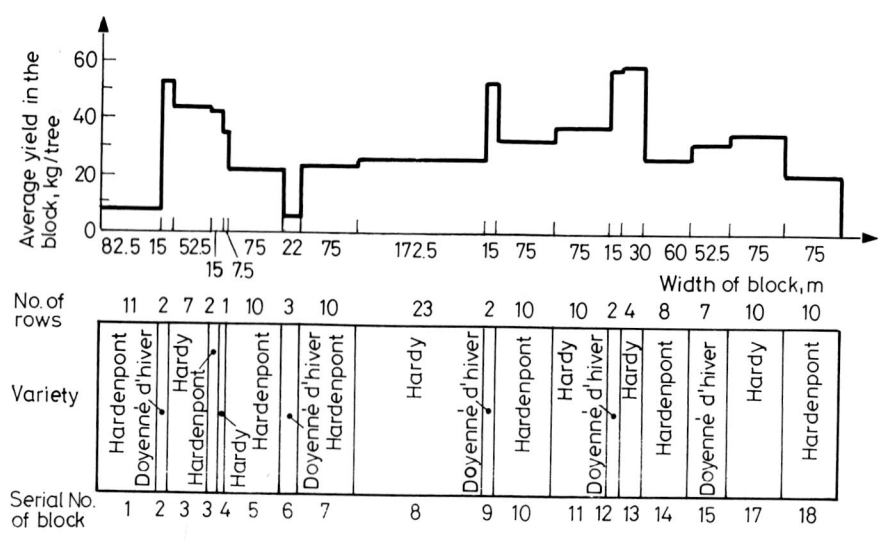

Fig. 8.4. Effect of width and arrangement of the cultivars on fruit yield in a pear orchard in Kutas in 1973 (after Nyéki and Soltész 1978b)

Table 8.16. Yield of Single Rows of 'Conference' and 'Butirra precoce Morettini' (B. p. M.) Pear Cultivars as a Function of Distance from the Pollinizer Cultivar (after Chiusoli 1966)

Distance from pollinizer (m)	Yield of the row (kg) Conference	Yield of the row (kg) B. p. M.	Index % Conference	Index % B. p. M.
3.5	212	263	100	100
7.0	194	143	92	54
10.5	192	81	91	31
14.0	159	63	75	24
17.5	73	56	34	21

Table 8.17. Effect of Distance from Pollinizer ('Schneiders Späthe Knorpelkirsche') on Fruit Set of 'Pándy' (Clone 'Kőrösi') Sour Cherry Cultivar (after Pejkic 1966)

No. of rows	Distance from pollinizer (m)	1963 No. of flowers	1963 fruit set (%)	1964 No. of flowers	1964 fruit set (%)	1665 No. of flowers	1665 fruit set (%)	Total No. of flowers	Means fruit set (%)
1	5	17,695	18.9	49.040	35.6	85,814	16.5	152,549	22.9
2	10	17,320	17.2	40,100	28.8	68,898	8.3	126,318	16.0
3	15	14,406	13.8	42,372	20.6	123,508	6.5	180,286	10.4
4	20	25,594	13.5	54,100	18.8	96,990	5.0	175,584	10.4
5	25	23,859	10.9	62,500	17.9	89,340	4.5	164,699	10.1
6	30	11,991	9.0	59,256	12.6	65,700	3.2	136,947	7.7
7	35	21,133	7.0	33,020	10.4	62,350	3.0	116,503	5.8

Table 8.18. Average Yields of 'Pándy' Sour Cherry Cultivar as a Function of Different Placings of Pollinizer Cultivars (1966 to 67) (after Bugarcic 1968)

Tree No.	Average yield/tree (kg) on both sides	Placing of pollinizers one side
1	31.0	18.5
2	40.0	23.0
3	23.5	25.0
4	24.5	23.5
5	22.5	21.5
6	19.5	22.0
7	23.0	22.0
8	25.0	19.5
9	23.0	18.0
10	31.0	14.5
11	22.0	16.0
12	22.5	14.0
13	26.5	16.0
14	35.0	16.0
15	25.5	13.5
16	22.0	14.5
Average (kg)	26.1	18.6
Index (%)	100.0	71.2

Note: year of planting: 1959; spacing: 5 x 5 m; rootstock: *Prunus avium*; distance from pollinizer row: 5 m; pollinizers: 'Májusi cseresznye', 'Olivet', 'Richmorency', 'Montmorency', 'Schattenmorelle', 'Dyehouse' and 'Umbra'

Fig. 8.5. Effect of the distance of the pollen donor ('Schneiders Späthe Knorpelkirsche') on fruit setting in 'Pándy' sour cherry at Mindszent, Hungary in 1974 (after Nyéki and Soltész 1977b)

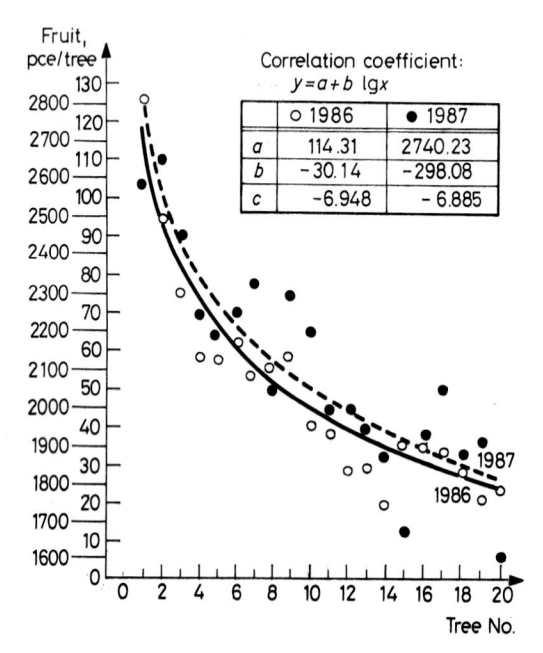

Correlation coefficient:
$y = a + b \lg x$

	○ 1986	● 1987
a	114.31	2740.23
b	-30.14	-298.08
c	-6.948	-6.885

Fig. 8.6. Effect of the distance from 'Rózsakajszi C 320' pollinizer on yield of 'Ceglédi óriás' apricot cultivar (after Szabó 1988)

is a wide variation of conditions depending on growing sites, fruit species, and different cultivars of a species. An *apple* cultivar that is fertile to a high extent, for instance, depends on the proportion and placing of pollinizers to a lesser extent (Kurennoy et al. 1984). Whereas the yield of 'Cox's Orange Pippin', an apple cultivar of a poor produc-

Table 8.19. Yield of 'Debreceni muskotály' Plum Cultivar Planted in the Border of the Orchard* (kg/tree) (after Tóth 1967)

Tree place	1961	1965	1966
7th row	\multicolumn{3}{l}{Pollinizer 'Besztercei'}		
6th row	30.7	38.2	45.6
5th row	23.4	28.6	37.0
4th row	19.9	21.5	26.7
3rd row	13.7	18.3	25.3
2nd row	11.6	15.9	24.8
1st row	8.2	13.8	22.7
LSD$_{5\%}$	6.9	7.0	3.4

* Distance between rows is 7 m.

Table 8.20. Effect of Pollinizer ('Golden Delicious') on Yield of 'Cox's Orange Pippin' Apple Cultivar (after Wertheim 1986)

Percent of pollinizer trees in the block	1973	1974	1975	1976	1977	1978	1979	1980	Average*
0	8.7	52.4	66.4	178.0	53.0	133.1	156.76	146.0	98.6a
10	11.4	54.3	84.7	182.1	46.7	138.0	167.8	152.2	106.8b
20	12.7	61.5	72.6	183.1	45.8	143.5	176.1	158.5	107.2b
60	19.7	69.6	91.9	187.0	36.7	147.5	207.7	168.5	115.3c
LSD$_{5\%}$									5.0

No. of fruits/tree

* Corrected for plot position. Values followed by the same letter do not differ significantly ($p = 0.05$).

Table 8.21. Effect of Pollinizer ('Golden Delicious') on Yield of 'Cox's Orange Pippin' Apple Cultivar (after Wertheim 1986)

Percent of pollinizer trees in the block	1973	1974	1975	1976	1977	1978	1979	1980	Average*
0	1.7	9.1	13.3	19.3	6.4	17.1	18.5	18.0	12.9a
10	2.2	9.1	14.3	19.2	5.3	16.6	19.0	18.6	13.2b
20	2.5	9.9	13.1	19.1	5.3	17.3	19.8	18.4	13.2b
60	3.4	10.2	15.1	19.8	4.7	17.7	21.5	20.3	13.9c
LSD$_{5\%}$									0.2

kg/tree

* Corrected for plot position. Values followed by the same letter do not differ significantly ($p = 0.05$).

tivity because of genetic reasons, strongly depends on its pollinizer's ('Golden Delicious') proportion and placing as illustrated by Wertheim's (1968) data in Tables 8.20 to 8.22. In the experiments the cultivar was planted alone, and in company with different proportion of pollinizers (10%, 20%, 60%).

Table 8.22. Effect of Pollinizer ('Golden Delicious') on Seed Content of 'Cox's Orange Pippin' Apple Cultivar (after Wertheim 1986)

Percent of pollinizer trees in the block	\multicolumn{7}{c}{No. of viable seeds/fruit}	Average*
	1973	1974	1975	1976	1977	1978	1979	
0	2.4 (2.3)	0.7 (1.3)	2.4 (1.3)	5.2 (0.3)	3.1 (0.4)	1.2 (4.1)	2.2 (1.5)	2.5a (1.7)a
10	2.4 (2.6)	0.9 (3.3)	2.8 (1.8)	6.0 (0.3)	4.0 (0.7)	1.6 (4.4)	2.7 (1.9)	2.9ab (2.2)b
20	2.4 (2.6)	0.8 (3.8)	2.9 (1.6)	6.1 (0.3)	4.2 (0.6)	1.1 (4.9)	2.9 (1.8)	3.0ab (2.2)b
60	2.6 (3.4)	0.8 (4.0)	3.4 (2.1)	6.8 (0.1)	4.5 (0.7)	1.4 (4.8)	3.3 (2.3)	3.2ab (2.5)c
LSD$_{5\%}$								0.6 0.1

* Values between parentheses refer to inviable seed. Corrected for plot position. Values followed by the same letter do not differ significantly ($p = 0.05$).

Table 8.23. Placement Variations of Apple Cultivars in Orchards

Placement	Author (year)
Every 3rd tree in every 3rd row is pollinizer	Gardner et al. (1952)
Max. distance from pollinizer 70 to 90 m	Kobel (1954)
Every 4th–5th row is pollinizer	Breviglieri (1960)
Few rows wide block	Maliga (1961)
4–5-row-wide block	Hoffman (1965),
4–5-row-wide block	Bayev (1967),
4–5-row-wide block	Fulford and Way (1967)
Max. 3–4-row-wide block	Krapf (1968)
Max. 4-row-wide block	Wertheim (1968)
Altering rows	Wertheim (1968)
Altering blocks of 3 to 4 rows	Greznitshenko (1969)
2–3-row-wide pollinizer block after every 5 to 6 rows	Greznitshenko (1969)
20-row-wide block	Pethö (1969)
4–5-row-wide block	Williams (1970)
At least one pollinizer for every tree	Tukey (1970)
Every 3rd tree in every 3rd row is pollinizer	Tukey (1970)
Max. 4-row-wide block, 1 to 3 pollinizers in the 5th row	Tukey (1970)
4–6-row-wide block ('Cox's Orange Pippin')	Williams and Wilson (1970)
Max. distance from pollinizer 20 to 25 m	Gautier (1971)
Altering max. 4-row-wide blocks	Gautier (1971)
Max. distance from pollinizer 70 m	Fritzsche (1972)
Every 3rd tree in every 3rd row is pollinizer	Teskey and Shoemaker (1972)
Max. 3-row-wide block	Teskey and Shoemaker (1972)
Max. 30–40 m wide blocks	Reichel (1972)
Max. 8–10-row-wide block	Blasse (1974)
Altering cultivar blocks	Blasse (1974)
Every 5th tree in every 5th row is pollinizer	Blasse (1974)
Max. distance from pollinizer 5 m	Popelyankov (1974)
Max. distance from pollinizer 15 m	Hugard (1975)
Max. 10-row-wide block	Blazek (1977)
Max. distance from pollinizer 100 to 150 m	Kurennoy (1977)
Pollinizer in every row	Parry (1978)
Every 8th–10th row is pollinizer	Stösser (1980)
Max. 4-row-wide block ('Delicious')	Gulino (1982)
Every 4th–5th row is pollinizer	Lalatta (1982a)
Pollinizer in every row	Lalatta (1982a)
Max. distance of pollinizer 25 m	Soltész (1982a)
At least one pollinizer for every tree	Mayer (1983)
Wide block is possible in the case of satisfactory pollination	Degrandi et al. (1984)
Max. 2-row-wide blocks (triploid cultivars)	Soltész (1986)

Table 8.24. Placement of Pear and Quince Cultivars in Orchards

Placement	Author (year)
Pear	
Pollinizer next to the tree	Brown (1943)
Every 3rd tree in every 3rd row is pollinizer	Gardner et al. (1952)
Max. distance from pollinizer 70–90 m	Kobel (1954)
Few rows wide block	Maliga (1961)
Altering rows, max. 2-row-wide block	Wertheim (1968)
Max. distance from pollinizer 15 m	Gautier (1971)
2–3-row-wide block	Teskey and Shoemaker (1972)
Every 3rd tree in every 3rd row is pollinizer	Childers (1975)
Max. distance from pollinizer 20 m	Soltész (1975)
Every 3rd tree in every 3rd row is pollinizer	Griggs and Iwakiri (1977)
Every 4th–5th row is pollinizer	Lalatta (1982*a*)
Pollinizer in every row	Lalatta (1982*a*)
Altering rows (2 cultivars)	Bellini (1987)
Few rows wide block (more than 2 cultivars)	Bellini (1987)
Quince	
Altering rows	Nyéki (1980)

An atypical effect of cultivar placing was observed in *walnut*: contrary to other fruit species, it is the trees nearest to the trees of the pollinizer that usually brought the least yield because of the high pollen concentration (Pór and Pórné 1990).

The amount of yield (kg/tree) in walnut cultivar 'Milotai 10' as a function of distance from pollinizer trees:

Distance (m)	1984	1985	1986
8	3.2	12.4	1.4
220	8.4	11.8	15.1

The effect of distance from pollinizer is to be established by means of a long series of observations because there are a lot of relevant factors (Soltész 1982*a*).

It was demonstrated in *apple* (Soltész 1982*a*, Wertheim 1986) and in *pear* (Soltész 1975, Wertheim 1984) that pollinizers may exert a negative indirect effect on flower buds. A high proportion of pollinizer cultivars resulted in so high a level of fruit set in the nearest trees to be pollinated and so great a seed content in certain years that exerted an unfavourable influence upon flower bud development.

Cultivar placement in an orchard should assure a sufficient extent of pollen supply even in unfavourable years. The same is true for the relation of cultivar placement with insect pollination. In a wide cultivar block foraging trips of bees may not ensure sufficient pollination or, parallel with an increasing distance from the pollinizer tree, a bee is more likely to turn back, or the pollen on its body is dispersed, resulting in a decreasing possibility for cross-pollination.

When one determines tree shape and the placing of pollinizers within rows, the activity and foraging behaviour of bees must also be considered. No pollen dispenser fixed on beehives will ensure reliable cross-pollination that substitutes an incorrectly planted orchard with very wide cultivar blocks or, with self-sterile cultivars, or perhaps for solid cultivar plantings.

Table 8.25. Recommended Cultivar Placement in Sweet and Sour Cherry Orchards

Placement	Author (year)
Alternative by 2 rows	Kamlah (1928a,b)
Max. distance from pollinizer 50 m	Kobel (1954)
Every 11th tree in every 2nd row is pollinizer	Ryubo and Mikutskis (1970)
1–2 rows of pollinizer after every 5 to 10–row-wide block	Ryabov and Ryabova (1970)
Max. distance from pollinizer 70 m (sweet cherry)	Fritzsche (1972)
Pollinizer in every row (sweet cherry)	Caillavet (1973)
Max. 4-row-wide block	Way (1974)
Every 3rd tree in every 3rd row is pollinizer	Way (1974)
Max. 2-row-wide block	Schaer and Schäfer (1975)
2–4-row-wide block	Childers (1975)
Alternative rows	Brózik et al. (in Brózik and Nyéki (1975)
Alternative rows	Nyéki et al. (1976)
Altering by 1–2 rows	Bargioni (1979)
Alternative rows (at right angles to wind direction) (sour cherry)	Maliga (1980)
Pollinizer in every row (if the direction of rows and wind are the same) (sour cherry)	Maliga (1980)
Max. 3-row-wide block	Koch (1981)

Table 8.26. Cultivar Placement in Other Stone Fruit Orchards

Placing	Author (year)
Plum	
Max. distance from pollinizer 70–90 m	Kobel (1954)
Few rows wide block	Maliga (1961)
Every 4th tree in every 4th row is pollinizer	Griggs and Hesse (1963)
2–3-row-wide block	Tóth (1967)
2–6-row-wide block	Childers (1975)
Max. distance from pollinizer 15 m	Keulemans (1980)
2-row-wide block	Tóth (in Nyéki 1980)
Pollinizer also in oriental plum rows	Bellini et al. (1982)
Altering 4-row-wide blocks	Gautier (1983)
Placing of pollinizers in square	Gautier (1983)
Alternating rows + every 5th tree is pollinizer in the rows of the main cultivar	Belmans and Keulemans (1985)
Max. 4-row-wide blocks	Szabó (1989)
Apricot	
Alternating rows (self-sterile cultivars)	Nyujtó (1980)
2–3-row-wide blocks (cultivars self-fertile to a low extent)	Nyujtó (1980)
6–8-row-wide blocks (self-fertile cultivars)	Nyujtó (1980)
Peach	
2-row-wide block	Childers (1975)
Almond	
Max. 3-row-wide block	Vansell and Griggs (1952), Pejovics (1976)
Alternating rows (2–4 cultivars)	Brózik (in Nyéki 1980)
Alternating rows (2–3 cultivars)	Barbier (1983)

Table 8.27. Cultivar Placement in Wind-Pollinated Fruit Species

Placement	Author (year)
Walnut	
Max. distance from pollinizer 100 m	Könemann (1943, cit. by Germain et al. 1973)
Max. distance from pollinizer 160 m	Impiumi and Ramina (1967)
6–8 pollinizer trees per hectare	Germain et al. (1975)
Max. distance from pollinizer 100 m	Bryner (1988)
Chestnut	
Max. distance from pollinizer 200–300 m	Breviglieri (1951)
3-row-wide block	Watanabe et al. (1964)
Max. distance from pollinizer 30–40 m	Solignat (1973)
Every 5th–6th row is pollinizer	Szentiványi (in Nyéki 1980)
Max. 80–100 m wide block (pollinizer also in block)	Szentiványi (in Nyéki 1980)
Hazelnut	
Max. 2-row-wide block	Romisondo (1963*a,b*)
Every 3rd tree in every 3rd row is pollinizer	Evreinov (cit. by Bordeianu 1967), Schuster (cit. by Thompson 1967)
Every 6th-10th bush in every row is pollinizer	Horn (1976)
5–7-row-wide blocks	Horn (1976)
Max. distance from pollinizer 25 m	Barbeau (1973)
Sea buckthorn	
Every 10th bush in every row is pollinizer	Porpáczy (1987)

Table 8.28. Recommended Cultivar Placement in Black Currant, Gooseberry and Strawberry Plantings

Placement	Author (year)
Black currant	
Alternating rows	Hilkenbäumer and Klämbt (1958), Porpáczy (in Nyéki 1980)
Max. 4-row-wide block between 2 pollinizers	Porpáczy (1987)
Gooseberry	
Every 20th bush in every 5th row is pollinizer	Harmat (1987)
Strawberry (with female flowers)	
Max. distance from pollinizer 10 m	Miculka (1965)

It is impossible to provide cultivar placement versions that can be applied anywhere at any time; these are to be worked out at a given site according to the requirements concerning the orchard to be designed. Tables 8.23 to 8.28 and Figs 8.7 to 8.9 provide useful data and suggestions available in the literature.

Cultivar placement in orchards is an area where the contradiction between favourable pollination and cultivar-specific cultivation becomes clear. Cultivar-specific growing technologies play an important role in the exploitation of fertility, and economical, envi-

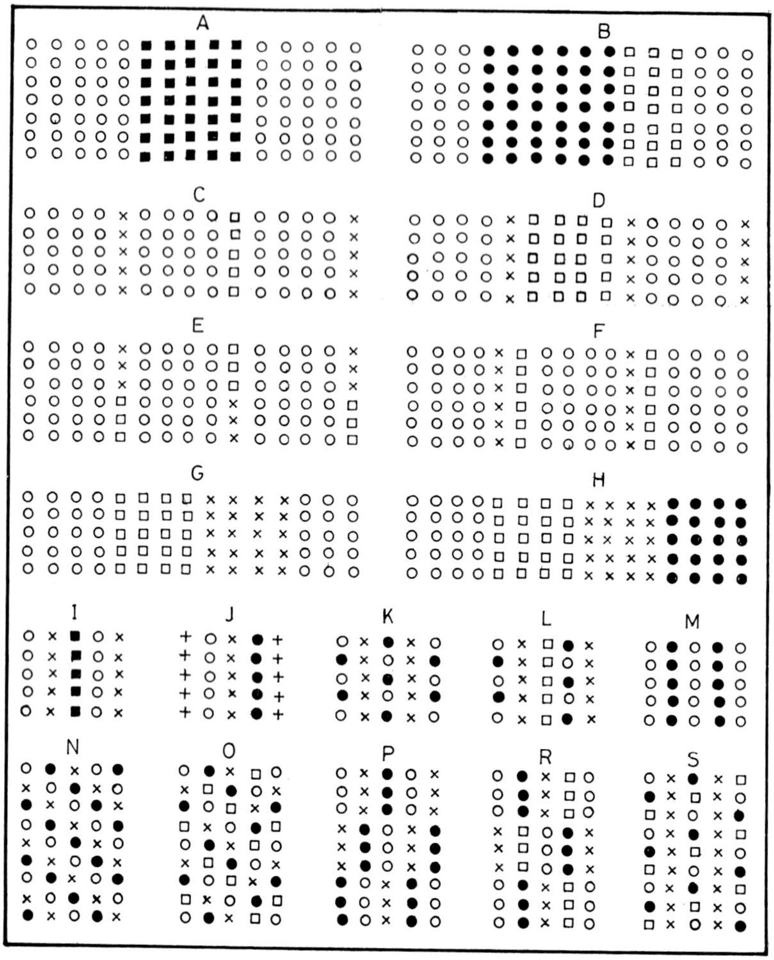

Fig. 8.7. Placement variations of cultivars (after Soltész 1982c)

ronment-preserving, integrated cultivation (Frick 1974, Dayton and Mowry 1977, Parry 1978, Meli 1981, Soltész 1981). Cultivar-specific production technology requirement can be satisfied to a high extent if a cultivar is planted in blocks alone, or more cultivars are planted in large blocks. However, the denser the pollinizers are placed, in neighbouring rows or alternately within rows, the more effective and reliable pollination is expected. These two factors are to be reconciled, that is, the best possible cultivar-specific technology is to be worked out that guarantees reliable and effective pollination.

The wider the cultivar blocks are, the less the number of spaces between rows of distinct cultivars is, that is, the more facility is given in cultivar-specific cultivation (Fig. 8.10). The creation of these advantageous wide cultivar blocks, however, is limited by the distance from pollinizer that is required because of cross-pollination. The same tendency presents itself in species such as *sweet cherry, quince, almond* and *black currant*, as in the self-sterile *sour cherry* cultivars.

Fig. 8.8. Placement of pollinizers along the roads in walnut and chestnut orchards (Soltész, original). (A) both cultivars are main marketable cultivars ■ O; pollinizer X; (B) marketable cultivar O; two pollinizers ■ X; block size in both cases 200 m × 200 m

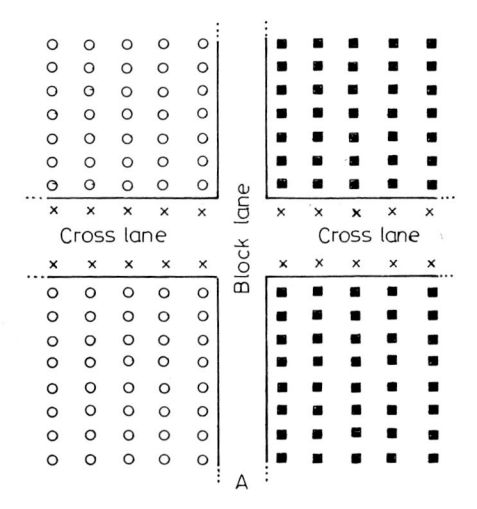

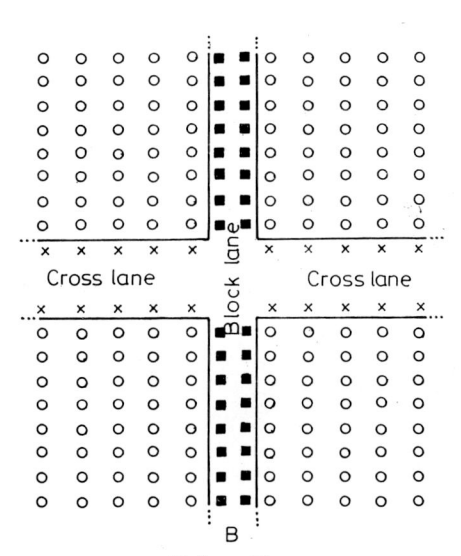

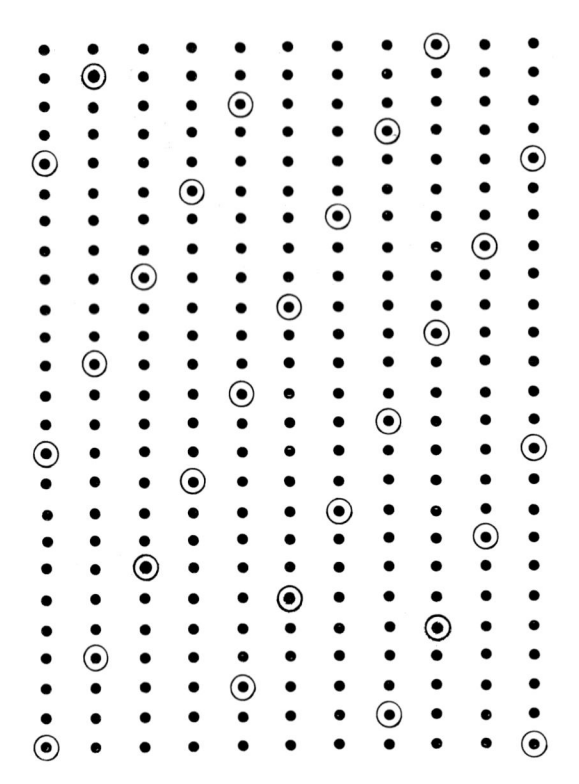

Fig. 8.9. Placement of bushes of dioecious sea buckthorn. Male flowers compromise 10% of the planting (after Porpáczy 1987)

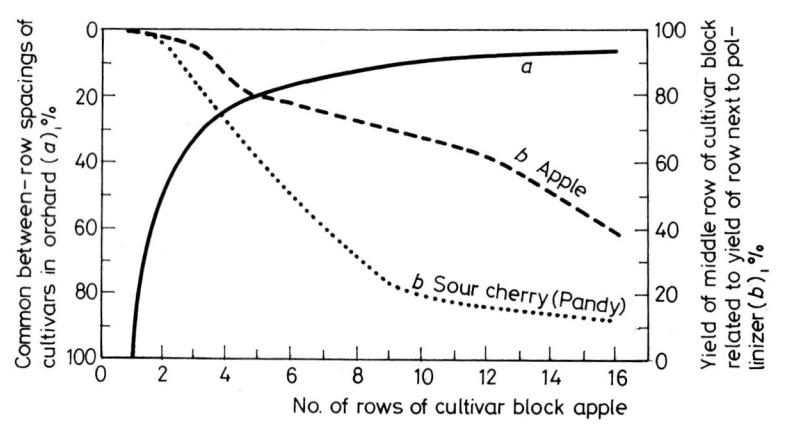

Fig. 8.10. Effect of width of cultivar blocks on yield and rate of common between-row spacings in apple and 'Pándy' sour cherry orchards (after Soltész in Nyéki 1980)

9. Insect Pollination of Fruit Crops
P. Benedek

9.1. The Possible Role of Wind in Pollination of Insect-Pollinated Fruit Species

Ancient horticulturists believed that wind pollinated every fruit species though they knew that blooming fruit trees were visited by insects. However, at that time, there were no reliable data on the role of the wind and insects in pollination.

Burchill (1963) and Langridge (1968) carried out experiments to identify pollen of fruit trees in the air. They found the most 60 pollen grains per m^3 near *apple* trees but the mean value was 20 to 25. The most apple pollen grains were observed between 10 and 14 h. In peach orchards, very few *peach* pollen grains were observed, and only between 10 and 12, or 10 and 14 h. The maximum was 20 grains per m^3 with a mean value of 5 grains per m^3. Similar pollen level was found near *sweet cherry* trees. The most sweet cherry pollen was observed during the night and early morning. These observations suggest that the wind cannot carry pollen very far, and hence it is unlikely that it is the wind that transfers the majority of pollen from the pollinizer tree to flowers of the cultivar to be pollinated.

Free (1964) designed an experiment to study the role of the wind and insects in pollination. He selected 16 uniform trees from apple trees. He divided them into four groups with four replicates in each group. Before the beginning of bloom all 16 trees were covered with netting. Under four cages small bee colonies and blooming branches of the pollinizer were placed near the blooming apple tree. The blooming branches were replaced with fresh ones every two or three days. Under the other group of cages blooming twigs were placed only coming from the pollinizer, which were also regularly replaced. There were no bees in these cages. Under the third set of cages small bee colonies were placed without blooming branches of the pollinizer cultivar. Under the last group of cages neither bees nor blooming twigs were placed. During the course of the experiment fruit set was measured at three points of time. Table 9.1 shows the results. Results show that the wind does not play a significant role in pollination.

Table 9.1. Role of Insects in Pollination of Apple Trees (after Free 1964)

Treatment	Fruit set (%)		
	after bloom	after first drop	after second drop
Blooming trees with branches of pollinizer cultivar and bees	33.8	19.6	10.7
Blooming trees with branches of pollinizer cultivar without bees	0	0	0
Blooming trees with bees without branches of pollinizer	2.6	0	0
Blooming trees without bees and branches of pollinizer	0.2	0.2	0

A later experiment of Free (1970) was also concerned with the role of the wind in pollination. With the aid of a wind-machine he blew air from the direction of a group of pollinizer trees toward *pear* trees of another cultivar. If the wind had played a prominent role, a significant effect should have arisen in this experiment. But no such effect was observed.

Stephen (1958) argued that the wind transmitted *pear* pollen at least to a distance of 0.75 miles. Stephen (1958) also reported that fruits that contained seeds, hence cross-pollinated, were sometimes found in solid cultivar blocks. This, however, cannot be regarded as an evidence for wind pollination in contrast to Stephen's opinion, but it may be derived from the special behaviour of pollinating insects or to be regarded as a result of self-fertilization.

According to Westwood et al. (1966), during bloom, about 50 wind-borne pollen grains get on the collective stigma surfaces of a flower cluster, whose total stigma area is 5 mm^2. They expressed the opinion that the wind might play a limited role in pear pollination. Smith and Williams (1967), however, reported that in *apple* trees isolated with cages (with small mesh screens) there was no fruit set despite the fact that pollen transmission by the wind had not been prevented by the cages. Wertheim (1968) observed a great extent of fruit set on pear limbs caged with muslin nets, but the rate of fruit set was lower than in the case of insect-pollinated flowers.

The debate on the role of the wind in pollination went on. Pisani and Ramina (1970) carried out experiments upon *apple*, *pear*, and *sweet cherry* trees and verified that the wind could transfer pollen to a relatively great distance from fruit trees when they are in the stage of full bloom. Furthermore, they verified that this pollen had arrived at stigmas and stuck on them. Later the same authors (Pisani and Ramina 1971) performed experiments with two apple cultivars, and verified the previous conclusion. On vaselined slides and on stigmas of flowering branches, even as far as 128 m from the orchard, they could find a significant amount of pollen grains. Furthermore, they imitated insect and wind pollination on blooming twigs caged with parchment bags. Insect pollination was imitated by the following method: pollen grains were transmitted to stigmas by touching with a brush. The wind was replaced with the method below: dropping compatible pollen grains on stigmas without any kind of contact. A significant difference had been observed between the extent of fruit set in the case of flowers pollinated by touch (10.6%) and that in the case of flowers pollinated with falling pollen grains (5.4%). They observed a much faster pollen tube growth in the case of flowers pollinated by touch; these pollen tubes had usually grown through the styles within 48 hours and reached the ovules, while it had taken more than 96 hours for the pollen tubes of flowers pollinated by pollen grain dropping to reach the ovules. Consequently, though these experiments had supported that *apple* flowers could be pollinated by the wind since pollen grains could arrive at stigmas, it is clear that wind pollination will not ensure a sufficient extent of fertilization.

Romisondo and Mé (1972a) carried out experiments upon *pear* trees. They received the most yield, the least extent of fruit drop, and the highest viable seed content in trees that had been closed in cages during bloom accompanied with small bee colonies, which were provided with compatible pollen of other cultivars. In the case of open pollination, they observed a lesser fruit set, a lower yield, more fruit drop, and less viable seeds. Finally, in trees pollinated with the aid of a fan which created a slight breeze and continuously blew compatible pollen grains, fruit set was even less, yield was lower, and much less viable seeds (90% empty seeds) were observed. The high rate of empty seedcoats can be interpreted as the occurrence of a higher level of parthenocarpy. Romisondo and

Mé (1972a) concluded that wind pollination was significant. Parthenocarpic fruit set, however, has nothing to do with wind pollination but it is a cultivar-specific capability. What has been verified by these experiments is that the pear cultivar under examination ('Abate Fetel') is strongly inclined to parthenocarpic fruit development but produces higher yield when insect pollinated.

If trees of a completely self-sterile cultivar are isolated with muslin nets, there sometimes develop fruits with fully developed seeds. According to Benedek et al. (1972), the explanation is the following: the bees, perceiving the untouched nectar content of the covered flowers, try to penetrate the muslin net, and a small amount of pollen gets from their bodies to stigmas of some flowers. This is prevented by the use of parchment bags. Parchment bag also prevents wind pollination due to its impenetrability. Occasionally, however, there are fruits with seeds, in parchment bags also which must result from pollination. The explanation is that tiny insects, thrips, pollen beetles and soldier beetles, can penetrate the parchment bags and carry pollen of another cultivar to stigmas on their bodies.

To sum up, though the possibility of wind pollination in the case of insect-pollinated trees is not excluded, the capability of wind-borne pollen for tube growth and fertilizing is poor. Hence, without insect pollination, self-sterile fruit trees will produce almost exclusively only parthenocarpic fruits, if the given cultivar is inclined to parthenocarpic fruit set.

9.2. Insect Pollinators of Fruit Species

The majority of temperate zone fruit species are self-sterile and hence require cross-pollination. Flowers are typically of entomophilous structure since they bear conspicuous petals, they are characterized by fragrance emission and produce nectar. The *walnut* and the *hazelnut* have an anemophilous flower structure.

The development of plants and their pollinating agents are the result of a long evolution. That is why the structure of flowers based upon wind pollination (anemophily) differs from that of those of entomophilous plants.

The greatest variation has come into existence among entomophilous plants due to evolution (Leppik 1957) because here the adjustment is reciprocal: it is not only the plant that adjusts to the pollinating agent in order to use it as possible (like in the case of wind pollination) but the plant and the pollinating agents (insects) adjust to each other reciprocally. On the one hand, the adjustment of the plant to the pollinating agent results in a more effective fertilization, and on the other hand, the pollinating insect strives for a better use of its feeding source.

The first real flower-visiting insects were beetles. Then flower visitation spread to butterflies and flies, and reached its peak with bees. A relevant factor of this development is that bees, but not other flower-visiting insects, feed their larvae exclusively with food obtained from flowers. Their connection with flowers is closer, their reciprocal adjustment is more intimate than in the case of other insects (Løken 1981).

The entomophilous flower structure of temperate zone fruit species is characterized by a radial symmetry and definite numbers while a spatial organization can be observed only in a few species. According to Leppik's (1957) classification, the majority of fruit species belong to the pleomorphic group (fruit trees, *raspberry*, *strawberry*), or the stereomorphic type, which has developed from pleomorphic types but has shorter or longer flower tube (*currant* species, *blueberry* species). In the pleomorphic flower type it is easy to obtain nectar and pollen while in simple stereomorphic flowers obtaining nectar is more difficult. Hence, it is surprising that flowers of fruit species, especially those that belong to the pleomorphic type, are visited by various insects of almost every type. According to observations of Solinas and Bin (1964), Benedek et al. (1972), Nye and Anderson (1974) and Boyle and Philogene (1983), flowers of fruit species are visited by pollinating insects of the distrophic, allotrophic, hemitrophic, and eutrophic groups (Fig. 9.1).

There is a group of insects belonging to the distrophic group, termed harmful pollinating insects. A few beetle species belong to this group that damage flowers, certain members of the Melolonthidae family (e.g. *Epicometis*, *Cetonia* species), other harmful beetles (*Omophlus* spp., *Attagenus* spp., *Meligethes* spp., *Cantharis* spp., etc.), and thrips (Thrypidae) complete the list. These insects flying from one tree to another, from one flower to another, might pollinate some flowers, but they may forage in the flower for a long time and usually devour not only pollen but other parts of the flower, the stigma and the stamens causing more harm than good by rare pollination. Fortunately, their quantity is usually negligible. They often appear in fruit trees, *raspberry* and *strawberry* but rarely visit *currant* and *blueberry* species whose flowers are stereomorphic, and more difficult to enter into.

The allotrophic group mainly consists of tenthredinid and ichneumonid wasps, chalcidoids and other hymenopterous insects not interested in pollen-gathering (Sphecidae, Vespidae, Formicidae), and beetles that eat nectar in their adult stage (Elateridae), but not flower parts. They are qualified as casual visitors of flowers as their lives do not depend on food that flowers can offer so their flower visitation does not comply with demands of flowers. Thus, their flower visitation is casual, their role in pollination is accidental.

Another group of insects, in which one can observe simple signs of adjustment to flowers (e.g. licking and sucking organs, special kinds of hair on the body), are the hemitrophic pollinators. Their lives partly depend on flowers because their adults can find their food exclusively, or at least in a high percentage, in flowers. Out of this group, flowers are visited by various flies (Syrphidae, Bombylidae, Bibionidae, Anthomyiidae, Tachinidae, Sarcophagidae, etc.), or by moths and butterflies and primitive bees (*Hylaeus* spp.), or parasitic bees (*Sphecodes* spp., *Nomada* spp.). Due to their simple adjustment, these insects are able to visit fruit species not only with pleomorphic but also stereomorphic flower structure. They sometimes play a significant role in pollination.

The fauna of insects pollinating fruit species is dominated by the eutrophic pollinators, that is, the apoids in respect of density as well as cultivar number. Flowers of fruit species are visited by apoids most groups of which fly in given blooming periods (Andrenidae, Halictidae, Anthophoridae, Megachilidae, Apidae). Colletidae and Melittidae, however, do not occur in orchards. In addition to *honeybees* (*Apis mellifera* L.), the most typical apoid visitors of fruit species are females of bumble-bees (*Bombus* spp.) as well as *Halictus*, *Lasioglossum*, and *Osmia* species.

Compared to other flower visitors, lives of apoids depend on flowers to a much greater extent because not only adults but also larvae live on products of flowers (pollen and nectar). They are characterized by a perfect adjustment to flower visitation, such as licking organ, lengthened proboscis to use up tubuliflorous plants, collecting hair and basket to gather and carry pollen. Thus they are the most effective pollinators.

The importance of pollinating insects is also affected by their amount and frequency. It is to be considered that fruit species bloom in the early part of spring, which is the beginning of the seasonal activity of wild pollinating insects. Hence, we can only rely on the activity of the insects lived through the winter, but they are few compared to the population of the same species that develops later in the season after a period of multipli-

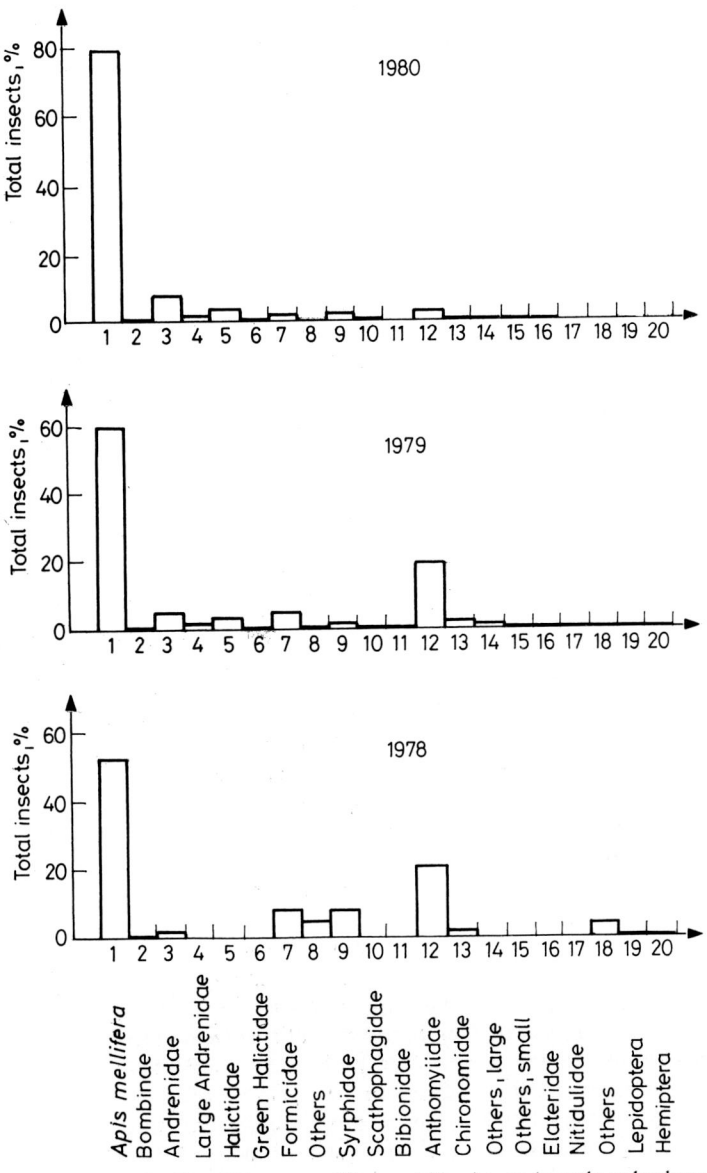

Fig. 9.1. Distribution of 20 groups of flower-visiting insects in apple orchards expressed as the percentage of total number of insects observed (after Boyle and Phylogéne 1983, modified)

cation. Unfortunately, the majority of wild pollinating insects do not work in the early period of spring, they appear later. Out of the numerous flight groups of wild bees, described by Benedek (1968), the first insects to appear, are those that start the multiplication of the population, belong to the short flight spring species, or the long flight bivoltine species, or the long flight continuously breeding species. Likewise, in the case

291

Table 9.2. Rate of Honeybees in Pollinator Insect Populations of Apple Trees (after Benedek et al. 1974)

Place of observation	Rate of honeybees	Author[*]
New Zealand	99	Roberts (1956)
USA, New York	95	Dyce (1958)
USA, Indiana	90	Davis (1925)
USA, New York	90	McDaniels and Heinicke (1929)
USA, New Jersey	90	Filmer (1941)
Egypt	89	Wafa and Ibrahim (1957)
England	87	Free (1966)
Soviet Union	85	Tsigankov (1953)
England	80	Hooper (1929)
Switzerland	80	Kobel (1942)
Germany	75	Zander (1936)
India, Bihar	72	Sharma (1961)
Canada, Ontario	60	Smith (1952)
Hungary		
Nyíregyháza	99	Benedek et al. (1972)
Budatétény	98	
Százhalombatta	97	
Debrecen	95	
Hajdúsámson	78	unpublished

[*] Sources are not included in the list of references, they can be found in Benedek et al. (1974).

of syrphid flies and other pollinators, we only rely on those "earliest individuals" that are first to appear after living through the winter.

These few insects may reach a sufficient density in single fruit trees or in fruit trees of small gardens but they are dispersed in large orchards where a great number of fruit trees begin to bloom at the same time.

Thus it is not at all surprising that in *apple* orchards, for instance, honeybees make up the overwhelming part of the pollinating insect population fro 60 to 99% but most usually 80 to 90% (Table 9.2).

In other fruit species, however, there may be different proportions. In the case of *sour cherry*, for instance, the role of wild pollinating insects is held to be negligible (Benedek and Martinovich 1971) because they are characterized by a negligible density among pollinating insects (0.7%). In the case of *raspberry*, there occur different flies, wasps and wild bees but the majority of pollinating population is still made up by honeybees (Benedek et al. 1974, McGregor 1970). In *strawberry*, honeybees made up 66% of the pollinating population while the proportion of dipterous species was 18% (Nye and Anderson 1974). According to experiments upon *pear* trees carried out by Benedek et al. (1974), the following insects take part in flower visitation in the following percentages: honeybees: only 21%, wild bees: 6%, beetles: 12%, dipterous species: 61%, an overwhelming part of visitations. In *currant*, the proportion of wild insects was also observed to be higher than that of honeybees (Benedek et al. 1974). These two types made up about the one-third of the population while flies and wasps belonged to the other two-thirds.

All in all, the overwhelming majority of the flower-visiting insect population are usually made up by honeybees but either the proportion of wild bees or that of dipterous species may be remarkable.

9.3. The Relative Efficiency of Insects in Pollination

Nye and Anderson (1974) intended to establish the pollination efficiency of insects visiting *strawberry* flowers on the basis of the amount of pollen carried on their bodies, body size, flightiness and their touching stamens and stigmas. Though this is a fairly subjective approach, it expresses some kind of relative pollination efficiency of insects because it depends on comparisons (Table 9.3).

Kendall and Solomon's (1973) method is much less subjective. They calculated the number of pollen grains and the number of apple pollen grains within the former, stuck on bodies of insects gathered in *apple* flowers. Insects of more than 70 species were observed (Table 9.4). Females of a few *Andrena* species and *Bombus terrestris* and *B. lucorum* workers bore much more pollen than honeybees (Group *A*). Honeybee workers were included in the same group (Group *B*) as wild bee females and some syrphid fly species. The groups of insects bearing less pollen (Groups *C–F*) contained mainly simpler pollinating insects, that is, fly species, beetles, and wasps above all. Although a few bee species, four out the 51, were also included in these groups, they were rather males or parasitic bees. These measurements have unquestionably proved that the most effective pollination is to be expected of bees and certain big syrphid flies. Klug and Bünemann (1983) executed similar experiments and they also found more pollen grains on certain insects (bumble-bees and a few andrenid bee species) than on honeybees.

Kendall (1973) examined the viability and compatibility of pollen carried on bodies of pollinating insects in order to get a more exact approximation of the pollination efficiency of insects. The author concluded that the pollen carried on bodies of insects, with the exception of a few male solitary bees and syrphid flies, was viable to the same extent as the pollen extracted from anthers just after their dehiscence.

Females of two solitary bees *Andrena haemorrhoa* and *A. jacobi* are more effective pollinators than honeybees because their visitations had resulted in better fruit set values than those of honeybees. Klug and Bünemann (1983) also argue for a better pollination efficiency in *Andrena* and *Halictus* species. They say that females of these species touch stigma in the case of every flower visited, whereas honeybees touch stigma only in the case of one-third of the visited flowers.

Table 9.3. Efficiency of Representative Strawberry Pollinators (after Nye and Anderson 1974)

Insect	Rating for loose pollen on body*	Rating in size	Rating for flightiness and action on flowers***	Efficiency rating****
Apis mellifera				5
pollen collectors	6	4	6	6
nectar collectors	4	4	5	4
Halictus ligatus				3.8
females	4	3	5	4.5
males	2	3	4	3
Odynerus dilectus	2	3	2	2
Phaenicia spp.	2	3	3	2.5
Eristalis tenax	4	4	4	4
Xylota (Syritta) pipiens	0.5	2	0.5	0.5

* Rated 0-6, the higher the number, the greater the pollinating efficiency; **1-4 small to larger; ***1-6 small to greater activity; ****loose pollen carried on the bees rated higher than other factors, the greater the figure the greater the efficiency.

Table 9.4. Species or Other Taxa Arranged According to the Mean Number of Fruit Pollen Grains per Individual (after Kendall and Solomon 1973, Modified))

Efficiency groups	Species	Total pollen grains geometric mean	S.D.	% fruit pollen
A	Hymenoptera: *Andrena pubescens* Ol.	29,970	2.2	81
	Hymenoptera: *Andrena haemorrhoa* (F.)	24,120	3.5	87
	Hymenoptera: *Andrena coitana* (Kirby)	16,870	2.2	97
	Hymenoptera: *Bombus terrestris* (L.), *B. lucorum* (L.)	19,010	3.4	85
	Hymenoptera: *Andrena jacobi* Perkins	18,490	3.7	71
B	Hymenoptera: *Psithyrus vestalis* (Geoff.)	7,240	–	80
	Hymenoptera: *Bombus pratorum* (L.)	10,720	2.4	51
	Hymenoptera: *Osmia rufa* (L.)	6,721	2.6	81
	Hymenoptera: *Bombus terrestris* (L.), *B. lucorum*	9,075	2.6	56
	Hymenoptera: *Bombus lapidarius* (L.)	11,050	3.4	40
	Hymenoptera: *Apis mellifera* L. (honeybee)	5,655	3.0	73
	Diptera: *Merodon equestris* (F.) (Syrphidae)	10,197	–	29
	Hymenoptera: *Halictus* spp.	3,940	3.5	75
	Diptera: *Myopa buccata* (L.) (Conopidae)	3,020	3.8	85
	Hymenoptera: *Andrena nana* Kirby	3,292	3.4	75
	Diptera: *Eristalis intricarius* (L.) (Syrphidae)	3,054	–	81
	Diptera: *Eristalis tenax* (L.)	3,571	2.7	67
	Hymenoptera: *Andrena haemorrhea* (F.)	2,343	1.5	96
	Hymenoptera: *Bombus pratorum* (L.)	3,350	–	61
	Hymenoptera: *Andrena chryosoceles* (Kirby)	3,658	3.6	53
	Hymenoptera: *Andrena wilkella* (Kirby)	2,109	3.9	66
C	Diptera: *Helophilus pendulus* (L.) (Syrphidae)	2,088	–	58
	Hymenoptera: *Andrena wilkella* (Kirby)	1,524	2.1	70
	Hymenoptera: *Nomada* spp.	1,460	2.4	59
	Hymenoptera: *Hoplocampa testudinea* (Klug)	955	4.4	85
	Diptera: *Eristalis pertinax* (Scop.)	10,720	2.8	7
	Coleoptera: *Melolontha melolontha* (L.)	1,095	–	70
	Hymenoptera: *Bombus agrorum* (F.)	4,635	3.2	12
	Diptera: *Leucozona lucorum* (L.) (Syrphidae)	5,160	–	11
	Neuroptera: *Chrysopa carnea* (Steph.)	1,007	1.1	53
	Diptera: *Pipiza* spp. (Syrphidae)	568	–	87
	Diptera: *Eristalis arbustorum* (L.)	2,370	2.3	20
	Diptera: *Rhingia campestris* Mg. (Syrphidae)	944	2.3	45
	Coleoptera: *Grammoptera ruficornis* (F.)	538	3.4	75
	Diptera: *Chilosia intonsa* (Loew) (Syrphidae)	600	–	47
D	Megaloptera: *Sialis lutoria* L.	520	–	39
	Coleoptera: *Pyrochroa serraticornis* (Scop.)	188	12.4	81
	Lepidoptera: *Aglais urticae* (L.)	1,535	–	10
	Diptera: *Rhagio scolopacea* (L.)	537	–	27
	Diptera: *Nephrotoma* sp. (Tipulidae)	273	2.2	49
	Diptera: *Bibio* spp.	226	15.3	55
	Hymenoptera: *Dolerus nigratus* (Müll.), *Aglaostigma aucuparia*	567	3.1	19
	Diptera: *Syrphus* spp.	364	10.4	30
	Hymenoptera: *Andrena nana* (Kirby)	297	–	33
	Diptera: *Platychirus* spp. (Syrphidae)	158	9.0	60

Table 9.4. cont.

Efficiency groups	Species	Total pollen grains geometric mean	S. D.	% fruit pollen
	Coleoptera: *Phyllobius pyri* (L.)	118	12.2	73
	Coleoptera: *Malachius bipustulatus* (L.)	85	–	86
	Diptera: *Melanostoma* spp. (Syrphidae)	195	10.0	33
E	Diptera: *Scopeuma stercorarium* (L.)	78	7.2	51
	Diptera: *Syritta pipiens* (L.) (Syrphidae)	108	2.9	10
	Diptera: *Muscina* sp., *Calliphora* sp., *Sarcophaga* sp.	302	10.8	10
	Hymenoptera: *Vespula vilgaris* (L.)	51	–	45
	Diptera: Stratiomyiidae	30	55.6	67
	Diptera: Tachinidae	71	10.8	25
	Diptera: *Dilophus* spp. (Bibionidae)	21	16.1	52
F	Dermaptera: *Forficula auricularia* L.	10	18.5	68
	Diptera: Empididae	150	7.1	5
	Lepidoptera: *Pieris napi* (L.)	8	–	62
	Diptera: Anthomyiidae	8	8.5	52
	Diptera: Sepsidae	4	3.0	95
	Diptera: Chloropidae	6	5.9	48
	Coleoptera: *Meligethes* spp. (Nitidulidae)	4	5.5	30
	Thysanoptera	1	2.7	47
	Diptera: Drosophilidae	1	2.5	55
	Diptera: *Sciara* spp. (Mycetophilidae)	1	2.8	29
	Coleoptera: *Adalia decempunctata* (L.) (Coccinellidae)	1	2.2	44
	Diptera: Chironomidae	1	1.4	0
	Diptera: Phoridae	0	–	0
	Diptera: Scatopsidae	0	–	0
	Diptera: Cecidomyiidae	0	–	0
	Coleoptera: *Oxytelus* spp. (Staphylinidae)	0	–	0
	Hymenoptera: Chalcididae, Cynipidae	1	–	0

In Kendall's (1973) experiments, however, fruit set values in flower clusters visited by bumble-bee queens was not higher than the values in the case of honeybees. The author rejects the traditional opinion that bumble-bees are better pollinators than honeybees. The explanation is that bumble-bees carry more extraneous pollen than honeybees and female solitary bees, as was pointed out by Free and Williams (1972) and Kendall and Solomon (1973), so they are not consistent flower visitors, that is, they visit different plants in the course of their foraging trips, which decreases their value as pollinators.

In addition to bees, only big-bodied syrphid flies, the *Eristalis* spp., and a conopid fly, the *Myopa* sp., bore compatible quantity of pollen on their body hair in a high percentage; they are also qualified as useful pollinators. According to Porter and Dibbens' (1977) experiment, pollination efficiency in bees is higher than that in dipterous insects. Under plastic tents, compared to self-pollination without insects, the yield could be increased only by 49% with blowflies whereas by 265% by honeybees.

9.4. The Role of Nectar and Pollen Production in Pollination

In attracting visiting insects, nectar and pollen production play a distinguished role among the entomophilous characteristics of flowers of fruit species. The quantity and quality of nectar and pollen production determine which species and which cultivars will be visited to what extent by foraging insects. Visitation, of course is in a close connection with success of pollination.

Nectar serves as a source of energy to pollinating insects. In wild insects this source of energy is required only to their own lives though in the case of apoids it is also required to a small extent to feed larvae. In honeybees the whole bee colony lives on the collected nectar and pollen. About two-thirds of a bee colony work inside the hive, and the field bees make up only one-third of the entire colony. The field bees are to ensure the source of energy of those working inside and they are to gather winter reserves for the whole colony. Bees visiting flowers of fruit trees and other plants gather much more nectar as they themselves need.

The attracting effect of different fruit species upon honeybees mainly depends on the calorific and nutritional value as well as on how easily the nectar can be gathered (Heinrich and Raven 1972). From an "energy exertion" point of view, flying is very "costly", so bees prefer the plants where they can gather nectar containing as much energy as possible with the use of as little energy as possible (Heinrich and Raven 1972, Pyke 1984). This general rule is also applicable to fruit species. The flower structure of fruit species, with the exception of currant and blueberry species, is quite simple, so to forage on flowers of fruit species is simple, and bees can reach the nectar quite easily due to the pleomorphic structure of the flowers. Currant and blueberry bear a simple stereomorphic flower structure, in which it is more difficult to reach the nectar, but the quantity of these species is less, hence they can be ignored in generalizations.

As the majority of fruit species have similar, relatively simple, flower structures, which fruit species and cultivars are preferred by honeybees basically depends on the nectar production of flowers and the sugar content of nectar. DeGrandi-Hoffman et al. (1991) also emphasized the importance of flower density. They stated that if there were more flowers on a unit of limb, then bees needed less energy to gather their nectar loads because they did not need to fly but only move from flower to flower.

Nectar production is affected by several factors, mainly the weather (Free 1970). In cool weather, or in a dry period, little nectar is produced, while in warm but not wet weather, when plants are provided with water well, there is a rich nectar production. The nectar produced does not evaporate in cloudy weather when humidity is high, but it is diluted, so its amount increases, whereas in dry weather, at high solar radiation, its amount decreases and its sugar content is condensed.

That is why different authors publish significantly differing data concerning the nectar content of flowers in fruit species (Free 1970). The differences in nectar determination are still increased by the fact that there is a periodicity of six hours in nectar production (Orosz-Kovács 1990).

Despite these factors yielding a wide variation, definite differences in nectar production could be demonstrated between different fruit species and cultivars (Simidchiev 1967, 1968, 1970, 1971a,b, 1972a,b, 1985, Benedek et al. 1989a, b, Szabó et al. 1989, Benedek et al. 1990a, 1991a, 1991b). Nectar production in *apple* is fairly rich compared to that in other species. The nectar production of *plum, sour cherry* and *apricot* trees is similar, and only somewhat less than that of apple trees, but more than that of *peach* trees. Finally, the nectar production of *pear* trees is even less. Though these differences

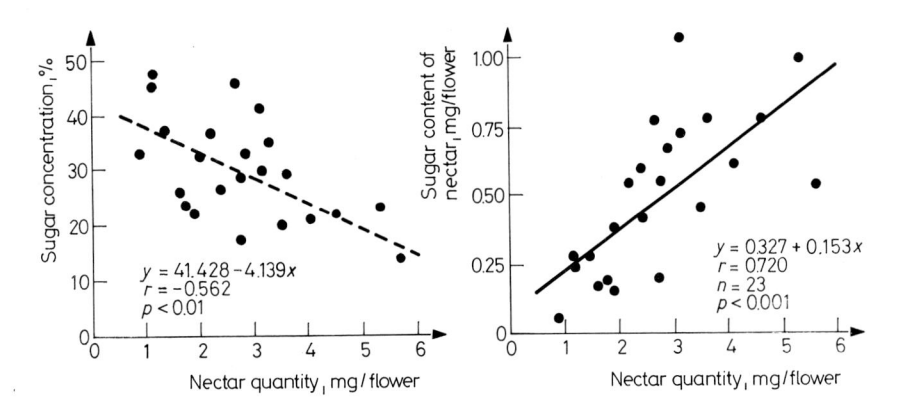

Fig. 9.2. Relationship between the nectar content of caged apple flowers and the sugar concentration (left) as well as the nectar content of flowers and the sugar content of nectar (right) (after Benedek et al. 1989a)

seem to be characteristic of the species, the influence of the weather and of the year, are usually so strong that they cover the differences caused by the species (Benedek et al. 1990a, b, 1991a,b). What has been said about nectar production holds true for the sugar concentration of nectar as well. Alhough there are less differences in sugar values and the cultivar-specific differences are more stable. It is a general observation that there is a negative correlation between nectar volume and sugar concentration, while there is a positive correlation between nectar volume and total sugar production. Figure 9.2 illustrates this point.

Bees bear foraging apparatuses specialized for pollen-gathering: certain apoids have foraging hairs on their abdominal sternum, while others have pollen baskets on their tibiae. The pollen of certain plants are more attractive to bees than that of others, whereas they discard the pollen of particular plant species (e.g. *raspberry*) from themselves instead of gathering it. It is not at all clear, however, what the attracting influence of different plant species depends on, and why the pollen of certain species is discarded instead of being gathered (Free 1970). With the exception of the raspberry, the pollen of fruit species pleases bees. As the earliest flowering spring plants, fruit species play an important role in feeding bee colonies.

Pollen productions of different fruit species are not the same. Among fruit trees, the pollen production of *apple* seems to be the richest, the pollen production of *sweet cherry* is less, and species, such as *peach, apricot, plum* and *sour cherry* are characterized by even lower extent of pollen production. The *pear* and the *black currant* produce very little pollen. The general pollen production trend does not hold true for every cultivar within a species. There are cultivars with very rich and very poor pollen production values within the same species (Godini 1981, Simidchiev 1985, Benedek et al. 1989a, b, Benedek et al. 1990a, b, 1991a,b, Szabó et al. 1989). Similar to nectar production, the pollen production of flowers also shows a fairly wide year-dependent variation. Though these differences are often greater than the demonstrated differences between cultivars, the pollen production values of cultivars of *almond* (Godini 1981), *apple* (Benedek et al. 1989), and *plum* (Szabó et al. 1989) are different. In contrast, the year-dependent variation in pollen production is so strong that it usually suppresses the cultivar-specific differences in *sweet cherry* and *sour cherry* (Benedek et al. 1990b) and *apricot* and *peach* (Benedek et al. 1991a,b).

To sum up, in most fruit species both nectar and pollen play an important role in attracting pollinating bees. In *pear*, however, it is the pollen that seems to be the more important attracting factor (Free 1970, Benedek et al. 1974). The role of pollen is also determinant in the case of *sweet cherry* and *sour cherry* though nectar also plays some role (Nyéki and Szabó 1989).

Flowers of certain *apricot* cultivars produce much more pollen than other cultivars. Thus in these cultivars pollen clearly plays a distinguished role in attracting pollinators (Benedek et al. 1991*b*). *Black currant* produces a limited amount of pollen (Free 1970), hence, here the nectar ensures the main attracting effect.

9.5. The Activity and Behaviour of Insect Pollinators in Different Fruit Species

On days with normal weather conditions pollinating insects begin to work in the morning, then their activity increases parallel with the rise in temperature, and reaches its peak in the early period of afternoon, when the temperature reaches its daily maximum, and then in the later period of afternoon the activity gradually decreases. However, there are remarkable differences between different pollinating insects.

Honeybees, for instance, begin to fly at about 8 a.m. during bloom of fruit species, and their daily activity comes to an end at about 17 to 18 in the case of good weather (Fig. 9.3). The most honeybees can be found in orchards between 13 and 14 p.m. (Free 1960*b*).

Dipterous insects fly only in a definitely good, warm (15 to 20 °C) and sunny weather (Boyle and Philogéne 1983). Thus, they are much more sensitive to weather. If the weather becomes cooler, their flower-visiting activity becomes less intensive, decreasing their pollination efficiency. Apoids, however, fly also under worse weather conditions. Honeybees, for instance, begin to fly at about as low a temperature value as 10 °C when the sun shines, and remain active in cloudy weather at 13 to 15 °C.

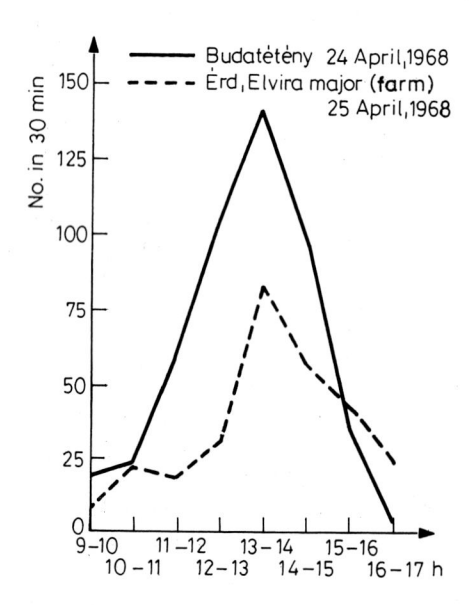

Fig. 9.3. Rhythm of daily activity of honeybees in apple trees (number of bees visiting 100 flowers in 30 min) (after Benedek et al. 1972)

Moreover, bumble-bees, and solitary and semisocial wild bees remain active at even lower temperatures so they eke out the decreased activity of honeybees, or partially substitute for honeybees (Paarmann 1977, Boyle and Philogéne 1983, Klug and Bünemann 1983). Big-bodied female bumble-bees remain active even in rain, when honeybee and wild bee activities come to a lull.

The daily activity of bumble-bees practically complements that of honeybees. On warm days, bumble-bees fly mainly during the evening and resume their activity in the morning (Paarmann 1977), while honeybees work during the day. On cool days honeybees, but not bumble-bees, cease to visit flowers; bumble-bees only postpone the daily peak of their activity to the afternoon. It very rarely occurs that the daily activity of bumble-bees coincides with that of honeybees.

In fruit species, wild bees forage primarily for pollen because pollen is required to feed their brood. Honeybees, however, are either pollen-gatherers, or nectar-gatherers, or gather both at the same time (Parker 1926). The type of behaviour strongly affects pollinating efficiency. Pollen-gatherer bees approach flowers from above and they must touch the stamens as well as the stigma in each flower visited because otherwise they cannot come at the pollen. Though they regularly sweep the pollen from their bodies into their baskets, pollen always remains stuck on the hairs of their bodies that may be enough for cross-pollination if it arrives at stigmas of flowers.

According to an observation by Free (1960*a*), when nectar-gatherer bees stop on anthers and push their proboscises and front parts of their bodies towards the nectary, they necessarily touch the stigma and pollinate it. Half of nectar-gatherer bees collect also pollen on the same flowers. Free (1960), however, argues that a bee follows a consistent behaviour in the course of a foraging trip.

In the case of honeybees, there is a fourth group besides the groups of nectar-gatherers, pollen-gatherers and nectar and pollen-gatherers. A bee can collect nectar in the following way: landing on a petal, it pushes its tongue towards the nectary from the side along the petals at the base of stamens. This way the nectar can be sucked from the flower without touching the stigma and the stamens, hence no pollination takes place (Robinson 1979). Brittain (1933) called attention to this type of behaviour called "side-workers", long ago. Later Free (1960) pointed out that the proportion of side-worker nectar-gatherers is a function of flower structure. If the filaments are short and relatively flexible, bees prefer approaching the flower from above, pollinating it. Whereas on flowers with long and rigid filaments, bees learn soon that nectar can be acquired also from the side. This is fairly usual in the case of cultivars with erect filaments because there is enough distance between stamens and petals. According to the data of Free and Spencer-Booth (1964*a*) concerning apple, 61% of nectar-gatherer bees approach flowers from above, 26% comes from the side, and only 13% chooses both ways to obtain nectar.

Hence, it is not at all surprising that different authors publish different data concerning the proportion of pollen-gatherer and nectar-gatherer bees in the case of different cultivars concerning the same fruit species (Free 1970, McGregor 1976). The proportion of pollen-gatherer bees is higher when the colony needs much pollen to increase brood (Hellmich and Rothenbuhler 1986).

Another main factor, which is also a variable, is the availability of pollen. Pollen production is different in various fruit species and cultivars (Benedek et al. 1989*a,b*, 1990*a,b*, 1991*a,b*, Szabó et al. 1989, DeGrandi et al. 1991, Sklanowska 1991). The pollen production of *black currant* flowers is very low. Bees cannot gather a sufficient amount of pollen load in the course of a foraging trip and the pollen of black currant does not really attract honeybees. Hence, the bees often sweep black currant pollen from their bodies (Free 1968a). Consequently, only 0 to 10% of honeybees, and 0 to 20% of female bumble-bees, visiting black currant gather pollen. Free (1968a) had set bee colonies near a block of flowering black currant plants and observed that only 1% of pollen gained by them had come from black currant flowers.

The effect of weather has major effect on the availability of pollen. On warm days, when pollen grains stick together to a lesser extent than on wet days. Hence pollen on warm days can be acquired more easily, and the number of pollen-gatherer bees increases parallel with rise in temperature (Free 1960a). The variation across days may be attributed to the fact that though the behaviour of bees shows a fairly constant tendency, pollen-gatherer bees are ready to change their behaviour when the amount of pollen becomes less, and they begin to gather nectar (Free and Spencer-Booth 1964a). In summary, the behaviour of honeybees on flowers of fruit species is affected by the available amount of pollen and nectar; meteorological factors, which also exert an immediate influence upon the first factor; the number of flower-visiting insects, which decrease the available amount of pollen and nectar; and the attracting effect of pollen and nectar upon the honeybees (Free 1970).

An important factor of the behavior of pollinating bees is the speed of flower visitation. The following circumstances impose influence upon this parameter: flower size and structure, types of behaviour and weather. Most data on the speed of flower visitation are concerned with honeybees (Free 1960b, 1970). Another sort of bees is either slower or faster than honeybees. Bumble-bees visit 2.5 times as many flowers as honeybees (18.7 flowers per minute vs. 7.1 flowers per minute). Anthophorid bees visit less flowers in a minute (15) but still twice as many as honeybees (Menke 1951). According to Løken (1958), bumble-bees are twice as fast as honeybees but the latter are twice as fast as small-bodied solitary bees. Paarmann (1983) argues that there may be a significant difference among bumble-bees. While workers of bumble-bees are only somewhat faster than honeybees, queens are twice or three times as fast.

Table 9.5. Foraging Speed of Honeybees as Affected by Their Foraging Behaviour on Flowers of Fruit Trees (after Free 1970)

Species	Nectar-gatherers	Pollen-gatherers	Unclassified	Author[*]
	flowers visited per min			
Apple	–	–	6.0	Sax (1922)
	–	–	6.0–9.0	Hutson (1926)
	–	–	7.1	Menke (1951)
	8.1	15.8	8.8	Rymashevskii (1956)
	6.7	10.9	–	Free (1960a)
	7.2	7.1	–	Free and Spencer-Booth (1964a)
Apricot	–	–	8.2	Rymashevskii (1956)
	5.6	7.9	–	Free (1960a)
Peach	2.7	4.0	–	Free (1960a)
Pear	–	–	16.0	Hutson (1926)
	9.1	10.2	–	Stephen (1958)
	7.7	6.6	–	Free (1960a)
Plum	3.8	10.7	–	Free (1960a)
Cherry	–	–	7.4	Rymashevskii (1956)
	6.0	6.7	–	Free (1960a)

[*] For references, see the original source (Free 1970).

In the case of honeybees, pollen-gatherers are always faster flower visitors than nectar-gatherers (Table 9.5). This is one of the reasons why pollen-gatherers are held to be more effective pollinators than nectar-gatherers. Another reason is that more pollen grains stick to the bodies of pollen-gatherers, and they always touch the stigma while visiting a flower (Free 1960*b*, 1970). Among flowers visited by pollen-gatherers, a fruit develops in 1.5 times to twice as many cases as in flowers visited by nectar-gatherers.

Nevertheless, the speed of flower visitation is not constant. Benedek and Prenner (in Benedek et al. 1974) demonstrated that the working speed of bees that were gathering pear pollen ranged from 3.1 to 14.5 flowers per minute, depending on air temperature ranging from 14 to 23 °C ($r = 0.42, p < 0.001$).

9.6. Factors Affecting the Distribution of Honeybees in Orchards and the Intensity of Their Flower Visitation

More honeybees visit those flowers that provide more food for them in an easily accessible way (Free 1960*b*, 1970). As fruit species are characterized by different pollen and nectar production, they attract honeybees to distinctly different extents. Bees are able to distinguish between cultivars (Free 1966), and if two cultivars are different in respect of providing them nutrition, the more favourable will be visited by more honeybees; moreover, the bees will become conditioned to this cultivar (Free 1963).

A favourable pollination of orchards primarily requires that the pollinizer and the cultivar to be pollinated should bear similar nutritional values to honeybees. Unless this requirement is satisfied, the majority of bees visiting the orchard will become conditioned to the cultivar more favourable to them. In a case like this, bees rarely transfer pollen from one cultivar to the other in spite of their high density, therefore they do not really help in cross-pollination. Although nectar and pollen production play as important role as pollen compatibility and coinciding blooming periods in cultivar arrangement, the available information on cultivar-specific features affecting bee pollination is still limited (Benedek et al. 1989*a,b*, 1990*a,b,* 1991*a,b*, Szabó et al. 1989, DeGrandi-Hoffman 1991, Sklanowska 1991). Only nectar production is an exception, on which topic information is available (Free 1970, McGregor 1976).

The unfavourable effect coming from differences in pollen and nectar production between cultivars in an orchard, however, may diminish under certain circumstances (Free 1970). If bee density increases, the large number of honeybees visiting the more favourable cultivar will soon gain so much pollen and nectar that the nutritional value of this cultivar, as well as its relatively great attracting effect, will diminish and become similar to that of the cultivar with lesser nectar and pollen production. In the case of a great bee density, this state of balance between the nutritional value of the originally "better" cultivar and that of the other will be achieved in a couple of hours. Nevertheless, if the original difference is too great, such an equalization will take place only at the end of the day, too late for good pollination. Thus the lost time, until the state of balance is achieved, are hours that would be valuable in respect of cross-pollination.

On warm days, when the humidity of the air decreases, honeybees gather the available amount of pollen in a short period. It may happen that until the state of balance is achieved the amount of pollen of the most favourable (pollinizer) cultivar will have become so little that an intensive cross-pollination in the remaining hours of the day will not be possible.

Another factor relevant to cross-pollination is the size of foraging areas of bees. To carry out pollination among neighbouring flowers, it suffices to fly from flower to flower, but cross-pollination requires bees to fly from tree to tree, moreover, from the tree of the pollinizer to the tree of the cultivar to be pollinated which may be several trees away.

Honeybees make six to eight foraging trips on a favourable day (Free 1970). In the course of a foraging trip, they strive to accumulate a load of pollen or nectar, which they carry into the hive at the end of the trip. How much time it takes to accumulate a load depends on how much pollen or nectar can be found in a flower. Thus the availability of pollen or nectar affects the length of the foraging trip. It is very difficult to investigate the entire trip of a bee so there are few pieces of information as to how many flowers a bee visits in the course of a foraging trip. Free (1960*a*), for instance, reports that pollen-gatherer honeybees visited 89 flowers on *apricot* trees, and 38 flowers on *pear* trees, whereas nectar-gatherers visited 76 *pear* flowers, and 82 *sweet cherry* flowers. Others also published similar data (Free 1970). According to Free and Spencer-Booth (1964*b*), the length of the trip covered is in positive correlation with the number of the flowers visited. While visiting 10 to 20 flowers, forager bees cover 1.5 m on average, whereas a trip of 2 m corresponds to the visitation of 30 to 40 flowers (Free and Spencer-Booth 1964*b*). The same authors observed that if the visited trees provided rich food, bees visited one or a few neighbouring fruit trees in the course of a foraging trip. According to Free (1960*a*, 1962), two apple trees are visited on average in the course of a foraging trip. If trees within rows are closer to each other than between-row-spacing, bees prefer flying from tree to tree in the same row rather than changing rows. If limbs of neighbouring trees touch each other, bees often fly from one tree to another, while they do not like to cover the long distance between trees whose canopies are far from each other (Free and Spencer-Booth 1964*a*). That is why bees rarely fly from one row to the next row in orchards where trees are planted in hedgerows but they cover long distances within rows.

The flying behaviour of bees plays a crucial role in the success of pollen transfer from pollinizer to main cultivar in orchards. Free (1962) and Free and Spencer-Booth (1964*b*) demonstrated that in *plum*, *apple* and *pear* orchards, farther from the pollinizer both the extent of fruit set and the amount of yield were less. Later, others produced further evidence in support of the negative correlation between fruit set and distance from pollinizer (Free 1970). Free (1962) designed experiments to reveal factors of the effect of flying behaviour. The author used eight trees surrounding a pollinizer tree and studied fruit set on the sides of trees towards the pollinizer tree and on the opposite sides. The results were 10.8 to 4.3% in *plum*, and 10.4 to 4.8% in *apple* in favour of the side towards the pollinizer. As for seed content, the mean results were 4.7 to 3.3 in *apple* also in favour of the side towards the pollinizer. DeGrandi-Hoffman et al. (1984), however, concluded to the opposite. In their experiments fruit set values did not show a decreasing tendency getting farther from the pollinizer cultivar. This fact can be attributed to hive-internal pollen transfer between bees, which comes from their social behaviour. That is why honeybees are held to be more valuable pollinators of self-incompatible fruit species than solitary bees.

These results clearly show that a bee works within a quite small area during the course of a foraging trip. The size of foraging area, however, is affected by several factors. It may spread to a large number of trees under unfavourable nutrient availability conditions, especially when nectar production is poor. In unfavourable weather bees visit less flower on a tree, but more trees than in good weather. The wind and the disturbing influence of other insects may also result in an increased foraging area. The foraging area is larger during the early and the late parts of the flowering period, and in the morning and just before the evening, than otherwise. Though bees usually work in the same area on two succeeding days, a bee's foraging area is 340 m^2 in a period of two days, but enlarges to 1016 m^2 in an eight-day-long period (Free 1970).

Kendall (1973) examined the compatibility of pollen carried on bodies of bees in orchards using two cultivars of *apples*, a dwarf bush tree and standard trees. The author observed that there was more compatible pollen stuck on bodies of bees where pollinizer trees had been interplanted in the row compared to those where pollinizers had been planted in discrete rows. This observation indicates, verified by others, that bees prefer transferring pollen from tree to tree within rows rather than transferring pollen from row to row.

Townsend et al. (1958) marked bees leaving their hive with fluorescent powder, which the bees left on flowers that they visited during the course of their foraging trips. Later the researchers found also wild bees stained with this powder. Thus these wild bees also visited the flowers, picked up the powder and spread it. This experiment verified that cross-pollination might be facilitated by a pollen transfer process. Honeybees, during the course of their foraging trips, pick up pollen that has been left on flowers by other bees in their overlapping foraging areas, and they spread it farther. This phenomenon thus facilitates cross-pollination (Free 1970).

Furthermore, a honeybee is able to transfer viable pollen from one of its foraging trips to another (Free 1966, Free and Durrant 1966). Pollen grains stuck on a bee during the course of a foraging trip may cross-pollinate flowers visited on a later trip. Stuck on bees, pollen remains viable for 48 hours (Latimer 1938), increasing the possibility for cross-pollination. This may be a reason for the phenomenon that sometimes there is a quite good yield in orchards where the placement of the main cultivar and the pollinizers is not advantageous to bees for transferring pollen directly (Free 1970).

Bee colonies send out foragers, whose number depends on the colony's demands for food and drink. The proportion of field bees, which work outside the hive, is usually about 30% (Free 1970), so the foraging area of a bee colony spreads to a large territory and under normal conditions is not limited to a single plant, or a single farm block, or a single orchard. Thus the foraging area of a whole bee colony makes up a multiple-square-kilometre territory though most foragers can always be found at a distance not more than some 100 m from the hives, especially in poor weather (Free 1970). The importance of this observation lies in the fact that, while in the hive bees are sweeping the pollen from themselves, pollen grains may be transferred to other bees that themselves have never visited the plant whose pollen they carry during their next foraging trip (Standhouders 1949, Lukoschus 1957, Karmo 1960, Free and Williams 1972*b*, Karmo and Vickerey 1987). This is supported by Free's (1966) observation that bees that were captured when leaving the hive or visiting a certain plant always carried on their bodies some pollen that had come from cultivars different from the cultivar that they were just visiting.

DeGrandi-Hoffman et al. (1984) examined, with scanning electron microscope, pollen grains found on forager bees captured on 'Delicious' ('Millespur' strain) and 'McIntosh' ('Macspur' strain) apple trees and collected from stigmas of fruit trees because this way not only the species but even the cultivar of the pollen could be identified. They reported that the bees visiting a certain apple cultivar had borne on their bodies pollen grains coming from numerous apple cultivars or other plant cultivars (Table 9.6). High percentage of the bees under examination bore non-apple pollen. In addition to the pollen of the cultivar they were just visiting, they carried also compatible apple pollen of

Table 9.6. Percentage of Foragers and of Stigmas with Compatible, Self and Non-Apple Pollen, from 'Delicious' and 'McIntosh' apples (after DeGrandi and Hoffman et al. 1984)

	Delicious	McIntosh
% foragers with:		
compatible pollen	100.0	100.0
self-pollen	84.2	87.7
non-apple pollen	89.5	93.3
No. of foragers	19	15
% stigmas with:		
compatible pollen	76.9	91.3
self-pollen	73.1	82.6
non-apple pollen	77.9	17.4
No. of stigmas	26	23

other cultivars suitable for pollination. The pollen combination observed on bees reflected the pollen combination observed on stigmas, that is, in addition to compatible pollen and their own pollen, on most stigmas non-apple pollen could be found as well.

DeGrandi-Hoffman et al. (1984) concluded, based on their results, that pollen transfer between bees inside the hive could serve as an explanation of the phenomenon that so many sorts of pollen grains could be found on bees visiting only apple trees. Furthermore, this can also serve as an explanation of the fact that certain researchers could observe significant extents of fruit set even in incompatible circumstances (Williams 1959, Free 1962, 1966, Free and Spencer-Booth 1964b, Maggs et al. 1971). The fact that bees can transfer pollen from trees they have not visited, can serve as an argument against Stephen's (1958) conjecture that the cross-pollinated fruits that he found in the middle of a solid pear block had been produced by wind pollination. The fact cannot be excluded that such fruit had been created as a result of the activity of bees.

As for the flower-visiting behaviour of wild bees in comparison with honeybees, wild bees visit flowers less systematically, insist on cultivars to a lesser extent, and more usually shift from a tree to another tree, or from a cultivar to another cultivar (Free 1970, Paarmann 1977, Klug and Bünemann 1983).

Foragers of honeybee colonies usually visit flowers of a single plant species, which must reflect their capability for communicating on good forages. Consequently, they gather the most of their food from the several best plant species, and neglect a lot of other plants at the same time (Free 1970). Synge (1947) examined foraging sources of neighbouring bee colonies, and found them to be different, and the plant assortment visited by their field bees was not necessarily the same. Such differences also facilitate a better cross-pollination of fruit species. Furthermore, bee colonies may differ in the intensity of their visiting a particular orchard and in the intensity of their gathering pollen. The value of intensity of gathering pollen may vary during the blooming period even with the same bee colony (Free 1959).

In blooming orchards a balanced fruit set and a uniform yield are to be expected if there is a uniform pollinator activity in the area. Wild pollinating insects fly into the area from outside thus there may occur a greater pollinator density in the outermost zone than in the middle of the orchard. The distribution of wild pollinating insects depends on the locality, which cannot be changed, at least within a single flowering period.

Honeybee density can be controlled since bee colonies can be moved to orchards. The distribution of honeybees has been a controversial topic for a long time.

It has often been reported that honeybee density is decreasing while getting farther from beehives (Free 1960b, 1970) though several bees visit plants very far from their hives. The distance-dependent distribution of bee density, however, is a function of several factors, primarily the weather. In good weather, for instance, the highest bee density can be found within a circle around the hives 200 to 300 m in diameter, whereas in unfavourable weather this diameter decreases to 100 to 150 m (Free 1970). The reason is that though honeybees are able to control temperature inside the hive to some extent due to their social life style, the foragers leaving the hive entirely depend on the effect of the weather. A warm and sunny weather free from wind is characterized by the most intensive visitation of flowering plants by bees, while clouds, cool air and strong wind hinder bees from flying. At a velocity of wind of 15 km/h, bee activity is strongly decreased, and at 30 km/h bees do not work at all. Bees begin to fly at a temperature of 8 °C but large crowds of bees begin to work at 12 to 25 °C. The most favourable temperature is above 18 °C (Free 1970, Benedek et al. 1974). Consequently, the limiting influence of weather can be compensated for by moving more bee colonies to the orchard.

Bee distribution also depends on whether the bee colony is stationary on the same location, or it has been moved to the new place. It has been known for long that if bee colonies are moved to an orchard, the bees of the colony remain near the hives on the first days and orientate themselves but spread their foraging areas only later (Free 1960b, 1970).

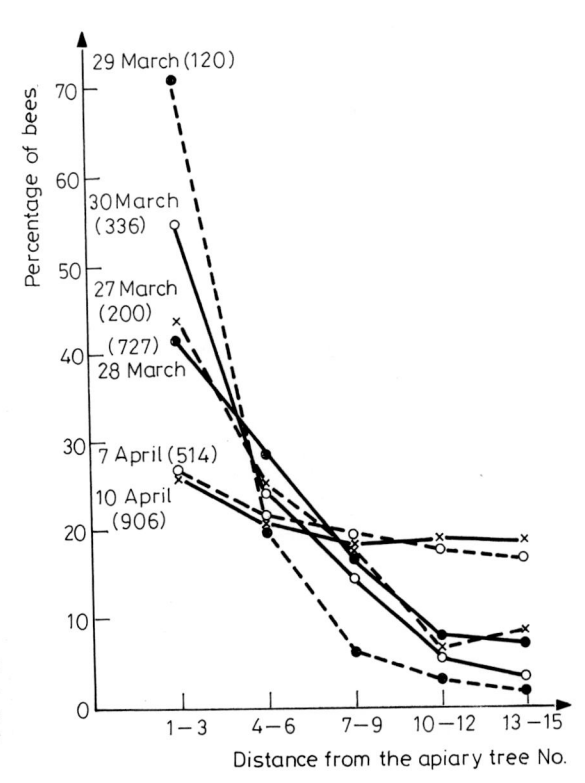

Fig. 9.4. Percentage distribution of bees on plum trees at different distances from honeybee colonies (figures in brackets are the mean numbers of bees per count) (after Free and Spencer-Booth 1963)

Free and Spencer-Booth (1963) placed an apiary in the middle of a *plum* orchard. At the beginning of full bloom (28 to 30 March), bee density decreased to a great extent with the distance from the apiary (Fig. 9.4); whereas at the end of full bloom (7 to 10 April) bee density became more uniform, that is, bees were flying farther from the apiary.

According to Levin (1961), foraging areas of bee colonies, and hence the distribution of honeybees in flowering orchards, are affected by neighbouring apiaries. Because of increasing competition, foraging areas contract towards neighbouring apiaries. The more bee colonies can be found in the neighbouring apiary, the stronger this effect is. Consequently, a uniform distribution of forager bees requires a well-spread distribution of bee colonies.

The relation between the placing of bee colonies and the distribution of bees has been discussed by several researchers (Free 1970). Experimental results suggest that bee colonies are to be placed singly and uniformly to achieve the most uniform bee density in the orchard, but an equidistant placing of groups of 15 to 20 bee colonies entails a similar result (Free 1970).

The statements above refer to situations when beehives are placed within the orchard or in its neighbourhood. Tsigankov (1953), however, studied the influence of the distance of the apiary from the orchard and its relation to flower visitation of bees. The results (Table 9.7) show that bee visitation quickly decreases parallel with an increasing distance between the orchard and the apiary. Within a distance of 500 m bee visitation remains intensive.

Honeybees visit flowering fruit species in order to gather some food. If there is a more favourable source of food available, they may shift to this "alternative source of food". Fruit species are in keen competition with weeds, the *dandelion*, for instance, because the bloom period of this weed coincides with that of most fruit trees (Free 1960, 1970, Benedek et al. 1974, McGregor 1976). If dandelions are widespread on the orchard floor field bees may visit dandelions instead of fruit trees (Free 1968b). *Stellaria* is another species that may distract bees from visiting fruit species and bushes, such as *Crataegus*, whose bloom periods coincide with those of fruit species, may compete for bee activity.

This distracting influence, however, may come from field crops as well. In the north, *rape*, may also attract bees. Blooming periods of fruit trees usually precede the blooming period of the rape, but in certain years, there may be a great overlap. In such years the majority of bees may be distracted from orchards by the rape. The early sown *mustard* is also a severe competitor to fruit species.

Table 9.7. Effect of Distance from Apiary on Flower Visitation of Honeybees (after Tsigankov 1953)

Distance between orchard and apiary (m)	No. of flower visitations on 100 flowers in 10 min	No. of flowers visited by bees in 10 min
100	45.75	130.25
500	34.95	113.55
1,000	15.35	50.55
1,500	9.50	29.00
2,000	0.90	3.40

Note: Observations were carried out in 17 apple cultivars.

Different fruit species attract bees to different extents. Among temperate fruit tree species, *sweet cherry*, *plum* and *apple* exert the most attractive influence upon honeybees. Though it may occur that the plum distracts bees from the apple, and the apple from the sweet cherry. However, this occurs in northern orchards only where bloom is condensed and only if these species are grown at the same locality. All other fruits are more attractive to honeybees than *pear*, *sour cherry*, or *apricot* trees. Distracting influence can be compensated for by moving even more bee colonies to the orchard, or, in the case of weeds, by their elimination.

Nevertheless, there is a keen competition within the orchard as well, between different fruit cultivars, for the attracting effect of particular cultivars may be different because of differences in their nectar and pollen production values (Free 1970). Benedek et al. (1989a,b) compared several *apple* cultivars with each other. They observed great differences in visitation by bees. They observed furthermore that the relative intensity of visitation by bees, that is, the relative attracting effect of cultivars, was primarily influenced by the amount of nectar, and not sugar concentration. Certain cultivars that produce little nectar can attract bees by their relatively rich pollen production. They also observed that flowers of apple cultivars bore properties that showed a wide cultivar-dependent variation. Bees are able to recognize flower structure and other cultivar-specific features (Free 1966) so they are also able to prefer visiting the more attractive cultivars, of course. As for other fruit species, the intensity of visitation by bees may be influenced by flower characteristics, such as flower density in *plum* (cultivars with lower flower density are visited by bees to a lesser extent; Szabó et al. 1989), or pollen production in *pear* (bees prefer a richer pollen production; Free 1970).

9.7. The Effect of Insect Pollination on Fruit Set, Yield and Fruit Characteristics

There are numerous experimental results in support of the conclusion that on branches caged during bloom, there will be less yield, if any, compared to open pollination. This is a widespread technique to reveal pollinating requirements of fruit species (Free 1960b, 1970, Benedek et al. 1974, McGregor 1976).

However, we know little about the question as to what will happen if a whole orchard lacks pollinating insects, and what is the relation between the number of pollinators and yield. The interpretation of the limited data is very difficult because of the influence of weather that exerts some influence upon bees as well as upon the plants. Now let us consider the few results obtained concerning this question.

In *apple* orchards with few bumble-bees and solitary bees, in a year when the weather was good, Bornus et al. (1976) received a yield of 35 tons per hectare with two bee colonies per hectare. Then they reduced the extent of visitation by bees by 40% by decreasing the number of bee colonies to 1.2 colonies per hectare, and the yield was reduced to 27 tons/hectare. A further decrease of visitation by bees by 20% to 0.8 colonies per hectare resulted in the yield being reduced to 23 tons per hectare.

Another way of examining the effect of bee pollination is to measure the effect of distance from apiary upon yield. Tsigankov (1953) indicated that when the apiary was at a distance of 1,000 m from the *apple* orchards, about half as many bee visitations occurred and almost 60% less yield was received compared to the case when bees were only 100 m from the orchards. When colonies were 500 m apart from the orchard, bee visitations were about 10% less but the amount of yield was almost 40% less. Yakovlev

Table 9.8. Yield of Fruit Trees as a Function of Distance from the Apiary Situated in the Border of the Orchard

Measurements	Sour cherry ('Pándy') (Benedek and Martinovich 1971)			Apple ('Jonathan') (Benedek et al. 1972)					
	Budatétény (Central Hungary)			Nyíregyháza (NE Hungary)			Százhalombatta (Central Hungary)		
Distance (m)	150	300	400	150	300	600	150	300	600
Fruit set (%)	45.6	30.2	17.0	66.9	61.6	46.1	14.5	12.0	5.2
Fruit weight from 500 flowers (g)	1,265	935	395	4,106	5,223	2,642	15,020	8,400	2,200
Mean fruit weight (g)	7.2	8.3	6.5	114	114	110	211	161	110

Measurements	Apple ('Jonathan') (Benedek et al. 1972)			Pear ('Bartlett') (Benedek et al. 1974)			
	Budatétény (Central Hungary)			Gersekarát (Western Hungary)			
Distance (m)	150	300	600	50	100	200	300
Fruit set (%)	13.2	11.5	6.1	26	16	17	11
Fruit weight from 500 flowers (g)	–	–	–	–	–	–	–
Mean fruit weight (g)	–	–	–	–	–	–	–

(1959) indicated that when bees were 300 m from the apple orchards, the yield was 5.1 tons/hectare but when the distance had been increased to 900 to 1,200 m, and then 1,500 to 1,900 m, the yield decreased to 2.7 and 2.3, tons per hectare, respectively. Pusztai et al. (1969) observed in a large apple orchard with an apiary outside of the orchard that the yield was three times as much as at 216 m than at 2,000 m from the apiary.

The distance-dependent decrease of fertilization and yield can also be observed if bee colonies have been placed at the side of the orchard (Benedek et al. 1971, 1972, 1974). Fruit trees are so sensitive that a significant effect can be demonstrated as near as at a distance of 200 to 300 m (Table 9.8). Decrease in yield is even more significant than decrease in fruit set. What is more important, parallel with decrease in yield, no increase in fruit size can be observed although these orchards were not irrigated, and that may account for the lack of fruit growth on trees with lighter crop. Benedek et al. (1974) measured yields of trees in a *pear* orchard to the north and to the south of the hives at distances of 100 and 200 m and received the following results: 62±6 and 43±6 kg per tree to the north respectively, and 98±6 and 24±4 kg per tree to the south, respectively. The decrease in fruit set and yield were consequences of the decrease of the intensity of bee visitation. Benedek and Martinovich (1971), and Benedek et al. (1972) argue that an increase in the intensity of bee pollination entails not only an increase in yield but that in fruit size, and hence fruit quality. This is also supported by another observation of Benedek et al. (1972): compared to isolated twigs, which bees cannot come at, open pollinated twigs brought not only more pieces of apple but such that were bigger and of a better quality. The same effect of bee pollination was also observed in *black currant* (Benedek, Porpáczy, and Virányi, cit. by Benedek et al. 1974). On open parts of branches, which bees were free to visit, a richer yield and berries of a better quality were obtained also in the case of a cultivar inclined to self-fertilization in a high percentage (40%).

As for fruit quality, we would like to cite a shared conclusion of several researchers (Free 1968, Nye and Anderson 1974, Blasse 1981, Svensson 1991) that bee pollination entails not only an increase in fruit set but, in *strawberry*, a decrease in the proportion of malformed berries, and hence an increase in the proportion of well-formed berries, that is, better quality and an increased market value.

Benedek and Bánk (cit. by Benedek et al. 1974) called attention to the influence of the length of bee pollination period. They restricted the bee visitation period and the yield decreased in this order: (*a*) most yield was obtained when flowers were open in the whole period, (*b*) yield was less when flowers were isolated with muslin from the fifth day of blooming, (*c*) no yield was obtained when flowers were isolated with muslin from the first day of blooming. Further detailed experiments (Benedek et al. 1989a,b) convincingly verified that restrictions on the effective period of bee pollination would drastically decrease the amount of yield. Consequently, rich yield is to be expected only when an intensive bee visitation covers the entire blooming period. If the effective period of bee pollination is reduced by any extent because of unfavourable weather conditions, a decrease in yield cannot be avoided.

Restrictions on the effective period of bee pollination have the same consequences on fruit set and yield in *plum* and *sour cherry* as in apple (Szabó et al. 1989, Benedek et al. 1990a,b). The effect is more explicit in self-sterile cultivars but is also observable in self-fertile cultivars.

Roversi and Ughini (1986) studied the effect of restrictions on bee pollination upon a *sweet cherry* cultivar, 'Flamengo'. The essence of their method is that they caged limbs for certain periods, and then exposed them to bee pollination for different periods of time varying day by day. Fruit set values were used to characterize the effect. The later they started this daily exposure to bee pollination in the bloom period (Fig. 9.5), and in the course of the day (Fig. 9.6), the lower the fruit set values were. In contrast, increase in the duration of daily exposures had a definitely increasing effect upon fruit set (Fig. 9.7). An increase of bee pollination period thus increases the extent of fruit set, while restrictions on bee pollination undoubtedly decrease it.

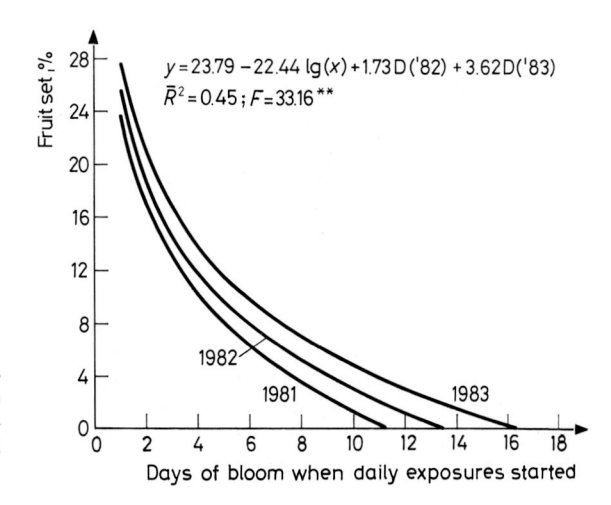

Fig. 9.5. Fruit set of sweet cherry flowers, cv. 'Flamengo', as affected by the start of daily exposures to bee pollination during bloom (after Roversi and Ughini 1986)

$y = 23.79 - 22.44 \lg(x) + 1.73 D(\text{'}82) + 3.62 D(\text{'}83)$
$\bar{R}^2 = 0.45; F = 33.16^{**}$

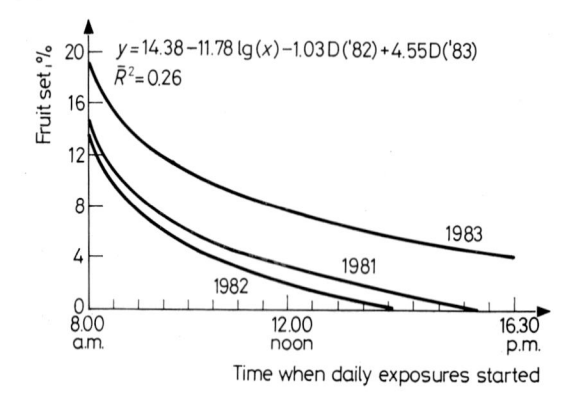

Fig. 9.6. Fruit set of sweet cherry flowers, cv. 'Flamengo', as exposed daily to bee pollination from different times during the day (after Roversi and Ughini 1986)

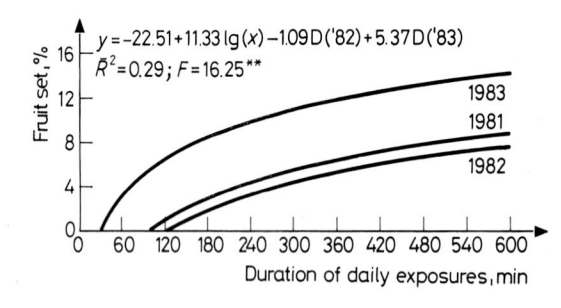

Fig. 9.7. Fruit set of sweet cherry flowers, cv. 'Flamengo', as affected by the start of daily exposures to bee pollination during the blooming period (after Roversi and Ughini 1986)

Benedek et al. (1989a) established that different apple cultivars react to restrictions on bee pollination to different extents. The sensitivity of the cultivars that they examined increased in the following order: 'Idared' was the least sensitive to restrictions on bee pollination, 'Jonnee' was a little more sensitive, 'Wellspur' and 'Jonathan' were even more sensitive, and 'Starkrimson Delicious' was extremely sensitive (Fig. 9.8). Thus, in the course of the arrangement of bee pollination in apple orchards, one should consider the sensitivity of the cultivar to restrictions on bee pollination. Thus the intensity of bee pollination depends not only on meteorological conditions but on certain features of plantation.

One might think that an increase in fruit load resulting from an increased intensity of bee pollination entails fruit size decrease. It is not the case, however, if yield and water limitations need not to be considered (Benedek 1980, Benedek et al. 1989a, Benedek and Nyéki 1990). Under conditions of intensive bee pollination, producing at least ten apples from 200 flowers, the individual weight of apples did not decrease, or scarcely decreased, in response to an increase in fruit load, in 'Idared', 'Jonnee' and 'Starkrimson Delicious'. Apple weight was significantly greater only when one to three apples had developed from 200 flowers (Fig. 9.9). It appears that when bee pollination is intensive, fruit weight is determined by other factors, such as water, nutrition, etc., than pollination.

Fruit weight in *apple* shows a significant correlation with the number of viable seeds (Fig. 9.10; Benedek et al., unpublished). A greater number of seeds is a result of a better fertilization, which can only be attributed to a more intensive bee pollination. The 'Abate Fetel' pear cultivar brings higher yield pollinated by insects than without their help (Romisondo and Mé 1972). Though individual fruit size shows a decreasing tendency in

Fig. 9.8. Relative sensitivity of apple cultivars to the limitation of the length of effective time of bee pollination. Sensitivity of cultivars is compared to cv. 'Jonnee' in both years (after Benedek et al. 1989a)

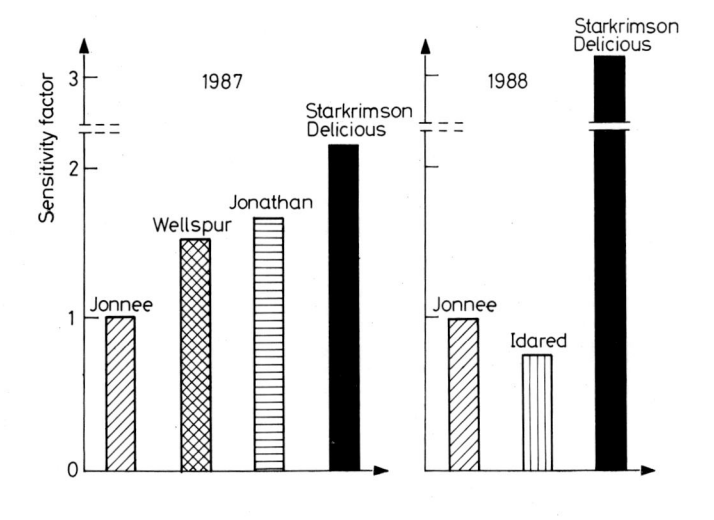

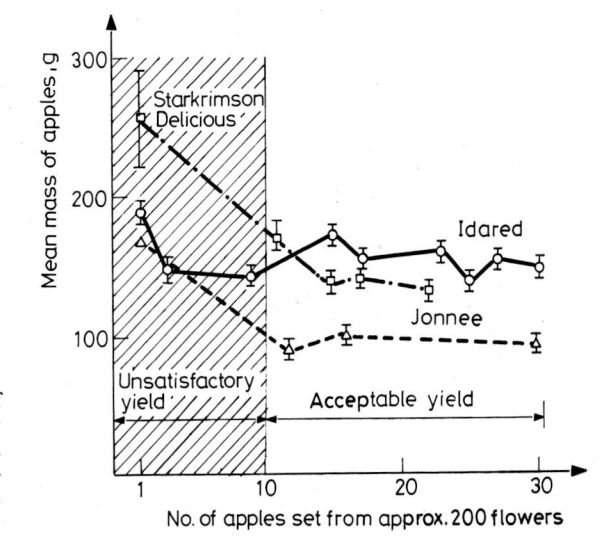

Fig. 9.9. Weight of apples in the case of intensive bee pollination as affected by the number of fruit set on branches with approximately 200 flowers (after Benedek et al. 1989a, Benedek and Nyéki 1990)

response to a greater fruit load, the decrease in fruit size is only 10%, which is much less than the percentage of the increase in fruit load. Thus it is true for the *pear*, too, that as a result of an intensive insect pollination, the negative influence of an increasing fruit load upon fruit size is more or less compensated for. Pears developed as a result of insect pollination contain more viable seeds.

In *raspberries* the intensity of bee visitation is extremely important (Oliveira et al. 1991). According to them, there is a close correlation between the frequency of bee visitation per flower, and the number of drupelets per berry, and berry size, in raspberry. The more times a flower is visited, the higher the number of drupelets is (Fig. 9.11), and the bigger the fruits are. Growth is limited, however. Five to six bee visitations per flower suffice for pollination and good development of fruits.

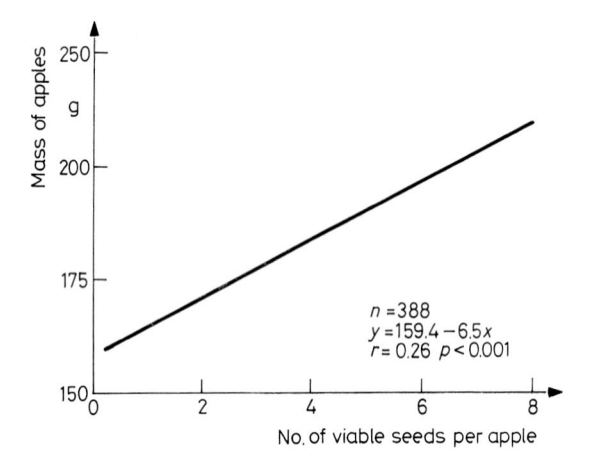

Fig. 9.10. Relationship between the number of viable seeds per apple and the weight of apples, cv. 'Idared' (Benedek et al., original)

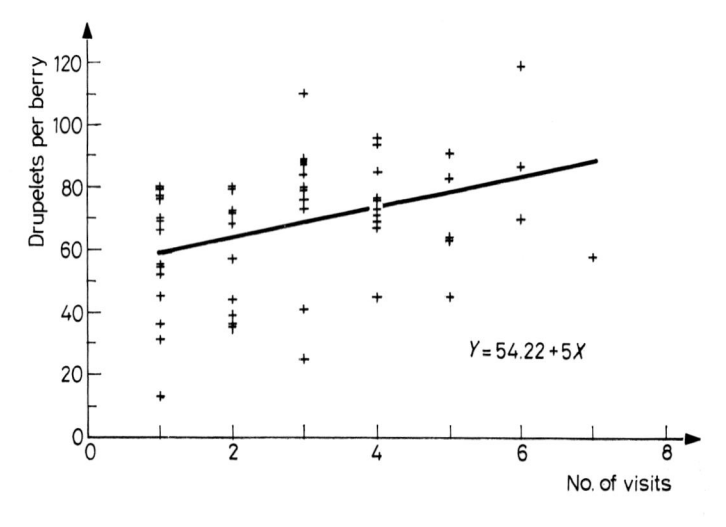

$n = 388$
$y = 159.4 - 6.5x$
$r = 0.26 \quad p < 0.001$

$Y = 54.22 + 5X$

Fig. 9.11. Relationship between the number of bee visits per flower and the number of drupelets per raspberry (after Oliveira et al. 1991)

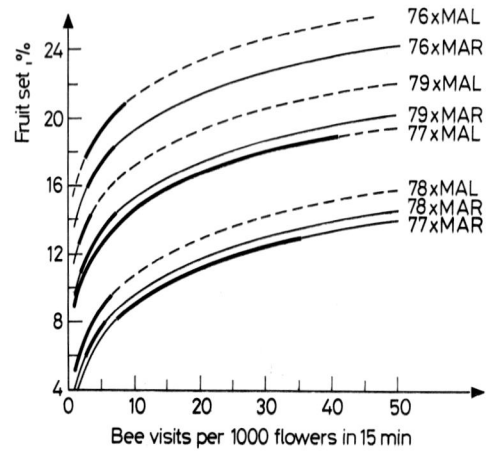

Fig. 9.12. Fruit set of sweet cherry flowers, cv. 'Flamengo', as affected by the frequency of bee visits. Thick line indicates the range of measurements (after Roversi et al. 1984)

Roversi et al. (1984) studied the effect of the intensity of bee visitation in *sweet cherry*. They demonstrated with three different cultivars that an increasing number of bee visitations per flower resulted in an increasing extent of fruit set. The effect is striking with one to ten visitations (Fig. 9.12), and then levels of saturation between 20 to 50 visitations.

In *plum*, a greater fruit load caused by an intensive bee pollination caused only a negligible decrease in fruit size and the solid content of fruits, but this decrease being within 15% was not relevant compared to the considerable increase in fruit load (Szabó et al. 1989).

9.8. Demand for Pollination in Fruit Species and Features of Their Insect Pollination

Apple. Without the aid of insects, there is no profitable yield (Free 1964). Moreover, getting farther away from the pollinizer, fruit set values are definitely decreasing (Free 1962). Though DeGrandi-Hoffman et al. (1984), and Palmer-Jones and Clinch (1967) argue against this statement, the dependence upon distance can be verified if not only when fruit set is taken into consideration but the amount of yield also indicates it. In a solid 'Granny Smith' block, Maggs et al. (1971) observed a highly significant correlation between the amount of yield per tree and distance from the pollinizer cultivar (Fig. 9.13). The distance of effective pollination proved to be as short as 12 m in their experiment.

Nevertheless, there are great differences between apple cultivars in respect of insect pollination. In addition to differences in nectar production (Free 1970, McGregor 1976), there are a lot of further factors that exert a crucial influence upon bee visitation and hence the success of pollination. These factors, such as flower colour, the relative position of stamens, stigmas, and petals, pollen production, pollen releasing capacity were pointed out upon a series of cultivars by Soltész et al. (1989) and Benedek et al. (1989a,b). According to daily nectar production, for instance, the cultivars under examination were distributed as follows: 28% was characterized by a good nectar production (at least 3 to 5 mg per day), 31% proved to be medium (2 to 3 mg), whereas the majority of cultivars (41%) produced little nectar (1 to 2 mg per day, or even less). The most widespread cultivars mostly belonged to the first two groups where pollen production was good or medium but 'Granny Smith', for instance, was included in the last group. Apple flowers produce 123,000 pollen grains on average. Pollen production in certain diploid cultivars is very poor and there are lots of non-viable pollen grains in cultivars such as 'Grimes Golden', 'McIntosh' and 'Melba', whereas other cultivars, such as 'Black Ben Davis' and 'Stark Delicious' are characterized by a fairly rich pollen production. Pollen grains of most widespread cultivars can easily be released but in other cultivars pollen grains are sticky, and hence they are difficult to be released (e.g. 'McIntosh', 'Melba').

'Delicious' apple cultivar, cultivated in the majority of growing areas in the USA, provides only 40 to 50% of its potential fertility (e.g. Brittain 1933, Preston 1949). According to some, the reason lies in flower structure, which influences the behaviour of bees (Roberts 1945, Preston 1949). Landing on the petals of flowers in the course of their foraging trips, bees do not touch the stigma (Free 1960a); hence, the poor fertility and yield of the 'Delicious' have something to do with the high proportion of side-worker bees, which work without pollinating flowers, because this cultivar group is characterized by rigid, erect stamens with great spacing between them (Robinson 1979, Robinson and

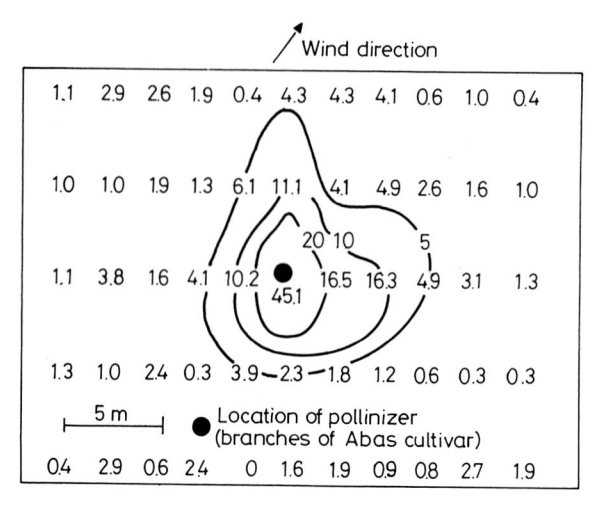

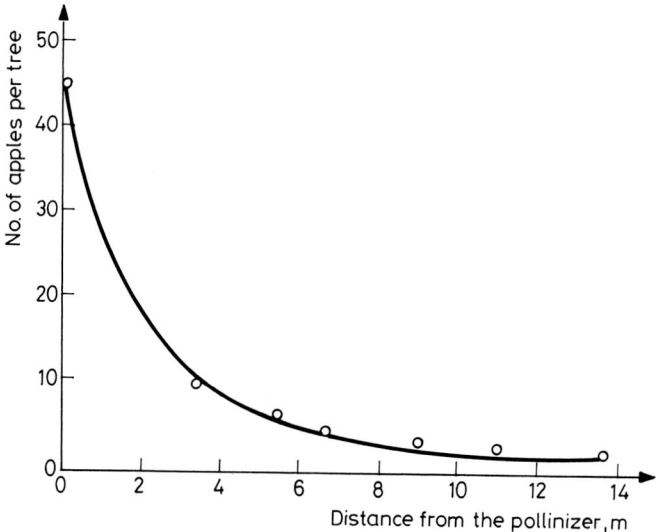

Fig. 9.13. Distribution of yield as expressed with the number of apples per tree in a solid block of 'Granny Smith' where pollinizer cultivar, 'Abas', is placed at a single spot. *Top*: distribution of yield at the site. *Bottom*: distribution of yield as a function of the distance from the pollinizer cultivar (after Maggs et al. 1971)

Fell 1981, Kuhn and Ambrose 1982). This character also appears in other cultivars. Robinson (1979) mentions 'Northern Spy' as such a cultivar, Ponomareva (1980) mentions three local cultivars, and Benedek et al. (1989b) mentions three further cultivars, 'Peasgood's Nonsuch', 'Spygold' and a local cultivar as cultivars with rigid, erect stamens. In the majority of apple cultivars, however, "side working" is relatively uncommon.

Bee visitation and the distribution of foraging behaviour are characteristic of a cultivar not only in the case of the 'Delicious' cultivar group (Benedek et al. 1984). When one makes observations on various cultivars at the same place and same time, one can determine the relatively effective bee visitation on the basis of the intensity of bee visitation and the proportion of side-worker nectar-gatherers (Fig. 9.14). In respect of effective bee visitation, 'Jonathan', 'Jonagold' and 'Mutsu' were eminent among 19 cultivars. Other cultivars had satisfactory number of bee visitations, but 'Gloster', 'Peasgood's Nonsuch', 'Spygold' and two local cultivars were characterized by a low effectivity of bee visitation. The 'Delicious' cultivar group also belongs to the low effectivity of bee visitation and low visitation can only be compensated for by having more than usual number of bee colonies in the orchard (Robinson and Fell 1981, Kuhn and Ambrose 1982).

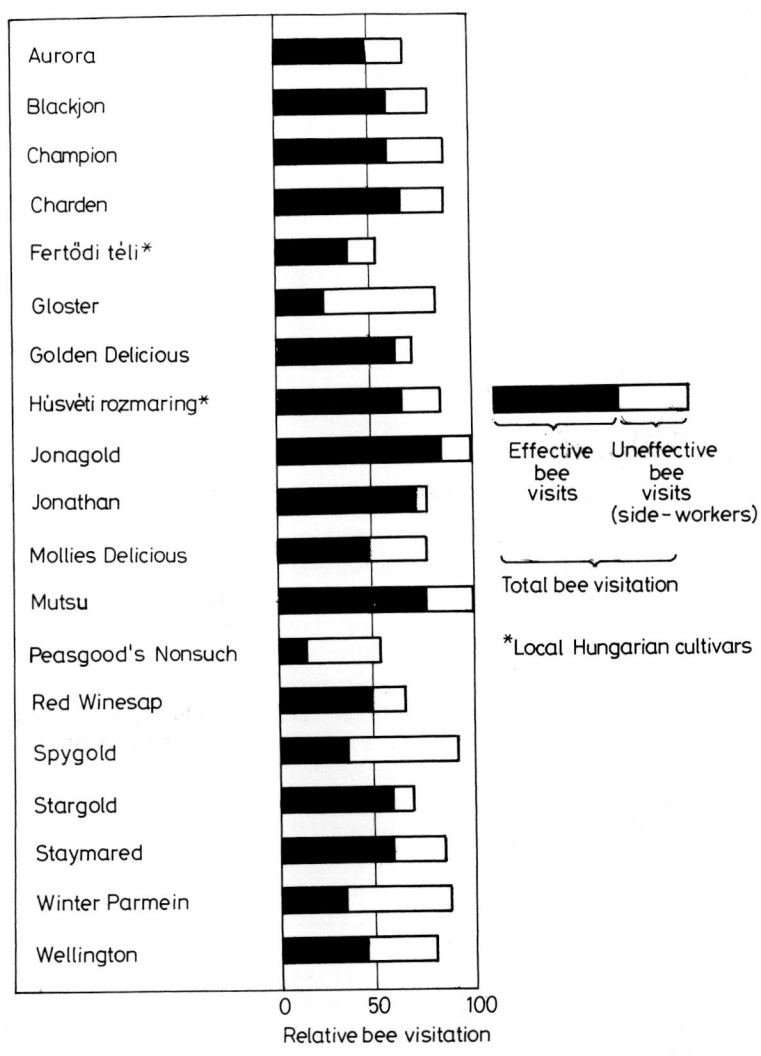

Fig. 9.14. Relative efficiency of honeybee visitation on the flowers of apple cultivars as affected by the proportion of side-worker nectar-gatherers (after Benedek et al. 1989a)

315

It is often suggested that ornamental *Malus* species (crab apple) should be used in apple orchards to serve as pollinizers because they produce much pollen and make cultivar-specific growing possible better than most other pollinizers. Kendall and Smith (1975) studied four ornamental *Malus* species and they determined that these species attracted bees during bloom to the same extent as eating cultivars. The number of bees visiting the species and cultivars was in proportion to the number of flowers; hence, more bee visitations were observed in the ornamental apple species because they had more flowers. Though in the course of visiting flowers, bees usually insist on a cultivar, those changing cultivars shift from an ornamental species to an eating cultivar or vice versa in a higher percentage than from an ornamental species to another ornamental species. Consequently, ornamental apple species seem to be suitable for pollinating apple cultivars, at least in respect of bee visitation.

Williams and Brain (1985) determined that the success of pollination in apple also depended on to what extent flowers in their different stages of development attracted bees. They think that such flowers are primarily popular with pollen-gatherer bees that are not far from, or have already achieved, the stage of entire anther dehiscence. More than 50% of bee visitations is directed to such flowers. Crab apple flowers just after opening, however, are rarely visited (7%). Bees take advantage of young flowers with anthers that have not dehisced only when gathering nectar.

Palmer-Jones and Clinch (1967, 1968) provided the following estimate of the sufficient number of bee visitations: 40 bee visitations per 30,000 flowers per minute were more than enough for a good yield. They examined six cultivars, with flower numbers ranging from 1,000 to 11,000 per tree. In their first series of experiments (Palmer-Jones and Clinch 1967), they observed no significant correlation between fruit set and mean seed content but later (Palmer-Jones and Clinch 1968) they found a significant correlation between the above mentioned parameters in three cultivars: 'Granny Smith' ($r = 0.93$), 'Kidd's Orange Red' ($r = 0.81$) and 'Golden Delicious' ($r = 0.40$).

Benedek et al. (1989a) calculated, on the basis of bee visitation values observed in apple trees, that in the orchards they studied a flower was visited five to twelve times on average in two days under favourable weather conditions. Furthermore, a lower frequency of bee visitation resulted in a lower extent of fruit set.

Free (1966) determined that every sixth flower visited by honeybees was to be expected to be fertilized. A statement of Petkov and Panov (1967) is in total harmony with this calculation because they say that fruit set values increase until six bee visitations per flower.

These data, however, are of little practical importance. The real question is always how many bee colonies are needed in an orchard. This is a topic under intensive discussion but suggestions, and aspects that suggestions are based upon, are usually so subjective. For such subjective information readers should consult the summaries of Free (1970), Benedek et al. (1974), McGregor (1970) and Tasei (1984).

In order to calculate bee colony requirement, Free (1970) offered the following reasoning: the pollination of apple trees in an area of 1 hectare theoretically requires 1,164 bee days. Since many bee visitations are directed to flowers that have already been pollinated, at least twice as many bee days, 2,328, are required practically. In the blooming period of fruit trees in Great Britain, bee colonies consist of about 10,000 bees, 33% of which, 3,300, fly out in favourable weather but only one-third of these, 1,100, visit fruit trees. Thus, 2.5 bee colonies per hectare are required to pollinate apple orchards. The author himself adds, however, that not only weather but other factors also have some influence upon the number of bee colonies necessary to pollination: for instance, the size and strength of colonies, the number of brood combs, the canopy shape of trees, the number of flowers, distance from neighbouring apiaries, etc. Hence, it is not surprising that there are so diverging suggestions concerning bee colony requirement in the literature.

Bornus et al. (1976) attempted to calculate the number of bee colonies necessary for pollination on the basis of an analysis of bee density, fertility, and yield. They say that in an orchard with 600 trees per hectare and 3,000 flowers per tree on average, 250 bee work-hours are required to achieve a yield of 50 kg per tree, that is, 30 tons per hectare. The authors argue that this result can be achieved with the aid of 1 bee colony per hectare. To achieve a higher level of crop, 55.5 kg per tree or 33 tons per hectare, however, 500 bee work-hours are required, which correspond to two bee colonies per hectare. As hiring bees is cheaper than other methods, the authors suggest that five bee colonies should be used per hectare since the efficiency of bee colonies may be decreased by unfavourable weather, or the competition with more attractive bee forages.

Benedek et al. (1989b) suggest that the initial number of bee colonies should be established on the basis of a differential analysis of characteristics of the orchard and the bee colonies (Table 9.9). They add, however, that the number of colonies recommended is only to be regarded as general approximations because of the influence of the factors that may drastically modify values concerning bee visitation and the efficiency of bee pollination in an orchard. These factors are summarized in Table 9.10.

After moving the bee colonies to the orchard, the number of bee visitations is to be checked immediately because this is the only way to obtain a real evaluation of the pollinating work of bees. The authors suggest that the grower should calculate the number of honeybees visiting flowers several times during bloom when the weather is suitable for bees to fly. If the number of bees is three to six per 50 flowers in 10 minutes, then the intensity of bee visitation is to be regarded as sufficient because this corresponds to a daily bee activity of eight hours in the two-day-long receptive period of flowers, supposing six to 12 bee visitations per flower. They stated that three to six bees per 50 flowers in 10 minutes were needed. This needs to be further qualified. Considering the fact, that to achieve a good yield, it suffices that fruits develop from as few as 5 to 10% of flowers, the bee visitation value mentioned above seems to be much more than enough for a sufficient yield. The sufficient level of bee visitation, 40 bee visitations per 30,000 flowers in a minute, corresponds to one bee visitation per 50 flowers in 10 minutes (Palmer-Jones

Table 9.9. Number of Bee Colonies Recommended for Optimum Pollination in Apple Orchards (after Benedek et al. 1990b)

Orchard type	strong (population = 20.000 bees) colonies	weak (population = 10.000 bees) colonies
Standard tree shape, middle-age orchard	1–2	2–4
Young (5–10-year-old) orchard with standard tree shape	0.5–1	1–1.5
Intensive orchard (free spindle, slender spindle)	1.5–3	3–6
Hedgerow	2–4	4–8

Table 9.10. Factors Affecting Number of Bee Colonies Required in Apple Orchards (after Benedek et al. 1990b)

Factors	Less colonies are needed	More colonies are needed
Weather	Warm and sunny weather with slight wind during bloom	Cool, cloudy and windy weather with rain
Features of the orchard	Young orchard	Intensive tree shape Wrong placing of pollinizer cultivars Strong effect of competing plants
Cultivar characteristics		Different pollen and nectar production of main and pollinizer cultivars Flower structure beneficial for side-worker nectar-gatherers Low rate of pollen-gatherer bees in wrong pollen producer cultivar Cultivar sensitive to restriction of intensive bee pollination Poor overlap of bloom
Apiary factors	Great quantity of bees coming from neighbouring apiaries Urging of pollen-gathering with sugar syrup feeding Artificial increase of brood in bee colonies	Pollinizer apiaries are placed outside the orchard Long distance Usage of weak colonies Too early or too late placing of colonies into the orchard

and Clinch 1967, 1968). Consequently, all things considered (Benedek et al. 1989a, Palmer-Jones and Clinch 1967, 1968), we suggest that the grower should increase the number of bee colonies if 50 flowers altogether are visited by less than one to two bees in 10 minutes. It is not necessary, however, to decrease a level of bee visitation that is higher than suggested above, because under intensive bee pollination conditions the quality of apple will not decrease if other requirements, such as nutrition and water, are satisfied (Benedek et al. 1989a, Benedek and Nyéki 1990).

This method that bee colony requirement should be adjusted by checking bee visitation, is supported by an observation by Chang et al. (1991) who found that in addition to managed bee colonies, also feral honeybees played an important role in pollinating apple orchards, and the number of feral bees could not be estimated really well. In apple orchards in Michigan, for instance, the density of feral honeybees is very high, and especially at peak flowering, orchards are swarming with them (Chang et al. 1991).

Certain wild bees are even better at pollinating apple than honeybees (Kendall and Solomon 1973, Paarman 1977). Nowadays they make much effort to artificially propagate early spring mason bees living in reeds in order to use them to pollinate apple orchards (Torchio 1991). They strive for the propagation and application of *Osmia cornifrons* in Japan, *O. lignaria propinqua* in the U.S.A., and *O. cornuta* and *O. rufa* in Europe. The most promising experimental results are related to the *Osmia lignaria*. The great advantage of mason bees is that they keep on working also when honeybees stop working because of unfavourable weather conditions. They have already been applied in commercial orchards. The *O. cornifrons* in Japan is intended to substitute for hand pollination applied by growers. In Europe, however, the propagation of *Osmia* species is still in an experimental stage. The most serious problem is how to free mason bees from nest parasites, which cause heavy losses of their populations.

Pear. Information available on demands of the pear for insect pollination is less clear-cut than those of apple though, as early as in the last century, it was observed that pear trees would not bring a good yield without insect pollination and required the planting of other cultivars in the orchard. Stephen (1958) observed a correlation between fertilization of pear trees in an orchard and the number of bees visiting flowers, which refers to the importance of pollination. Another relevant observation (Benedek et al. 1974) is that the amount of yield of pear shows a decreasing tendency in response to the increasing distance from the apiary. At a distance of 300 m from the apiary, the yield was only half as much as at a distance of 50 m from the apiary. All these data clearly show the importance of bee pollination under conditions where pears do not set parthenocarpic fruit.

The insects visiting and pollinating pear flowers are the same as those in apple (Solinas and Bin 1964) but the proportion of honeybees is somewhat lower while the proportion of dipterous insects is higher (Benedek et al. 1974). This situation may be derived from the fact that pear flowers attract bees to a lesser extent as they produce less nectar of a lower sugar concentration (Free 1970, Simidchiev 1970).

Hence, in pear, honeybees prefer gathering pollen (Free and Smith 1961). The proportion of pollen and nectar-gatherers, however, shows a variation across days as well as across parts of the day. Much pollen sticks on bodies of nectar-gatherers, too, and they sweep it into their baskets. Free and Smith (1961) observed that in bee colonies moved to pear orchards 47 to 91% of pollen grains carried into the hive had come from pear. Thus, though as a source of nectar, the pear is not very attractive to bees, it attracts them as a source of pollen but even pollen production is less in pear than in other fruit species (Percival 1955). In addition, there are male-sterile pear cultivars that produce a negligible amount of pollen. 'Magness' is one of them. According to Caron (1973), trees of this cultivar are much less often visited by bees than trees of the pollinizers in the same orchard. Hence, in a weather unfavourable to bees, flowers of 'Magness' are not visited at all. Because of poor pollen production, only 7% of bees visit flowers of this cultivar carried a pollen load.

There are great differences also in nectar production between different pear cultivars. Simidchiev (1970) studied several cultivars and found in certain cultivars nectar production was ten times as great as in other cultivars, and in some nectar concentration was thrice as high as in others. Differences in nectar production have some influence on the intensity of bee visitation. That is why flowers of 'Winter Nellis' are half as often visited by bees as those of 'Packham's Triumph' (Langridge and Jenkins 1975).

In pear, numerous but contradictory, data and suggestions are available concerning the ideal bee colony density (McGregor 1976). Benedek et al. (1974) argue for the placing of 2.5 to 3 bee colonies per hectare because of the relatively weak attracting influence of pear flowers and competition with plants distracting pollinators. It is necessary also in the case of pear to check the intensity of bee pollination and compare it with the expected visitation from number of colonies placed originally. The authors suggest that the number of bee colonies moved to the orchard should be increased if a flower is visited by less than 1.2 to 2 bees on average in a day.

Quince. Reviewing data available in the literature, McGregor concluded that the quince required the help of pollinating insects. Quince flowers attract honeybees very much (Simidchiev 1967). Quince nectar is characterized by a sugar content of 40 to 50%, and nectar production reaches its peak intensity at noon. Bees, however, gather not only nectar but also pollen on quince flowers. Simidchiev (1984) reports 5% of bees to forage only pollen, 11% only nectar, and the majority (84%) both pollen and nectar. Considering the strong attracting effect of the quince upon bees, one bee colony per hectare is to be moved to orchards.

Sweet cherry. Insects play a crucial role in pollination of sweet cherry flowers (Free 1970, McGregor 1976, Tasei 1984). Restrictions on bee pollination result in a decrease in fruit set and yield (Langridge and Goodman 1973, Roversi and Ughini 1986), and a total exclusion of bee pollination totally prevents trees from producing fruit (Benedek et al. 1990).

In the pollinating insect population, usually honeybees make up the majority but the population also includes a number of wild bees and flies, sometimes in a higher percentage than honeybees (Frilli and Barbattini 1980, Frilli et al. 1983). The number of wild bees (*Andrena* and *Halictus* species), and bumble-bees (*Bombus* spp.) primarily depends on characteristics of the region around the orchard, that is, the existence of places suitable for their nesting and the amount of further sources of food. The number of dipterous insects, whose main representatives are the syrphids, is fairly variable, and though their percentage might sometimes exceed even that of honeybees (Frilli et al. 1983), their pollinating activity is of little importance (Frilli and Barbattini 1980).

Sweet cherry flowers produce little nectar and much pollen (Benedek et al. 1990). Péter (1972) argued that more nectar was produced though he found limited nectar production in flowers of early flowering sweet cherry cultivars compared to flowers of late flowering ones. Sugar concentration showed a negative correlation with the amount of nectar produced. The annual weather causes a wide variation in flower size, anther number, pollen production, and the relative position of stamens and stigmas. Year-dependent variations are so great that they often suppress cultivar-specific differences (Benedek et al. 1990). Filaments are quite erect but occurrence of side-worker bees is not typical. Sweet cherry shows such a very wide variation in flower characteristics that Frilli et al. (1983) could not point out significant differences in the bee visitation among the three cultivars they examined.

Honeybees gather pollen and nectar from flowers of sweet cherry trees. Benedek et al. (1990) report the following proportion: 32% of honeybees gathers pollen, 47% is nectar-gatherer, and 21% forages both pollen and nectar, though this proportion may vary during the blooming period as well as in the course of a day (Free 1960*b*, 1970).

In sweet cherry trees, fruits can be found at a height between 2.5 and 5 m, which corresponds to the middle section of canopies of traditional large trees (Frilli and Barbattini 1980). According to Barbattini et al. (1983), the reason is that this is the zone of the trees that honeybees are fond of and visiting often (Fig. 9.15).

As for the optimum bee colony density for pollination, many suggestions are available in the literature but they are contradictory (McGregor 1976). There is a suggestion that one bee colony should be sufficient per acre (about 2.5 bee colonies per hectare) but according to another suggestion, twice as many bee colonies, five bee colonies per hectare, are required. These suggestions have been made but fruit set and yield in relation to bee colony density have scarcely been studied. Skrebtsova and Jakovlev (1959) report that placing 3.8 bee colonies per hectare in the orchard, they received a fruit set of 15%, whereas with the aid of 2.8 colonies, a fruit set of 13% was received. Considering that the desirable fruit set level should be 50% and sweet cherry trees have a huge number of flowers, McGregor (1976) suggests that as many as five bee colonies should be placed in an area of 1 acre. Furthermore, each colony should contain bees 3 to 4 kg in weight and brood 600 cm² in area. This means 12 bee colonies per hectare, which is more than any previous suggestion.

Frilli and Barbattini (1980) used five to six bee colonies per hectare. There is still no precise estimate of the optimum bee colony density for pollination because, as is pointed

Fig. 9.15. Intensity of bee visitation at different sections of the canopy of cherry trees of traditional shape (after Barbattini et al. 1983)

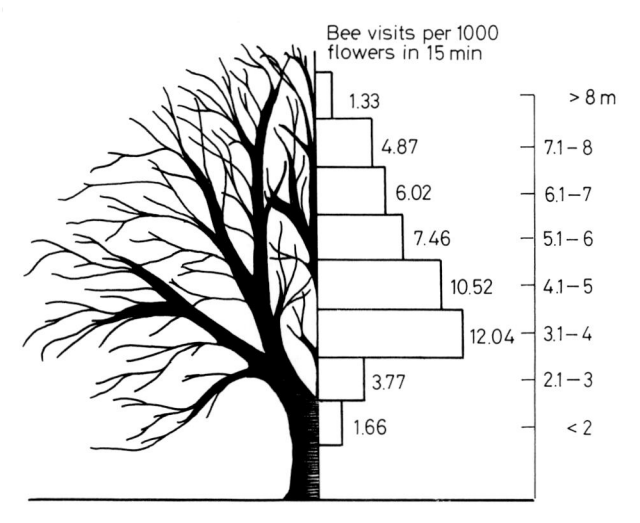

Bee visits per 1000 flowers in 15 min	Height
1.33	> 8 m
4.87	7.1 – 8
6.02	6.1 – 7
7.46	5.1 – 6
10.52	4.1 – 5
12.04	3.1 – 4
3.77	2.1 – 3
1.66	< 2

out by McGregor (1976) as well, further research is required to reveal the relation between bee visitation, the sufficient level of fruit set and bee colony density.

In addition, the optimum bee colony density is affected by external factors also in sweet cherry, primarily by the temperature and competitor plants. In a temperature domain between 14 and 20 °C, each increase by 4 °C in temperature makes bee visitation on sweet cherry trees almost four times as intensive (Barbattini et al. 1983). Cloudiness and wind hinder bees from visiting flowers to some extent but, according to the authors, it is the temperature that has an overriding effect and may stop working of bees entirely. The negative influence of weather can be compensated for by moving (even) more bee colonies to the orchard though fruit set values in sweet cherry show an increasing tendency in response to an increase in the effective period of bee visitation (Roversi and Ughini 1986). Compared to the sweet cherry, other fruit species are less competitive and distract honeybees from apple, sour cherry and pear orchards (Free 1970, Benedek et al. 1974).

Sour cherry. In Europe the importance of insect pollination had not been accepted for long because they used to think that wind pollination sufficed. Although self-fertile cultivars set fruit pollinated with their own pollen, they still need an agent for transferring pollen from anther to stigma (McGregor 1976). Self-acting fertilization is avoided in flowers. In Europe, however, they used to think that self-fertile sour cherry cultivars required no insect pollination. Blasse (1978) pointed out the falsity of this opinion with an experiment where bees were excluded from trees of a self-fertile sour cherry cultivar; the trees that had been fertilized without insects brought a negligible yield compared to trees of the same cultivar that had been insect-pollinated (Table 9.11). Benedek et al. (1990*a*) also provided evidence in support of the indispensable role of insect pollination in sour cherry cultivars. Moreover, they verified that not only a total exclusion of bee pollination but even partial restrictions of pollination caused a decrease in yield. Although the self-fertile cultivars were sensitive to partial restrictions and/or to a lesser extent than self-sterile sour cherry cultivars (Table 9.12). Another observation, due to Eaton (1962), is also against restrictions on insect pollination. The author reports that in sour cherry orchards, bees placed in the orchard for a day only, after the first flowers opened, resulted in a decrease in yield. Considering these data, it is not surprising that

Table 9.11. Effect of Insect Pollination on Fruit Set and Yield of Self-Fertile 'Schattenmorelle' Sour Cherry Cultivar (after Blasse 1978)

			Yield	
Treatment	Replication	Fruit set after bloom (%)	No. of fruits	
			at harvest	compared to No. of flowers (%)
Caged trees	tree #1	2.4	201	1.2
	tree #2	1.6	100	0.9
Open pollinated trees with intensive bee visitation	tree #1	40.9	3,110	33.3
	tree #2	60.0	1,008	15.9
	tree #3	38.6	4,146	24.9

Table 9.12. Effect of Restriction of Bee Pollination on Fruit Set of Sour Cherry Cultivars (after Benedek et al. 1990a)

	Fruit set (%)			
Cultivar	total restriction	partial restriction caged on the		open pollination
		4th	6th	
		day of bloom		
Pándy meggy (self-sterile)	0	1.0	1.5	4.8
Cigánymeggy (self-sterile)	11.1	26.7	31.0	32.1
Újfehértói fürtös (self-sterile)	10.0	32.3	32.1	31.4

McGregor (1976) regards it as an unfortunate custom that in North America sour cherry growers return the bees hired to pollinate flowers after two to three favourable days.

Flowering sour cherry trees are visited by representatives of insect groups similar to those visiting apple and sweet cherry trees. Nevertheless, as sour cherry trees bloom even earlier, the amount of wild pollinators is usually even less than in apple orchards. Benedek and Martinovich (1971), reported the proportion of wild pollinating insects to be only 0.7%.

The result of bee pollination can be measured either in an increase in fruit set, or in a decreasing tendency of fruit set and yield in response to an increasing distance from the apiary (Benedek and Martinovich 1971). This latter influence may be suppressed by the influence of gatherers of neighbouring apiaries (Benedek et al. 1990a).

According to the experiments of Benedek et al. (1990a) upon 12 sour cherry cultivars, the majority of bees (64.8%) gathered nectar, and 34% gathered exclusively pollen. One-third of nectar-gatherers also gathered pollen in the course of their foraging trips. The authors did not observe significant differences between the cultivars but they attributed this fact to the unfavourable weather. They reported 34 bee visitations per 100 flowers in 10 minutes on average. Side-worker nectar-gatherers were not observed, because sour cherry flowers contain reclining filaments, which obstruct side working.

Simidchiev (1971b) concluded that nectar production in sour cherry was peculiar to individual cultivars. Daily nectar production ranged from 1.1 to 12.2 mg depending on cultivars, with a sugar content ranging from 16 to 38%. In experiments of Benedek et al. (1990a), who studied a large number of cultivars, nectar production showed a cultivar-dependent variation ranging from 0.6 to 11.2 mg per day, with sugar concentration values between 30 and 54%. According to their experimental results, daily nectar production may range to 18 mg (with a sugar concentration 20 to 38%), while at a low nectar production (0.5 to 2 mg) sugar concentration may range to 60%.

These observations are in accordance with Simidchiev's (1971) conclusion that nectar production is in inverse proportion to the sugar concentration of nectar. Simidchiev (1971) argues that a sour cherry flower produces nectar during its entire life, that is, through a several-day-long period, reaching the peak of nectar production on about the third day after the flower opened. Nectar production is varying during the course of a day but there is no such part of the day when no nectar is secreted. Orosz-Kovács et al. (1989) pointed out a daily periodicity of six hours in nectar secretion. In clones of 'Pándy' sour cherry cultivar group, three classes were formed according to daily nectar production maximums. The six-hour-long periods began at 3, 4, and 6 a.m. in the three classes. Later they (see Orosz-Kovács 1990) discovered a positive by significant correlation between the fertilization of flowers and the daily maximums of nectar secretion, or rather, between the point of time of the first daily maximum (3, 4, or 6 a.m.) and fruit set (Fig. 9.16). The reason is that in the three classes, considering the six hours periodicity, the first daily maximum plus the first and second six hours coincide with daily periods of large-scale bee flight. The class that is the best in this sense is the one with a first maximum at 3 a.m. because the further two maximums are at 9 and 15 h that are favourable to bees, and there are only two unfavourable secretion periods at 3 and 21 h. The most unfavourable is the class where the first maximum is at 6 o'clock early in the morning because this way only a single maximum, the one at noon, coincides with a period of large-scale bee flight.

'Pándy' sour cherry and other self-sterile cultivars produce more nectar than 'Cigánymeggy', a self-fertile cultivar (Parnia et al. 1979, Benedek et al. 1990a). Regard-

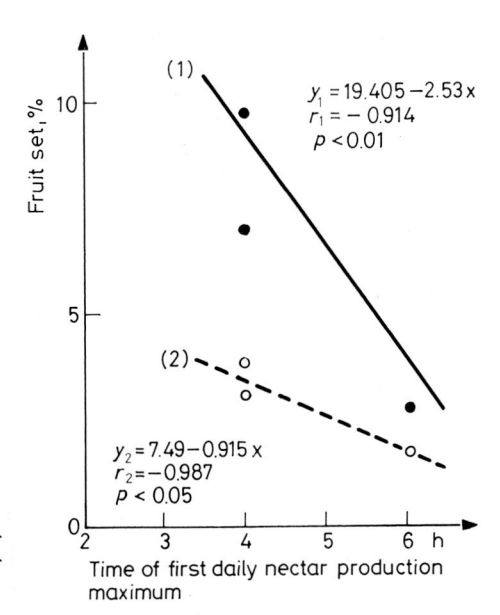

Fig. 9.16. Relationship between the time of the first daily nectar secretion maximum and the fruit set of flowers in case of sweet (1) and bitter (2) clones of cv. 'Pándy' sour cherry (after Orosz-Kovács 1990)

$y_1 = 19.405 - 2.53x$
$r_1 = -0.914$
$p < 0.01$

$y_2 = 7.49 - 0.915x$
$r_2 = -0.987$
$p < 0.05$

Time of first daily nectar production maximum

less of the role of nectar production in sour cherry, Benedek et al. (1990*a*) still consider the rich pollen production, which is similar to that of the sweet cherry, to be the main attracting factor of sour cherry.

Competition with plants distracting pollinators is more relevant to the sour cherry than to the sweet cherry because sour cherry trees attract honeybees to a lesser extent than apple and sweet cherry trees (Free 1960*b*, 1970, Benedek et al. 1974).

As for the optimum number of bee colonies, Benedek et al. (1974) suggest four to five bee colonies per hectare, whereas McGregor (1976) suggests 12. Both authors, Benedek et al. (1974), as well as McGregor (1976), emphasize that bee colonies should be distributed in blooming sour cherry orchards in small groups, that is, as uniformly as possible.

Plum. Pollen transfer is mostly carried out by honeybees though during bloom there are also wild bees and dipterous insects on plum trees. Undoubtedly, however, honeybees are the most important due to their good pollinating efficiency as well as their huge number, because plum trees require a large-scale pollinating insect population to be sufficiently pollinated (McGregor 1976). Szabó et al. (1989) studied some plum cultivars in two consecutive years and reported the proportion of honeybees in the flower-visiting insect population to be 80% and 95%.

The effect of bee pollination is illustrated by the observation that fruit set values of plum trees show a definitely decreasing tendency in response to an increasing distance from the apiary (Free 1962, Szabó et al. 1990). Restrictions on the effective period of bee pollination exert a significant influence upon fruit set values in the case of self-fertile as well as self-sterile cultivars (Szabó et al. 1989). Naturally, self-sterile cultivars are much more sensitive to this influence than self-fertile ones (Fig. 9.17, Table 9.13). In self-sterile cultivars, it is indispensable that there must be an intensive bee pollination through the whole blooming period. In self-fertile cultivars, however, it may happen that as soon as the third or fifth day of the blooming period, a sufficient amount of fruit (20%) will have been set; in a case like this, further bee visitations may increase the level of fruit set to a negligible extent. It has been unquestionably verified, however, that without bee pollination even trees of self-fertile cultivars will set a sufficient amount of fruit.

Honeybees mostly gather pollen from plum flowers. Simidchiev (1968) reports the proportion of pollen-gatherers to be 40%, but the majority of nectar-gatherers, 58.2% of all bees, gathered pollen as well, and bees that gather exclusively nectar made up only 1.2%. On the basis of their experiments upon different plum cultivars, Szabó et al. (1990) stated that the majority of bees, more than 50% but sometimes 80 to 86%, were pollen-gatherers, and the proportion of nectar-gatherers was about 30%, sometimes slightly more but usually somewhat less. Bees that are pollen and nectar-gatherers at the same time make up a proportion of about 20%. The proportion of bees belonging to different behaviour types, however, widely varies depending on the time of observation. Szabó et al. (1989) say that at the beginning of bloom the proportion of pollen-gatherers is higher but later, at the end of bloom, the proportion changes in favour of nectar-gatherers. Sometimes side-worker nectar-gatherers can also be observed on plum flowers but their proportion never exceeds 10%.

Changes in the behaviour of bees must be connected with the pollen and nectar production of flowers. As for pollen, the decreasing tendency is natural since with time pollen is being collected and the remaining pollen is less attractive to the bees. As for nectar, Simidchiev (1968) argues that though there is a continuous decrease in nectar production during the course of bloom, sugar concentration is increasing parallel with the decrease of the amount of nectar, and this makes nectar more attractive to bees. Simidchiev

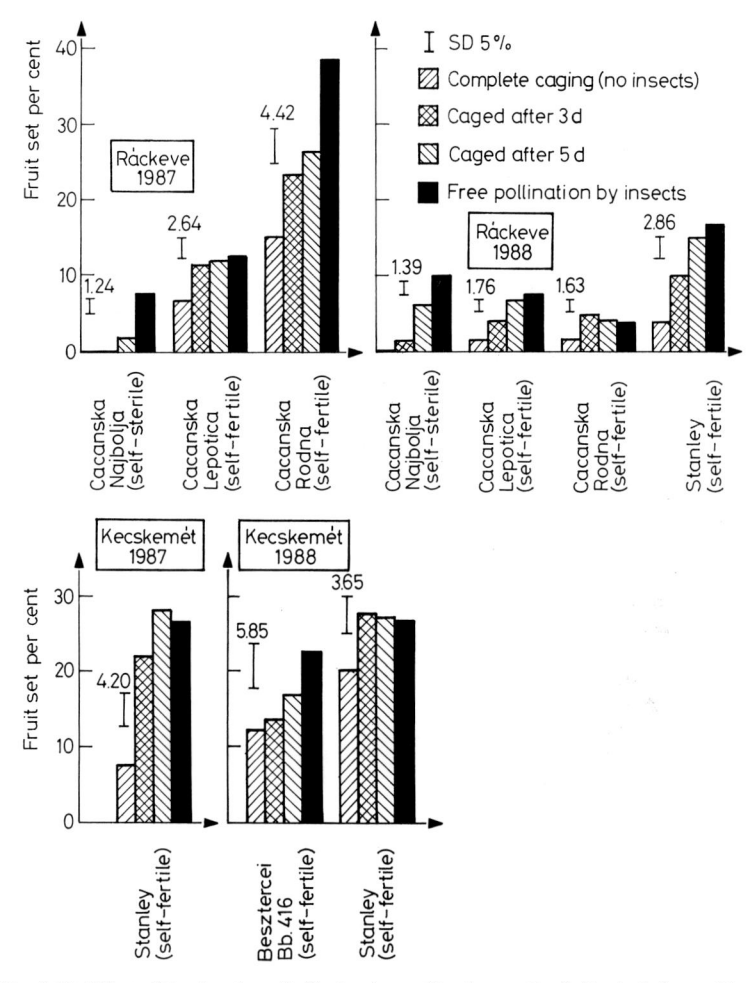

Fig. 9.17. Effect of the duration of effective bee pollination on the fruit set of plum cultivars at two different sites in two consecutive years (after Szabó et al. 1989)

(1968) adds that the nectar production of flowers is a cultivar-specific property. The nectar content of a flower ranged from 0.05 to 7.4 mg depending on the time and the cultivar.

Szabó et al. (1989) mention the extreme values 0.3 and 6.3 mg, and they also observed cultivar-dependent differences in nectar production in plums. Furthermore, they observed different nectar amounts at different times. In their measurements the sugar content of nectar ranged from 16.5 to 56.7%, while Simidchiev (1968) received the extreme values 5.5 and 61.5%. Sugar concentration is also cultivar-specific. Szabó et al. (1989) report that more pollen was produced in self-fertile plums than in self-sterile ones. Pollen production, however, is not only cultivar-specific but depends on years and growing sites as well.

As for the optimum number of bee colonies for pollination, researchers agree that one bee colony suffices in a plum orchard of 1 acre (2.5 colonies per hectare), according to

Table 9.13. Effect of Bee Pollination on the Fruit Quality of Plum (Szabó et al. 1989)

Site cultivar	Year	\multicolumn{5}{c}{Fruit weight (g)}
		1	2	3	4	LSD$_{5\%}$
Ráckeve						
Tsathsanska naybolia	1987	0	0	36.10	32.76	1.57
	1988	0	31.75	33.18	31.95	4.28
Tsathsanska lepotica	1987	27.13	29.94	27.36	27.13	1.27
	1988	28.87	33.80	33.70	30.00	4.94
Tsathsanska rodna	1987	15.54	15.69	15.87	14.82	0.67
	1988	29.80	26.93	26.20	27.23	4.81
Kecskemét						
Stanley	1987	23.0	25.1	23.3	21.4	–
Stanley	1988	30.29	29.17	29.76	27.87	–

Site cultivar	Year	\multicolumn{5}{c}{Soluble solids (%)}
		1	2	3	4	LSD$_{5\%}$
Ráckeve						
Tsathsanska naybolia	1987	–	–	12.55	12.23	0.17
	1988	–	14.55	14.34	13.90	1.04
Tsathsanska lepotica	1987	13.37	13.30	12.73	13.05	0.18
	1988	12.91	13.17	13.13	12.89	1.45
Tsathsanska rodna	1987	12.02	11.58	11.68	11.25	0.11
	1988	17.30	17.68	18.15	17.40	1.68
Kecskemét						
Stanley	1987	14.73	13.93	13.66	14.13	–
Stanley	1988	12.34	12.68	12.98	13.52	–

1 = caged before bloom; 2 = caged on the fourth day of bloom; 3 = caged on the sixth day of bloom; 4 = open pollination

McGregor's (1976) review. Szabó et al. (1989) performed experiments upon orchards where there were four to five, and eight to ten, bee colonies per hectare. The greater bee colony density produced an exaggerated extent of fruit set, while the lesser bee colony density, four to five colonies per hectare, mostly ensured a sufficient fruit set (15 to 20%), and with 2.5 colonies per hectare insufficient fruit set values were observed only in self-sterile cultivars. All in all, self-fertile cultivars do not require more than 2.5 colonies per hectare while in self-sterile ones four to five bee colonies are required.

In plum orchards, we need not worry about the distracting influence of other fruit species because plum trees can distract bees even from apple trees (Free 1960*b*, 1970) but competition with flowering weeds (e.g. dandelion) is not to be ignored.

Apricot. Murneek (1937) established that in both self-sterile and self-fertile cultivars, a sufficient level of pollination required the help of insects. McGregor (1976) is somewhat more precise. Self-sterile cultivars require insect pollination, whereas the activity of flower-visiting insects is helpful in self-fertile cultivars.

Only limited data are available in the literature concerning insect pollination in apricot. Stark (1944) claims that the main pollinators of apricot trees are bees. Other insects casually take part in pollination. Benedek et al. (1991b) performed a several-year-long series of experiments upon 30 apricot cultivars in several growing areas in order to study bee pollination in apricot and to reveal cultivar-dependent factors affecting pollination. They concluded that parameters of bee visitation depended on the site and time. When the weather was such that bees could fly, a flower was visited 0.6 to 17.4 times a day. However, the main characteristic was the low level of bee visitation. Because of this weak bee activity, they could not point out reliable differences between the apricot cultivars. The distribution of bees according to their activity also varied. Though the majority of bees (40 to 60%) usually gathered pollen, including those gathering pollen exclusively as well as those gathering both pollen and nectar. Occasionally, the nectar-gatherers made up the majority, but the proportion of side-worker nectar-gatherers was usually low (14 to 25%). It was rare that the proportion of side-workers was higher but it could reach 40 to 80%.

Flower size and stamen number show a wide variation in apricot depending on years and growing sites (Benedek et al. 1991b). This statement is also true for stamen position and pistil type, though pistil position is somewhat more stable than stamen position. As for stamens, in certain cases erect and extrorse types were observed, while in other cases the reclining and mixed types were typical. As for pistil position, there is a variation across years, but the pistil is typically situated at or above the level of the whorl of stamens, and this is mostly true in certain years and in given growing sites. The variation of stamen position may exert a strong influence on bee pollination but the typically highly situated pistil facilitates the success of pollination. The varying position of stamens is of much importance because in the case of erect stamens side-worker bees are able to get nectar without pollinating the flower. In addition to late spring frosts, the changing position of stamens might be another cause of the great variation in fruit set in apricot.

Benedek et al. (1991b) observed the nectar content of apricot flowers to vary. Nectar weight per flower ranged from 0.28 to 11.85 mg, but the majority of data fell between 2 and 3 mg. These data agree with Péter (1972) that the nectar content of flowers shows a remarkable variation in the course of a day. Significant cultivar-dependent differences in nectar content of flowers, however, could not be observed. The influence of the sugar content of nectar upon bee pollination is at least as important as the amount of nectar, because bees prefer nectar with a greater sugar content. The sugar content ranges between 7.5 to 63.9% (Benedek et al. 1991b) and 5.4 to 58.2% (Péter 1972). Apricot nectar is characterized by a high sugar content in comparison with other fruit species.

Pollen grains are about 50 μm in diameter, and diameter values show a quite little variation. Contrary to this, pollen production values per anther vary between the extreme values 500 and 5,900 grains, which corresponds to 14,000 to 166,000 pollen grains per flower. The pollen production of apricot flowers is much less than that of apple flowers, somewhat less than in sweet cherry flowers, and is similar to that of sour cherry and plum flowers (Benedek et al. 1989a,b, 1990a,b, Szabó et al. 1989).

Experimental results suggest that, though the apricot belongs to the fruit species first to bloom in spring, when the activity of honeybees and other pollinating insects is usually still weak, its flowers are perfectly suited to attract pollinating insects. Apricot flowers are medium in size, contain numerous filaments with high pollen content. The sugar content of its nectar is high, and often very high. Compared to other fruit species that bloom early in spring, the nectar and pollen production of apricot flowers as well as the sugar content of apricot nectar are quite good. On most occasions, because of the early bloom, a large-scale bee activity is hindered. Bee pollination in apricot requires an intensive and more thorough research in the future. It seems to be likely that in apricot, similar to other fruits, bee pollination improves the quantity and quality of the yield. Finally, we mention a suggestion concerning the optimum bee colony density. Corner et al. (1964) suggest that one bee colony should be placed for each acre (2.5 colonies per hectare) in apricot orchards.

Peach. Chandler (1951) pointed out that pollen was to be applied to the stigma of peach by insects in order to achieve a sufficient degree of fertilization.

Randhawa et al. (1963) argue that the honeybee is the most important pollinator of peach trees. According to Romisondo and Marletto (1972), honeybees make up 87% of the flower-visiting insect population. Not only their number but also their working capacity exceeds that of other pollinators. They work more steadily and are disturbed by momentary changes in weather to a lesser extent. The other pollinators of peach belong to the groups of wild bees and beetles (*Tropinota* sp.). Among the wild bees, *Anthophora* and *Bombus* species, but not *Halictus* and *Xylocopa* species, play a major role in peach pollination.

Under favourable weather conditions, flowers are visited by insects from 9 a.m. to 17 p.m., with a daily maximum from 11 to 15. Within this period of time, wild insects mostly work only a few hours around noon, it is only honeybees that are active through the entire period. Simidchiev (1985) regards the visitation of flowers of peach trees by honeybees in good weather as very intensive. Benedek et al. (1991*a*) provide detailed data on flower visitation. One hundred flowers were visited by bees 4 to 107 times in 15 minutes in favorable meteorological circumstances. Bees visited clingstone cultivars the most. They considered seven-hour-long effective bee flying period per day which is possible during peach bloom but often decreased because of the changeable early spring weather. Using this seven-hour period, 100 peach flowers were visited 110 to 3,000 times, that is, one flower was visited one to 30 times on a favourable day. This is a high frequency of bee visitation, which must exert a favourable influence upon fruit set and yield.

Romisondo and Marletto (1971) made a comparison in peach between three kinds of pollination. They studied self-pollination on twigs caged with double bags, with parchment inside and linen outside. In caged trees they imitated wind pollination (anemophily) with ventilators, and compared the results to those with open pollination. Fruit set values were some 80% with anemophily and entomophily, and only 20% less with self-pollination. More but smaller fruits had resulted from entomophily, and the extent of fruit drop was lower than in the other treatments. The sugar content of insect-pollinated fruits was higher at harvest. Because of the great fruit set, fruit thinning was to be carried out in the orchard. McGregor (1976) also refers to this problem but argues that fruit thinning is a less serious problem than an insufficient fruit set because of poor pollination.

In peach, honeybees display a flower-visiting behaviour similar to that in other fruit trees. In peach, the majority of bees (65%) gather nectar from peach flowers, and the proportion of those gathering exclusively pollen is 31% (Benedek et al. 1991*a*). One-third of nectar-gatherers gather pollen as well, and another one-third, performs side-worker nectar-gathering, without pollination. This proportion of types of behaviour, however, shows a wide variation at different times. Pollen-gatherers range from 0 to 69%, nectar-gatherers without pollen load from 0 to 75%, nectar-gatherers with pollen load from 0 to 83% and side-worker nectar-gatherers from 0 to 38%. The flower-visiting speed of honeybees is 4.2 to 6.4 flowers per minute (Simidchiev 1972).

The peach is characterized by an intensive bee visitation because, few other sources of food are available for bees in its blooming period and, the nectar production of flowers is good. The daily nectar production ranges from 2.2 to 6.4 mg, with a sugar content ranging from 16.2 to 26.9% (Simidchiev 1972). Benedek et al. (1991a) report a higher extent of daily nectar production, 5 to 46 mg per flower. They concluded on the basis of data on 26 cultivars, that significant differences between cultivars could not be identified because of the strong influence of external factors upon nectar production. Sugar content ranges from 28% to 52%. Nectar weight and sugar concentration correlate negatively. Simidchiev (1972) accepts the large influence of weather but regards the extent of nectar production as a cultivar-specific property. Later the same author (Simidchiev 1985) argues that there is a positive correlation between air temperature and nectar content and temperature and sugar concentration.

Pollen production is more stable than nectar production (Simidchiev 1985). Pollen production in peach is 3 to 19 kg per hectare, depending on cultivars. According to Benedek et al. (1990a), there are 450 to 2,800 pollen grains in an anther. On the basis of experiments upon 16 cultivars, they stated that pollen production was somewhat less in freestone cultivars (47,000 grains per flower), than in clingstone cultivars and nectarines (55,000 grains per flower). There are great year-to-year variations, therefore, it is impossible to determine cultivar-specific numbers.

Due to this large nectar and pollen production, peach orchards are attractive to honeybees. Hence, peach is visited by bees even from great distances (McGregor 1976). Nevertheless, there are few data on bee colony density requirement of peaches. Generally, one colony per 5 acres (0.5 colonies per hectare) is suggested. This seems to suffice due to the strong attracting influence of peach flowers and the lack of competitors. In the neighbourhood of inhabited areas, if there are apiaries, it is not always necessary to move apiaries to orchards.

Almond. Kester and Griggs (1959) point out that to achieve a maximum yield, every flower should be cross-pollinated so that they can set fruit. Each flower that has not set fruit, is a loss.

The almond blooms very early. Hence, during its bloom, air is usually cool and cloudy, so the activity of insects is limited to the hours around noon and early in the afternoon. Because of this early flowering, there are still no wild pollinating insects, with the exception of a few female bumble-bees (*Bombus*) that have lived through the winter and a few other insects of *Andrena* and *Halictus* species. They might pollinate flowers of almond trees standing alone but the pollination of orchards is to be expected only from honeybees. Although almond trees (would) require a large-scale honeybee visitation (McGregor 1976). Because of the early bloom, honeybee colonies are still weak and bee density is also often insufficient.

With rare self-fertile almond cultivars pollination may not be important. Moleas and Serio (1981) claim that there is no positive correlation between insect pollination and fruit set in the case of the self-fertile 'Filippo Ceo' cultivar. Some authors assume that insect pollination is advantageous in self-fertile sour cherry (Benedek et al. 1990) and plum (Szabó et al. 1989) cultivars. Therefore, it is likely to be advantageous also in almond. Demands of self-fertile almonds for insect pollination thus require further research.

Honeybees gather pollen as well as nectar from almond flowers. DeGrandi-Hoffman et al. (1991) observed no significant differences between five almond cultivars in nectar production. Nectar content per flower ranged from 0.53 to 1.13 ml, with a sugar content 5.8 to 38.5%. In one of two years less nectar was found in flowers with higher sugar concentration (34.9 to 38.5%), whereas in the other year more nectar was observed that parallelled with a much less sugar concentration (5.8 to 6.1%). Pollen production was 0.55 to 1.2 mg per flower, with cultivar-dependent differences in one year, but not the other. When nectar and pollen production were calculated per m of branch nectar was 9.2 to 33.3 ml and pollen: 20.5 to 71.5 mg. There were differences between cultivars in nectar and pollen per m of branches, as a consequence of differences in flower density but the differences in the two succeeding years were not consistent.

Almond flowers produce less nectar and pollen in comparison with flowers of other species but at the time almonds bloom almost no other food is available for bees, therefore, almond is quite attractive to them. Griggs and Iwakiri (1960) observed 150 to 200 honeybees per tree, which is held to be sufficient. DeGrandi-Hoffman et al. (1991) counted much less bees, 2 to 10.9 per tree, with differences between cultivars, though these differences were not consistent in two consecutive years.

McGregor (1976) suggests that at least three strong bee colonies should be placed to an acre (seven to eight colonies per hectare). These numbers seem to be large but under the poor early spring meteorological conditions, even so many bees cannot guarantee a sufficient extent of fruit set in every year. An advantage of early blooming, however, is that competition with plants distracting bees can be ignored.

Chestnut. The question as to whether insects or the wind play a determining role in the pollination of chestnut flowers has not been decided yet. It is a fact that the inflorescence of the chestnut shows anemophilous as well as entomophilous features at the same time. Porsch (1950) and Breviglieri (1955) regard the following features as entomophilous: attracting flower odour, fresh yellow staminate flowers, great extent of nectar secretion, sticky pollen. The anemophilous features are the following: an extremely great amount of staminate flowers, the lack of an attractive colour and odour in pistillate flowers, poor nectar secretion, small stigma surface and very rich pollen production.

Breviglieri (1951) thought that in chestnut insects also helped to shed pollen. Later the same author (Breviglieri 1955) claimed that the wind and insects carried out pollination together. Nevertheless, the question as to whether insects or the wind play a determining role in the pollination of chestnut flowers has not been decided yet (Chapa 1984). It is to be regarded as a verified fact, however, that the role of insects is not to be ignored (McKay and McGregor 1974, Manino et al. 1991). On the basis of an analysis of flower structure and characteristics, Porsch (1950) concluded that the chestnut flower used to be a primitive entomophilous flower, which is undergoing a secondary transformation and eventually will become a wind-pollinated flower.

Chestnut trees bloom much later than other fruit trees. Hence, the composition of insect population is different from what can be found on other fruit trees. At the time of chestnut bloom most insects have reached the maximum of their population growth. Chestnut flowers are attractive to primarily coleopterous insects. Porsch (1950) observed 103 insect species on chestnut flowers. Half of them were coleopterous insects but the flower-visiting population was quite miscellaneous since these insects belonged to six insect orders, 26 families, and 80 genera. Among the insects, wild bees and honeybees are the most important in addition to beetles. Manino et al. (1991) report that honeybees make up about half of the insect population visiting chestnut flowers. In addition to honeybees, the relative abundance of butterflies, syrphids, and beetles are worth mentioning. Manino et al. (1991) stated that the move of chestnut-growing areas from the traditional mountainous environment towards areas cultivated more intensively changed the composition of the flower-visiting insect population. McGregor (1976) thinks that it is worth moving bee colonies to chestnut orchards but McKay and McGregor (1974) mention that honeybees tend to disregard chestnut for the sake of more attractive bee forages.

Strawberry. Automatic self-fertilization in strawberry is prevented because at the beginning of the receptive period of the stigma the anthers already shed their own pollen, which facilitates cross-pollination. Furthermore, stamens are so situated that at anther dehiscence pollen grains are shed only on a part of the many stigmas of the flower, hence some stigmas remain unpollinated, and the maximum berry size requires that every pistil be fertilized. In primary flowers there are 400 to 500 stigmas and achenes, and in later flowers there are 200 to 300, and in the last flowers 80 to 200. Although the number of fruitlets may be less than that of stigmas. Because of flower structure, if unpollinated stigmas remain malformed berries will develop. Consequently, good yield and fruits of uniform appearance require insect pollination (Free 1970, McGregor 1976, Pouvreau 1984).

Strawberry flowers are visited by honeybees and wild insects. The latter are mainly early spring wild bees (*Andrena*, *Halictus*, *Bombus* species) and dipterous insects (Syrphidae) but strawberry flowers are sometimes visited by beetles, wasps, thrips, butterflies, and even few, representatives of other insect groups (Nye and Anderson 1974, McGregor 1976, Oliveira et al. 1991). Nye and Anderson (1974) observed 105 insect species on strawberry flowers in Utah (U.S.A.), which belonged to 35 families. In Canada, only 45 insect species were observed (Oliveira et al. 1991). In Utah, honeybees and syrphids, in this order, made up the two largest groups in the entire flower-visiting population. In Canada, also honeybees occupied the first position (56%) but wild bees were the second (25%) instead of syrphids, which had a proportion of 10%. Petkov (1965) indicated that the proportion of honeybees may range even higher depending on the year and the annual weather because in his experience honeybee proportion may range from 45% to 78%. He observed seven to 19 honeybees in a m^2 in an hour in favourable weather, and five to 11 other insects. The high proportion of honeybees in strawberries has been also mentioned by others (Free 1968a, McGregor 1976). Nevertheless, honeybees seem to work only in favourable weather (Free 1968a, Blasse 1981). When the weather is bad, pollination is performed by other insects, primarily wild bees and flies, with an efficiency that is lower than that of honeybees. Pollination of strawberry is facilitated by the fact that in bad weather, flowers may retain their receptivity for as long as ten days but on average the receptive period may last seven days. Regardless of the length of receptivity of stigmas, the most suitable period for pollination falls between the first and the fourth day (McGregor 1976) after flower opening.

Free (1968a) reported that in strawberry several nectar-gatherers gathered also pollen, while others deliberately gathered pollen. Nectar-gatherers touch stamens as well as stigmas in almost every flower but sometimes land on petals and approach the nectary from the side, without touching the stigmas. Pollen-gatherers, however, scrabble about for pollen while walking around on the stamens or standing on the stigmas in the middle of the flower. In the course of this activity, pollen grains get on the stigmas from their bodies. Honeybees visit 6.7 strawberry flowers on average in a minute (Petkov 1965). Free (1968a) estimated that a bee spends 10 s in a flower-visiting six flowers per min. Chagnon et al. (1989) reported that the first bee visitation on a flower lasts 24 s, while the second visitation lasts only 8 s, and the further visitations are even shorter.

It is generally held that the frequency of malformed fruits is higher without insect pollination, that is, insect pollination results in an increase in the number of well-formed fruits, which improves market value (Free 1968c, Moore 1969, Nye and Anderson 1974, Blasse 1981, Svensson 1991). Furthermore, insect pollination increases fruit size (Moore 1969, Nye and Anderson 1974), and fruit set (Free 1968c). However, there is no agreement on the effect of insect pollination upon yield. Nye and Anderson (1974), and Svensson (1991) claim that this effect is negligible, whereas Moore (1969) and Blasse (1981) report a significant decrease in yield in the case of the exclusion of insect pollination. This decrease is 32 to 71% (Moore 1969), or 15 to 50% (Blasse 1981).

Skrebtsova (1957) studied the effect of the frequency of bee visitations. With less bee visitations smaller berries developed. After 16 to 20 bee visitations, the mean fruit weight was 5.36 g, and after 21 to 25 visitations, 8.13 g. Consequently, insect pollination is useful to strawberry so bee colonies are to be moved to flowering strawberry parcels in order to increase yield. As fruit size depends on the number of pollinated stigmas in flowers, and the flowers have many stigmas, an over-pollination is impossible in strawberry and careful growers use bee colonies. Some other species flowering at the same time may be more attractive to bees than strawberries (McGregor 1976), which only can be compensated for by applying more bee colonies. McGregor (1976) suggests five to ten colonies per 1 acre (12 to 25 colonies per hectare).

In addition to open field cultivation, strawberries are cultivated also under cover. When they began to grow strawberries under glass or plastic, the proportion of malformed fruits was very high (Matsuka and Sakai 1989). Then the use of pollinating bee colonies became widespread, and the strawberry has brought good yield, perfect fruits since then even under cover.

Raspberry and blackberry. Fertilization and demands for pollination in raspberry and blackberry cultivars show a wide variation. In flowers the stamens are situated in the outer circles. Hence, the anthers that dehisce before the stigmas have become receptive can touch only the outermost stigmas, so the majority of stigmas would remain unpollinated without insects.

Shanks (1969) designed experiments in order to check a usual opinion of fruit growers that the wind played a crucial role in pollination of raspberry and blackberry. The author had a continuous air current blown towards a few caged raspberry bushes but this "wind pollination" entailed no increase in yield compared to bushes that have not received artificial air current.

Raspberry and blackberry flowers produce abundant nectar (Free 1970). Raspberry flowers produce 1.4 mg sugar a day, and secrete 13 mg nectar in the entire period of flowering. In blackberry, the daily nectar content is 4 to 6 mg with a sugar concentration of 20 to 40%. The nectary is fairly liable to weather so nectar content and its sugar concentration are also influenced by weather. There are lots of stamens in flowers of both fruit species, and as modern cultivars are bisexual, pollen production is abundant as well. Hence, they strongly attract pollinating insects though there are few data on insect visitation of these species.

Free (1970) reports that Hymenoptera, Diptera, and Coleoptera species are also visiting the flowers of raspberry and blackberry. Benedek et al. (1974) emphasize the large number of honeybees, and the many *Andrena* and *Bombus* species. McGregor (1976) also states that honeybees are the best pollinators of raspberry and mentions that blackberry cultivars are also visited by crowds of honeybees. Blackberry flowers are visited also by bumble-bees and other wild bee species but in the plantings their amount is never sufficient, hence McGregor (1976) suggests that honeybees should be placed near raspberry or blackberry plantings.

While gathering nectar from raspberry or blackberry, a bee is positioned on the petals or the stamens and pushes its head between the filaments and the stigma towards the nectary (Oliveira et al. 1991). Free (1968a) reports that while gathering nectar, the bee is going around along the nectar glands and its body is touching the stigmas. Abundant pollen is being shed on nectar-gatherers while working but there are few bees that deliberately gather only pollen on raspberry. Moreover, only a certain percentage of nectar-gatherers put this pollen in their baskets, the majority discard it. Oliveira et al. (1991) reported that the majority of honeybees (66%) gathered exclusively nectar, and those gathering both nectar and pollen made up the minority (34%). The proportion of bees carrying a pollen load is not constant because they are more likely to carry pollen in the afternoon than before noon. Free (1968a) studied the steadiness of bees. He marked bees that had gathered either nectar or pollen and was checking them through two days. Seventy percent of bees had insisted on their original behaviour.

Honeybees spend 9 s on a flower while gathering pollen, and 8 s while gathering only nectar. About 50 flowers are visited in the course of a foraging trip, which lasts 7 min (Free 1968a).

The behaviour of wild bees seems to differ from that of honeybees in that more bumble-bees (*Bombus* spp.) carry pollen loads than honeybees. Bumble-bees work in raspberry somewhat faster than honeybees.

There are no data on the bee visitation of the blackberry but, considering the similarities in flower structure, it is likely that the differences are negligible.

As for the effect of insect pollination, it is an old observation that caged raspberry and blackberry bushes produce less fruits, and the proportion of imperfect berries increases in the crop (Free 1970, McGregor 1976). The reason for the imperfect fruit is that the development of a perfect blackberry fruit requires that all styles, or at least their majority, have received an effective pollination (Shoemaker and Davis 1966). For raspberry, Couston (1963) states essentially the same. Shanks (1969) reported a decrease by 10 to 27% in fruit weight in parcels without bees, compared to parcels visited by bees (*see also* Oliveira et al. 1991). Wieniarka (1987) also remarked that the decreased yield in parcels without bees and the decreased fruit weight could be derived from fewer drupelets, which was the result of an insufficient pollination.

Eaton et al. (1968) pointed out that in order to grow more raspberry fruits (large berries with numerous drupelets), the grower was to repeat pollination at least through four days. However, they performed their experiments with hand pollination. Oliveira et al. (1991) checked these results on the basis of the influence of the number of effective bee visitations. They established a significant positive correlation between the number of bee visitations and the number of drupelets, and between bee visitations and fruit weight. The correlation can be made even more precise by considering the behaviour of bees also. The number of drupelets and fruit weight were significantly higher when bees had gathered both pollen and nectar than when they had visited flowers only in order to gather nectar. The number of drupelets and fruit weight reach their maximum at an accumulated bee visitation time of about 150 s per flower. A large increase in number of drupelets and fruit weight is observable to an accumulated bee visitation time about 50 s, and over this point the further increase is slow and little. Consequently, Oliveira et al. (1991) emphasized that five to six bee visitations per flower are enough for a sufficient pollination, that is, to achieve a favourable level of fruit set and a maximum yield. Contrary to these facts, no data are available concerning the optimum bee colony density in raspberry or blackberry plantings. Hence, we suggest that in large plantings a few bee colonies should be placed per hectare at the beginning of bloom, and then the number of bee colonies should be modified on the basis of the frequency of bee visitation. What we know (Shanks 1968, Oliveira et al. 1991) is that a flower is to be visited five to six times in four days. Five to six bee visitations are to be ensured in two days (Oliveira et al. 1991). Hence, supposing a seven-hour daily bee visitation, nine bee visitations should be observed on 100 flowers in 15 min. Unless the extent of bee visitation achieves this level, further bee colonies should be placed in the orchard.

There are no records on competition with other plants. Nevertheless, raspberry and blackberry flowers are attractive to bees so the probability of distracting influence is negligible.

Gooseberry. Yield in *gooseberry* cultivars is a function of insect pollination, as is pointed out by Offord et al. (1944). They report that no fruits had been developed from 736 self-pollinated flowers, while 280 fruits had resulted from 621 flowers with allogamy.

Insects seem to play a crucial role in the achievement of the maximum gooseberry yield (Hughes 1961, Free 1970, McGregor 1976). There are no data on the pollinating insect population. According to our own observations, gooseberry flowers are mostly visited by honeybees but different wild bee species can also be observed on gooseberry flowers. Where the local pollinating insect population is insufficient, including honeybees and wild bees, bee colonies are worth using in order to achieve the maximum yield.

Currants. The *red currant* produces fruit steadily and uniformly so its insect pollination has not been investigated intensively. Practically, there are no experimental data on red currant. Hughes (1961) only mentions that the red currant produces more fruit if honeybees are in its neighbourhood.

The *black currant*, however, causes much problems because its cultivars have different capabilities for self-fertilization. This led to the idea that insect pollination should be studied in this species.

Black currant flowers are visited by honeybees and wild bees. Black currant flower structure is different from that of other fruit species discussed so far. In this bell-shaped, primitive stereomorphic flower, beetles and other primitive flower-visiting insects cannot get to the nectar. Nevertheless, the proportion of honeybee visitations, without an artificial increase of their number, is usually less than in other fruit species. Bumble-bees are also important pollinators (Free 1968*d*). In black currant, Benedek et al. (1974) report that honeybees make up only one-third of the flower-visiting insect population of the black currant, the others are wild insects. Among the wild insects 19% were bumble-bees, 25% were other wild bees (*Andrena, Osmia* species), 9% were wasps, and 47% were syrphid flies. Clinch and Faulke (1976), however, thought that the most usual visitors of black currant flowers were honeybees. Furthermore, they mention bumble-bees, other wild bees (*Lasioglossum* spp.), flies (hover flies, drone flies), and certain butterflies, but not nocturnal insects. The data thus are contradictory, which refers to the influence of environment upon the black currant visiting population.

Pollinators visit flowers for the sake of the quite abundant amount of nectar and pollen (Free 1970).

While a bee is visiting a black currant flower, it is pushing its proboscis forward between the stigma and the stamens towards the nectary. According to Free (1968*a*), it is rare that honeybees and wild bees gather exclusively pollen. The proportion of pollen-gatherers is 0 to 10 or 0 to 22%, depending on the day. In other circumstances, Clinch and Faulke (1980) observed that 1 to 26%, and 1 to 50% of the honeybee population were pollen-gatherers in two succeeding years.

Bee colonies moved to black currant orchards gathered only 1% of the accumulated pollen from black currant bushes. Honeybees spend 14 s on average on a flower, while bumble-bees spend a much shorter time, 5 to 6 s. On a bush, honeybees visit five flowers, while bumble-bees visit much more, nine to 13. Within an inflorescence, bumble-bees visit several neighbouring flowers. Thus, bumble-bees are as valuable pollinators as honeybees. Another argument for this statement is that in the flowering period of black currant, which is earlier than that of most fruit species, the weather is often bad, resulting in a low bee density. Bumble-bees remain active in cool weather that honeybees cannot bear (Paarmann 1977). Benedek et al. (1974) observed in black currant that the number of honeybees and their flower visitations were less with distance of the colonies. At a distance of 600 m, half as many bees were found compared to 150 m. Wild pollinating insects, however, were characterized by a more or less uniform distribution in the area of black currant planting.

Without placing many honeybee colonies to the black currant planting, honeybees do not necessarily make up the majority of the black currant visiting insect population. In contrast to other fruit species, in black currant wild insects usually make up the majority of visiting population. It is not surprising, therefore, that there is no significant change in yield depending on distance from the apiary (Benedek et al. 1974, Bergfeldt 1980). Nevertheless, several data suggest that honeybees exert a favourable influence upon the fruit set and yield of the black currant. Bushes or branches that pollinating insects are free to visit always bring a more considerable yield than those caged, which insects cannot visit (Free 1970, Benedek et al. 1974, McGregor 1976, Pouvrean 1984). In order to study the pollinating efficiency of honeybees, Schanderl (1956) enclosed honeybees with caged bushes and compared the yield obtained from bushes caged without honeybees. The yield of the bushes caged with bees was 10.9 and 17.3 times more than that of bushes caged without bees. Free (1968d) demonstrated that a cross-pollination by insects (bees) improved both the quantity and the quality of yield. Another positive influence of insect pollination is that as a result of an intensive bee pollination, the extent of fruit drop after flowering will decrease (Zakharov 1959, Bergfeldt 1980). Furthermore, the flower-visiting activity of honeybees has an increasing influence upon seed content and berry weight. On bushes where honeybees are excluded, berries are smaller with fewer seeds, which increases the fruit drop rate (Bergfeldt 1980).

Considering the favourable influence of bee pollination, it is generally held that the achievement of the maximum yield and good fruit quality requires growers to place honeybees to large black currant plantings (Free 1970, Benedek et al. 1974, McGregor 1976, Bergfeldt 1980, Pouvrean 1984).

As for the optimum bee colony density, Skrebtsova's (1959) data on its relation with fruit set values are mentioned first. A density of 0.5 bee colonies per hectare corresponded to a fruit set of 53%, while a density of six bee colonies per hectare corresponded to a much higher fruit set, 88%. More than six bee colonies, however, were not worth placing to the orchard, as fruit set values ceased to increase with higher bee density, supposedly because the plant could not produce more. On the basis of these data, Benedek et al. (1974) suggest that three to five bee colonies should be placed per hectare. Nevertheless, competition with other plants must not be ignored since blooming orchards in the neighbourhood and field crops, such as rape, may distract the majority of bees. Furthermore, the often cool and unfavourable weather during bloom of the black currant decreases the pollinating activity of bees, which can be compensated for by placing even more honeybee colonies to the planting. In spite of Skrebtsova's (1959) above-mentioned observations, McGregor (1976) regards demands of the black currant for auxiliary pollination as not very strong though accepts that the achievement of a large yield does require the application of bee colonies because the native pollinating insect population is usually insufficient.

Blueberry. There are miscellaneous and often contradictory data on fertilization conditions and insect pollination in different blueberry species. Although a good yield requires a good pollination because blueberry species are able to achieve a fruit set of 100%, and a sufficient yield requires a fruit set of 80% in *highbush blueberry* and at least 50% in *lowbush blueberry*. Flower structure is relevant to pollination. Blueberry species bear tube-shaped flowers with petals grown together. The filaments can be found inside, but the stigma is almost twice as long as filaments. As only its top being receptive, automatic self-pollination is excluded. Petals of the *cranberry* are free. Its filaments make up a tube, and the stigma rises above it. When pollen is being shed, the stigma is still short, and only later, when it has risen above the stamens, becomes receptive. Hence, self-acting self-pollination is also excluded in this species. Flower structure thus requires insects to take part in pollination.

The *Vaccinium* berries are visited by crowds of wild pollinating insects in their forest environment, depending on the size of the area. The flower structure is available almost only for bees so the most usual flower-visiting insects are bumble-bees (*Bombus* spp.) and other wild bees (*Andrena, Colletes, Halictus, Nomada* species), and accidentally some syrphid, bombylid, and calliphorid flies (Free 1970). The proportion of bumble-bees often exceeds that of honeybees but certain researchers argue that bumble-bees are less reliable pollinators than honeybees because of their changeable density. In areas with a widespread blueberry population, or where blueberry is cultivated intensively, the density of wild pollinating insects is much lower than in relatively undisturbed areas, so especially at large production sites, honeybees are indispensable as pollinators (Free 1970). Different authors (Free 1970, McGregor 1976) agree that the application of honeybees is necessary for a good yield. Apiarists move their bee colonies to *Vaccinium* berry plantings with pleasure because these species give a good honey yield.

Free (1970) provides a good review concerning flower-visiting behaviour of bees. *Bombus* species are much faster flower-pollinating insects, visiting 10 to 20 flowers per min, than solitary bees and honeybees which visit only five flowers per min. With their long tongues, Bombus species can easily get to the nectary, situated at the bases of filaments. Honeybees cannot get to the nectary with their shorter proboscises in certain species. In these species, honeybees select flowers with wide ringent corollas, where they can force their heads inside in order to reach the nectary. However, this procedure is impossible in the case of species with narrow ringent corollas. In species with wide ringent corollas, bees often work without touching the stigma, therefore not every bee visitation entails pollination. It may also occur in some *Vaccinium* species with long flowers that certain bumble-bees with short proboscises, or carpenter bees (*Xylocopa* spp.) bite a small slit with their strong mandibles at the base of the flower to gather nectar. Honeybees can also use these holes, neglecting pollination. Helms (1970) thought that these small slits were made by honeybees but McGregor (1976) refused to accept this idea. We agree with McGregor (1976) because it is well known (Eickworth and Ginsberg 1980) that on papilionaceous species with long tube-shaped flowers are firm and honeybees' mandibles are not strong enough to bite such flower tubes (Benedek et al. 1974).

Honeybees gather nectar primarily from blueberry. The proportion of pollen-gatherers in *Vaccinium* species is generally only 1 to 2%. The cultivars differ from each other in flower tube structure as well as nectar production. Hence, honeybees prefer certain cultivars to others. Bumble-bees, however, visit different cultivars without selection.

As for the influence of bee pollination upon yield, it has been demonstrated (Free 1970, McGregor 1976) that in caged bushes, isolated from bees, yield is less and of poorer quality, compared to bushes where bees are free to visit. Berries resulting from cross-pollination are larger, contain more seeds, and ripen earlier.

McGregor (1976), reviewing the miscellaneous suggestions concerning the application of honeybees to pollination (0.5 to 10 colonies per acre = 1.2 to 25 colonies per hectare), concluded that richer yield, better fruit quality, and earlier harvest, are all due to bee pollination. They are all advantageous to the grower, and therefore, the highest possible bee density should be achieved. Hence, the optimum of five to ten colonies per acre (12 to 25 colonies per hectare) should be placed to *Vaccinium* plantings, but at least one to two colonies per acre (2.5 to 5 per hectare) are absolutely necessary. Another useful suggestion is (Filmer and Marucci 1963) that lowbush blueberry populations at full bloom require one bee per sq yard (1.1 bees per sq m). Below this density, the native pollinating insect population should be supplemented placing bee colonies. Howel et al. (1972) pointed out that there was a correlation between yield and the time of the introduction of bee colonies relative to the flowering period. An earlier introduction entailed a richer yield and larger berries. Hence, bees should be introduced not later than at 25% of full bloom.

Nowadays bumble-bees can also be propagated artificially. Readers who are interested in methods of propagation and the application of bumble-bees to pollinating *Vaccinium* species are referred to Johansen (1967).

Wood (1971) worked with the mixture of two sexually incompatible cultivars of *Vaccinium*. He increased bee colony density up to eight per acre but could not achieve an increase in yield, berry quality and perfect seed content. Hence, in *Vaccinium* populations, as is suggested also by others (Free 1970, McGregor 1976), it is worth growing more compatible species or cultivars together.

9.9. The Placing and Management of Bee Colonies

In order to pollinate fruit species, bee colonies should bee placed near or inside orchards. When and how should this task be done? We discuss this topic on the basis of reviews by Free (1976), Shaer and Schäfer (1975), Benedek et al. (1983, 1989b) and Jay (1986).

The time of the introduction of hives is relevant to the bee visitation of the orchard. Free et al. (1960) report that much more foragers visit the orchard from those hives placed in the orchard after the beginning of bloom than from those that were placed there earlier. Though this difference may be equalized under weather conditions favourable for blooming, when bloom is short, nevertheless, the time of introduction may be of great importance. Benedek et al. (1972) demonstrated that bee colonies moved to the orchard before bloom were characterized by a lower pollinating efficiency (lower fruit set values) than bee colonies moved to the orchard just at the beginning of bloom. Howel et al. (1972) verified the same in the case of highbush blueberry.

The explanation of this phenomenon is that first bee colonies orient themselves in the new place with the aid of their scout bees, and the foragers follow the scouts and visit plants flowering at the time they were placed in the orchard and not the trees that start bloom later. Williams and Brain (1985) verified this idea in *apple*; moreover, they pointed out differences in cultivars. Flowers of 'Cox's Orange Pippin', for instance, are visited by bees after petal fall more intensively than flowers of 'Golden Delicious'. If a plant begins to flower later, scouts discover it also later and it takes them more or less time to make the foragers of the colony use the new forage subject. Consequently, bee colonies should be moved to the orchard just at the beginning of bloom. The disadvantageous influence of a too early introduction can only be compensated for by the increase of the number of bee colonies, whereas a too late introduction results in decreased fruit set and lower yield.

Another important question is to where bee colonies should be placed. Remember if bee colonies are placed at a greater distance from the orchard, less field bees will visit it (Tsigankov 1953), and lower fruit set values are to be expected in negative correlation with distance from the hive (Tsigankov 1953, Yakovlev 1959, Pusztai et al. 1969). Thus, bee colonies should be placed as near to the orchard as possible. Placing inside the orchard is preferred to placing outside, and if the former is impossible, then the apiary should be placed at the edge of the orchard. Nevertheless, placing apiaries too near to each other may cause drifting (exchanging populations), which can be avoided by keeping an isolating distance over 600 m (Boylen-Pett and Hoopingarner 1991) between hive placements.

Bee density definitely begins to decrease at a distance of 100 to 200 m, but definitely decreases at a distance of 300 m (Free 1960*b*, 1970, Benedek et al. 1974). This is reflected in a decrease in fruit set and yield as well (e.g. Free 1962, Free and Spencer-Booth 1964, Benedek and Martinovich 1971, Benedek et al. 1972, 1974, Bergfeldt 1980, Szabó et al. 1989). Thus bee colonies should be placed singly and scattered in the area of the orchard. Apiarists, however, are not ready to do so because looking after bee colonies, indispensable also through the bloom period would require too much time.

As for the placing of bee colonies, the best compromise, initiated by several authors (Free 1970, Benedek et al. 1974, McGregor 1976, Benedek et al. 1983, 1990*b*) is that bee colonies should be placed in groups consisting of ten to 20 hives in equidistant distribution with a distance of 200 to 300 m inside the orchard. This way it is possible to look after bee colonies, and because of the overlapping foraging areas of bees of different hive groups, a uniform bee visitation is to be expected. Furthermore, even under unfavourable weather conditions, when bees are concentrated around the hive in a circle of 150 m in diameter, a sufficient extent of bee visitation is ensured (Free 1970).

A further question is as to what kind of bee colonies should be hired. Strong ones, of course, but what does "strong" mean? The precise definition of a strong colony is fairly difficult. Considering that fruit species (with the exception of chestnut) bloom early in spring, when bee colonies have just lived through the winter and are still before the annual population multiplication, "strong bee colonies" cannot mean the same as later in summer, when they achieve the peak of their annual period of growth. There are two widespread parameters to characterize the strength of bee colonies. The one is the total number of the bee population, which can be expressed either by the number of combs occupied by bees, or the number of bees in the colony. The other is the size of brood. In the bloom periods of fruit species, a colony of 10,000 bees, is to be regarded as strong, but in summer this number may exceed 20,000. Todd and Reed (1970) argue that a really appropriate parameter to express the pollinating value of bee colonies is the size of brood because an unsealed brood requires pollen as a food since it contains undeveloped larvae; and this demand stimulates bees of the colony to gather pollen. Whereas large bee colonies with a little brood prefer gathering nectar, so pollination requires bee colonies with large brood. McGregor (1976) regards colonies as suitable for pollination in which at least four to eight frames are occupied with brood. Further generalizations are impossible because in different countries different hive types are used, and it may also occur that two to three bee colonies live together in a hive. Furthermore, colonies that move from the south cannot be regarded as colonies which just passed the winter.

Another old question is how to increase the pollinating efficiency of bee colonies in order to ensure a sufficient extent of bee pollination of orchards. Istomina-Tsvetkova and Skrebtsov (1965) suggest that bee colonies should be rotated because in a new place at the beginning bees work immediately near the hive, and their foraging area spreads out only gradually (Free, 1960*b* 1970). It must be noted, however, that after relocation bee colonies tend to visit the plant that they visited earlier if it is available (Free 1959). Jay (1986) calls attention to disadvantages of rotation of bee colonies. Rotation requires a continuous monitoring of the foraging activity of bees, and the disorientation of the removed bees may cause losses, decrease reserved food, and require much money and work.

Naturally, apiarists know tricks that the fruit grower does not have to know. For instance, hives should be carried with specially designed vehicles (e.g. Shaer and Schäfer 1975), or they are worth storing in simple, temporary buildings. At least it should be ensured that hives are placed in the shadow of trees that they do not overheat during the warm hours around noon.

To increase pollinating efficiency of bee colonies, increasing pollen-gathering seems to be a suitable method, because pollen-gatherers are usually better pollinators than nectar-gatherers (Free 1970). What exerts an increasing influence upon the proportion of pollen-gatherers is the colony's demand for pollen, and a large or increasing brood, requiring much pollen (Free 1967). Hence, inserting extra brood into the hive from another colony would increase pollen-gathering. In practice, however, this method is senseless (Jay 1986) because it requires extra bee colonies. Pollen-gathering can be stimulated by creating a lack of pollen, also. The frames that contain the accumulated pollen can be removed, for instance, or pollen traps can be placed at the entrance of the hive in order to pick pollen loads from baskets of bees returning home (Free 1970). Though pollen traps can increase the proportion of pollen-gatherers, using this method for a long period decreases the size of brood, which is not favourable. Providing pollen or pollen supplement for bee colonies, however, encourages brood rearing, which results in an increased proportion of pollen-gatherers (Free 1965c). Another method to increase pollen-gathering is to feed sugar syrup to bee colonies (Free and Spencer-Booth 1961, Free 1965c). Increase in pollen-gathering stimulated by sugar feeding seems to be independent of brood rearing. Rather, it seems that the influence of feeding sugar changes bee foraging behaviour. Because of the availability of nectar substitute, sugar, some nectar-gatherers change their behaviour and begin to gather pollen (Free 1965a). Thus, feeding sugar syrup is a good method to make bee colonies increase their pollen-gathering activity, which entails an increase in the pollinating efficiency as well as the proportion of pollen-gatherers in the species to be pollinated (Free 1970). It seems that the success of the method of feeding perfumed sugar syrup to bee colonies, suggested mostly in Russian articles, which serves the purpose of directing bees to certain cultivars, is simply due to the sugar syrup, without perfuming (Free and Spencer-Booth 1961, Free 1965a). Free (1958b) points out that scout bees of bee colonies fed with perfumed sugar syrup look for flowering plants and communicate with foragers about its location but no increase in yield can be observed. Some further experiments supported this opinion but others served evidence in support of the conjecture that there is an influence upon fruit set and yield (Jay 1986). All in all, directing bees with perfumed sugar syrup is not a reliable method but feeding sugar itself is undoubtedly useful.

Another method to increase the pollinating efficiency of bee colonies is to fix hive inserts and pollen dispensers on the hive so that bees can get out of the hive only by penetrating the apparatus and have pollen stuck on their bodies. Legge (1976) provides a detailed comprehensive survey on this method. Although several bees may load pollen into their baskets in the dispenser and turn back to the hive (Griggs and Iwakiri 1960), others do carry some pollen to the flowers (Legge 1976). The necessary amount of pollen can be gathered with pollen traps by the help of bees. There are several types of pollen traps. A modern type is described by Cook (1985), which is suitable for continuously gaining pollen, perhaps with the purpose of pollination. Szalay et al. (1983) report that the (apple) pollen gathered this way, can be stored at a temperature of −76 °C for three months, then at −196 °C for nine months, and 60 months, while it entirely retains its activity, it germinates on stigma, and sets fruit. Nevertheless, hive inserts and pollen dispensers have not become widespread in practice but Jay (1986) argues that they are valuable pollination aids and will become more common.

Another approach is to increase the attracting influence of the orchards to be pollinated. Free (1965), for instance, sprayed plants with sugar syrup. This method increased the number of bees visiting these plants but they gathered the syrup instead of the pollen and the nectar of the flowers. Other researchers sprayed plants with honey solution or sugar syrup but influence upon yield was rarely observed (Jay 1986).

Spraying flowering plants with "Beeline", a food supplement, is based upon the same idea. The results are contradictory, the influence of this material upon bee visitation and yield is not significantly stronger than that of simple water (Jay 1986). Williams and Jefferies (1980) report that in apple "Beeline" was not effective in a weather unfavourable to bees but in good weather spraying increased bee visitation. Rajotte and Fell (1982) analysed another bee-attracting product, "Beelure". No influence was observed so this method is not promising.

9.10. Concluding Remarks

It is clear on the basis of this review that wind does not play an important role in the pollination of temperate zone fruit species with entomophilous flower structure, with the exception of the *chestnut, walnut* and *hazelnut*. These latter species are in a transitional position between insect-pollinated and wind-pollinated flowers. The other fruit species with entomophilous flower structures cannot bring a sufficient yield without the aid of insects. More or less pollen grains carried by the wind might reach the stigma but a good pollen germination requires insects to touch the stigma surface, which start certain enzymes functioning. The help of insects is indispensable not only in self-sterile but in partially or completely self-fertile fruit species as well because the flower structure of fruit species excludes a self-acting self-pollination.

Fruit species are visited by species of numerous insect groups. Among them, the apoids are the most effective pollinators. As fruit species bloom early in the spring, among wild pollinating insects only some that have lived through the winter are already active. Thus in large orchards their density is variable and insufficient. Good pollination requires the use of honeybees, which usually make up the majority of pollinating insect population without any kind of artificial intervention. Nevertheless, as the presence of native pollinators is usually not sufficient for the achievement of optimum fruit set and yield, an auxiliary introduction of bee colonies to the orchard is required.

The pollinating efficiency of bee colonies shows a close relation with the flower-visiting behaviour of honeybees and the influence of factors affecting the distribution of honeybees. These factors and the influence of bee pollination upon fruit set and yield should be known by fruit growers, so that they can apply bee colonies for pollination to be successful.

During the last decades they tried out some artificial pollinating methods but it turned out that the use of honeybees was simpler and more effective. The pollinating efficiency of bee colonies can be increased artificially though one can achieve the optimum fruit set and yield without artificial increasing methods, only by ensuring the sufficient bee colony density in the orchard, which is a cultivar-dependent parameter. These facts can lead us to the conclusion that nowadays bee pollination is an indispensable factor of intensive fruit growing.

10. Cited Literature

Abbott, D. L. (1959). *Ann. Rep. Long Ashton* 1958, p. 52–56.
Abbott, D. L. (1970). In: Luckwill, L. C. and Cuttings, C. V. (eds), *Physiology of Tree Crops.* Academic Press, New York, pp. 65–80.
Abdalla, M. M. F. and Hermsen, J. G. Th. (1972). *Euphytica* 21:32–47.
Abdulkadyrov, S. K., Batyrkhanov, Sh. G. and Dzhabaev, B. R. (1972). *Trudy Dagest. Sel'skokh. Inst.* 22:58–71.
Abou-El-Nasr, N. M. A. and Stösser, R. (1989). *Angew. Botanik* 63:33–42.
Abramov, N. A. (1955). *Izv. Akad. Nauk. SSSR* 1:53–65.
Aeppli, A. (1984). *Schweiz. Zeitsschr. Obst-Weinbau* 120(4):102–109.
Afify, A. (1933). *Pomol. Hort. Sci.* 11:113–119.
Albertini, A. (1980). *Frutticoltura* 2:5–64.
Albertini, A. (1981). *L'Informatore Agrario* 37 (26):16281–16301.
Albertini, A. (1982). In: Bargioni, G. (ed.), *Giornate Frutticole Veronesi il Ciliegio,* Verona, pp. 83–102.
Alexander, M. P. (1969). *Stain Technology* 44:117–122.
Alfaro, J. F., Griffin, R. E., Keller, J., Hanson, G. R., Anderson, J. L., Aschroft, G. L. and Richardson, E. A. (1974): *Trans. Amer. Soc. Agr. Eng.* 17: 1025–1028.
Al-Jaru, S. and Stösser, R. (1983). *Angew. Botanik* 57:371–379.
Almeida, C. R. (1949). *Anis. Inst. Sup. Agronomia* 16:51–71.
Almeida, C. R. and Marques De (1945): *Anais Inst. Sup. Agron Univ. Téc. Lisb.* 15: 7–13.
Andreies, N., Serboiu, L. and Thiesz, R. (1981). *Lucrari Stiintifice* 9:95–110.
Angelov, T. (1966). *A "Lippay János" Tud. Ülésszak előadásai,* Budapest 2:301–311.
Angelov, T. (1975). Prouchvaniya verhu nyakoy biologichni osobennosti i sortoviya seostav pri dulata. Dissertation. Plovdiv.
Angelov, T. (1981). *Gartenbau* 28 (3):85–87.
Anvari, S. F. and Stösser, R. (1978a). *Z. Pflanzenzüchtg.* 81:333–336.
Anvari, S. F. and Stösser, R. (1978b). *Mitt. Klosterneuburg* 28:23–30.
Anvari, S. F. and Stösser, R. (1981). *Mitt. Klosterneuburg* 31:24–30.
Anvari, S. F. and Stösser, R. (1984). *Mitt. Klosterneuburg* 34:221–225.
Apostol, J-né, Brózik, S. and Nyéki, J. (1977). *Bot. Közl.* 64 (4):267–272.
Ashworth, E. N. (1984). *Plant Physiol.* 74:862–865.
Ashworth, E. N. and Rowse, D. J. (1982). *HortScience* 17:790–791.
Avanzi, S., Tagliasacchi, A. M., Forino, L. M. C., Loiero, M. and Filiti, N. (1980). *Riv. Ortoflorofruttic.* 64:111–124.
Babaleanu, P. (1938). Diss. der Friedrich-Wilhelms-Univ., Berlin. Verlag Gebr. Borntraeger.
Babóné Moharos, G. (1974). *Kert. Egy. Közl.* 38:205–213.
Bach, F. (1928a). *Gartenbauwiss.* 1:358–374.
Bach, F. (1928b). *Gartenbauwiss.* 1:615–618.
Badescu, Gh., Badescu, L. and Sotiriu, D. (1981). *Lucrari Stiintifice* 9:47–57.
Bagni, N. and Gerola, F. M. (1978). *Riv. Ortoflorofruttic.* 62:309–327.
Baillod, M. and Mottier, P. Ph. (1966). *Agric. Romande* 3:29–32.
Baldini, E. (1949). *Riv. Ortoflorofruttic.* 33:1–2.
Baldini, E. (1950). *Riv. Ortoflorofruttic.* 34(11–12):200–208.

Baldini, E. and Pisani, P. L. (1961). *Riv. Ortoflorofruttic.* 45:640–663.
Baldini, E. and Scaramuzzi, F. (1980). *Il Susino. Da Frutticoltura* Fanni 80. Manuale REDA, Bologna, pp. 9–74.
Barbattini, R., Frilli, F., Roversi, A. and Ughini, V. (1983). *Redia* 66:343–363.
Barbeau, G. N. (1973). *La Pomologie Française* 1, Paris.
Barbier, E. (1983). *L'Arboriculture Fruitière* 30(348):32–36.
Barbier, E. (1985). *L'Arboriculture Fruitière* 379:39–42.
Bargioni, G. (1978). *Riv. Ortoflorofruttic.* 62(4):383–402.
Bargioni, G. (1979). *L'Informatore Agrario* 35(5):4431–4467.
Bargioni, G. (1982). Biologia fiorole e di fruttificazione. In: *Il ciliegio dolce.* Ed. Agricole, Bologna, pp. 59–99; 139–181.
Barrett, C. and Ariaumi, T. (1952). *Proc. Amer. Soc. Hort. Sci.* 59:259–263.
Bartz, M. and Stösser, R. (1989). *Gartenbauwiss.* 54:132–137.
Basso, M. (1972). *Riv. Ortoflorofruttic.* 4.
Basso, M. and Natali, S. (1967). *Yugosl. vočarstvo* 1:21–31.
Basso, M. and Natali, S. (1972). *L'Agricultura Italiana* 3:154–165.
Batjer, L. P., Schomer, H. A., Newcomer, E. J. and Coyier, D. L. (1967). Agr. Handb. No. 330. U. S. Dep. Agr., pp. 1–47.
Bayev, H. (1976). *Ovohstartstvo* 14(12): 15–18.
Beakbane, A. B., Chapelow, H. C. and Grubb, N. H. (1935). *Ann. Rep. East Malling Res. Stn. 1934*, pp. 100–114.
Beilani, L. M. and Bell, P. R. (1986). *Ann. Bot.* 58:563–568.
Bellini, E. (1973). *L'Informatore Agrario* 29(14):12077–12111.
Bellini, E. (1975). *Riv. Ortoflorofruttic.* 59(3):210–221.
Bellini, E. (1978). *L'Informatore Agrario* 2:43–49.
Bellini, E. (1980). *L'Informatore Agrario* 48:1–23.
Bellini, E. (1987). *L'Informatore Agrario* 43(1):43–49.
Bellini, E. and Bini, G. (1978). *Riv. Ortoflorofruttic.* 62(4):403–417.
Bellini, E., Liverani, A., Nicotra, A. and Sansavini, S. (1982). *Cent. Oper. Ortofruttic.* pp. 39–55.
Belmans, K. (1986). *Fruit Belge* 54(413):53–76.
Belmans, K. and Keulemans, J. (1985). *Erwerbsobstbau* 27(9):213–216.
Benedek, P. (1968). *Acta Phytopathol. Acad. Sci. Hung.* 3:59–71.
Benedek, P. (1980). Insect pollination of fruit trees. In: Nyéki, J. (ed.), *Gyümölcsfajták virágzásbiológiája és termékenyülése* (Bloom Biology and Fertility of Fruit Cultivars), Mezőgazdasági Kiadó, Budapest, pp. 101–110 (in Hungarian).
Benedek, P. and Martinovich, V. (1971). *Kertgazdaság* 3:37–42.
Benedek, P. and Nyéki, J. (1990). Abstracts of Contributed Papers (XXIIIrd Int. Horticult. Congress, Florence, August 27–September 1, 1990), 1:457 (No. 1984).
Benedek, P., Martinovich, V. and Dévai, Gy. (1972). *Kertgazdaság* 4(4):50–58.
Benedek, P., Manninger, S. and Virányi, S. (1974). *Pollination of Crops with Honeybees.* Mezőgazdasági Kiadó, Budapest, p. 199 (in Hungarian).
Benedek, P., Soltész, M. and Nyéki, J. (1983). *Proc. 29th Int. Congr. Apiculture (Budapest),* Apimondia Press, Bucharest pp. 289–290.
Benedek, P., Nyéki, J. and Lukács, Gy. (1989a). *Kertgazdaság* 21(3):8–26.
Benedek, P., Soltész, M. and Nyéki, J. (1989b). *Kertgazdaság* 21(6):41–64.
Benedek, P., Nyéki, J. and Szabó, Z. (1990a). *Kertgazdaság* 22(5):1–23.
Benedek, P., Soltész, M. and Nyéki, J. (1990b). *Kertgazdaság* 22(1):1–19.
Benedek, P., Nyéki, J. and Szabó, Z. (1991a). *Kertgazdaság* 23(1):40–58.
Benedek, P., Nyéki, J. and Szabó, Z. (1991b). *Kertgazdaság* 23(2):27–39.
Benko, B. (1967). *Biol. Plant.* 9:263–269.
Berezenko, N. P. (1963). *Sadov. Vinogr. Vinod. Moldavii* 18(8):20–23.
Bergamini, A. and Ramina, A. (1968). *Atti Conv. Naz. Studi Nocciuolo* 18: 1–16.
Bergamini, A. and Ramina, A. (1971). *Riv. Ortoflorofruttic.* 55:484–491.
Bergfeldt, G. (1980). *Sverig. Lantbruksuniv. Institut Tradgardsv. Rapport* 9:1–42.

Bernhard, R., Delmas, H. G. and Sanfourche, G. (1951). *Ann. Amel. des Plant.* 1(2):179–207.
Bernier, G. (1970). *Can. J. Bot.* 49:803–819.
Bespetshalnaya, V. V. (1981). *Sadov. Vinogr. Vinod. Moldavii* 12:49–50.
Bidabé, B. (1955). *Annal. Amelior. des Plant.* INRA, Recherches d'Arboriculture Fruitière, pp. 249–253.
Bini, G. (1972). *Riv. Ortoflorofruttic.* 4:299–307.
Bini, G. and Bellini, E. (1971). *Riv. Ortoflorofruttic.* 4:322–337.
Blaja, D. (1962). *Via şi Lividá. Bucureşti* 11(3):48–53.
Blasse, W. (1976). *Blühen und Früchten beim Obst.* VEB. Dtsch. Landw. Verlag, Berlin.
Blasse, W. (1978). *Gartenbau* 25:111–112.
Blasse, W. (1981). *Gartenbau* 28:337–338.
Blasse, W. (1986a). *Gartenbau* 33:308–309.
Blasse, W. (1986b). *Blühen und Früchten beim Obst.* VEB. Deutsch. Landw., Berlin, p. 117.
Blasse, W. and Barthold, F. (1970). *Arch. Gartenbau* 18:125–138.
Blasse, W. and Eger, J. (1978). *Arch. Gartenbau* 25:317–322.
Blasse, W. and Hoffmann, S. (1989). *Arch. Gartenbau* 37(3):169–232.
Blauhorn, W. and Schimmelpfeng, H. (1986). *Obstbau* 11(4):161–164.
Blazek, J. (1977). *Metodika pro zavádeni vysledku do praxe.* 10:1–30.
Blazek, J., Kloutvor, J. and Drobková, R. (1974). *Sbornik Uvti-Genetika a Slechténi* 10(2):139–145.
Bodi, I. (1981a). *Lucrari Stiintifice* 9:292–300.
Bodi, I. (1981b). *Lucrari Stiintifice* 9:119–125.
Bodi, I. (1981c). *Lucrari Stiintifice* 9:323–331.
Bordeianu, T., Lupescu, F., Mihaescu, G. R. and Popa, V. (1955). Lucrari Sec. Stiin. Inst. Agr., pp. 161–172.
Bordeianu, T., Constantinescu, N., Péterfi, St., Stefan, N. and Anghel, G. H. (1963). *Pomológia. Republicii Populare Romine.* I. Editura Academiei Republicii Populare Romine.
Borhidi, A. (1993). *A zárvatermők fejlődéstörténeti rendszere* (Phylogenetic System of Angiosperms). Janus Pannonius Univ., Pécs, p. 566.
Bornus, L., Jablonsi, B. and Król, S. (1976a). *Pszczelnicze Zesyty Naukowe* 20:1–20.
Bornus, L. et al. (1976b). *Pszczelnicze Zesyty Naukowe* 20:21–39.
Botez, I., Onigoaie, M. and Straulea, M. (1960). *Lucrari Stiintifice* pp. 411–414.
Bouché-Thomas, E. (1953). *La Méthode Bouché-Thomas*, Angers.
Boylan-Pett, W. and Hoopingarner, R. (1991). *Acta Horticult.* 288:111–115.
Boyle, R. M. D. and Philogene, B. J. R. (1983). *J. Hort. Sci.* 58:355–363.
Bradt, O. A., Hutchinson, A., Leuty, S. J. and Ricketson, C. L. (1978). *Min. Agri. Food, Ontario* 430:1–112.
Braniste, N. and Amzár, V. (1986). *Lucrari Stiintifice* 11:123–132.
Branscheidt, P. (1931). *Gartenbauwiss.* 4:387–427.
Branscheidt, P. (1933a). *Gartenbauwiss.* 7:546–566.
Branscheidt, P. (1933b). *Gartenbauwiss.* 8(1):45–76.
Branzanti, E. C. (1964). *Riv. Ortflorofruttic.* 5:424–427.
Branzanti, E. C., Cristoferi, G. and Zocca, A. (1965). *Riv. Ortoflorofruttic.* 90(1):79–84.
Branzanti, E. C., Cirillo, A., Cobianchi, D., Curzel, G. C., Damiano, C., Fideghelli, C., Manzo, P., Monastra, F., Nicotra, A. and Rosati, P. (1974). *Frutticoltura* 38(10–11):13–28.
Branzanti, E. C., Cobianchi, D., Faedi, W. and Sansavini, S. (1978). *Riv. Ortoflorofruttic.* 5:501–508.
Braun, J. (1984). *Narben- und Griffelstruktur sowie die im Griffel eingelagerten Kohlenhydrate und ihre Bedeutung für Pollenschlauchwachstum und Fruchtansatz beim Kernobst.* Univ. Hohenheim, Stuttgart, p. 85.
Braun, J. and Stösser, R. (1985). *Angew. Botanik* 59:53–65.
Braun, J., Neubeller, J. and Stösser, R. (1986). *Gartenbauwiss.* 51:1–6.
Breviglieri, N. (1951). *Centro di Studio sul Castagno* 21:15–49.
Breviglieri, N. (1955). *Centro di Studio sul Castagno* 25:5–25.

Breviglieri, N. (1958). *Frutticoltura* 20(5):433–458.
Breviglieri, N. (1960). *L'Italia Agricola* 97:1179–1208.
Breviglieri, N. and Baldassari, T. (1956). *Ann. Speriment. Agr.* 10:1345–1382.
Brewbaker, J. L. (1957). *J. Heredity* 48:271–277.
Brittain, W. H. (1933). *Bull. Dep. Agric. Canada New. Ser.* 162:198.
Brittain, W. H. (1935). *J. Econ. Ent.* 28:553–559.
Brittain, W. H. and Eidt, C. C. (1933). *Canad. J. Res.* 9:307–333.
Brown, A. G. (1940). *J. Pomol.* 18:68–73.
Brown, A. G. (1943). *J. Pomol. Hort. Sci.* 20:107–110.
Brown, A. G. (1955). *John Innes Inst. Ann. Rep.* 1954:7–8.
Brózik, S. (1962). *Kísérletügyi Közl.* 2:111–114.
Brózik, S. (1967). *Szőlő- és Gyümölcstermesztés* 3:93–127.
Brózik, S. (1969). *Szőlő- és gyümölcstermesztés* 5:59–95.
Brózik, S. (1971). MÉM. 1970. évi főbb kutatási eredményei, Budapest, pp. 132–143 (in Hungarian).
Brózik, S. and Nyéki, J. (1971). *Szőlő- és Gyümölcstermesztés* 6:43–73.
Brózik, S. and Nyéki, J. (1975). *Gyümölcstermő növények termékenyülése* (Fertility of Fruit-Producing Plants). Mezőgazdasági Kiadó, Budapest, p. 234.
Brózik, S. and Nyéki, J. (1979). *Tag. Ber. Akad. Landwirtsch. Wiss. DDR, Berlin* 174:137–149.
Brózik, S., Nyéki, J. and Soltész, M. (1980). Sour Cherry. In: Nyéki, J. (ed.), *Gyümölcsfajták virágzásbiológiája és termékenyülése* (Bloom Biology and Fertility of Fruit Cultivars). Mezőgazdasági Kiadó, Budapest pp. 229–233 (in Hungarian).
Brózik, S., Jr., Nyéki, J. and Dunai, J. (1978). *Bot. Közl.* 65:253–264.
Bryner, W. (1988). *Schweiz. Z. Obst- und Weinbau* 124 (10):271–273.
Bryant, L. R. (1935). *Tech. Bull. N. H. Agric. Exp. Stn* 61:40.
Bubán, T. (1965). *Kísérletügyi Közl.* 3:113–119.
Bubán, T. (1967). *Arch. Gartenbau* 15:129–148.
Bubán, T. (1969). *Bot. Közl.* 56:251–256.
Bubán, T. (1980a). In: Nyéki, J. (ed.), *Gyümölcsfajták virágzásbiológiája és termékenyülése* (Bloom Biology and Fertility in Fruit Cultivars). Mezőgazdasági Kiadó, Budapest, pp. 10–24.
Bubán, T. (1980b). Újabb kutatási eredmények a gyümölcstermesztésben (New Results in Fruit Production Research). *Gyümölcs és Dísznövény Kutatóintézet Publ., Budapest* VIII:29–39.
Bubán, T. (1981). *Acta Horticult.* 120:113–117.
Bubán, T. (1992). In: Timon, B. (ed.), *Őszibarack* (Peach). Mezőgazdasági Kiadó, pp. 125–129.
Bubán, T. and Faust, M. (1982). *Horticult. Rev.* 4:174–203.
Bubán, T. and Hesemann, C. U. (1975). 2nd Int. Symp. Plant Growth Reg., Oct. 21–24, Sofia, Bulgaria., Proceeding.
Bubán, T. and Hesemann, C. U. (1979). *Acta Bot. Acad. Sci. Hung.* 25:53–62.
Bubán, T. and Zeller, O. (1974). In: Timon, B. (ed.), *Őszibarack* (Peach). Mezőgazdasági Kiadó, Budapest, pp. 97–107.
Bubán, T., Seitz, U. and Seitz, U. (1974). *Acta Bot. Acad. Sci. Hung.* 20:215–220.
Bubán, T., Zatykó, I. and Gonda, I. (1979). *Kertgazdaság* 11(5):17–31.
Bubán, T., Klement, Z., Bodnár, G. and Turi, I. (1982). *Gartenbauwiss.* 47:212–217.
Budig, H. (1960). *Bess. Obstb. Wiesbaden* 15(5):75–76.
Bugarcic, V. (1968). Symp. über Kirschen und Kirschenbau. Inst. für Obstbau und Gemüsebau der Universität. 25–28, June, Bonn, pp. 56–60.
Bulatovic, S. (1961). *Arhiva za poljopriv. nauka* 14(45):1–8.
Bullock, R. M. and Overly, F. L. (1949). *Proc. Amer. Soc. Hort. Sci.* 54:125–132.
Bumbac, E. (1970). Einfluss der klimatischen Bedingungen auf die Morphogenese beim Pfirsich, *Tagungsber. Deutsch. Akad. Landwirtschaftswiss., Berlin* 99:47–58.
Burchill, R. T. (1963). *Ann. Rep. East Malling Res. Stn, 1962*, pp. 109–111.
Burgos, L., Egea, J. and Dicenta, F. (1989). 9th Int. Horticult. Symp. Apricot Cult., July 9–15, Caserta, Italy, Proceedings.

Burmistrov, A. D. (1972). *Yagodnye kultury*. Kolos, Leningrad.
Caillavet, H. (1973). Société Coopérative de Recherches et d'Expérimentations Agricoles des Pyrénées-Orientales, pp. 1–26.
Calabrese, F., Fenech, L. and Raimondo, A. (1984). *Frutticoltura* 46(5):27–30.
Callan, N. W. and Lombard, P. B. (1978). *J. Amer. Soc. Hort. Sci.* 103(4):496–500.
Canisius, A. I. and Woudenberg, G. J. (1959). *Fruitteelt.* 49:603.
Cappellini, P. and Limongelli, F. (1981). *Ann. Ist. Sperm. Fruttic.* 12:47–55.
Capucci, C. (1959). *Frutticoltura* 4.
Caron, D. M. (1973). *Fruit Varieties* J. 27(4):81–83.
Carrera, M. (1982). *Acta Hort.* 124:157–163.
Cassani, V. (1981). *Frutticoltura* 9:57–68.
Catlin, P. B. and Polito, V. S. (1989). *HortScience* 24:1003–1005.
Chagnon, M., Gingras, J. and Oliveira, D. de (1989). *J. Econ. Ent.* 82:1350–1353.
Champagnat, P., Come, D. et al. (1986). *Acta Hort.* 179(1):117–176.
Chan, B. and Cain, J. (1967). *Proc. Amer. Soc. Hort. Sci.* 91:63–68.
Chandler, W. H. (1942). *Deciduous Orchards*. Lea and Febiger, Philadelphia.
Chandler, W. H. (1951). *Deciduous Orchards*. Lea and Febiger, Philadelphia.
Chandler, W. H. (1957). *Deciduous Orchards*. Lea and Febiger, Philadelphia.
Chang, S. V. and Hoopingarner, R. A. (1991). *Acta Hort.* 288:239–243.
Chapa, J. (1984). Pollinisation du chataignier. In: Pesson, P. and Louvaux, J. (eds), *Pollinisation et production végétales*. INRA, Paris, pp. 182–194.
Chaparro, J. and Sherman, W. B. (1988). *HortScience* 23:753–754.
Child, R. D. (1967). *Ann. Rep. Long Ashton Res. Stn, 1966*, pp. 115–120.
Childers, N. F. (1975). *Modern Fruit Science. Horticultural Publications.* Rutgers, The State University, New Brunswick–New York, pp. 134–143.
Chiriac, St., Roman, R. and Rudi, E. (1981): *Lucrari Stiintifice* 9:155–164.
Chittenden, F. J. (1911). *J. Roy. Hort. Soc.* 37:350–361.
Chiusoli, A. (1966). *Frutticoltura* 18(2):101–103.
Chollet, P. (1965). Thesis présentée a la Faculté des Sciences de l'Université de Rennes. Série C. No. d'Ordre:39. No. de Série:18.
Chollet, P. (1976). *Le Fruit Belge* 44:11–16.
Christensen, J. V. (1970). *Tiddskr. Planteavl.* 74:44–72.
Christensen, J. V. (1973). *Erwerbobstbau* 15(8): 123–125.
Christensen, J. V. (1974). *Tiddskr. Planteavl.* 78:303–312.
Christensen, J. V. (1977). *Tiddskr. Planteavl.* 81:148–156.
Church, R. M. and Williams, R. R. (1983). *J. Hort. Sci.* 58:169–172.
Church, R. M., Cooke, B. K. and Williams, R. R. (1983*a*). *J. Hort. Sci.* 58:161–163.
Church, R. M., Morgan, N. G., Cooke, B. K. and Williams, R. R. (1983*b*). *J. Hort. Sci.* 58:165–168.
Church, R. M., Williams, R. R., Andrews, L. (1983*c*). *J. Hort. Sci.* 58:349–353.
Church, R. M., Copas, L. and Williams, R. R. (1984). *J. Hort. Sci.* 59:161–164.
Clinch, P. B. and Faulke, J. (1976). *New Zeal. J. Expt. Agricult.* 4:399–402.
Cobianchi, D. and Sansavini, S. (1984). *L'informatore Agr.* 5:85–85.
Cobianchi, D., Faedi, W., Rivalta, L. and Battelli, T. (1978*a*). *Riv. Ortoflorofruttic.* 62:552–562.
Cobianchi, D., Faedi, W., Turci, E. and Cicognani, C. (1978*b*). *Riv. Ortoflorofruttic.* 62:563–571.
Cociu, V. (1981). *Lucrari Stiintifice* 9:309–315.
Cociu, V. and Bumbac, E. (1968). *Anal. Inst. Cerc. Pentru Pomicult. Pitești* 1:25–42.
Cociu, V. and Bumbac, E. (1973). Manuscript from Inst. Fruit Production, Pitești.
Cociu, V. and Gozob, T. (1960–1961). *Lucrari Stiintifice* 4:571–580.
Cociu, V. and Gozob, T. (1962). *16th Int. Hort. Congr. Duculot. Gembloux.* 1:477.
Cociu, V. and Tudor, A. (1981). *Lucrari Stiintifice* 9:59–67.
Cociu, V., Gozob, T., Rudi, E., Amzar, V. and Micu, C. (1981). *Lucrari Stiintifice* 9:273–284.
Constantinescu, N. (1939). *Horticultura rom.* 17(9–10):2–4.
Cook, V. A. (1985). *Grower* 103:20–22.

Corner, J., Lapis, K. O. and Arraud, J. C. (1964). *Min. Agr. Victoria, B. C., Apiar. Circ.* 14:1–18.
Costa, G. and Grandi, M. (1982). *Centro Oper. Ortofruttic.*, Ferrara, pp. 74–76.
Couston, R. (1963). *Scott. Beekeep. Bladg.* 40:196–197.
Coutaud, J. (1954). *Ann. Inst. Nat. Rech. Agron.* 25:233.
Crabbé, J. J. (1981). *Acta Hort.* 120:167–172.
Crabbé, J. J. (1984). *Acta Hort.* 146:113–120.
Crabtree, C. D. and Westwood, M. N. (1976). *J. Amer. Soc. Hort. Sci.* 101:454–456.
Crane, M. B. (1923). *J. Pomol. Hort. Sci.* 3:67–84.
Crane, M. B. (1925). *J. Genet.* 15:301–322.
Crane, M. B. (1927/28). *J. Pomol. Hort. Sci.* 6:157–166.
Crane, M. B. and Brown, A. G. (1937). *J. Pomol. Hort. Sci.* 15:86–116.
Crane, M. B. and Brown, A. G. (1942). The fertility rules in fruit planting. In: Lawrence, W. C. D. (ed.), *The Fruit, the Seed and the Soil.* Oliver and Boyd, Edinburgh and London.
Crane, M. B. and Brown, A. G. (1954). The fertility rules in fruit planting. In: Lawrence, W. C. D. (ed.), *The Fruit, the Seed and the Soil.* Oliver and Boyd, Edinburgh and London.
Crane, M. B. and Brown, A. G. (1955). *Sci. Hort.* 11:53–55.
Crane, M. B. and Lawrence, W. J. C. (1929). *J. Pomol. Hort. Sci.* 7:276–301.
Crane, M. B. and Lawrence, W. J. C. (1931). *Rep. Proc. IXth Int. Hort. Congr., London.* 1930:100–116.
Crane, M. B. and Lewis, D. (1942). *J. Genet.* 43:31–43.
Crescimanno, F. G. (1960). *Riv. Ortoflorofruttic.* 85:13–23.
Cresti, M., Ciampolini, F. and Sansavini, S. (1980). *Sci. Horticult.* 12:327–337.
Cresti, M., Ciampolini, F. and Sansavini, S. (1985). *Riv. Ortoflorofruttic.* 69:49–62.
Crisosto, C. H., Miller, A. N. and Lombard, P. B. (1990). *HortScience* 25 (4):426–428.
Crossa-Raynaud, P. (1955). *Ann. Serv. Bot. Agr. Tun.* 28.
Crossa-Raynaud, P. (1957). *Ann. Serv. Bot. Agr. Tun.* 30:33–43.
Crossa-Raynaud, P. (1984). Quelques productions fruitère dépandant d'une pollinisation anémogame: noyer, noisetier, olivier, palmier-dattier, pistachier. In: Pesson, P. and Louveaux, J. (eds), *Pollinisation et production végétales.* INRA, Paris, pp. 162–180.
Crossa-Raynaud, P., Soleille, B., Martines-Tellez, J. and Jraidi, B. (1984). *Proc. Conv. Internaz. Pesco*, Verona-Ravenna-Campania, pp. 143–155.
Csapody, V. and Tóth, I. (1982). *A Colour Atlas of Flowering Trees and Shrubs.* Akadémiai Kiadó, Budapest.
Cummings, M. B., Jenkins, E. W. and Dunning, R. G. (1936). *Vermont Agr. Expt. Stn Bull.* 408.
Cummins, J. N. and Norton, R. L. (1974). *Plant Sci. New York's Food and Life Sci. Bull.* 41:1–15.
Czigankov, S. K. (1953). *Sad i Ogorod* 5:9–11.
Dale, A. (1988). *Crop Res.* 28(2):123–135.
Dániel, L. (1962). *Kísérletügyi Közl.* 2:23–44.
Daubeniy, H. A. (1969). *Can. J. Plant. Sci.* 49:511–512.
Davary Nejad, Gh. H. (1992). Almafajták virágzásbiológiája, termékenyülése, társítása (Flower Biology, Fertility and Combination of Apple Varieties). PhD Thesis, Magyar Tudományos Akadémia, Budapest.
Davary Nejad, Gh. H. and Nyéki, J. (1990). *"Lippay János" Tudományos Ülésszak előadásainak és posztereinek összefoglalói.* Kertészeti Szekció, KÉE. Kiadványai, Budapest, pp. 144–145.
Davis, L. D. and Tufts, W. P. (1941). *Calif. Agr. Ext. Serv. Circ.* 122:87.
Dayton, D. F. and Mowry, J. B. (1977). *HortScience* 12(5):434.
Degman, E. S. and Auchter, E. C. (1935). *Proc. Auer. Soc. Hort. Sci.* 32:213–220.
DeGrandi-Hoffman, G., Hoopingarner, R. and Baker, K. (1984). *Bee World* 65:126–133.
DeGrandi-Hoffman, G., Thorp, R. and Eisikowitch, D. (1991). *Acta Hort.* 288:299–302.
Deidda, P. and Pisanu, G. (1968). *Studi sassar Sez. III*, 16:315–324.
Dennis, F. G., Jr. (1976). *J. Amer. Soc. Hort. Sci.* 101:629–633.
Detjén, L. R. (1945). *Bull. Del. Univ. Agric. Exp. Stn* 257:1–24.
Deveronico, L. and Marro, M. (1980). *Frutticoltura* 42:37–40.
Deveronico, L. and Marro, M. (1982). *La ric. sci.* 110:165–171.

Deveronico, L., DiGiambattista, N. and Marro, M. (1982). *Riv. Ortoflorofruttic.* 62:1–17.
DeVries, D. P. (1967). *Euphytica* 16:177–182.
DeVries, D. P. (1968). *Euphytica* 17:207–215.
Dibuz, E. (1989). Vth Symposium of the Hungarian Plant Anatomy. Abstract, p. 37.
Dibuz, E. (1991). VIth Symposium of the Hungarian Plant Anatomy. Abstract, pp. 30–31.
Dibuz, E. (1993). Körtefajták rendszerezése morfológiai tulajdonságaik alapján (Systematization of pear cultivars according to their morphological characteristics). PhD Thesis. Hung. Acad. Sci., Budapest.
Donk, J. A. W. M., van der (1974). In: Linskens, H. F. (ed.), *Fertilization in Higher Plants.* North-Holland, Amsterdam, pp. 279–283.
Dorsey, M. J. (1919). *J. Genet.* 4:417–488.
Dozier, W. A., Jr., Griffey, W. A. and Burgers, H. F. (1980). *HortScience* 15:743–744.
Drescher, W. and Engel, G. (1978). *Erwerbsobstbau* 20:91–92.
Ducom, P. (1968). *Pomol. Franç.* 10:5–7.
Duganova, E. A. and Hrolikova, A. H. (1977). *Sadov. Vinogr. Vinod. Moldavii* 32:56–60.
Duhan, K. (1944). *Gartenbauwiss.* 18:253–265.
Durner, E. F. and Grianfagna Th. J. (1990). *HortScience* 25:1222–1224.
Dwyer, R. E. P. and Bowman, F. T. (1933). *Agric. Gaz. N. S. W.* 44:516–526.
Dzheneyev, S. Yu. (1958). *Agrobiologia* 3:129–131.
Dzieciol, W. (1967). Ziawiska ksenii nasion i metaksenii owocow kilku odmion jabloni Praca doktorska wykonona w Instytucie Sadownictwa w Skierniewicah (masynopis).
Eaton, G. W. (1959*a*). *Can. J. Bot.* 37:1203–1205.
Eaton, G. W. (1959*b*). *Can. J. Plant. Sci.* 39:466–476.
Eaton, G. W. (1962). *Mich. State Hort. Sci. Ann. Rep.* 92:102–104.
Eaton, G. W. and Jamont, A. M. (1965). *Proc. Amer. Soc. Hort. Sci.* 86:95–101.
Eaton, G. W., Daubeny, H. A. and Norman, R. C. (1968). *Canad. J. Plant. Sci.* 48:342–344.
Efimov, V. A. (1963). *Izvestiya TSHA* 3:148–154.
Egea, J., Burgos, L., Garcia, J. E. and Egea, L. (1991). *J. Hort. Sci.* 66:19–25.
Eickworth, G. C. and Ginsberg, H. S. (1980). *Ann. Rev. Entomol.* 25:421–446.
Einset, O. (1932). *N. Y. State Agric. Exp. Stn Bull.* 617:1–13.
Einset, O. (1934). *Gartenbauwiss.* 9:157–158.
Elek, Lné (1966). *Kert. Szől. Főisk. Közl.* 30:91–105.
Elek, Lné (1974). *Kert. Egy. Közl.* 38:163–174.
Eliseeva, E. P. (1970). *Naukh. Trudy Kursk. Sel'skokh. Inst.* 6:23–27.
Ende, H., van den (1976). *Sexual Interactions in Plants.* Acad. Press, London–New York–San Francisco, pp. 1–26 and 143–157.
Enikeev, H. K. (1973). *Sel'skokh. Biol.* 8:370–373.
Engel, G. (1960). *Gartenbauwiss.* 25:67–106.
Erson, L. A. and Hrolikova, A. H. (1970). *Trud. Gos. Nikit. Botan. Sada* XLV:169–182.
Esaulova, J. N. (1960). *Agrobiologia* 2:268–270.
Eti, S. and Stösser, R. (1987). *Angew. Botanik* 61:505–519
Ewert, R. (1906). *Landw. Jahrb.* 35:259–287.
Ewert, R. (1907). *Die Parthenokarpie oder Jungfernfrüchtigkeit der Obstbäume und ihre Bedeutung für den Obstbau.* Parey Verlag, Berlin–Hamburg.
Ewert, R. (1909). *Landw. Jarb.* 38:767–839.
Ewert, R. (1926). Pflanzenphysiologische und biologische Forschungen im Obstbau. Land. Jahrb., pp. 759–785.
Ewert, R. (1929). *Blüten und Früchten der insektenblütigen Garten- und Feldfrüchte unter dem Einfluss der Bienenzucht.* Verlag Neumann J. Neudamm, Leipzig.
Fabbri, A., Ramina, A. and Albertini, A. (1983). *L'Informatore Agrario* 39:27555–27557.
Faccioli, F. (1981). *Scelte varietali in frutticoltura,* Ferrara, pp. 92–94.
Facteau, T. J. and Chestnut, N. E. (1983). *HortScience* 18:717–718.
Faccioli, F. and Marangoni, B. (1978). *Riv. Ortoflorofruttic.* 62:584–596.
Faccioli, F. and Regazzi, D. (1972). *Riv. Ortoflorofruttic.* 1:33–45.

Facteau, T. J. and Rowe, R. E. (1977). *J. Amer. Soc. Hort. Sci.* 102:95–96.
Facteau, T. J. and Rowe, R. E. (1981). *J. Amer. Soc. Hort. Sci.* 106:77–79.
Faedi, W., Rosati, P. (1974a). *Frutticoltura* 36:5–17.
Faedi, W., Rosati, P. (1974b). *Frutticoltura* 36:41–48.
Faedi, W., Rosati, P. (1975a). *Frutticoltura* 37:11–15.
Faedi, W., Rosati, P. (1975b). *Frutticoltura* 37:25–35.
Faust, M. (1989). *Physiology of Temperate Zone Fruit Trees.* John Wiley and Sons Inc., New York.
Faust, M., Zimmerman, R. and Zwet, T., van der (1976). *HortScience* 11:59–60.
Faust, M., Liu, D., Millard, M. M. and Stutte, G. W. (1991). *HortScience* 26:887–890.
Fedtzenkova, G. A. (1970). *Sborn. Trud. Aspir. Molod. Naukh. Sotrud. Leningrad,* 15:551–558.
Fernquist, I. B. (1961). *Kungl. Skogs. Lautbr. Akad. Tiddskr.* 100:357–396.
Fideghelli, C. and Cappelini, P. (1978). *Riv. Ortoflorofruttic.* 62:372–380.
Fideghelli, C. and Monastra, F. (1978). *L'Informatore Agrario* 34:3733–3749.
Filmer, R. S. and Marcucci, P. E. (1963). Proc. 31st Ann. Blueberry Open House. N. J. Agr. Expt. Stn, pp. 14–21.
Fletcher, S. W. (1908). *Proc. Amer. Soc. Hort. Sci.* 4:29–40.
Florin, R. (1924). *Sver. Pomol. För. Arsskr.* 1:1–34.
Fogle, H. W. (1977). *Fruit Var. J.* 31(4):74–75.
Fogle, H. W., Snyder, J. C., Baker, H., Cameron, H. R., Cochran, L. C., Schomer, H. A. and Yang, H. Y. (1973). *Agriculture Handbook.* 442. USDA, Washington.
Forde, H. I. and Griggs, W. H. (1972). *Pollination and Blooming Habits of Walnuts.* Agricultural Extension, University of California.
Forde, H. I. (1975). Walnuts. In: Janick, J. and Moore, J. N. (eds): *Advances in Fruit Breeding.* Pardue Univ. Press, West Lafayette, Ind.
Frecon, J. (1980). *Fruit Var.* 34:18.
Free, J. B. (1958a). *Anim. Behav.* 6:219–223.
Free, J. B. (1958b). *Bee World* 39:221–230.
Free, J. B. (1959). *J. Agric. Sci.* 53:1–9.
Free, J. B. (1960a). *J. Anim. Ecol.* 29:385–395.
Free, J. B. (1960b). *Bee World* 41:141–151.
Free, J. B. (1962). *J. Hort. Sci.* 37:262–271.
Free, J. B. (1963). *J. Anim. Ecol.* 32:119–131.
Free, J. B. (1964). *Nature (London)* 201:726–727.
Free, J. B. (1965a). *J. Apicult. Res.* 4:85–88.
Free, J. B. (1965b). *J. Apicult. Res.* 4:61–64.
Free, J. B. (1965c). *J. Agric. Sci.* 64:167–168.
Free, J. B. (1966). *J. Hort. Sci.* 41:91–94.
Free, J. B. (1967). *Anim Behav.* 15:134–144.
Free, J. B. (1968a). *J. Appl. Biol.* 5:157–168.
Free, J. B. (1968b). *J. Appl. Ecol.* 5:169–178.
Free, J. B. (1968c). *J. Hort. Sci.* 43:107–111.
Free, J. B. (1968d). *J. Hort. Sci.* 43:69–73.
Free, J. B. (1970). *Insect Pollination of Crops.* Acad. Press, London, pp. 544.
Free, J. B. and Durrant, A. J. (1966). *J. Hort. Sci.* 41:87–89.
Free, J. B. and Nuttall, P. M. (1968). *Nature (London)* 218:982.
Free, J. B. and Smith, M. V. (1961). *Bee World* 41:11–12.
Free, J. B. and Spencer-Booth, Y. (1963). *J. Hort. Sci.* 38:129–137.
Free, J. B. and Spencer-Booth, Y. (1964a). *J. Hort. Sci.* 39:78–83.
Free, J. B. and Spencer-Booth, Y. (1964b). *J. Hort. Sci.* 39:54–60.
Free, J. B. and Williams, I. H. (1972a). *J. Appl. Ecol.* 9:627–634.
Free, J. B. and Williams, I. H. (1972b). *J. Appl. Ecol.* 9:609–615.
Free, J. B., Free, N. W. and Jay, S. C. (1960). *J. Econ. Ent.* 53:69–70.
Frick, F. (1974). *Erwerbsobstbau* 16:119–120.

Friedrich, G. and Preusse, H. (1983). *Obstbau in Wort und Bild*. Neumann Verlag, Leipzig.
Frilli, F. and Barbattini, R. (1980). *L'Informatore Agrario* 36:11619–11625.
Frilli, F., Roversi, A., Barbattini, R. and Ughini, V. (1983). *L'Informatore Agrario* 39:28307–28315.
Fritzsche, R. (1972). *Obstbau*. Schweizerischer Verband der Ingenieur-Agronomer, Zollikofen.
Fulford, R. (1966). *Ann. Bot. N. S.* 29:167–180.
Fulford, R. (1970). *Proc. 18th Int. Hort. Congr.* 4:143–150.
Fulford, R. and Way, D. W. (1967). *Ann. Rep. East. Malling Res. Stn, 1966*, pp. 103–105.
Furukawa, Y. and Bukovac, M. J. (1989). *HortScience* 24:1005–1008.
Gagnard, J. (1954). *Fruits et Primeurs, Casablanca*, No. 253.
Galletta, G. J. (1983). Pollen and seed management. In.: Moore, J. N. and Janick, J.: *Methods in Fruit Breeding*. Purdue University Press, West-Lafayette, pp. 23–25.
Gardner, V. R. (1913). A preliminary report of the pollination of the sweet cherry. *Oreg. Agr. Exp. Stn Bull.*, pp. 116–140.
Gardner, V. R., Merill, T. A. and Toenjes, W. (1949). *Mich. Agr. Expt. Stn Spec. Bull.* p. 358.
Gardner, V. R., Bradford, F. C. and Hooker, H. D., Jr. (1952). *The Fundamentals of Fruit Production*. McGraw-Hill, New York–Toronto–London.
Gasser, Ch. S. (1991). *Ann. Rev. Plant Physiol. Plant Molec. Biol.* 42:621–649.
Gautier, M. (1971). *L'Arboriculture Fruitière* 18(209–210):20–27.
Gautier, M. (1973). *L'Arboriculture Fruitière* 20(238):14–20.
Gautier, M. (1974). *L'Arboriculture Fruitière* 21(241):23–29.
Gautier, M. (1977). L'établissement de la plantation. *L'Abricot fruit.* 276:33–44.
Gautier, M. (1981). *L'Arboriculture Fruitière* 28(329–330):27–36.
Gautier, M. (1983). *L'Arboriculture Fruitière* 30(349):29–36.
Gerin, G. (1972). *Riv. Ortoflorofruttic.* 56(2):113–117.
Germain, E. (1983). *Proc. Conv. Int. Nocciuolo,* Avellino, pp. 47–55.
Germain, E., Jalinat, J. and Marchau, M. (1973). Biologie florale du noyer (*Juglans regia* L.) *B. T. I.* 282. 5(63):661–673.
Germain, E., Jalinat, J. and Marchau, M. (1975). Divers aspects de la biologie florale du noyer. In: Bergougnoux F. and Grospierre, P. (eds), *Le noyer*. Paris.
Ghena, N. and Braniste, N. (1978). *Horticultura, Bucureşti.* 27(9):20–28.
Ghosh, S. P. (1970). *Dokl. Moskow. Sel'skokh. Akad. K. A. Timiryazeva*, 165:47–51.
Godini, A. (1981). *Riv. Ortoflorofruttic.* 65:135–141 and 173–178.
Goldwin, G. (1990). *Grower* 113(5):31–32.
Golikova, N. A. (1969). *Sel'skokh. Biol.* 4:940–942.
Golubinsky, I. N., Samorodov, V. N., Kekalo, V. I. and Glazkov, A. N. (1977). *Ukr. Bot. Zh.* 34:577–582.
Golubinsky, I. N., Samorodov, V. N. and Berezenko, N. P. (1979). *Ukr. Bot. Zh.* 35:73–81.
Gornevsky, V. (1976). *Grad. Lozar. Nauka* 13:20–26.
Goryaczkowski, W. (1926). Doswiodczalmictwo Rolmioze t. II., oz I.
Gorter, C. J. and Visser, T. (1958). *J. Hort. Sci.* 33(4):217–227.
Got, N. (1958). *L'abricotier.* 3me édition. La maison rustique, Paris.
Gough, R. E. (1983). *HortScience* 18:934–935.
Gozob, T., Rudi, E. and Amzar, V. (1979). *Lucrari Stiintifice* 8:37–51.
Gozob, T., Rudi, E., Micu, C. and Amzar, V. (1981). *Lucrari Stiintifice* 9:301–308.
Götz, G. (1970). *Süss- und Sauerkirchen*. Eugen Ulmer Verlag, Stuttgart.
Granger, R. L. (1982). *Acta Hort.* 124:43–50.
Grasselly, Ch. and Olivier, G. (1976). *Ann. Amél. des Plant.* 26:107–112.
Greznitshenko, A. G. (1969). *Biologiya razvitiya plodovy rastenii*. Vysshaya Shkola, Moscow.
Gribanovsky, A. P. (1970). *Trudy cent. gen. Leb. I. V. Mitshurina* 10:256–259.
Griggs, W. H. (1948). Pollination requirements of fruits and nuts. *Calif Agric. Exp. Service Circ.* 424.
Griggs, W. H. and Hesse, C. O. (1963). *Div. Agr. Sci. Univ. Cal.* 8.
Griggs, W. H. and Iwakiri, B. T. (1954). *Hilgardia* 22:643–678.

Griggs, W. H. and Iwakiri, B. T. (1960). *Proc. Amer. Soc. Hort. Sci.* 75:115–128.
Griggs, W. H. and Iwakiri, B. T. (1977). *California Agriculture* 31:8–12.
Griggs, W. H., Iwakiri, B. T. and Claypool, L. L. (1957). *Proc. Amer. Soc. Hort. Sci.* 78:74–84.
Griggs, W. H., Martin, G. C. and Iwakiri, B. T. (1970). *J. Amer. Soc. Hort. Sci.* 95:243–248.
Griggs, W. H., Beutel, J. A. and Iwakiri, B. T. (1972). *Riv. Ortoflorofruttic.* 1:246.
Grigorian, V. (1974). *Ann. Amél. des Plant.* 25:415–420.
Grochowska, M. (1968). *Bul. Acad. Pol. Sci. Cl. V.* 16:581–583.
Grochowska, M. and Karasewska, A. (1974). *Proc. 19th Int. Hort. Congr.* 18:618.
Grochowska, M. and Karasewska, A. (1976). *Fruit Sci. Rep.* 3:5–16.
Grochowska, M. and Karasewska, A. (1978a). *Acta Hort.* 80:181–185.
Grochowska, M. and Karasewska, A. (1978b). *Acta Hort.* 80:457–464.
Grubb, N. H. (1949). *Cherries.* Crosby Lockwood and Sons Ltd., London.
Gruner, A. M. and Medvezhova, S. S. (1977). *Sadovodstvo* 5:30–31.
Grupce, R. (1966). *Ann. Fac. Agric. Sylvic. Univ. Skopie Agric.* 19:65–75.
Guerrero-Prieto, V. M., Vasilakis, M. D. and Lombard, P. B. (1985). *HortScience* 20:913–914.
Guerriero, R. (1982). Atti del Convegno "Prospettive per l'Agricoltura Collinare Fiorentina", 1982. 01:27–28. Florence, pp. 93–116.
Gulino, F. (1982). *Frutticoltura* 44(6–7):158–162.
Gushtshin, M. F. (1972). *Sadovodstvo* 6:29.
Gustafson, F. G. (1942). *Proc. Natn. Acad. Sci. USA* 38:131–133.
Gylcan, R. and Askin, A. (1989). Int. Symp. Apricot Cult. July 14–17, Salerno, Italy. *Acta Horticult.* (in press).
Gylcan, R. and Askin, A. (1990). *Abstr. 23th Int. Hort. Congr.* 2:4334.
Gyuró, F. (1959). *Kert. Szől. Főisk. Közl.* 23:135–142.
Gyuró, F. (ed.) (1977). *Gyümölcsfajták társítása* (Association of Fruit Varieties). Mezőgazdasági Kiadó, Budapest.
Gyuró, F., Soltész, M. and Nyéki, J. (1976). *Kertgazdaság* 8(1):1–14.
Harmat, L. (1987). *Köszméte* (Gooseberry). Mezőgazdasági Kiadó, Budapest.
Hartman, F. O. and Howlett, F. S. (1954). *Res. Bull. Ohio Agric. Exp. Stn,* p. 745.
Hassibb, M. (1966). *Arch. Gartenbau* 14:277–287.
Heinrich, B. (1975). *Ann. Rev. Ecol. Syst.* 6:139–170.
Heinrich, B. and Raven, P. H. (1972). *Science* 176:597–602.
Helms, C. W. (1970). *Can. J. Zool.* 48:185.
Hellmich, R. L. and Rothenbuhler, W. C. (1986). *Apidologie* 17:13.
Hendrickson, A. H. (1918). *Proc. Amer. Soc. Hort. Sci.* 15:65–66.
Herbst, W. and Rudloff, C. (1939). *Gartenbauwiss.* 13:286–317.
Herbst, W. and Weger, N. (1940). *Forschungsdienst, Org. d. Landwirtschaft,* pp. 518–525.
Herrero, M., Cambra, M. and Felipe, A. J. (1977). *An. Inst. Naci. Invest. Agr.* 7:99–103.
Hesemann, C. U. and Bubán, T. (1973). *Histochemie* 36:237–246.
Heslop-Harrison, J. (1975). *Ann. Rev. Plant Physiol.* 26:403–425.
Heslop-Harrison, J. (1976). *Rep. East Malling Res. Stn, 1975,* pp. 141–157
Hibbard, A. D. (1933). *Proc. Amer. Soc. Hort. Sci.* 30:140–142.
Hilkenbäumer, F. and Buchloh, G. (1954). *Gartenbauwiss.* 1:7–21.
Hilkenbäumer, F. and Klämbt, H. D. (1958). *Gemüse-, Obst- und Gartenbau* 46:11–12.
Hill, H., Cottingham, D. G. and Williams, R. R. (1967). *J. Hort. Sci.* 42:319–338.
Hiratsuka, S. and Tezuka, T. (1980). *J. Jap. Soc. Hort. Sci.* 49:57–62.
Hiratsuka, S. Hirota, M. Takahashi, E. and Hirata, N. (1985). *J. Jap. Soc. Hort. Sci.* 53:377–382.
Hladik, F. (1972). *Vedecké Práce Ovocn.* 4:145–156.
Hlava, B., Pospisil, F. and Stary, F. (1983). *Pflanzen.* Artia, Prague.
Hoad, G. V. (1978). *Acta Horticult.* 80:93–103.
Hoad, G. V. and Donaldson, S. (1977). *Ann. Rep. Long Ashton Res. Stn,* pp. 39–40.
Hoad, G. V., Ramirez, H. R. and Gaskin, P. (1977). *Ann. Rep. Long Ashton Res. Stn,* p. 39.
Hoffman, M. B. (1965). *Corn. Ext. Bull.* 1146:2–8.

Hoffmann, H. (1887). *Phänologische Untersuchungen.* Universitätsprogramm zum 25. 8. 1887, Giessen.
Hoffmann, K. (1962). *Angewandte Meteorologie* 4:146–154.
Hooper, C. H. (1912). *J. Ray. Hort. Soc.* 38:238–248.
Hooper, C. H. (1932). *J. S. E. Agric. Coll.* 30:244–246.
Horavka, B. (1961). *Biol. Plant.* 3:137–139.
Horn, E. (1976). *Dió – Mandula – Mogyoró – Gesztenye* (Walnut—Almond—Hazelnut—Chestnut). Mezőgazdasági Kiadó, Budapest.
Horn, J. (1927). *Mezőgazdaság és Kertészet* 7:107–108.
Hortobágyi, T. (1962). *Növénytan* (Botany), Vol. 1. Tankönyvkiadó, Budapest.
Hortobágyi, T. (1968). *Növénytan* (Botany), Vol. 2. Tankönyvkiadó, Budapest.
Hortobágyi, T. (1986). *Agrobotanika* (Agrobotany). Mezőgazdasági Kiadó, Budapest.
Höstermann, D. (1924). *Angew. Botanik* 6:232–242.
Howell, G. S., Kilby, M. W. and Nelson, J. W. (1972). *Hort. Sci.* 7:129–131.
Hruby, K. (1963). *Biologia Plantarum* 5:124–128.
Huet, J. (1974). *Acta Hort.* 34:193–198.
Huet, J. and Lemoine, J. (1972). *Physiol. Vég.* 10:529–545.
Hugard, J. (1975). *La Pomol. Franç.* 17:63–78.
Hugard, J. (1978). *Le Fruit Belge* 381:11–32.
Husz, B. (1942). *Magy. Kir. Kert. Akad. Közl.* 8:128–133.
Husz, B. (1943). *Magy. Kir. Kert. Szől. Főisk. Közl.* 2:60–86.
Iliev, P. (1985). *Rastenieveodni Nauki* 22:65–73.
Impiumi, G. and Ramina, A. (1967). *Riv. Ortofloro-fruttic.* 51(6):538–543.
Istomina-Tsvetkova, K. P. and Skrebtsov, M. C. (1964). *Pkhelokodsvo (Moscow)* 41:205–222.
Ivan, I., Minorin, N., Micu, I., Pattantyus, K., Modoran, D. and Iliuta, I. (1981a). *Lucrari Stiintifice* 9:285–291.
Ivan, I., Modoran, D., Pattantyus, K. and Iliuta, I. (1981b). *Lucrari Stiintifice* 9:89–94.
Ivan, I., Minoiu, N., Micu, I. and Modoran, D. (1981c). *Lucrari Stiintifice* 9:317–322.
Jackson, D. I. and Sweet, G. B. (1972). *Horticult. Abstr.* 42:9–24.
Jacob, F., Jager, E. J. and Ohmann, E. (1981). *Kompendium der Botanik.* G. Fischer Verlag, Jena.
Jaovani, A. (1973). *Rapp. Activ. Lab. Arbo. Fruit.* pp. 1–20.
Jaumien, F. (1968). *Acta Agrobot.* 21:75–106.
Jay, S. C. (1986). *Ann. Rev. Entomol.* 31:49–65.
Jaynes, R. A. (1964). *Silvae Genetica* 13:146–154.
Jefferies, C. J. (1977). *Stain Technol.* 52:277–283.
Jefferies, C. J. (1979). *HortScience* 14:229.
Jefferies, C. J. and Belcher, A. R. (1974). *Stain Technol.* 49:199–202.
Jefferies, C. J. and Brain, P. (1984). *Planta* 160:52–58.
Jefferies, C. J., Atwood, J. G. and Williams, R. R. (1982a). *Sci. Hort.* 16:147–153.
Jefferies, C. J., Brain, P., Stott, K. G. and Belcher, A. R. (1982b). *Plant, Cell and Environment* 5:231–236.
Johansen, C. (1967). *Coll. Agric. Coop. Ext. Serv. Washington State Univ.*, EM. 2262:1–3.
Johansson, L. and Callmar, G. (1936). *Sver. Pomol. Foren. Arsskr.* 1:1–29.
Jona, R. (1967). *Riv. Ortoflorofruttic.* 51:544–553.
Jona, R. (1986). Hazelnut. In: Monselise, S. P. (ed.), *CRC Handbook of Fruit Set and Development.* CRC Press, Boca Ratow, Fl. pp. 193–216.
Kamlah, H. (1928a). *Kühn.-Arch.* 19:133–195.
Kamlah, H. (1928b). *Gartenbauwiss.* 2:10–45.
Kandaurova, E. F. (1985). *Plodoovoshchoye Hozyaystvo, Moscow.* 10:38–40.
Kandaurova, E. F., Krylova, V. V. and Smykov, V. K. (1973). *Selektsiya i sortoizucheniye plodovyh i yagodnyh kultur., Kisinev,* pp. 102–121.
Kapil, R. N. and Bhatnagar, A. K. (1975). *Phytomorphology* 25:334–368.
Karatsharova, L. P. (1977). *Podbor opyliteley Sadovodstvo* 5:19–20.

Karamysheva, V. I. and Blinova, E. N. (1976). *Trudy Prikl. Bot. Genet. Selektsii, Leningrad* 57:34–38.
Karmo, E. A. (1960). *Rep. Nova Scotia Fruit. Grower Assoc.* 97:125–128.
Karmo, E. A. and Vickery, V. R. (1987). *Canad. Beekeep.* 13:163.
Karnatz, A. (1960). Untersuchungen über Parthenokarpie bei Apfeln und Birnen. Thesis, Berlin.
Karnatz, A. (1962). *Erwerbsobstbau* 4:31–33.
Karnatz, A. (1963). *Der Züchter* 33:249–259.
Karnatz, A. (1971). *Gartenbauwiss.* 36:525–533.
Kárpáti, Z. (1969). *A növények világa* (The World of Plants), Vol. II. Gondolat Könyvkiadó, Budapest.
Kárpáti, Z. and Terpó, A. (1968). *Növényrendszertan* (Plant Taxonomy). Mezőgazdasági Kiadó, Budapest.
Kavetzkaya, A. and Tokar, L. O. (1963). *Bot. Zh.* 4.
Kellerhals, M. (1986). *Schweiz. Zschr. Obst- und Weinbau* 122:319–327 and 363–371.
Kendall, D. A. (1973). *J. Appl. Ecol.* 10:842–853.
Kendall, D. A. and Smith, B. P. (1975). *J. Appl. Ecol.* 12:465–471.
Kendall, D. A. and Solomon, M. E. (1973). *J. Appl. Ecol.* 10:627–634.
Kerékné-Vida, P. (1981). *Kertgazdaság* 13(3):53–60.
Keremidarska, S. (1968). *Grad. Lozar. Nauka* 5:3–18.
Kester, D. E. (1966). A review of almond varieties. *Cal. Agr. Ext. Serv.* AXT–215.
Kester, D. E. (1981). Almonds. In: Jaynes, R. A. (ed.), *Nut Tree Culture in North America.* North Nut Growers Assoc., Hamden, Connecticut.
Kester, D. E. and Asay, R. (1975). Almonds. In: Janik, J., Moore, J. N. (eds), *Advances in Fruit Breeding.* Purdue University Press, West Lafayette, pp. 387–419.
Kester, D. E. and Bradley, M. V. (1976). *J. Amer. Soc. Hort. Sci.* 101:490–493.
Kester, D. E. and Griggs, W. H. (1959). *Proc. Amer. Soc. Hort. Sci.* 74:206–213.
Keulemans, J. (1980). *Fruit Belge* 48:117–121.
Keulemans, J. (1984). *Acta Hort.* 149:103–108.
Kevan, P. G. and Barker, H. G. (1983). *Ann. Rev. Entomol.* 28:407–453.
Kim, C. H. (1946). An Inquiry into the Factors Affecting the Shape of Barlett Pears Fruits with Special Reference to Xenie, Metaxenie and Pollination. Doctoral Thesis, Oregon State College.
Kim, S. K., Lagerstedt, H. B. and Daley, L. S. (1985). *HortScience* 20:944–946.
King, J. R. (1965). *Bull. Torrey Bot. Club.* 92:270–287.
Kirakosian, A. M. and Beketovskaya, A. A. (1970). *Biol. Zh. Arm.* 23(8):84–89.
Kirby, E. G. and Smith, J. E. (1974). In: Linskens, H. F. (ed.), *Fertilization in Higher Plants.* North-Holland, Amsterdam, pp. 127–130.
Kirtvaya, E. K. (1965). *Sadovodstvo* 5:33–34.
Klug, M. and Bünemann, C. (1983). *Acta Hort.* 139:59–64.
Knight, R. L. (1963). *Abstract Bibliography of Fruit Breeding and Genetics to 1960.* Malus and Pyrus. Commonwealth Agricultural Bureaux, Farnham Royal, VK.
Knight, R. L. (1969). *Abstract Bibliography of Fruit Breeding and Genetics to 1965 Prunus.* Comm. Agric. Bur. East Malling. Maidstone, Kent.
Kobel, F. (1931). *Schweiz. Zeitschr. Obst- und Weinb.* 45:129–132.
Kobel, F. (1954). *Lehrbuch des Obstbaus auf physiologischer Grundlage.* Springer, Berlin.
Kobel, F. and Sachov, T. H. (1929). *Landw. Jahrb. Schweiz* 43:1036–1064.
Kobel, F. and Steinegger, P. (1933). *Landw. Jahrb. Schweiz* 47:973–1018.
Kobel, F. and Steinegger, P. (1934). *Landw. Jahrb. Schweiz* 48:741–768.
Kobel, F., Steinegger, P. and Anliker, J. (1938). *Landw. Jahrb. Schweiz* 52:564–595.
Kobel, F., Steinegger, P. and Anliker, J. (1939). *Landw. Jahrb. Schweiz* 53:160–191.
Koch, H. J. (1979). *Obstproduktion* 4:3–4.
Koch, H. J. (1981). *Gartenbau* 28:213–215.
Koleshnikov, M. A. (1953). Kultura tseresni i vishni na Kuban. Krasnodar.
Koleshnikov, M. A. (1959). *Sad i Ogorod* 4.

Kollányi, L. (1976). A fajtahibridizáció lehetősége a málna és szedermálna nemesítésében (The possibility of cultivar hybridization in raspberry and bramble improvement). PhD Thesis, Magyar Tudományos Akadémia, Budapest.
Kollányi, L. (1990). *Málna* (Raspberry). Mezőgazdasági Kiadó, Budapest.
Koloteva, N. I. and Zhukov, O. S. (1984). *Bul. Nauk. Inf. Central. Ord. Trud. Krasn. Znam. Gen. Lab. Mitsurina* 41:28–32.
Kordon, R. Ja. (1934). *Aiva. Bull. Appl. Bot. Genet. Plant Breeding. Kishinev.*
Kostina, K. F. (1926–1927). *Zap. Gos. Nikit. Bot. Sada* 1:3.
Kostina, K. F. (1928). *J. Gov. Bot. Gard. Nikit. Yalta* 10:1–86.
Kostina, K. F. (1970). *Trudy. Gos. Nikit. Bot. Sada* 45:7–17.
Koul, A. K., Singh, A., Singh, R. and Wafai, B. A. (1985). *Euphytica* 34:125–128.
Kovács, S. (1966). *Kertészet és Szőlészet* 24:7.
Kovács, S. (1968). *Agrártud. Közl.* 27:573–581.
Kovács, S. (1977). *Nyári gyümölcsök termesztése* (Production of Summer Ripening Fruits). Mezőgazdasági Kiadó, Budapest.
Kozma, P. (1950). *Szőlészeti Kutató Intézet Évk.* 10:49–78.
Kozma, P. (1963). *A szőlő termékenységének és szelektálásának virágbiológiai alapjai* (Flower Biological Bases of Fertility and Selection of Grape). Akadémiai Kiadó, Budapest.
Krapf, B. (1966). *Zschr. Obst- und Weinbau* 102:127–133 and 153–163.
Krapf, B. (1967). *Schweiz. Zschr. Obst- und Weinbau* 103:266–268.
Krapf, B. (1968). *Erwerbsobstbau* 10:14–16.
Krapf, B. (1969). *Zschr. Obst- und Weinbau* 105:268–269.
Krapf, B. (1971). *Schweiz. Zschr. Obst- und Weinbau* 107:3–9 and 49–53.
Krapf, B. (1972). *Schweiz. Zschr. Obst- und Weinbau* 108:341–347.
Krapf, B. (1976). *Mitt. Eidg. Forschungs. Obst-, Wein- und Gartenbau. Wadenswill.* 30:1–24.
Krapf, B. and Bryner, W. (1977). *Schweiz. Zschr. Obst- und Weinbau* 113:156–158.
Krapf, B., Theiler, R. and Odermatt, Th. (1972). *Schweiz. Zschr. Obst- und Weinbau* 108:23–29.
Krause, W. (1973). *Grower* 80:367–368.
Kravtshenko, L. M. (1955). *Agrobiologia* 2:115–117.
Kronenberg, H. G. (1979). *Neth. J. Agric. Sci.* 27:131–135.
Kronenberg, H. G. (1985). *Neth. J. Agric. Sci.* 33:45–52.
Krumbholz, G. (1932). *Gartenbauwiss.* 6:404–424.
Krümmel, H. (1932). *Gartenbauwiss.* 6:262–302.
Krümmel, H. (1933). *Kühn-Archiv* 38:202–222.
Kuhn, E. D. and Ambrose, J. T. (1982). *J. Amer. Soc. Hort. Sci.* 107:391–395.
Kurennoy, V. N. (1977). *Sadov. Vinogr. Vinod. Moldavii* 1:59–61.
Kurennoy, N. M., Babuk, V. J. and Plugar, I. M. (1984). *Sadov Vinogr. Vinod. Moldavii* 4:57–58.
Kursakov, G. A. and Dubovitzkaya, L. A. (1974). *Bull. Nauk. Inf. Cent. Ord. Trud. Krasn. Znam. Genet. Lab. I. V. Mitshurina* 21:22–27.
Lacey, H. J., Fulford, R. M. and Smith, J. G. (1976). *Rep. East Malling Res. Stn, 1975,* pp. 65–66.
Laczy, G. (1943). *Borászati Lapok* 34:202.
Lake, J. V. (1956). *J. Hort. Sci.* 31:244–257.
Lalatta, F. (1982a). *Atti del Convegno "Nuovi orientamenti per la coltura del melo nel Veronese".* Verona, pp. 125–144.
Lalatta, F. (1982b). *La ric. Sci. Biol. della riprod. Roma* 110:487–493.
Lalatta, F., Marro, M. and Sansavini, S. (1978a). *Riv. Ortoflorofruttic.* 62(4):350–371.
Lalatta, F., Marro, M. and Sansavini, S. (1978b). *Atti Sem. Fertilità delle piante de frutto.* C. N. R. Reg. Emilia - Romagna. S. O. I. Bologna.
Lalatta, F., And Sansavini, S. (1983). Le scelte varietali nella melicoltura di montagna. Proc. Conv. sa Scelte Varietali e Rinnovamento della Frutticoltura Montana, 22–23 April 1983. Bologna, pp. 5–18.
Lamb, R. C. (1948). *Proc. Amer. Soc. Hort. Sci.* 51:313–315.
Lamb, R. C. (1966). *Cornell Ext. Bull. New York State Call. Agri.* 1168:1–4.

Lamb, R. C. and Stiles, W. C. (1983). Apricots for New York State. *New York's Food and Life Sci. Bull.* 100:1–4.
Lange, W. D. (1979). *J. Hort. Sci.* 34:87–89.
Langridge, D. F. (1968). *Aust. J. Exp. Agr. Anim. Husb.* 9:549–552.
Langridge, D. F. and Goodman, R. D. (1973). *Aust. J. Exp. Agr. Anim. Husb.* 13:193–195.
Langridge, D. F. and Jenkins, P. T. (1975). *Aust. J. Exp. Agr. Anim. Husb.* 15:105–107.
Lapins, K. O. (1970). *Fruit Var. Hort. Dig.* 24:19–20.
Lapins, K. O. (1971). *Can. J. Plant Sci.* 51:252–253.
Lapins, K. O. and Schmid, H. (1976). *Inf. Div. Can. Dep. Agr. Ottawa Publ.* 1471:1–19.
Larsen, F. E. and Fritts, R. (1987). *Sci. Hort.* 30(4):283–288.
Latimer, L. P. (1938). *Proc. Amer. Soc. Hort. Sci.* 34:16–18.
Layne, R. E. C. (1967). *Fruit Var. Hort. Digest.* 21:28–32.
Layne, R. E. C. (1983). Hybridization. In.: Moore, J. N. and Janick, J. (eds), *Methods in Fruit Breeding*. Purdue University Press, West Lafayette, Ind., pp. 48–65.
Lech, W. (1976). *Acta Agr. et Silv. Ser. Agr.* 16:81–93.
Lee, C. L. (1980). Pollenkeimung. Pollenschlauchwachstum und Befruchtungsvernältnisse bei *Prunus domestica*. Thesis, Fakult. Gartenb. Landeskult., Univ. Hannover, p. 120.
Lee, C. L. and Bünemann, G. (1981). *Erwerbsobstbau* 23:52–55.
Legave, J. M. (1978). *Ann. Amél. Plant.* 28:533–607.
Legge, A. P. (1976). *Bee World* 57:159–167.
Legge, A. P. and Williams, R. R. (1975). *J. Hort. Sci.* 50:279–281.
Le Lezec, M. and Babin, J. (1979). *Pépin. Hort. Maraichers* 195:59.
Le Lezec, M., Babin, J., Michelesi, C., Lespinasse, M., Masseron, A., Tronel, C. and Chartier, A. (1981). *L'Arboriculture Fruitière* 331:3–12.
Lemaitre, R. (1978). *Le Fruit Belge* 46:89–91.
Leppik, E. E. (1957). *Evolution* 11:466–481.
Lespinasse, Y. and Salesses, G. (1973). *Ann. Amél. Plant.* 23:381–386.
Levin, M. D. (1961). *Ins. Soc.* 8:195–201.
Levitzkaya, L. L., Kotoman, E. M. (1980). *Sadav. Vinogr. Vinod. Moldavii* 7:21–24.
Lewis, C. I. and Vincent, C. C. (1909). *Oregon Agr. Exp. Stn Bull.*, p. 104.
Lewis, D. (1942). *J. Pomol.* 20:40–41.
Lewis, D. (1942). *Proc. Roy. Soc. B.* 131:12–26.
Lewis, D. (1948). *Heredity* 2:219–236.
Lewis, D. (1949). *Heredity* 3:339–355.
Lewis, D. (1956). *Genet. Plant Breed.* 9:89–100.
Lewis, D. and Crowe, L. K. (1954). *Heredity,* 8:357–363.
Lewis, D. and Modlibowska, I. (1942). *J. Genet.* 43:211–222.
Li, S. H., Bussi, C., Hugard, J. and Clanet, H. (1989). *Gartenbauwiss.* 54:49–53.
Lier, P, van (1967). *Fruitteelt.* 57:465–466.
Limongelli, F. and Cappellini, P. (1978). *Riv. Ortoflorofruttic.* 6:623–631.
Lin, C. H., Chan, L. R., Lin, S. H. and Lee, C. H. (1987). *Gartenbauwiss.* 52:200–204.
Linder, R. (1974). In: Linskens, H. F. (ed), *Fertilization in Higher Plants*. North-Holland, Amsterdam, pp. 325–326.
Liwerant, J. (1966). *L'Acad. d'Agr. France* 52:187–193.
Lobanov, G. A. and Basina, I. G. (1969). *Sadovodstvo* 5:31.
Logintsheva, A. G. (1958). *Agrobiologia* 6:125–126.
Løken, A. (1958). *Proc. Xth Int. Congr. Entomol. (1956),* 4:961–965.
Løken, A. (1981). *Bee World* 62:130.
Lomakin, E. N. (1974). *Sel'skokh. Biol.* 9:151–152.
Lombard, P., Westwood, M. and Thompson, M. (1971). *Proc. Ore. Hort. Soc.* 62:31–36.
Lombard, P., Williams, R. R., Stott, K. G. and Jefferies, C. J. (1972). Proc. Symp. Pear Growing, 1972, pp. 265–279.
Lombard, P., Hull, J. and Westwood, M. (1980). *Fruit Var. J.* 34(4):74–83.
Lombard, P., Guerrero-Prieto, M., Thompson, M. (1983). *Ann. Rep. Ore. Hort. Soc.* 74:121–127.

Looney, N. E., Kamienska, A., Regge, R. L. and Pharis, R. P. (1978). *Acta Hort.* 80:105–114.
Löschnig, H. J. and Passecker, D. F. (1954): *Die Marille (Aprikose) und ihre Kultur.* Österreichischer Agrarverlag, Wienna.
Lott, R. V. and Simons, R. K. (1968). *Hortic. Res.* 1.
Lucke, M. and Lech, W. (1974). *Proc. 19th Int. Hortic Congr.* 1:451.
Lucka, W. A. (1959). *Fruit Var. Hort. Digest*, 14(2):25.
Luckwill, L. C. (1959). *J. Linn. Soc. Bot. G. B.* 56:294–302.
Luckwill, L. C. (1970). In: Luckwill, L. C. and Cuttings, C. V. (eds), *Physiology of Tree Crops.* Acad. Press, New York, pp. 237–254.
Luckwill, L. C. (1974). *Proc. 19th Int. Hort. Congr.* 3:237–245.
Luckwill, L. C. (1977). Pestic. Chem. 20th Century Symp. 1976. *Amer. Chem. Soc. Symp. Ser.* 37:293–304.
Lukoschus, F. (1957). *Z. Bienenforsch.* 4:3–21.
Lutri, I. (1935). *Ital. Agr.* 72:139–155.
Mackowiak, M. (1974). *Rolnicz. Komisii Nauk. Lesnych* 37:183–192.
Maggs, D. H., Martin, G. J. and Needs, R. A. (1971). *Aust. J. Exp. Agr. Anim. Husb.* 11:113–117.
Magness, J. R., Overly, F. L. and Luce, W. A. (1931). *Wash. Agr. Exp. Stn Bull.*, p. 249.
Magyar, Gy. (1935). *Magy. Kir. Kert. Tanint. Közl.* 1:22–26.
Mainland, Ch. M. (1985). *Acta Hort.* 165:29–35.
Maliga, P. (1942). *Magy. Kir. Kert. Akad. Közl.* 8:3–5.
Maliga, P. (1944). *Kert. Szől. Főisk. Közl.* 10:287–319.
Maliga, P. (1946). A gyümölcsfélék termékenysége és terméketlensége (Fertility and infertility of fruit crops). In: Mohácsy M. (ed), *A gyümölcstermesztés kézikönyve* (Handbook of Fruit Production). Pátria, Budapest.
Maliga, P. (1948). *Agrártud. Egy. Kert- és Szőlőgazdaságtud. Kar. Közl.* 12:1–7.
Maliga, P. (1952). *Agrártud. Egy. Kert- és Szőlőgazdasági Tud. Kar Közl.* 16:27–49.
Maliga, P. (1953*a*). *Magy. Tud. Akad. Agrártud. Oszt. Közl.* 3:177–215.
Maliga, P. (1953*b*). *Kert. Szől. Főisk. Közl.* 17:25–41.
Maliga, P. (1956*a*). *Acta Agr. Acad. Sci. Hung.* 6:287–305.
Maliga, P. (1956*b*). A cseresznye- és a meggy fenológiai szakaszai és termékenyülése (Phenological phases and fertilization of cherry and sour cherry). In: Mohácsy, M. and Maliga, P. (eds), *Cseresznye- és meggytermesztés* (Cherry and Sour Cherry Growing). Mezőgazdasági Kiadó, Budapest.
Maliga, P. (1957). *Magy. Tud. Akad. Agrártud. Oszt. Közl.* 12:96–100.
Maliga, P. (1961). *Kert. Kut. Int. Évk.* 4:33–76.
Maliga, P. (1966*a*). *Szőlő- és Gyümölcstermesztés* 1:87–97.
Maliga, P. (1966*b*). *Szőlő- és Gyümölcstermesztés* 1:111–131.
Maliga, P. (1970). *Agrártud. Közl.* 29:247–253.
Maliga, P. (1980): Az új meggyhibridek termékenyülése (Fertility of sour cherry hybrids). In: Nyéki, J. (ed.): *Gyümölcsfajták virágzásbiológiája és termékenyülése* (Bloom Biology and Fertility of Fruit Cultivars). Mezőgazdasági Kiadó, Budapest, pp. 223–228.
Manandhar, D. N. and Lawes, G. S. (1980). *Orchardist New Z.* 53:269–272.
Manaresi, A. (1953). *Riv. Frutticoltura* 15:78–104.
Manino, A., Patetta, A. and Marletto, F. (1991). *Acta Hort.* 288:335–339.
Manzo, P. (1956). *Riv. Ortoflorofruttic.* 40:166–171.
Marcucci, M. C. and Filiti, N. (1984). *Gartenbauwiss.* 49:28–32.
Marcucci, M. C. and Visser, T. (1983). *Sci. Hort.* 19:311–319.
Marenaud, L. and Desvignes, J. C. (1965). *C. R. Acad. Agr. Fr.* 51(11):782–790.
Marro, M. (1976). *Riv. Ortoflorofruttic.* 60:184–198.
Marro, M. (1982). *La ric. sci.* 110:125–132.
Marro, M. and Deveronico, L. (1978). Com pre. Semin. "Fertilità delle piante de frutto", Bologna.
Marro, M. and Lalatta, F. (1978). *Proc. Fertility of Fruit Trees. Bologna, Dec. 15.* 1978, pp. 637–642.
Marro, M. and Lalatta, F. (1982). *La ric. sci.* 110:323–327.

Marro, M. and Ricci, A. (1962). *Riv. Ortoflorofruttic.* 46:450–457.
Marschall, E. R., Johnston, S., Hoosman, H. D. and Wells, H. M. (1929). *Mich. State Spec. Bull.* 188:38.
Martinez-Tellez, J. J., Monet, R. and Crossa-Raynaud, P. (1982). *L'Arboriculture Fruitière* 29:39–43.
Mascarenhas, J. P., Terenna, B., Mascarenhas, A. F. and Rueckert, L. (1974). In: Linskens, H. F. (ed.), *Fertilization in Higher Plants.* North-Holland, Amsterdam, pp. 137–143.
Matsuka, M. and Sakai, T. (1989). *Bee World* 70:55–61.
Matthews, P. (1966). *Grower* 65:840–841.
Matthews, P. (1970). Eucarpia. Proc. Angers. Fruit Breeding Symp., pp. 307–316.
Matthews, P. and Dow, K. P. (1969). Incompatibility groups: Sweet cherry (*P. avium*). Appendix III. In: Knight, R. L.: *Abstract Bibliography of Fruit Breeding and Genetics to 1965—Prunus.* Comm. Agric. Bur. East Malling Maidstone, Kent, pp. 540–544.
Mattusch, P. (1968). Untersuchungen über einige für den Anbau von Süßsskirschen zu Brennzwecken wichtige Kriterien. Thesis, Universität Hohenheim.
Mattusch, P. (1970). *Erwerbsobstbau* 12:47–49.
Mayer, D. F., Johansen, C. A. and Lunden, J. D. (1983). *HortScience* 24(3):510–512.
McGranahan, G. and Forde, H. I. (1985). *Amer. Soc. Hort. Sci.* 110:692–696.
McGregor, S. E. (1976). *Insect Pollination of Cultivated Crop Plants. Agricultural Handbook*, No. 496. Agr. Res. Serv., U. S. D. A., Washington.
McKay, J. W. (1969). Chestnuts. In: Jaynes, R. A. (ed.), *Handbook of North America Nut Trees.* Northern Nut Grow. Assoc. Knoxville, Tennessee.
McKay, J. W. and McGregor, S. E. (1974). *Amer. Bee J.* 114:336–354.
McKenzie, A. (1971). *Orch. New Zealand* 44:175–181.
McLagen, J. F. A. (1933). *Plant Physiol.* 8:395–423.
McLaren, G. F. Fraser, J. A. and Grant, J. E. (1992). *The Orchardist* 9:20:23.
Medeira, M. C. and Guedes, M. E. (1989). Proc. 9th Symp. Apricot Cult., 1989, Caserta, Italy, *Acta Hort.* 293:311–318.
Medeira, M. C. and Guedes, M. E. (1990). *Brotéria Genetica* 11:171–186.
Medeira, M. C., Maia, M. I. and Moreira, A. C. (1991). *Adv. Horticult. Sci.* (in press).
Mehlenbacher, S. A. and Thompson, M. M. (1991). *HortScience* 26:442–443.
Meli, T. (1981). *Erwerbsobstbau* 23:158–162.
Mellenthin, W. M., Wang, C. Y. and Wang, S. Y. (1972). *HortScience* 7:557–559.
Menke, H. F. (1951). Proc. XIVth Int. Beekeep. Congr., p. 11.
Mesnil, G. (1987). *Rev. Horticult.* 278:23–27.
Miculka, B. (1965): *Ovoc. Zelin.* 4:106–107.
Middlebrook, W. J. (1915–1916). *J. Agricult.* 22:418–433.
Mihatsch, H. and Schumann, G. (1971). *Arch. Gartenbau* 19:371–378.
Milella, A. (1959). *Riv. Ortoflorofruttic.* 43:322–332.
Miller, A. N., Lombard, P. B., Westwood, M. N. and Stebbins, R. L. (1990). *HortScience* 25:176–178.
Milovankic, M. (1968). *Letopis naučnih radova Poljoprivrednog fakulteta u Novom Sadu* 12:117–133.
Milovankic, M. (1972). *Yug. vočarstvo* 17–18:173–178.
Milovankic, M. (1974). *Jul. God. Zborn. Zemliodelsko- Sumarskiot Fak. Univ. Skopie* 25:39–44.
Milutinovic, M. (1973). *Sborn. Semidelsk. Shumarsk. Fak. Univ. Skopie* 25:93–102.
Milutinovic, M. (1975). *Yug. vočarstvo* 8:15–23.
Misic, P. (1969). *Archiv. Polyoprv. Nauke* 22(78):41–49.
Misic, P., Todorovic, R. R., Lekic, N. K. and Pavlovic, C. V. (1977). *Nauka u praksi* 7:141–144.
Misic, P., Todorovic, R., Lekic, N., Pavlovic, V. and Vinterhalter, D. (1979). *Nauka u praksi* 9:225–236.
Misota, Iné (1973). *Kertgazdaság* 5(4):17–26.
Mittempergher, L. and Roselli, G. (1966). *Riv. Ortoflorofruttic.* 91:555–567.
Mittempergher, L., Aradski, M. and Fideghelli, C. (1965). *Riv. Ortoflorofruttic.* 49:171–185.

Modic, D. (1974). Raziskave tzvetenia in interfertilnosti med sortami leske v predalpskem obmocyu SR. Sloveniye. (*Corylus avelana, C. maxima* in niihovi krizanci). Izvirno raziskovalno delo, pp. 83–99.
Modlibowska, I. (1945). *J. Pomol. Horticult. Sci.* 21:57–89.
Moffet, A. A. (1931). *J. Pomol.* 9:100–110.
Mohácsy, M. Porpáczy, A., Sr., Kollányi, L. and Szilágyi, K. (1965). *Szamóca, málna, szeder.* (Strawberry, raspberry, blackberry). Mezőgazdasági Kiadó, Budapest.
Mohácsy, M., Maliga, P. and Mohácsy, M. (1967). *Az őszibarack* (The Peach). Mezőgazdasági Kiadó, Budapest.
Moleas, T. and Serio, G. (1981). *L'Informatore Agrario* 37:18462–18470.
Molnár, L. and Stollár, A. (1971). *Acta Agr. Acad. Sci. Hung.* 20:47–53.
Monastra, F. and Fideghelli, C. (1978). *Frutticoltura* 15:11–14.
Monastra, F., Proto, D., Fideghelli, C., Grassi, G., Della Starada, G., Magliano, V. and Pennone, V. (1984). *L'Informatore Agrario* 40:29–54.
Montalti, P. and Selli, R. (1984). *Riv. Frutticolt. Ortofloricolt.* 9–10:57–62.
Möhring, A. (1942). *Dtsch. Obstbau Ig.* 57:142.
Montesinos, E. and Vilardell, P. (1991). *Phytopathology* 81:113–119.
Moore, J. N. (1969). *J. Amer. Soc. Hort. Sci.* 99:362–364.
Morettini, A. (1943). *Nord* 13:18–22.
Morettini, A., Baldini, E., Scaramuzzi, F., Bargioni, G. and Pisani, P. L. (1962). Monografia delle principali cultivar del pesco. C. N. R., Florence.
Muhturi-Stylianidu, E. (1979). *Agr. Res.* 3:64–73.
Müller-Thurgau, H. (1898). *Landw. Jahrb. Schweiz.* pp. 135–205.
Mullins, C. A. and Lockwood, D. W. (1979). *Fruit Var. J.* 33:94–96.
Murawski, H. (1959). *Züchter* 29:72–78.
Murneek, A. E. (1937). *Midwest Fruitman* 10:8–9.
Murneek, A. E. (1954*a*). *Mon. Agr. Exp. Stn Bull.,* p. 622.
Murneek, A. E. (1954*b*). *Proc. Amer. Soc. Hort. Sci.* 64:573–582.
Mussano, L., Radicati, G. Me. L. and Vallania, R. (1983*a*). *Proc. Conv. Internaz. Nocciuolo, Avellino,* pp. 305–308.
Mussano, L., Radicati, G. Me. L. and Sacerdote, S. (1983*b*). *Proc. Conv. Internaz. Nocciuolo, Avellino,* pp. 321–325.
Muth, F. and Voigt, G. (1928). Ber. Lehraust. Gesenheim für 1927, Berlin.
Müller, W. (1976): *Schweiz. Zschr. Obst.- und Weinbau* 112:344–351; 381–386.
Naeumann, W. D. (1964). *Erwerbsobstbau* 6:64–67.
Nagasawa, K. Ohno, M. and Tsukahara, K. (1972). *Techn. Bull. Fac. Hort. Chiba Univ.* 20:1–8.
Nagy, P. (1957). *Kert. Kut. Int. Évk.* 2:261–280.
Nagy, P. (1960). *Kísérletügyi Közl.* 3:27–45.
Nebel, B. R. (1930). *N. Y. Agr. Exp. Stn Tech. Bull.* 170:1–16.
Nebel, B. R. (1932). *Proc. 6th. Int. Congr. Genet.* 2:140–141.
Nebel, B. R. (1936*a*). *J. Pomol.* 14:203–204.
Nebel, B. R. (1936*b*). *J. Hered.* 27:345–349.
Nebel, B. R., Kertész, Z. I. (1934). *Gartenbauwiss.* 9:45–64.
Nebel, B. R. and Trump, I. J. (1932). *Proc. Nat. Acad. Sci.* 18.
Nedev, N. and Stefanova, A. (1979). *Grad. Loras. Nauka* 16:3–9.
Nesterov, Ya. S. (1952). *Sad. Ogorod Moskva* 1:30–32.
Nesterov, Ya. S. (1956). *Agrobiologia* 4:132–135.
Nesterov, Ya. S., Anisimova, G. G. and Savelev, N. I. (1972). *Vestnik Sel'skokhoz. Nauki* 10:114–118.
Neumann, U. (1962). *Arch. Gartenbau* 10:11–23.
Nicotra, A., Damiano, C., Cobianchi, D., Moser, L. and Faedi, W. (1976). *Frutticoltura* 38(5):5–22.
Ninkovski, I. (1989). *Nektarin a sima héjú őszibarack* (Nectarine, the Smooth-Skinned Peach). Mezőgazdasági Kiadó, Budapest.

Noll, F. (1902). Sitz. Ber. Nied. Ges. Nat. Heilk. Bonn., pp. 149–162.
Nybom, H. (1986). *J. Hort. Sci.* 61(1):49–55.
Nye, W. P. and Anderson, J. (1974). *J. Amer. Soc. Hort. Sci.* 99:40–44.
Nyéki, J. (1970). Körtefajták termékenyülési viszonyainak elemzése (Analysis of fertilization conditions in pear cultivars). (Doctoral Thesis). Kertészeti Egyetem, Budapest.
Nyéki, J. (1971). *Szőlő- és Gyümölcstermesztés* 6:75–87.
Nyéki, J. (1972a). *Kísérletügyi Közl.* 1–3:13–27.
Nyéki, J. (1972b). *Acta Agr. Acad. Sci. Hung.* 21:75–80.
Nyéki, J. (1973a). *Kísérletügyi Közl.* 1–3:25–42.
Nyéki, J. (1973b). *Acta Agr. Acad. Sci. Hung.* 22:207–210.
Nyéki, J. (1973c). *Kert. Egy. Közl. 1972.* 36:147–154.
Nyéki, J. (1974a). *Acta Agr. Acad. Sci. Hung.* 23:93–99.
Nyéki, J. (1974b). *Acta Agr. Acad. Sci. Hung.* 23:85–93.
Nyéki, J. (1974c). Meggyfajták virágzása és termékenyülése (Flowering and Fertilization of Sour Cherry Cultivars). PhD Thesis. Hung. Acad. Sci., Budapest.
Nyéki, J. (1974d). *Kert. Egy. Közl.* 38:149–159.
Nyéki, J. (1975). *Kert. Egy. Közl.* 39:49–56.
Nyéki, J. (1976). *Agrártud. Közl.* 35:377–386.
Nyéki, J. (1976). Metaxénia hatások (Effects of metaxenia). In: Gyuró, F. (ed.), *Körte* (Pear). Mezőgazdasági Kiadó, Budapest.
Nyéki, J. (1977). *Acta Agr. Acad. Sci. Hung.* 26:282–289.
Nyéki, J. (1978). *Kertgazdaság* 10(1):31–38.
Nyéki, J. (1980). *Gyümölcsfajták virágzásbiológiája és termékenyülése* (Bloom Biology and Fertility of Fruit Cultivars). Mezőgazdasági Kiadó, Budapest.
Nyéki, J. (1989). *Csonthéjas gyümölcsűek virágzása és termékenyülése* (Flowering and fertility in stone fruits). DSc. Thesis. Acad. Sci. Hung., Budapest. pp. 1–36.
Nyéki, J. (1990). A gyümölcstermő növények virágzása, megporzása és termékenyülése (Flowering, pollination and fertilization of fruit crops). In: Gyuró, F. (ed.) *Gyümölcstermesztés* (Fruit Growing). Mezőgazdasági Kiadó, Budapest.
Nyéki, J. and Ifjú, Z. (1975). *Bot. Közl.* 62(4):271–285.
Nyéki, J. and Soltész, M. (1977a). "Lippay János" Tudományos Ülésszak Előadásai. May 27–28, 1975. I., pp. 490–516.
Nyéki, J. and Soltész, M. (1977b). *Acta Agr. Acad. Sci. Hung.* 26:87–89.
Nyéki, J. and Soltész, M. (1978a). *Acta Agr. Acad. Sci. Hung.* 27:72–74.
Nyéki, J. and Soltész, M. (1978b). *Acta Agr. Acad. Sci. Hung.* 27:271–277.
Nyéki, J. and Tóth, F. (1976). *Bot. Közl.* 63:165–176.
Nyéki, J., Brózik, S. and Ifjú, Z. (1974). *Bot. Közl.* 62:103–107.
Nyéki, J., Soltész, M. and Brózik, S. (1976). *Kertgazdaság* 8(2):31–38.
Nyéki, J. Brózik, S. and Ifjú, Z. (1980). *Kertgazdaság* 12(2):37–50.
Nyéki, J., Szabó, Z., Tóth, F-né and Pete, A. (1985). *Kertgazdaság* 17(2):35–63.
Nyujtó, F. (1958). *A Duna-Tisza közi Mezőgazd. Kísérl. Int. Évi Jel.* (Annual Report), Cegléd, Hungary, pp. 153–154.
Nyujtó, F. (1966). Kertészet és Szőlészet Tudományos Tanácsadó. Csonthéjasok, pp. 5–9.
Nyujtó, F. (1967). *A mezőgazdasági kutatások 1966. évi főbb eredményei.* (Important 1966 results of agricultural research). Ministry of Agriculture, Budapest. pp. 110–115.
Nyujtó, F. (1971). *Szőlő- és Gyümölcstermesztés* 4:89–107.
Nyujtó, F. (1978). A review of the varieties applicable in apricot growing. In: *Varieties, Cultural Practices and Possibility of Mechanical Harvest in Stone-Fruits.* Publication of the Research Institute for Fruit and Ornamental Plants, Budapest, pp. 23–31.
Nyujtó, F. (1980). A kajszibarack (Apricot). In: Nyéki, J. (ed.), *Gyümölcsfajok virágzásbiológiája és termékenyülése* (Flowering Biology and Fertilization of Fruit Cultivars). Mezőgazdasági Kiadó, Budapest pp. 248–266.
Nyujtó, F., Brózik, S., Nyéki, J. and Brózik, S., Jr. (1982). *Acta Hort. Agr. Acad. Sci. Hung.* 121:159–165.

Nyujtó, F., Brózik, S., Nyéki, J. and Brózik, S., Jr. (1983). *Acta Agr. Acad. Sci. Hung.* 32:46–58.
Nyujtó, F., Brózik, S., Brózik, S., Jr. and Nyéki, J. (1985). *Acta Agr. Acad. Sci. Hung.* 34:65–72.
Nyujtó, F. and Tomcsányi P. (1959). *A kajszibarack és termesztése* (Apricot Production). Mezőgazdasági Kiadó, Budapest, p. 330.
Offord, H. R., Quick, C. R. and Moss, U. D. (1941). *J. Agric. Res.* 68:65–71.
Ohno, M., Takahashi, E., Nakai, J. and Kawano, Y. (1961). *Techn. Bull. Fac. Horticult. Chiba Univ.* 9:75–81.
Okályi, I. and Maliga, P. (1956). *Gyümölcstermesztés* (Fruit Production), Vol. 2. Mezőgazdasági Kiadó, Budapest, p. 377.
Oliveira, D., de., Gingras, J. and Chagnom, M. (1991). *Acta Hort.* 288:415–419.
Oliveira, D., de., Savoie, L. and Vincent, C. (1991). *Acta Hort.* 288:420–426.
Ollivier, H. (1967). *Pomol. Franç.* 9:7–13.
Oprea, St. (1981). *Lucrari Stiintifice* 9:77–87.
Oprea, St. and Palocsay, R. (1977). Realizari si perspective in ameliorazea marolni la statiunes experimentala horticola Chy-Napoca. In: *Realizari in ameliorarea pomilor şi arbustilor fructiferi din Romanica.* Editura Ceres, Bucureşti.
Oratovsky, M. T. (1935). Perehresne zapilenia i samozadileniya kistochkovih plodovih. Kiev.
Oratovsky, M. T. (1940). Samozapilennaya i missortove zbirnik pac Melitopolskoi zonalnoy nauko doslidnoy plodoyagodn. Kiev.
Orero, J. (1974). *Cerezos. Viveros Orero. Catalogo General* 84:24–25.
Orosz-Kovács, Zs. (1990). *Kertgazdaság* 22(5):24–31.
Orosz-Kovács, Zs., Gulyás, S. and Halászi, Zs. (1989). *Acta Bot. Hung.* 35:237–244.
Orosz-Kovács, Zs., Nagy-Tóth, E., Csatos, A. and Szabó, A. (1990). *Bot. Közl.* 77:127–132.
Orosz-Kovács, Zs., Gulyás, S. and Kaposvári, F. (1992). *Acta Biol. Szeged* 38:47–55.
Ostapenko, V. I. and Zhukov, O. S. (1985). *Sadov. Vinogr. Vinod. Moldavii* 5:20–21.
Osterwalder, A. (1907). *Landw. Jahrb. Schweiz* 36.
Osterwalder, A. (1909). *Landw. Jahrb. Schweiz* 38.
Osterwalder, A. (1910). *Landw. Jahrb. Schweiz* 39:917–998.
Overcash, J. P. (1965). *Proc. Assoc. Southern Agr. Workers 62nd Conv.*, Texas, pp. 175–176.
Paarmann, W., von (1977). *Z. angew. Ent.* 84:164–178.
Paarman, W. (1980). *Erwerbsobstbau* 22 (3):44–47.
Palara, U., Sansavini, A. and Benati, R. (1985). *Riv. Fruttic. Ortofloricoltura* 47:9–20.
Palazon, L. J. (1965). *El almondro y su cultivo.* Ediciones Mundi-Prensa, Madrid.
Palmer-Jones, T. and Clinch, P. G. (1967). *New. Zeal. J. Agr. Res.* 10:143–149.
Palmer-Jones, T. and Clinch. P. G. (1968). *New Zeal. J. Agr. Res.* 11:149–154.
Palocsay, R. (1961). *Zöldség és gyümölcstermesztési kísérleteim* (Vegetable and Fruit Growing Trials). Private edition, Bucharest.
Parfitt, D. E. and Ganeshan, S. (1989). *HortScience* 24:354–356.
Parker, R. L. (1926). *Mem. Cornell Agr. Expt. Stan* 98.
Parnia, C., Ritiu, C., Ivan, J. and Burlói, N. (1979). *Lucrari Stiintifice* 8:223–237.
Parry, M. S. (1978). *Le Fruit Belge* 46(381):62–66.
Papp, J. (1984). *Bogyósgyümölcsűek* (Berry Crops). Mezőgazdasági Kiadó, Budapest, p. 324.
Paskevic, V. V. (1930). *Bull. Appl. Bot. Genet. Plant. Breed.* Suppl. 49:1–204.
Paunovic, S. A. (1971). *Yug. vočarstvo* 17–18:109–122.
Pavlov, O. A. (1973). *Sadov. Vinogr. Vinod. Moldavii* 28:13–15.
Pearce, S. G. and Preston, A. P. (1954). *Ann. Rep. East Malling Res. Stan, 1953,* pp. 133–137.
Pejkic, B. (1966). *Rev. Res. Work. Fac. Agr. Univ., Beograd.* 14:1–8.
Pejkic, B. (1968). *J. Sci. Agricult. Res.* 21:63–68.
Pejkic, B. (1969). *Yug. vočarstvo* 7:115–123.
Pejkic, B. (1971). *Yug. vočarsvo* 17–18:391–403.
Pejkic, B. (1972). *Vitolnost zenskog gametofita u visnye Kereske (P. cerasus L.).* Poljoprivredni fakultet. Zemun, Beograd.
Pejkic, B. (1973). *Genetika* 5:265–274.
Pejkic, B. and Dokic, A. (1968). *J. Sci. Agr. Res.* 21:67–77.

Pejkic, B. and Jovovic, V. (1968). *J. Sci. Agr.* 21:37–47.
Pejkic, B. and Popovic, R. (1973). *Archiva poljopriv. Nauka* 26:44–52.
Pejovics, B. (1966). "Lippay János Tudományos Ülésszak" előadásai. *Kert. és Szől. Főisk. Kiadv.* 2:227–242.
Pejovics, B. (1968). A mandula egyes biológiai és termesztési sajátosságai (Biological and Production Characteristics of Almond). PhD Thesis. Hung. Acad. Sci., Budapest.
Pejovics, B. (1976). Mandula (Almond). In: Horn, E. (ed.), *Dió, mandula, mogyoró, gesztenye* (Walnut, Almond, Hazelnut and Chestnut). Mezőgazdasági Kiadó, Budapest.
Percival, M. S. (1955). *New Phytol.* 54:353–368.
Percival, M. S. (1965). *Floral Biology.* Pergamon Press, Oxford.
Perfileva, Z. N. (1982). *Bullet. Gosud. Nikit. Botan., Sada* 48:54–57.
Perry, J. L. and Moore, J. N. (1985). *HortScience* 20(4):338–339.
Péter, J. (1972). *Agr. Egy. Keszthely, Mosonm. Mezőg. Kar Növ. Tansz. Közl.* 15:5–32.
Peterson, R. H. and Taber, H. G. (1987). *HortScience* 22:953.
Pethő, F. (1969). *Almatermesztés* (Apple Growing). Mezőgazdasági Kiadó, Budapest.
Petkov, V. (1965). *Grad. Lozar. Nauk.* 2:421–431.
Petkov, V. G. and Panov, V. (1967). *Proc. XXIst Int. Beekeep. Congr. (College Park, Md.).* Apimondia Press, Bucharest, pp. 432–436.
Pheasant, J. (1985). *Fruit Grower* 6:20–21.
Phillips, H. A. (1922). *N. Y. Agr. Exp. Stn Mem.* 59:1377–1416.
Philp, G. L. (1947). *Calif. Agr. Ext. Circ.* 46.
Pimienta, E., Polito, V. S. and Kester, D. E. (1983). *J. Amer. Soc. Hort. Sci.* 108:643–647.
Pisani, P. L. (1962). *Riv. Ortoflorofruttic.* 6:1–17.
Pisani, P. L. (1978). *Riv. Ortoflorofruttic.* 61(5):311–329.
Pisani, P. L. and Ramina, A. (1970). *L'Informatore Agrario* 26:1217–1219.
Pisani, P. L. and Ramina, A. (1971). *Riv. Ortoflorofruttic.* 55:51–59.
Pisani, P. L., Ramina, A., Zocca, A. and Cristoferi, G. (1970). *L'Agricoltura Italiana* 70:371–382.
Plancher, B. (1985). *Erwerbsobstbau* 11:271–275.
Plancher, B. and Dördrechter, H. (1983). *Erwerbsobstbau* 25:80–84.
Plock, M. (1966). *Erwerbsobstbau* 8.
Pogorelov, P. F. (1970). *Sborn. Nauk. Rabot. Sibirskiv NII* 15:137–141.
Polunin, O. (1976). *Trees and Bushes of Europe.* Oxford University Press.
Ponomareva, E. G. (1980). *Dokl. THSA* 266:13–19.
Popelyankov, G. (1974). *Ovoshtarstvo* 6:30–34.
Pór, J. and Pór, Jné (1990). *Kertgazdaság* 6:9–14.
Porpáczy, A. (ed.) (1964). *A korszerű gyümölcstermelés elméleti kérdései* (Theoretical Questions of Modern Fruit Growing). Mezőgazdasági Kiadó, Budapest.
Porpáczy, A., Jr. (1974). *Proc. XIXth Int. Hort. Congr., Warsawa,* 1. A., p. 366.
Porpáczy, A. Jr. (1987). *Ribiszke, áfonya, bodza, fekete berkenye* (Currant, Vaccinium, Elderberry and Sorbus). Mezőgazdasági Kiadó, Budapest.
Porsch, O. (1950). *Oest. Bot. Z.* 97:269–321.
Porter, J. and Dibbens, R. (1977). *Horticult. Industry* 9:633–634.
Postweiler, K., Stösser, R. and Anvari, S. F. (1985). *Sci. Horticult.* 25:235–239.
Potter, J. M. S. (1956). New varieties of fruit. In: T. Wallace and Buch, R. G. W. (eds), *Modern Commercial Fruit Growing.* Country Life Ltd., London, pp. 330–347.
Poulsen, G. B. (1983). *Tidsskr. Planteavl.* 87:33–38.
Pouvreau, A. (1984). Culture de petits fruits. In: Pesson, P. and Louveaux, J. (eds), *Pollinisation et production végétales.* INRA, Paris, pp. 373–392.
Pratt, Ch. and Einset, J. (1955). *Amer. J. Bot.* 42:637–645.
Preston, A. P. (1949). *Ann. Rep. East Malling Res. Stan, 1948.* pp. 44–47.
Pritts, M. (1989). *Amer. Fruit Grower* 109:14–15.
Pugliano, G. and Forlani, M. (1985). *Acta Hort.* 192:384–400.
Pusztai, J. et al. (1969). *Pollination Experiments with Apples and Red Clovers in County Zala.* Zala megyei Nyomda, Nagykanizsa (SW-Hungary), p. 47 (in Hungarian).

Pyke, G. H. (1984). *Ann. Rev. Ecol.* 15:523–575.
Quarta, R. and Bunalti, R. (1983). *Abstr. Int. Symp. Flowering and Fruit Set in Fruit Trees, Aug. 25–26.*
Quarta, R. and Bunalti, R. (1984). *Acta Hort.* 149:85–94.
Radulescu, C. (1971). *Resultate ale cercetarilor privind sortimentul de visin din Romania.* CIDAS, pp. 151–164.
Rajotte, E. G. and Fell, R. D. (1982). *HortScience* 17:230–231.
Ramina, A. (1969). *Riv. Ortoflorofruttic.* 53:3–12.
Ramina, A. (1970). *L'Agricoltura Italiana* 70:383–397.
Randhawa, G. S. and Nair, P. K. R. (1960). *Ind. J. Hort.* 17:83–101.
Randhawa, G. S., Yadar, I. S. and Nath, N. (1963). *Ind. J. Agr. Sci.* 33:129–138.
Rapillard, Ch. (1981). *Rev. Suisse Viticult. Arboricult. Horticult.* 13:301–303.
Rasmussen, P. M. (1984). *Tidsskr. Planteavl.* 88:193–202.
Rasmussen, K., Kold, E. and Christensen, J. V. (1983). *Tidsskr. Planteavl.* 87:505–514.
Raseira, M. C. B. and Moore, J. N. (1986). *HortScience* 21:1367–1368.
Raseira, M. C. B. and Moore, J. N. (1987). *HortScience* 22:216–218.
Rawes, A. N. (1922). *J. Hort. Soc.* 47:8–14.
Redalen, G. (1977). *Meldinger fra Norges Landbonkshgskole* 56:1–10.
Redalen, G. (1981). *Gartenbauwiss.* 46:223–227.
Redalen, G. (1984a). *Acta Hort.* 149:71–75.
Redalen, G. (1984b). *Gartenbauwiss.* 49:212–217.
Reichel, M. (1964a). *Physiol. Probl. Obstbau* 65:35–43.
Reichel, M. (1964b). *Kühn-Arch.* 78:268–333.
Reichel, M. (1972). *Arch. Gartenbau* 20:427–444.
Reinecke, O. S. H. (1930). *Union South Afr. Dep. Agr. Sci. Bull.* 90:92.
Reino, A., Giorgio, V. and Godini, A. (1986). *Riv. Frutticolt. Ortofloricolt.* 12:61–63.
Roberts, R. H. (1922). *Wisconsin Agr. Exp. Stn Bull.* 344.
Roberts, R. H. (1945a). *Proc. Amer. Soc. Hort. Sci.* 46:87–90.
Roberts, R. H. (1945b). *Amer. Fruit Grower* 65:16.
Robinson, W. S. (1979). *J. Amer. Soc. Hort. Sci.* 104:596–598.
Robinson, W. S. and Fell, R. D. (1981). *HortScience* 16:326–328.
Roemer, K. (1966). *Mitt. Obstbauversuchsring Altes Land* (Jork) 21:248–254.
Roemer, K. (1968–1970). *Mitt. Obstvers. Jorh., Hannover 1968.* 23:222–239; *1970.* 25:2909–2915, 242–249, 259–309, 354–361.
Roh, L. M. (1929a). *Arbetein der Mleewer Gartenbauversuchs-station* 23:1–250.
Roh, L. M. (1929b). *Obstbau* 23:250.
Roh, L. M. (1929c). *Dokl. Opyt. Stanc. Mleev.,* pp. 1–99.
Rom, R. C. and Arrington, E. H. (1966): *Proc. Amer. Soc. Hort. Sci.* 88:239–244.
Roman, R. (1981). *Lucrari Stiintifice* 9:145–153.
Roman, I. and Blaja, D. (1984). *Stat. Cerc. Prod. Pomic. Tirgu Jiu.,* pp. 86–88.
Roman, R. and Radulescu, C. (1986). *Polenizatorii soiurilor noi de prun recent create in tara* (Private edition).
Romisondo, P. (1963a). *Riv. Ortoflorofruttic.* 88:202–215.
Romisondo, P. (1963b). *Frutticoltura.* 11–12:1–9.
Romisondo, P. (1963c). *Ann. Fac. Sci. Agr. Univ., Studi di Torino. 1962–1963* 2:25–56.
Romisondo, P. (1977). *Riv. Ortoflorofruttic.* 61:227–302.
Romisondo, P. (1978). *Riv. Ortoflorofruttic.* 62:423–429.
Romisondo, P. and Limongelli, F. (1978). *Riv. Ortoflorofruttic.* 62:655–661.
Romisondo, P. and Marletto, F. G. (1972). *Wiss. Bull. Apimondia 1972,* pp. 162–172.
Romisondo, P. and Me, G. (1972). *Wiss. Bull. Apimondia 1972,* pp. 185–190.
Rosati, P. and Faedi, W. (1974). *Frutticoltura.* 36:23–33.
Rosati, P. and Gaggioli, D. (1987). *Acta Hort.* 212(2):379–390.
Roversi, A. and Ughini, V. (1986). *Ann. Fac. Agr. U.C.S.C. (Piacenza)* 26:189–203.
Rubin, B. A. (1968). *Fiziologia Sel'skoh. rastenii.* 10:35–61.

Rudenko, I. S. (1958). *Sad i Ogorod* 8:52–54.
Rudloff, C. F. and Schanderl, H. (1935). *Gartenbauwiss.* 9:501–508.
Rudloff, C. F. and Schanderl, H. (1950). *Die Befruchtungsbiologie der Obstgewachse.* 3rd Ed., Stuttgart.
Ruiz, V. S. (1977). *Acta Hort.* 69:235–241.
Rünger, W. (1971). *Blütenbildung und Blütenentwicklung.* Paul Parey, Berlin, Hamburg.
Russell, S. D. (1991). *Ann. Rev. Plant Physiol. Plant Mol. Biol.* 42:189–204.
Ryabov, I. N. (1930). *Nikit. Opyt. Bot. Sada.* 14.
Ryabov, I. N. (1975). *Trudy Nikit. Bot. Sada,* 67.
Ryabov, I. N. and Kantsherova, V. P. (1970). *Trudy Gos. Nikit. Bot. Sada* 4:155–159.
Ryabov, I. N. and Ryabova, A. N. (1970). *Trudy Gos. Nikit. Bot. Sada* 4:53–153.
Ryabova, A. N. (1961). *Vinogr. Sadovodstvo Kryma* 6.
Ryabova, A. H. (1970). *Trudy Gos. Nikit. Bot. Sada* 45:37–51.
Ryabova, A. H. (1977). *Izuchenie i onedrenie v proizvostvo novyh sortov plodovyh, dekoratiny i tekhnicheskikh rasteniy.* Yalta. 137:23–39.
Sachs, R. M. (1977). *HortScience* 12:220–222.
Sanagian, M. B. (1968). *Techn. Comm. Int. Soc. Hort. Sci.* 11:57–64.
Sandsten, E. P. (1909). *Res. Bull. Wisconsin Agr. Exp. Stn* 4:149–172.
Sanfourche, G. (1972). *Proc. Conv. del Ciliegio, Verona. Cam. Comm.* IAA, pp. 143–154.
Sansavini, S. and Bassi, D. (1977): *Acta Hort.* 75:73–85.
Sansavini, S. and Lane, W. D. (1983). *Cent. Oper. Ortofrutti. Ferrara,* pp. 55–57.
Sansavini, S., Costa, G., Credi, R., Grandi, M., Bindi, V. and Monti, C. (1980). *Riv. Ortoflorofruttic.* 64:563–577.
Sansavini, S., Grandi, M. and Atti, P. (1981*a*). *Scelte varietali in Frutticoltura, Ferrara,* pp. 23–37.
Sansavini, S., Zambrini, M., Costa, G. and Cavicchi, C. (1981*b*). *Scelte varietali in Frutticoltura, Ferrara,* pp. 95–102.
Sárkány, S. (1969). *A növények világa* (The World of Plants) Vol. 1. Gondolat Kiadó, Budapest.
Sárkány, S. and Szalai, I. (1966). *Növénytani Praktikum* (Botanical Teaching Exercises) Vol. I. Tankönyvkiadó, Budapest.
Sartorius, R. (1990). Anatomische, histologische und cytologische Untersuchungen zur Samenentwicklung bei der Walnuss (*Juglans regia* L.) unter besonderer Berücksichtigung der Apomixis. Thesis, Univ. Stuttgart-Hohenheim, p. 123.
Sartorius, R. and Stösser, R. (1992). *Erwerbsobstbau* 34:98–106.
Sartorius, R., Stösser, R. and Anvari, S. F. (1984). *Angew. Botanik* 58:307–318.
Saure, M. (1967*a*). *Erwerbsobstbau* 9.
Saure, M. (1967*b*). *Mitt. Klosterneuburg* 22:219–223.
Saure, M. (1975). *Erwerbsobstbau* 17:188–189.
Savio, A. (1970): *CTFL-DOKUMENTS* 26(2):1–14.
Scagel, R. F. et al. (1984). Plants. *An Evolutionary Survey.* Wadsworth Publishing Company, Belmont, California.
Scaramuzzi, F. (1951). *Ann. Sper. Agr.* 5:543–557.
Schaer, E. (1952). *Swiss Plum and Damson Varieties.* Buchverlag Verhandsdruckerei of Bern.
Schaer, E. and Schafer, H. (1975). *Schweiz. Zschr. Obst- und Weinbau* 111:148–154.
Schanderl, H. (1932). *Gartenbauwiss.* 6:192–239.
Schanderl, H. (1934*a*). *Gartenbauwiss.* 8:135–145.
Schanderl, H. (1934*b*). *Züchter* 1:6–12.
Schanderl, H. (1937). *Gartenbauwiss.* 11:297–318.
Schanderl, H. (1955*a*). *Mitt. Obstvers. Altes Landes* 9:271–277.
Schanderl, H. (1955*b*). *Zeitschr. Pfl. Züchtung* 34:255–306.
Schanderl, H. (1956). *Gartenbauwiss.* 3:284–291.
Schanderl, H. (1965). *Erwerbsobstbau* 7:149–153.
Schanks, H. (1969). *J. Apic. Res.* 8:19–21.

Schauz, R. (1989). Anatomische, histologische und cytologische Untersuchungen zur Samenentwicklung bei Steinobst unter besonderer Berücksichtigung der Haustorien. Thesis, Univ. Stuttgart–Hohenheim. p. 106.
Schauz, R., Stösser, R. and Anvari, S. F. (1989). *Angew. Botanik* 63:455–469.
Schmedlak, J. (1965). *Arch. Gartenbau* 13:497–513.
Schmidt, M. (1940). *Züchter* 12:281–289.
Schmidt, M. (1954). *Arch. Gartenbau.* 2:355–384.
Schmidt, S. (1978). *Arch. Gartenbau* 26:177–187.
Schmidt, M. (1982). *Erwerbsostbau* 24(1):6–9.
Schnelle, F. (1955). *Pflanzenphänologie.* Akad. Verlag., Leipzig.
Schoemaker, J. S. and Davis, R. M. (1966). *Circ. Florida Agr. Expt. Serv.* 294:20.
Schossig, S. (1959). *Gartenbau* 64:105–106.
Schultz, J. H. (1948). *Proc. Amer. Soc. Hort. Sci.* 51:171–174.
Schuster, C. E. (1922). *Oregon Agric. Exp. Stn Circ.* p. 27.
Schuster, C. E. (1925). *Oregon Agr. Exp. Stn Bull.* 212:1–40.
Schuster, C. E. (1961). *Oregon Stn Univ. Exp. Bull.* 628:9–10.
Schwerdtfeger, G. (1978). *Gartenbauwiss.* 43:145–156.
Sedov, E. H. (1958). *Agrobiologia* 3:131–135.
Seilheimer, M. and Stösser, R. (1982). *Gartenbauwiss.* 47:49–55.
Seyfert, F. (1955). *Angew. Meteorol.* 2:183–190.
Shaw, J. K. (1911). *Mars. Agr. Exp. Stn Ann. Rep.* 190:177–245.
Sheljahudin, A. (1960). *Kert. Kut. Int. Évk.* 4:123–128.
Sheljahudin, A. and Brózik, S. (1965). *Acta Agr. Acad. Sci. Hung.* 4:253–260.
Sheljahudin, A. and Brózik, S. (1966). *Acta Agr. Acad. Sci. Hung.* 15:187–198.
Sheljahudin, A. and Brózik, S. (1967). *Acta Agr. Acad. Sci. Hung.* 16:63–74.
Sherman, W. B. and Janick, J. (1964). *Fruit Var. Hort. Digest.* 18:37–38.
Sherman, W. B., Knight, R. J. and Lyrene, P. M. (1978). *HortScience* 13:162.
Shimura, I., Yasuno, M. and Otomo, C. (1971). *Jap. J. Breed.* 21:77–80.
Shing, K. C. and Feng, Y. F. (1936). *J. Agr. Assoc. China* 149:1–13.
Shitt, P. G. and Metlickij, Z. A. (1940). *Plodovodstvo M. Sel'hozgiz.* 658.
Siegfried, D. (1973). *Urania Pflanzenreich. Höhere Pflanzen*, Vols 1–2. Urania Verlag, Leipzig, Jena, Berlin.
Silbereisen R. (1970). *Proc. Angers Fruit Breeding Symposium, Versailles.* Station de Recherches Fruitiers, Angers, pp. 79–97.
Silbereisen, R. (1982). *Erwerbsobstbau* 24:96–103.
Silbereisen, R. and Sherr, F. (1969). *Obst und Garten* 88:39–42.
Simidchiev, T. (1967). *Nauch. Trud. Vissh. Selskostop. Inst. Vasil Kolarov* 12:241–253.
Simidchiev, T. (1968). *Nauch. Trud. Vissh. Selskostop. Inst. Vasil Kolarov* 17:165–172.
Simidchiev, T. (1970). *Nauch. Trud. Vissh. Selskostop. Inst. Vasil Kolarov* 19:73–87.
Simidchiev, T. (1971a). *Nauch. Trud. Vissh. Selskostop. Inst. Vasil Kolarov* 20:71–86.
Simidchiev, T. (1971b). *Nauch. Trud. Vissh. Selskostop. Inst. Vasil Kolarov* 20:87–97.
Simidchiev, T. (1972a). *Gradinar. Lozar. Nauka* 9:25–32.
Simidchiev, T. (1972b). *Nauch. Trud. Vissh. Selskostop. Inst. Vasil Kolarov* 21:97–103.
Simidchiev, T. (1985). *Rastenievad. Nauki* 23:110–117.
Simons, R. K. (1965). *Proc. Amer. Soc. Hort. Sci.* 87:55–65.
Simons, R. K. (1974). *J. Amer. Soc. Hort. Sci.* 99:69–73.
Simons, R. K. and Chu, M. C. (1968). *Proc. Amer. Soc. Hort. Sci.* 92:37–49.
Simons, R. K. and Lott, R. V. (1963). *Proc. Amer. Soc. Hort. Sci.* 83:88–100.
Sinska, I., Grochowska, M. and Lewak, S. (1973). *Bull. Acad. Pol. Sci. Cl. V.* 21:291–295.
Sisler, G. and Overholser, P. (1943). *Proc. Amer. Soc. Hort. Sci.* 43:29.
Sklanowska, K. (1991). *Acta Hort.* 288:452–457.
Skrebtsova, N. D. (1957). *Pchelovodstvo* 34:34–26.
Skrebtsova, N. D. (1959). *Pchelovodstvo* 36:26–27.
Skrebtsova, N. D. and Jakovlev, A. S. (1959). *Pchelovodstvo* 36:25–26.

Smirnov, A. G. (1974). *Sb. Nauk. Rab. Vses. Nauk. Int. Sadov I. V. Mitshurino* 19:233–240.
Smith, B. D. and Williams, R. R. (1967). *Rep. Agr. Hort. Res. Stn Univ. Bristol 1966,* pp. 120–125.
Smith, B. D. (1970). Natural pollen transfer. In: Williams, R. R. and Wilson, D. (eds), *Towards Regulated Cropping.* Grower Books, London, pp. 29–40.
Smykov, V. K. (1974). *Kultura abrikosa v neoroshaemyh usloviiah Moldavii* Izd. Stii, Kishinev.
Socias i Company, R., Kester, D. E. and Bradley, M. V. (1976). *J. Amer. Soc. Hort. Sci.* 101:490–493.
Socias i Company, R. and Felipe, A. J. (1987). *HortScience* 22:113–116.
Soenen, Ir. A., Paternotte, E. and Verheyden, C. (1978). *Acta Hort.* 46(381):33–55.
Sogomonian, S. A., Kalatsian, V. G. and Minasian, L. G. (1974). *Biol. Zh. Armenii, Yerevan* 1:32–87.
Sokolova, E. P. (1951). *Dokl. Akad. Nauk. SSSR.* 81:997–940.
Soldatov, I. V. (1982). *Bul. Glavn. Bot. Sada, Moscow* 123:80–84.
Solignat, G. (1958). *Ann. Amél. des Plant.* 1:31–58.
Solignat, G. (1966). *Ann. Amél. des Plant.* 16:71–80.
Solignat, G. (1973). *Bull. Techn. Inf.* 280. 5:425–450.
Solignat, G. and Chapa, J. (1975). *La biologie florale du Châtaignier.* I. N. V. U. F. L. E. C. 22. rue Bergere, Paris.
Solinas, M. and Bin, F. (1964). *Riv. Ortoflorofruttic.* 48:479–496.
Soltész, M. (1975). Körtefajták termőképességének vizsgálata. (Studies on productivity of pear cultivars). Doctoral Thesis. Kertészeti Egyetem, Budapest.
Soltész, M. (1977). Technológiai követelmények az ültetvények fajtaösszetételének és -elhelyezésének meghatározásakor (Technological requirements at the combination and placing of cultivars in plantations. In: Gyuró, F. (ed.), *Gyümölcsfajták társítása* (Association of Fruit Cultivars). Mezőgazdasági Könyvkiadó, Budapest, pp. 20–28.
Soltész, M. (1980). A gyümölcsültetvények faj- és fajtatársításának alapjai (Bases of association of fruit species and cultivars in orchards). In: Nyéki, J. (ed.), *Gyümölcsfajták virágzásbiológiája és termékenyülése* (Bloom Biology and Fertility in Fruit Cultivars). Mezőgazdasági Kiadó, Budapest, pp. 83–100.
Soltész, M. (1981). *Gyümölcs-Inform* 3:25–28.
Soltész, M. (1982a). Almaültetvények fajtatársítása (Cultivar arrangement in apple orchards). PhD Thesis. Hung. Acad. Sci., Budapest.
Soltész, M. (1982b). "Lippay János" Emlékülés és Tudományos Ülésszak Előadásai. April 28, 1982. Budapest, I. pp. 572–580.
Soltész, M. (1982c). *Kertészet és Szőlészet* 31(44):6–7.
Soltész, M. (1985). *Kert. Élelmiszerip. Egy. Közl.* 49:83–92.
Soltész, M. (1986a). *Kert. Élelmiszerip. Egy. Közl.* 50:135–143.
Soltész, M. (1986b). *Gyümölcs-Inform* 8:141–145.
Soltész, M. (1986c). "Lippay János" Tudományos Ülésszak Előadásai, Kertészeti és Élelmiszeripari Egyetem, Budapest. I. pp. 420–428.
Soltész, M. (1987). *Kert. Élelmiszerip. Egy. Közl.* 51:113–123.
Soltész, M. (1989a). *A kert* 4:13–14.
Soltész, M. (1989b). Az almafajták virágzása, termékenyülése és az ültetvények fajtatársítása (Flowering, fertilization and arrangement of apple cultivars). In: Gvozdenovic, D. (ed.), *Intenzív almatermesztés a homokon* (Intensive Apple Growing on Sandy Soils). Fórum, Novi Sad, pp. 77–79.
Soltész, M. (1992). A virágzásfenológiai adatok és összefüggések hasznosítása az almaültetvények fajtatársításában (The use of phenological information in determining cultivar combinations in apple). DSc Thesis. Hung. Acad. Sci., Budapest.
Soltész, M. and Nyéki, J. (1982). Variety combinations in pear orchards. In: van der Zwet, T. and Childers, N. F. (eds), *The Pear.* Hort. Publ., Gaimesville, Florida.
Soltész, M., Nyéki, J. and Benedek, P. (1983). *Proc. 29th Int. Congr. Apiculture.* Apimondia Press, Bucharest, pp. 286–289.

Soltész, M., Bartha, Cs. and Tomcsányi, P. (1984). "Lippay János" Tudományos Ülésszak előadásai, Budapest, pp. 577–588.
Soó, R. (1965). *Fejlődéstörténeti növényrendszertan* (Evolutionary Plant Taxonomy). Tankönyvkiadó, Budapest.
Soyanov, A. and Gormesky, V. (1984). *Grad. Lozar.* Nauka 21(1):3–9.
Soyla, M. and Lüdders, P. (1988). *Gartenbauwiss.* 53(6):253–257.
Speigel-Roy, P. and Alston, F. H. (1982). *J. Horticult. Sci.* 57:145–150.
Spence, S. and Couvillon, G. A. (1975). *J. Amer. Soc. Hort. Sci.* 100:242–244.
Srivastava, R. P. and Singh, L. (1970). *Abstr. Hort. Abstr. 1971* No. 5874.
Stadler, J. D. and Srydom, D. K. (1967). *S. Afr. J. Agr. Sci.* 10:831–840.
Stancevic, A. S. (1967). *Yug. vočarstvo* 1:21–31.
Stancevic, A. S. (1969). *Yug. vočarstvo* 9:1–15.
Stancevic, A. S. (1970). *Eucarpia. Angers Fruit Breeding Symposium.* INRA, pp. 297–305.
Stancevic, A. S. (1971). *Yug. vočarstvo* 16:1–28.
Stancevic, A. S. (1972). *Arch. za polioprivedne nauke* 25(90):85–99.
Stancevic, A. S. (1974). *Proc. 19th Int. Hort. Congr., Warsawa,* Vol. 1, A:442.
Stancevic, A. S. (1975). *Yug. vočarstvo* 31–32:25–31.
Standhouders, P. (1949). *Meded. Dir. Tuinb.* 12:821–830.
Stanley, R. G. and Linskens, H. F. (1974). *Pollen: Biology, Biochemistry and Management.* Springer Verlag, Berlin.
Stark, A. L. (1944). *Farm and Home Sci.* 5(4):5–6.
Steinborn, G. (1983). *Erwerbsobstbau* 25:188–190.
Steinborn, G. (1985). *Erwerbsobstbau* 27:173–175.
Stephen, W. P. (1958). *Oregon Agr. Expt. Stn Techn. Bull.,* pp. 42–43.
Stoll, K. and Krapf, B. (1973). *Obst- und Weinbau* 10:242–246.
Stösser, R. (1966a). Befruchtungsbiologische und embryologische Untersuchungen bei der Süsskirsche (*Prunus avium* L.). Thesis, Landw. Hochschule, Stuttgart–Hohenheim. p. 112.
Stösser, R. (1966b). *Erwerbsobstbau* 8:211–214.
Stösser, R. (1967). *Erwerbsobstbau* 9.
Stösser, R. (1979a). *Erwerbsobstbau* 21:124–126.
Stösser, R. (1979b). *Obst- und Gartenbau* 98:233–235.
Stösser, R. (1980a). *Angew. Botanik* 54:319–327.
Stösser, R. (1980b). *Z. Pflanzenzüchtg.* 84:30–34.
Stösser, R. (1980c). *Obst- und Gartenbau* 99:185–188.
Stösser, R. (1982). *Z. Pflanzenzüchtg.* 88:261–264.
Stösser, R. (1983a). *Angew. Botanik* 57:173–179.
Stösser, R. (1983b). *Obst- und Gartenbau* 102:258–260.
Stösser, R. (1984). *Erwerbsobstbau* 26:110–115.
Stösser, R. and Anvari, S. F. (1978). *Gartenbauwiss.* 43:157–162.
Stösser, R. and Anvari, S. F. (1981). *Gartenbauwiss.* 46:154–158.
Stösser, R. and Anvari, S. F. (1982). *Sci. Horticult.* 16:27–29.
Stösser, R. and Anvari, S. F. (1983). *Acta Hort.* 139:13–22.
Stösser, R. and Hartmann, W. (1982). *Mitt. Klosterneuburg* 32:124–130.
Stösser, R. and Neidhart, M. (1975). *Erwerbsobstbau* 17:137–139.
Stösser, R. and Neubeller, J. (1980). *Gartenbauwiss.* 45:97–101.
Stott, K. G. (1972). *J. Hort. Sci.* 47:191–198.
Streitberg, H. (1975). *Gartenbau* 22:178–180.
Streitberg, H. and Handschack, M. (1983). *Arch. Gartenbau* 31:91–103.
Sullivan, D. T. (1965). *Proc. Amer. Soc. Hort. Sci.* 87:41–46.
Surányi, D. (1977a). *Bot. Közl.* 64:125–133.
Surányi, D. (1977b). *Bot. Közl.* 64:259–265.
Surányi, D. (1978a). *Acta Agr. Acad. Sci. Hung.* 27:247–257.
Surányi, D. (1978b). *Bot. Közl.* 65:89–97.
Svensson, B. (1991). *Acta Hort.* 288:260–264.

Swartz, H. J. Geyer Anne, S., Powell, L. E. and Shung-Hui Cynthia, Lin (1984). *J. Amer. Soc. Hort. Sci.* 109:745–749.
Swayne, G. (1824). *Royal Hortic. Soc. Trans.* 5:208–212.
Swingle, W. T. (1928). *J. Hered.* 19:257–268.
Synge, A. D. (1947). *J. Anim. Ecol.,* 16:122–138.
Szabó, L. Gy. (1980). *A magbiológia alapjai.* (Bases of Seed Biology). Akadémiai Kiadó, Budapest.
Szabó, Z. (1989). Európai és japán szilvafajták virágzása, termékenyülése, társítása (Flowering Fertilization and Combinations of European and Oriental Plum Cultivars). PhD Thesis, Hung. Acad. Sci., Budapest
Szabó, Z. and Nyéki, J. (1989). *9th Int. Symp. "Apricot Culture". Caserta, Italy.* Abstracts pp. 45–64.
Szabó, Z., Nyéki, J. and Benedek, P. (1989). *Kertgazdaság* 21(1):53–70.
Szalai, Z., Farkas, J. and Barnabás, B. (1983). *Állatteny. Takarm. Kutatóközp. Közl. 1983,* pp. 371–374.
Szentiványi, P. (1976a). In: Horn, E. (ed.), *Dió – Mandula – Mogyoró – Gesztenye* (Walnut—Almond—Hazelnut—Chestnut). Mezőgazdasági Kiadó, Budapest, pp. 9–84.
Szentiványi, P. (1976b). In: Horn, E. (ed.), *Dió – Mandula – Mogyoró – Gesztenye* (Walnut—Almond—Hazelnut—Chestnut). Mezőgazdasági Kiadó, Budapest, pp. 271–320.
Szentiványi, P. (1990). *Acta Hort.* 284:251.
Szentiványi, P. (1992). A héjasgyümölcsűek termesztésének korszerűsítésében elért főbb kutatási eredmények. Doktori értekezés (Main Results of Research on Development of Nut Growing). DSc Thesis, Hung. Acad. Sci., Budapest.
Szilágyi, K. (1975). *Szamóca* (Strawberry). Mezőgazdasági Kiadó, Budapest.
Tabuenca, M. C. (1964). *Ann. Aula Dei,* 7.
Tabuenca, M. C. (1965): *Ann. Aula Dei,* 8.
Tabuenca, M. C. (1968). *Ann. Aula Dei,* 9.
Tabuenca, M. C. (1972). The effect of temperature on flowering date in cherry varieties. *Proc. 2nd Conv. del Ciliegio, Verona. Camera di Comm. I. A. A.,* 1972.
Tamás, P. (1959). *Züchter* 29:78–91.
Tamás, P. (1963). *Züchter* 33:202–306.
Tamássy, I. and Nyéki, J. (1976). *Acta Agr. Acad. Sci. Hung.* 25:450–455.
Tamássy, I., Nyéki, J. and Migend, D. (1975). *Kert. Egy. Közl.* 39:87–98.
Tarnavschi, I. T., Bordeianu, T., Radu, I. F., Bumbac, E. and Botez, M. (1963). *Lucrar. Grad. Bot.* 1:343–363.
Tasei, J. N. (1984). Arbes fruitières des régions tempérées. In: Lesson, P. and Louveaux, (eds), *Pollinisation et production végétales.* INRA, Paris, pp. 349–372.
Taylor, R. H. (1918). The Almond in California. *Berkeley Bull.* 297.
Terpó, A. (1968). Az ivaros szaporítás és szervei (Sexual reproduction and its organs) In: Kárpáti, Z., Görgényi, L-né and Terpó, A.: *Kertészeti Növénytan* (Horticultural Botany) *I. Növényszervezettan.* Mezőgazdasági Kiadó, Budapest, pp. 228–318.
Terpó, A. (1987). *Növényrendszertan az ökonombotanika alapjaival* (Plant Taxonomy and Its Econobotanical Basis), Vols 1–2, Mezőgazdasági Kiadó, Budapest.
Terpó, A., Brózik, S., Sr., Nyéki, J., Apostol, Jné and Pozvai, E. (1978a). *Bot. Közl.* 65:51–60.
Terpó, A., Nyéki, J. and Pozvai, E. (1978b). *Bot. Közl.* 65:39–40.
Terrettaz, R. (1978). *Rev. Suisse Viticult. Arboricult., Horticult.* 10:147–150.
Terziyski, D. and Stefanova, A. (1981). *Grad. Lozar. Nauka* 18:3–8.
Teskey, B. J. E., Shoemaker, J. S. (1972). *Free Fruit Production.* The Avi Publishing Company, Westport, Connecticut.
Theiler, R. (1985). *Acta Hort.* 169:63–72.
Thibault, B. (1971). Centre Techn. Interprofes. d. Fruits et Legumes-Dokuments, Paris, pp. 41:1–3.
Thibault, B. (1979). *Proc. Eucarpia Symp. Tree Fruit Breeding. Angers,* pp. 47–58.
Thompson, M. (1967). *Proc. Nut Growers Society Oregon and Washington* 53:31–36.
Thompson, M. (1971). *Proc. Nut Growers Society Oregon and Washington* 56:73–79.
Thompson, M. M. and Liu, L. J. (1973). *J. Amer. Soc. Hort. Sci.* 98:193–196.

Thompson, P. A. (1963). *Nature* 200(4902):146–148.
Timon, B. (1970). A fenológiai fázisok meghatározása, mint a technológia alapja a nagyüzemi gyümölcstermesztésben (Determination of phenological phases in large-scale fruit production). Doctoral Thesis, Kertészeti Egyetem, Budapest.
Todd, F. E. and Reed, C. B. (1970). *J. Econ. Ent.* 63:148–149.
Tomcsányi, P. (1979). *Információk a gyümölcsfajtákról* (Information on Fruit Cultivars). Mezőgazdasági Kiadó, Budapest.
Toptsiski, S. and Mihaylov, C. (1975). *Grad. Lozar. Nauka* 12:13–21.
Torchio, P. F. (1991). *Acta Hort.* 288:49–61.
Townsend, G. F., Riddell, R. T. and Smith, M. V. (1958). *Can. J. Plant Sci.* 38:39–44.
Tóth, E. (1957). *Kert. Kut. Int. Évk.* 2:11–129.
Tóth, E. (1966). *Kert. és Szől. Tud. Tanácsadó. Ministry of Agriculture*, pp. 12–13.
Tóth, E. (1967). *Szőlő- és Gyümölcstermesztés* 3:129–150.
Tóth, E. (1968). *Acta Hort.* 10:233–240.
Tóth, E. (1969). Szilvafajták öntermékenyülésének vizsgálata (Self-fertility in plum cultivars). Doctoral Thesis, Kertészeti Egyetem, Budapest.
Tóth, E. and Surányi, D. (1980). *Szilva* (Plum). Mezőgazdasági Kiadó, Budapest.
Tóth, E. and Tóth, Ené (1959). *Kert. Szől. Főisk. Évk.*, pp. 161–168.
Tóth, E., Tóth, Ené and Szilágyi, K. (1966). *Szőlő- és Gyümölcstermesztés* 1:105–113.
Toyama, T. K. (1980). *Fruit Var. J.* 34:2–4.
Trefois, R. (1966). *Le Fruit Belge* 54 (413):3–42.
Tromp, J. (1968). *Acta Bot. Neerl.* 17:212–220.
Tromp, J. (1976). *Sci. Hort.* 5:331–338.
Tromp, J. (1990). *Ann. Rep. 1990, Res. Stn Fruit Growing*, p. 65.
Tsukanova, Z. G. (1974). *Sadov. Vinogr. Vinod. Moldavii* 29:18.
Tudor, A. (1981). *Lucrari Stiintifice* 9:229–239.
Tufts, W. P. (1919). *Univ. Calif. Bull.* 306:337–366.
Tufts, W. P. and Hansen, C. J. (1933). *Proc. Amer. Soc. Hort. Sci.* 30:134–139.
Tufts, W. P. and Philp, G. L. (1922). *Univ. California, Berkeley Bull.* 346.
Tufts, W. P. and Philp, G. L. (1925). *Calif. Agr. Expt. Stn Bull.* 373.
Tufts, W. G., Hendrickson, A. H. and Philp, G. L. (1926). *Mem. Hort. Soc. N. Y.* 3:171–174.
Tukey, H. B. (1933). *Bot. Gaz.* 3.
Tukey, H. B. (1956). *Proc. Amer. Soc. Hort. Sci.* 68:32–43.
Tukey, H. B. (1970). *Dwarfed Fruit Trees*. MacMillan Company, New York.
Tustin, S. D., Hirst, P. M. and Warrington, I. J. (1988). *J. Amer. Soc. Hort. Sci.* 113:693–699.
Ubatshina, E. G., Pobbubnaya-Arnoldi, V. A. and Enikeev, H. K. (1976). *Bul. Glavn. Bot. Sada* 101:85–94.
Vahl, E. (1956). *Mitt. Obstbauversuchsring Altes Land (Jork).* 11:173–175.
Vahl, E. (1959). Weitere Ergebnisse der Befruchtungsversuche bei Süsskirschen. *Mitt. Obstbauversuchsring Altes Land (Jork).* 14:87–88.
Vahl, E. (1965). *Erwerbsobstbau* 7:126–129.
Vasil, I. K. (1974). In: Linskens, H. F. (ed.), *Fertilization in Higher Plants*. North-Holland, Amsterdam, pp. 105–118.
Vasilakis, M. D. and Porlingis, I. C. (1984). *HortScience* 19:659–661.
Vasilakis, M. D. and Porlingis, I. C. (1985). *HortScience* 20:733–735.
Vasiliev, V. N. and Rodeva, V. (1978). *Grad. Lozar. Nauka* 15:3–8.
Vazart, B. (1955). *Bull. Soc. Bot.* 102:405–443.
Vértessy, J. and Nyéki, J. (1974). *Acta Phytopathol Acad. Sci. Hung.* 9(1–2):17–22.
Virdi, B. V. and Eaton, G. W. (1969). *Can. J. Bot.* 47:1891–1893.
Visser, T. (1955). *Meded. Landb. Hogesch. Wageningen.* 51:1–68.
Visser, T. (1973). *J. Amer. Soc. Hort. Sci.* 98:26–28.
Visser, T. (1984). *Acta Hort.* 149:109–116.
Visser, T. and Marcucci, M. C. (1984). *Euphytica* 33:699–704.
Visser, T. and Oost, E. H. (1982). *Euphytica* 31:305–312.

Visser, T. and Scharp, A. A. (1967). *Euphytica* 16:109–121.
Visser, T. and Verhaegh, J. J. (1987). *Gartenbauwiss.* 52:13–16.
Visser, T., Verhaegh, J. J., Marcucci, M. C. and Uijtewahl, B. A. (1983). *Euphytica* 32: 57–64.
Vitanov, M. (1963). *Izv. Inst. Ovoch.* 4:23–31.
Vondracek, J. (1962). *Rostlinná Vyroba, Praha* 8:1249–1256.
Vondracek, J. (1969). *Gartenbauwiss.* 34:149–157.
Vondracek, J. (1972). *Zahradnitz*, pp. 235–244.
Vondracek, J. (1975). *Acta Hort.* 48:23–34.
Vondracek, J. and Kloutvor, J. (1976). *Zahradnictvi, Praha* 3:11–17.
Vukolova, A. M. (1960). *Sadov Vinogr. Vinod. Moldavii* 2:24–25.
Waite, M. B. (1894). *U.S.D. Agr. Div. Veget. Pathol. Bull.* 5:3–86.
Waite, M. B. (1898). *Yearb. U. S. Dept. Agr.*, pp. 167–180.
Watanabe, S. (1984). *Bull. Yamagata Univ. Agr. Sci.* 9:326–349.
Way, R. D. (1968). *Proc. Amer. Soc. Hort. Sci.* 92:119–123.
Way, R. D. (1971). *Search Agr. New York St. Agr. Exp. Stn* 1:1–84.
Way, R. D. (1973). *Plant Sci. Pomol.* 8:2–16.
Way, R. D. (1974). *New York's Food and Life Sci. Bull. New York St. Agr. Exp. Stn* 37:1–6.
Way, R. D., Livermore, K. G. and Aldwinkle, H. S. (1982). *New York's Food and Life Sci. Bull. New York State Agr. Exp. Stn* 99:1–3.
Weber, H. J. (1984). *Obstbau* 9:60–65.
Weger, N., Herbst, W. and Rudloff, C. F. (1940). *Wiss. Abh. Reichsamt. Wetterdienst.* 7:3–28.
Weinbaum, S. A. (1985). *Sci. Hort.* 27:295–302.
Weinbaum, S. A. and Simons, R. K. (1974*a*). *J. Amer. Soc. Hort. Sci.* 99:266–269.
Weinbaum, S. A. and Simons, R. K. (1974*b*). *J. Amer. Soc. Hort. Sci.* 99:311–314.
Weinbaum, S. A., Urin, K., Micke, W. C. and Meith, H. C. (1980). *HortScience* 15:78–79.
Weinberger, J. H. (1956). *Proc. Amer. Soc. Hort. Sci.* 67.
Welkerling, E. M. L. (1954). *Rev. Fac. Cienc. Agr. Univ. Nac. Cuyo.* 4:35–39.
Wellensiek, S. J. (1977). *Acta Hort.* 68:17–27.
Wellington, R. (1921). *Proc. Amer. Soc. Hort. Sci.* 18:28–29.
Wellington, R. (1926). *Mem. Hort. Soc. N. Y.* 3:165–170.
Wellington, R., Stout, A. B., Einset, O. and Alstyne, L. M. (1929). *New York St. Agr. Exp. Stn Geneva Bull.* 577:1–54.
Wertheim, S. J. (1968). *Meded. Dir. Tuinb.* 31:438–447.
Wertheim, S. J. (1984). *Int. Soc. Hort. Sci.* 1:211–219.
Wertheim, S. J. (1986). *Gartenbauwiss.* 51:63–68.
Werzilov, W. F., Plotnikova, I. V. and Alexandrova, W. A. (1978). *Acta Hort.* 80:175.
Westwood, M. N. and Grim, J. (1962). *Proc. Amer. Soc. Hort. Sci.* 81:103–107.
Westwood, M. N., Stephen, W. P. and Cordy, C. B. (1966). *Hort. Sci.* 1:28–29.
Whelan, E. D. P., Hornby, C. A. and Eaton, G. W. (1968). *Can. J. Genet. Cytol.* 10:819–824.
White, L. M. (1979). *Agricultural Meteorology* 20:189–204.
Wieniarska, J. (1987). *Fruit Sci. Rep.* 14:65–70.
Wilcox, W. F. (1962). Flower Biology and Seed Formation. In: Kappert, H. and Rudorf, W. (eds), *Handbuch der Pflanzenzüchtung*, 2nd Ed Parey, Berlin, pp. 637–645.
Williams, R. R. (1959). *Rep. Long. Ashton Res. Stn, 1958*, pp. 61–64.
Williams, R. R. (1963). *J. Hort. Sci.* 38:52–60.
Williams, R. R. (1965). *J. Hort. Sci.* 40:31–41.
Williams, R. R. (1966*a*). *Rep. Long Ashton Res. Stn, 1965*, pp. 128–135.
Williams, R. R. (1966*b*). *Rep. East Malling Res. Stn, 1941*, pp. 29–35.
Williams, R. R. (1966*c*). *Ann. Rep. Long Ashton Agr. Hort. Res. Stn*, pp. 112–114.
Williams, R. R. (1966*d*). *Ann. Rep. Long Ashton Res. Stn, 1965*, pp. 136–138.
Williams, R. R. (1970*a*). In: Luckwill, L. C. and Cutting, C. V. (eds), *Physiology of Tree Crops*. Acad. Prees, London–New York, pp. 193–208.
Williams, R. R. (1970*b*). The effect of supplementary pollination on yield. In: Williams, R. R. and Wilson, D. (eds), *Towards Regulated Cropping*, Grower Books, London, pp. 7–10.

Williams, R. R. (1975a). *Commercial Grower* 4131:479–480.
Williams, R. R. (1975b). *Grower* 11:86–87.
Williams, R. R. and Brain, P. (1985). *J. Hort. Sci.* 60(1):25–28.
Williams, R. R. and Church, R. M. (1983). *J. Hort. Sci.* 58:337–342.
Williams, R. R. and Corton, D. C. (1990). *J. Amer. Soc. Hort. Sci.* 115(2):207–212.
Williams, R. R. and Jeffries, C. J. (1980). *Rep. Long. Ashton Res. Stn 1980,* p. 27.
Williams, R. R. and Legge, A. P. (1979). *J. Hort. Sci.* 54:67–74.
Williams, R. R. and Maier, M. (1977). *J. Hort. Sci.* 52:475–483.
Williams, R. R. and Wilson, D. (1970). *Towards Regulated Cropping.* Grower Books, London.
Williams, R. R., Church, R. M. and Legge, A. P. (1979). *J. Hort. Sci.* 54(1):75–79.
Williams, R. R., Brain, P., Church, R. M. and Flook, V. A. (1984). *J. Hort. Sci.* 59:337–347.
Williams, R. R., Child, D. V., Copas, L. and Holgate, M. E. (1987). *J. Hort. Sci.* 62:291–294.
Willing, H. (1960). *Arch. Gartenbau* 8:561–594.
Wilson, C. L. and Loomis, E. W. (1967). *Botany.* Holt, Rinehart and Wiston, New York–Chicago– San Francisco– Toronto–London.
Winter, F. et al. (1974). *Lucas' Anleitung zum Obstbau.* Verlag Eugen Ulmer, Stuttgart.
Wociór, S. (1976a). *Roczn. Nauk. Roln.* 101:7–16.
Wociór, S. (1976b). *Roczn. Nauk. Roln.* 101:63–76.
Wociór, S. (1976c). *Postepy Nauk Rolniczych, Warsawa* 23(2):47–55.
Wociór, S. T., Mitrut, T., Opiat, Z. and Wolosiu, J. (1976). *Roczn. Nauk. Roln.* 101:47–62.
Wood, G. W. (1971). *Hort. Sci.* 6:413.
Wood, M. N. (1937). *Calif. Agric. Ext. Serv. Circ.* 103:87.
Wood, M. N. and Tufts, W. P. (1938). *Almond Facts* 2:6–7.
Yakovlev, A. S. (1959). *Pehelovodstvo* 36:22–25.
Yedrov, A. A., Zinim, G. V. and Dunaeva, L. A. (1982). *Bot. Sada, Yalta* 49:68–72.
Yoshida, Y. (1965). *Bull. Hort. Res. Stn Morioka, Ser. C,* 3:1–11.
Young, E. and Werner, D. J. (1985). *J. Amer. Soc. Hort. Sci.* 110(3):411–413.
Zaderbauer, E. (1926). *Fortschr. Landw.* 1:8–9.
Zakharov, G. A. (1958). *Pehelovdstvo* 35:29–33.
Zatykó, I. (1974). *Gyümölcstermesztés* 1:5–22.
Zatykó, J. M. (1962). In: Porpáczy, A. (ed.), *A korszerű gyümölcstermelés elméleti kérdései* (Theoretical Implications of Modern Fruit Growing). Mezőgazdasági Kiadó, Budapest, pp. 426–429.
Zeller, O. (1954). *Angew. Botanik* 28:178–191.
Zeller, O. (1955). *Angew. Botanik* 29:69–89.
Zeller, O. (1960a). *Obstbau* 78:1–6.
Zeller, O. (1960b). *Angew. Botanik* 34:110–120.
Zeller, O. (1960c). *Z. Pflanzenzüchtg.* 44:175–214, 243–278.
Zeller, O. (1961). *Obstbau* 79:7–12.
Zeller, O. (1964a). *Tagungsber.* 65:53–60.
Zeller, O. (1964b). *Maataloustieteellinen Aikakauskirja* 36:85–105.
Zeller, O. (1964c). In: Kreeb, K. (ed.), *Beiträge zur Phytologie.* Verlag Eugen Ulmer, Stuttgart, pp. 1–17.
Zeller, O. (1968). *J. Sci. Agr. Soc. Finland* 40:150–169.
Zeller, O. (1969). *Angew. Botanik* 43:159–173.
Zeller, O. (1973). *Gartenbauwiss.* 38:327–342.
Zeller, O. (1983). *Blütenknospen; verborgene Entwicklungsprozesse im Jahreslauf.* Verlag Urachhaus, Stuttgart, p. 246.
Zenina, V. V. (1967). *Sborn. Hanch. Rab. Veses. Mitshurina* 12:159–165.
Ziegler, J. (1879). *Über phänologishe Beobachtungen und thermische Vegetations-Konstanten.* Zwei Vorträge. Jahrester. d. Snkenbeng. Naturforsch, Ges, Frankfurt.
Zielinski, Q. B. and Thompson, M. M. (1967). *Proc. Amer. Soc. Hort.* 91:187–191.
Zwintzscher, M. (1962). *Mitt. Klosterneuburg* 12:125–134.
Zych, C. C. (1965). *Amer. Soc. Hort. Sci.* 86:307–312.

Index

achene 19, 23, 78, 247
achlamydeus 72
acidic phosphatase enzyme's activity 43
acropetal anthesis order 63, 119
actinomorph flower 57, 73
activity of bees 155, 281, 304, 306, 327
　　– – insect pollinators 290, 298, 299, 329, 335
additional pollination 255
affinity coefficient 159
allotrophic pollinating insect 290
alternately bearing trees 4
andrenid bee species 290, 293
androecium 55, 72, 74
androfertile 47
anemophilous flower 289, 330
anomalies in embryo development 180, 181
anther 5, 15, 16, 19, 21, 33, 37, 74, 76
　　– opening 38
anthesis 19, 21, 23, 33, 35, 39, 40, 45, 46, 47, 49, 53, 93, 107, 108, 109, 110, 111, 117, 118, 132, 133, 136, 138, 142, 147, 157, 160, 164, 168, 169, 170, 171, 174, 181
apex-like structure 26
apex of the bud 9, 10
apical dominance 3
apical meristem 3, 26
apocarpic gynoeceum 78
apochlamydeus 72
apogamy 178, 179, 180
apomictic embryo development 179
apomictic progenies 178
apomixis (somatic reproduction) 177, 179, 185
apospory 178, 179
apparatuses for pollen-gathering 297
apple inflorescence 54, 62, 110
archesporium 8, 9, 21, 27, 31, 33
autogamic pollination 187
　　– – by hand 209
autogamy (natural self-pollination) 191, 213
auto-incompatible 209
autoregulation of fertility 157
axillary bud (axial bud) 6, 15, 16, 21, 29

bearing capacity 3
bearing potential 3
bearing shoot (generative shoot type; types of bearing shoot) 117, 118, 122, 123
bee colony 296, 300, 303, 304, 305, 306
　　– colonies per hectare 307, 315, 316, 317, 319, 320, 324, 325, 328, 329, 330, 332, 336
　　– placement in the orchard 337
　　– pollination 199, 315, 327, 328, 335
Beeline and Beelure (food supplements) 340
beginning of bloom (start of blooming) 80, 84, 87, 106, 107, 114, 122, 125, 131
bell-flower shape 70, 114
big syrphid flies 295
binucleate embryo sac 53
bisexual flower 72
blackberry fruit 77
bloom (blooming, bloom period, flowering) 4, 5, 7, 19, 57, 80, 83, 168
　　– dynamics 110, 126, 128
　　– order of fruit-producing species 80
　　– phenology (floral phenology) 80, 82, 119, 126
　　–, simultaneous (simultaneous blooming; joint blooming; blooming together; coinciding blooming) 97, 98, 99, 100, 101, 103, 112, 123, 125, 126, 131, 262, 263, 264, 301
　　– time groups (blooming group, bloom groups) 97, 98, 99, 110, 115, 130, 131
blooming phenogram 126, 127, 128
brachyblast 5
bract 57, 58, 59, 60, 63, 72, 78
bracteole 29, 59, 63, 78
branching system 3
bud burst 14, 110, 116, 117, 120
　　– drop in December 32
　　– mutation 230, 232, 239
bumble-bees 290, 293, 295, 299, 300, 307, 320, 329, 332, 333, 334, 336
bunch flower 60
bunched inflorescence 68

bur 58, 210
butterflies as pollinators 289, 290, 330, 331, 334

callose 37, 158
callyx 5, 6, 55, 61, 72
capability for fertilizing 227, 228
– of self-fertilization 186, 192
carpel 61, 76, 165
catkin 6, 26, 29, 57, 59, 144, 168
cavity 76, 78, 169
central meristem 10
centrifugal anthesis order 63, 118
centripetal order of bloom 63, 118
chalaza 45, 49, 53, 163, 165, 169, 177
chalazogamy 163
chalcidoid wasp insect 290
changes caused by metaxenia 86, 269, 270, 271, 273
characteristics of parthenocarpy in pear cultivars 247
characterizing the fertility and compatibility relations 182
chasmogamy 153
cleistogamy 133, 138, 139, 144, 148, 153
clonal geitonogramy 154, 207
coenocarpic fruit 76
coinciding blooming 301
cold effect (cold requirement, chilling requirement, chilling effect) 8, 82, 86, 104, 105, 106, 114, 120, 131
coleopterous insects 330
collecting of nectar 155
column-like formation 10
compatibility between closely related cultivars 234
compatible pollen tube 171, 182
compatible pollination 158
connecting tissue 40
corolla 6, 55, 61, 63, 72, 73
cotyledon primordia 173, 174, 177
course of blooming 11, 125
crimson stigma 57
cross-fertilization 258, 259, 264
cross-incompatibility 183
cross-pollination (allogamy) 133, 145, 153, 158, 162, 170, 172, 185, 186, 189, 193, 198, 199, 211, 212, 213, 215, 216, 217, 218, 221, 222, 223, 224, 225, 226, 281, 288, 289, 299, 301, 303, 304
cultivar combination (cultivar arrangement; cultivar-composition; combining cultivars; mixed plantation) 80, 98, 216, 217, 222, 223, 228, 230, 235, 238, 239, 257, 258, 261, 262, 263, 264, 265, 266, 273, 274, 301

cultivar features affecting bee pollination 301
– placement in orchard 282, 283, 284, 285, 286
– -specific production (cultivar-specific technology; variety-specific growing) 186, 283, 284, 316
cultural value of pollinizer 260
cyme 137
cymose 59, 63, 70, 72, 119, 138
cytokinesis 36, 74, 76, 174, 176

damage of flower primordia in autumn 30, 31, 32
degeneration of embryo sac 50, 180, 181
degradation of the papilla 39, 40, 172
dehiscence of anthers 35, 74, 138, 139, 140, 141, 142, 143, 144, 146, 147, 148, 149, 152, 155, 251, 316
– – pollen 126, 139
dermatogen cell 10
determining the time of blooming 125
diagram of blooming 128
dichasium (double node branchings) 57, 59
dichogamy 40, 96, 98, 144, 145, 146, 147, 148, 151, 154, 264
diclinous flower 71, 72, 96
dictyosome 35, 53
differentiation during the winter months 30
– of tissues 9
dimer ovary 76
dioecious 66, 71, 72, 98, 264
dipterous insect pollinators 292, 295, 298, 319, 320, 324, 331
distance from pollinizer 157, 264, 265, 274, 313
distribution of flower-visiting insects 291
distrophic pollinating insect 290
diverged form of anthesis order 63, 118
DNA and RNA levels of nuclei 11, 35
dormancy 3, 33, 86, 104, 106, 107, 108, 111, 120, 168, 263
double calyx 72
double dichogamy 152
double fertilization (amphimixis) 162, 163, 266
double-segmented carpel 76
drupelets 258, 259, 274
duration of bloom (length of bloom) 80, 81, 93, 94, 96, 110, 112, 119, 123, 125, 126, 155

early blooming fruit species 80, 81
ecological factors (ecological conditions; environmental condition; environmental pollu-

tion) 6, 8, 80, 84, 86, 111, 155, 170, 171, 172, 185, 186, 188, 189, 190, 192, 193, 196, 197, 198, 199, 201, 202, 204, 216, 217, 223, 242, 243, 272, 296, 298, 300, 305, 318
end of bloom 84, 93, 125
effective pollination period (EPP) 121, 132, 133, 138, 171, 172, 173, 261
effect of cultivation on bloom-time 121, 122, 123, 124
– – distance from apiary 307, 319, 324, 334
– – – – pollinizer 274, 275, 276, 277, 278, 279, 280, 281, 286
– – insect pollination on fruit characteristics 307
– – – – – – set 307
– – sugar syrup 339, 340
efficiency of bee pollination 317
effective bee visitation 315, 333
effective period of bee pollination 309, 315, 321, 324
effective pollinators 290
egg cell (zygote formation) 49, 53, 54, 163, 171, 181
emasculation 251, 252
embryo 4, 7, 165, 167, 175, 177, 178, 266, 271
– development 173, 177, 184
– in the globular stage 19
– sac 26, 132, 163, 164, 165, 169, 170, 171, 174, 178, 179, 180, 181
endocarp 61, 69, 79
endogenous growth substance 6
endogenous hormone 4
endoplasmic reticulum 35, 40, 41, 53, 158
endosperm 8, 165, 167, 173, 174, 176, 177, 179, 182, 266, 267, 271
endothecium 33
entomophilous flower 289, 296, 330
eutrophic pollinating insect 290
evaluation of cross-combinations 219, 220, 221
exine (external wall of the pollen) 35, 36, 158, 159
exocarp 69, 78

factors affecting fruit set results 216
– – number of bee colonies 318
– – pollination 153
– inducing or affecting parthenocarpy 242
fauna of insect pollinating 290
female flower (inflorescence) 6, 26, 29, 49, 57, 59, 60, 72, 96, 98, 104, 138, 144, 151, 152, 157, 168, 181, 182, 228
feral honeybees 318
fertilization 5, 33, 38, 45, 53, 54, 132, 158, 164, 168, 170, 171, 172, 181, 187, 190, 192, 194, 196, 199, 203, 205

fertilization conditions 185
– groups 265
filaments of stamens 33, 74, 76, 299, 335
filiform apparatus of synergids 53
fleshy calyx 72, 79
fleshy mesocarp 78
fleshy receptacle 78, 79
flower bud 3, 57, 68, 71
– – development 3, 4, 5, 6, 7, 8, 86, 111, 122
– – initiation 3, 4, 5, 6, 7, 8, 10, 15, 21
– circles 73
– drop 104
– -forming characteristics 21
– induction 6, 7, 8
– initiation in autumn 6
– isolation 249, 250
– meristem 9, 10, 15, 21
– primordium 3, 6, 7, 13, 14, 20, 27
– visitation by pollinators 289, 290, 292, 300, 306
– -visiting insects 289, 292, 328, 330
flowering tendencies on long shoots 117
fluctuation of self-fertilization 187
foraging areas of bees 302, 303, 305, 306
foraging trip 295, 299, 302, 303
fruit abscission 7
– drop 121, 173, 181
– set 38, 39, 121, 156, 157, 162, 164, 170, 171, 172, 178, 185, 186, 189, 191, 193, 195, 198, 200, 202, 204, 205, 206, 207, 213, 214, 216, 217, 220, 221, 222, 223, 224, 225, 226, 228, 243, 261, 262, 269, 270, 281, 288, 302, 309
– -setting potential 5
full bloom 15, 40, 84, 87, 90, 91, 99, 101, 102, 115, 123, 125, 126, 128, 131, 165, 171, 174, 179, 180, 288
funiculus 45, 165, 180
functional female flower 213
functional value of flower 5

gametophytic incompatibility system (GSI) 40, 157, 158
geitonogamy 154, 191, 207, 209, 213
generative shoot types 86
gibberellin 4, 9, 35
glomerule 26, 29
Golgi bodies 40, 41
groups of blooming time 97
gynoecium 39, 55, 72, 74

hand pollination 157
haploid chromosome number 33
haploid generation 47

haploid microspore 35
haploid spore 158
haustorium 164, 165, 166, 177
heat requirements of blooming 86, 107, 108, 109, 110, 111, 114, 120
height above sea level 8, 80, 111, 112, 114, 144
hemitrophic pollinators 290
heterogeneous orchard 217
heteromorphic flower (inflorescence) 71, 138, 152
heterosis effect 268
heterotactic 59
histological differentiation 9, 10, 12
histological transformation 10
homogamous flower 40, 168
homogamy 98, 144, 147, 148, 150, 264
homogeneous plantation (plantation without pollinizers) 211, 212, 214
honeybees 290, 292, 295, 296, 298, 299, 300, 301, 303, 319, 320, 324, 328, 329, 330, 331, 332, 334, 336
hormone balance 3
hormone degradation 4
hormonal interaction 9
hull 58, 60
husk 58
hypsophyll 59, 67, 71, 72

ice-nucleation active (INA) 31
ichneumonid wasp insect 290
incompatibility 40, 157, 229, 230
 – group 158, 230, 231, 235, 236, 237, 238, 262
incompatible pollen tube 171, 172, 182
indicator pollen in fertilization 267
induced blooming periodicity 110
induced periodicity of pollen shedding 139
inflorescence primordium 5, 13, 15, 16, 19, 22, 28
information concerning blooming 257
inhibitory factors 4, 9, 157
insect pollination 126, 142, 143, 153, 287, 288, 289
integrated cultivation 284
integument 26, 46, 163, 178, 181, 182
intercompatibility 219
inter-incompatibility 158, 216, 228, 229, 230, 232, 234, 235, 237, 238, 239, 240
intensity of bee visitation 308
internal integuments 45
intine (internal wall of the pollen) 35, 158
isolator effect 251

labile blooming cultivars 115, 116
late blooming fruit species 80

lateral bud 5, 10
 – – of long shoot 3, 117, 118, 120, 122
lateral flowers of the short shoot 5
lethal pollen factors 36
linear tetrade 51
long flight bivoltine insects 291
longevity of embryo sac 54
 – – ovule 45, 172

macrospore tetrade 49
macrosporogenesis (female gametophyte formation) 47, 48, 49, 180, 182
male flower (inflorescence) 6, 26, 29, 57, 59, 66, 72, 96, 98, 104, 138, 144, 151, 152
male gametophyte 35
male sterility 36, 37, 47, 203, 205, 206, 211, 213, 226, 263, 319
Malus pollinizer 316
market value of parthenocarpic fruit 244
marking the individual flowers 126
mason bees 318
maturity of the pistil 26
megaspore cell 169
 – mother cell 47, 180
meiosis 8, 29, 33, 34, 36, 47, 178
meiotic division 33, 157
mentor pollen method 159
meristematic growing point 3
meristem block 10
metaxenia 265, 266, 268, 274
methods for examination of pollen viability 182
microgametophyte 35
micropilar area 179
micropyle 45, 47, 53, 161
microspore 35
microsporogenesis (pollen formation) 33, 34, 36, 157
minimum level of simultaneous blooming 265
mitochondria 35, 41, 53
mitosis 163, 170, 171, 179
mitotic activity 9, 10, 15
mitotic division 162
mixed bud (compound bud) 57, 60, 67, 71
mixed inflorescence 26
mixed plantation 261, 263
mezocarp 69
monocarp flower 76
monoclinous species 76
monoecious 66, 71
monomerous ovary 76
morphogenetic phase 9
morphological changes 10, 15
morphological differentiation 9, 10, 12, 15, 174
mother cell of embryo sac 46, 47, 49, 50

mother cell of macrospore 47, 48
multiple embryo sac 53
multiple pollination 132, 156, 261
mutants 115, 117
Myrobalan pollinizer 203, 263

natural autogamic pollination 154, 207, 208
natural parthenocarpy 241, 245
necessity of cross-fertilization 258
nectar concentration 155, 319
    – -gatherer bees 296, 299, 300, 301, 302, 319, 320, 322, 324, 327, 328, 331, 332, 336, 339
    – production (nectar content) 35, 40, 261, 289, 296, 297, 301, 302, 307, 313, 319, 320, 322, 324, 325, 327, 328, 329, 332
nectary 299, 332, 334, 336
nitrogenous reserve 4
nucellar embryony 179
nucellus 26, 45, 47, 48, 49, 52, 53, 69, 163, 167, 168, 169, 171, 177, 178, 179, 180, 181, 182
nucleohistone inhibition 10
    – level 11
number of cultivars in orchard 265
    – – drupelets per berry 311
    – – flowers in the inflorescence 8, 19, 26, 29, 57, 59, 60, 71, 119, 138
    – – fruits per 100 burs 228
    – – seeds in the fruit (viable seed content) 77, 221, 222
nut 29
nutrient diversion hypothesis 10
nutrient supply 86

obligatory self-pollination 144
obturator 45, 46, 169
occurrence of inter-incompatibility 230
opened endocarp 61
open pollination 118, 199, 209, 211, 213, 216, 220, 222, 244, 255, 307
organogenetic activity 10
organ specification at floral apex 9
ovary 15, 21, 31, 39, 40, 45, 46, 75, 78, 163, 164, 165, 169, 170, 181, 182, 184, 185
overlap of stigma receptivity and pollen shedding 144, 148, 150, 151
ovule 15, 18, 21, 25, 27, 45, 46, 47, 49, 61, 75, 76, 160, 164, 169, 170, 171, 173, 176, 181, 182, 184

papilla 39, 40, 45
papillar epidermis 76

papillary cells 76
parasitic bees 290, 293
parthenocarpic fruit set 190, 191, 289
parthenocarpy 185, 189, 190, 210, 241, 242, 246, 247, 248, 254, 273, 288, 289
parthenogenesis 178, 179
partially self-fertile 186
peaks of pollen shedding 35
pear inflorescence 63
peculiar stigma position 138
perianth 72, 73
periclinal cell wall 10, 165
periclinal division 33, 47
perigonium 55
perigynous ovary 79
period between bud burst and bloom 110
    – – pollination and fertilization 169
periodicity of nectar production 296, 323
    – – nectar secretion 35
peroxidase enzyme's activity 43
petal 7, 18, 57, 61, 73, 126, 138
petaloid calyx 72
Pflock-Stadium 10, 13, 15
phases of flower formation 9
phenophases of bloom 84
physiological condition 6, 185
physiological differentiation stage 7, 9
pioneer pollen method 159
pistil 5, 7, 15, 16, 19, 22, 35, 38, 40, 47, 63, 76, 78, 168, 170
    – -point of the fruit 69
pistillate flower 210, 211, 264
pith-rib meristem 10
pith tissue 32
placement of cultivars in orchard (locating cultivars) 154, 274, 280, 281, 302, 303
placenta 21, 25, 45
placental axis 45
placental tissue 181
placentation 75, 76
placing of bee colonies 305, 306, 308, 320, 321, 337, 338
    – – pollinizer 226
    – – – within rows 281
plasmodial organism 32
plastochron 9
pleomorphic flower type 289, 290, 296
pollen 15, 21, 35
    – adhesion 76, 136, 156
    – basket of bees 290, 296, 319
    – collection and storage 249
    – combination by bees on stigma 304
    – concentration (quantity of pollen on stigma) 157, 281

pollen disperser 267, 281, 339
- formation 18
- -gatherer bees 299, 300, 301, 302, 316, 319, 320, 322, 324, 327, 328, 331, 333, 334, 336, 339
- -gathering insects 290
- germination 35, 36, 38, 39, 45, 132, 155, 156, 168, 169, 170, 171, 172
- mother cell 8, 21, 29, 33, 34
- production 144, 228, 296, 297, 299, 301, 307, 319, 320, 324, 327, 328, 329
- shedding 126, 138, 139, 140, 141, 142, 143, 144, 145, 146, 148, 154, 155, 157, 228
- tetrade 19, 22, 33, 34, 37
- transfer between bees 302, 304
- trap 339
- tube 36, 45, 53, 132, 158, 159, 163, 164, 168, 169, 170, 172, 230
- - growth (pollen tube development) 35, 38, 40, 45, 46, 132, 136, 137, 156, 157, 159, 160, 161, 162, 163, 169, 170, 171, 172, 183, 184, 271, 289

pollinating efficiency of insects 293, 294, 295, 298, 299
pollinating population 290, 292, 320, 334
pollination 39, 253
- period 155
- system 259, 261, 275
- with pollen mixture 215, 217
pollinizer (pollen donor variety; pollinizing partner; male partner; pollen giving cultivar) 100, 101, 125, 145, 154, 156, 185, 186, 190, 191, 195, 196, 207, 210, 215, 219, 220, 221, 222, 223, 224, 225, 226, 228, 229, 232
- placement 274
- proportion 226
- selection 259
polymerous ovary 76
polyploid 47, 66, 230
porogamy 163
position of stamens and stigmas 320, 327
preliminary methodological studies 10
primary endosperm 163, 164, 165, 166
primary thickening 21
primitive bees 290
primitive entomophilous flower 330
primitive stereomorphic flower 334
primordium differentiation 9
proembryo 23, 173, 174, 176
proportion of pollinizer 228, 260, 261, 262, 263, 264, 274, 278, 279
prospective lateral bud 15

protandrous 26, 168, 264
protandry 144, 146, 147, 150, 151
protein synthesis 9, 36
proterandric 33
protogynous 168, 264
protogyny 144, 145, 147, 148
pseudo berry 62, 77, 78, 79
pseudo-compatibility 159
pseudodrupa 79
pseudo fruit 62
pseudogamy 178, 179, 180
pyrenarium 79

raceme 119
racemose 60, 162
radial symmetry of the proembryo 173
realization of flower induction 9
real stone fruit 77
real vascularization 31
receptacle 19, 32, 61, 62, 71, 72, 78, 247
receptivity of ovule 47
- - the pistil 26
- - sexual organs 148, 149, 151, 152
- - stigma 40, 45, 132, 133, 134, 135, 136, 137, 138, 144, 148, 149, 152, 157, 168, 171
reciprocal fertilization 216
reciprocal inter-incompatibility 230
relative attracting effect of cultivars 296, 298, 300, 301, 307, 319
relative beginning of blooming 87, 93
relative bloom order 85, 87
relative bloom time 85, 86
relative full bloom time 91, 93
relative growth rate 5
relative order of full bloom 92, 93
relative pollination efficiency 293, 315
re-organization of the meristem 9
rhythm of blooming (blooming rhythm; flower rhythm) 125, 126, 128
- - secretion activity 40
rich pollen supply (high pollen concentration) 264
RNA synthesis 15
rootstock effect 8, 54, 86, 110, 120, 121, 147, 180
rose-flower shape 70
rose hip 77, 78, 79

S alleles 230, 231, 235, 238
scape 63
sclereid 78
sclerenchymatization 79
secondary embryo sac 163, 180
secondary endosperm 165, 181

secondary flower induction 10
secondary polyandria 72
secondary polyploid 217
secondary shoot 7
second auxin peak 4
second blooming (second flowering) 116, 261
secretory activity of stigma 132, 136, 138, 146, 158
seedless fruit 4, 221
selection of pollinizer 226
self-compatible 159, 162, 172, 199
self-fertile (autofertile) 135, 136, 145, 147, 154, 155, 160, 169, 172, 186, 188, 190, 191, 192, 193, 195, 196, 197, 198, 199, 200, 201, 202, 203, 204, 205, 206, 209, 210, 211, 212, 213, 230
self-fertile mutants 195
self-fertilization (autogamy) 148, 153, 154, 185, 187, 188, 189, 190, 191, 192, 193, 194, 195, 198, 199, 200, 202, 203, 204, 205, 206, 207, 209, 210, 211, 212, 213, 288
– – groups of fruit species 214
self-incompatibility 154, 159, 172, 183, 187, 192
self-incompatible pollen tube 185
self pollen tube 172
self-pollination (idiogamy) 41, 144, 153, 154, 156, 158, 159, 160, 162, 170, 172, 186, 189, 190, 194, 195, 199, 201, 211, 212, 213, 215, 217
self-regulation of reproduction (autoregulation) 228
self-sterile (autosterile) 101, 145, 154, 160, 170, 178, 185, 186, 187, 188, 189, 190, 191, 192, 193, 194, 195, 196, 198, 199, 200, 201, 202, 203, 204, 205, 209, 211, 212, 230, 239, 289
semi-compatible pollen 156
sepal 10, 15, 24, 25, 30, 31, 60, 63, 72, 73, 139
separation of the archesporium cells 33
sexual organs of the flowers 33
sexual reproduction 53
shape of fruit 164
– – pollen 35
shoot growth period 7
short flight spring insects 291
short shoots 6, 7, 97
shuck 58, 59
side-worker bees 299, 313, 320, 327
single cultivar orchard 185
single sex flower 59
size of the pollen 35
small bee colonies 288
solid cultivar block 186, 288

solitary bees 293, 295, 299, 300, 307, 336
somatic parthenogenesis 178, 180
special pollinizers 261
sporogenic tissues 36
sporogenous cell 47
sporophytic incompatibility system (SSI) 40, 157, 158, 169
sporopollenin 35
stable blooming 131
stable bloom marker cultivar 131
stages of crossing 249
stamen 5, 29, 33, 40, 61, 63, 72, 74, 76, 144, 154, 299
– position 154
staminate flower 210
start of bloom (beginning of bloom) 112, 114, 120
stenospermocarpic 221, 241
stereomorphic flower type 289, 290, 296
sterile flower 54
stigma 26, 39, 75, 76, 126, 132, 154, 168, 169, 299
– papillae 45, 158
– receptivity 132
stigmatic secretion 35, 40, 132, 136, 137, 155
stigmatoid tissue 40
stone accessory 77
– cell 78
strawberry fruit 77
structure of inflorescence 118, 119
– – stigmas 138
subdermatogen cell 10
subepidermal cells 47
sugar content of nectar 296, 297, 307, 319, 320, 322, 324, 327, 328, 329, 332
symmetry of flowers 73
synergids 53, 54, 168, 180, 181

tapetum 33, 36, 37, 158
tenthredinid wasp insect 290
tepal 59, 72, 78
terminal bud 5, 6, 12, 13, 14
tetrade cell 47, 49
tetrade stage 35
tetraploid cultivar 159, 178, 191
theca of the anther 33
thrip pollinators 289, 290, 331
tiny insect 289
transmitting tissue of the style 40, 42, 43, 45, 172
tricellular pollen 35
trimerous ovary 76
triploid 35, 47, 48, 54, 135, 156, 160, 161, 178, 179, 180, 188, 189, 191, 217, 218, 219, 220, 221, 230, 232, 244, 261
types of compatibility 158

types of fruit 77
   – – parthenocarpy 241

umbrella 72
– -like inflorescence in raspberry 68
unbalanced number of chromosomes 220
unicellular papilla 39
unidirectional incompatibility 230
unilateral incompatibility 159
unisexual flower 66, 154, 264
uppermost bud 3

value of the inflorescence 10
vascular bundle 32, 61, 165, 176, 180, 182
vegetation period 3, 5, 8, 9
vegetative apex 12, 15, 16, 18, 20, 22, 24

vegetative bud 9, 10
vegetative meristem 9
viability of embryo sac 189
virus infection 120, 121, 186

wild bees 291, 292, 299, 303, 304, 318, 320, 328, 330, 331, 332, 334, 336
wild pollinating insects 290, 291, 292, 296, 304, 322, 329, 336
wind pollination (anemophily) 138, 153, 287, 288, 289, 304, 321, 328, 330, 332

xenia 265, 266, 267
xenogamy 154

zygomorphic flower 57, 73

Pál Benedek, PhD, DSc

Birth: 1943

Present post: full professor and chairman of Zoology Department, Pannon University of Agricultural Sciences, Faculty of Agricultural Sciences
H-9201, Mosonmagyaróvár, Vár 4, Hungary

Main fields of research:

- Insect pollination of cultivated crop plants (structure of pollinating insect populations, their behaviour on crop plant flowers, its effect on the yield) and the effect of pesticides on pollinating insects as vell as bee-safe plant protection technologies
- Ecology and population dynamics of insect pests, their effects on crop plants, and warning forecasts, on their outbreaks and damage
- Scientific publications: over 210 research papers, books and book chapters

Tamás Bubán, PhD, DSc

Birth: 1938

Present post: senior research fellow
Research Station for Fruit Growing
H-4244, Újfehértó, Vadas-tag 2, Hungary

Main fields of research:

- Cropping potential: flower bud formation, fertilization and fruit set in fruit trees
- Nitrogen nutrition and soil management in the integrated apple production

Erzsébet Dibuz, PhD

Birth: 1944

Present post: assistant professor
University of Horticulture and Food Industry
College Faculty of Horticulture
Department of Natural Science
H-6000, Kecskemét, Erdei Ferenc tér 1–3, Hungary

Main fields of research:

- Systematization of pear cultivars based on their morphological characteristics
- The susceptibility of pear species and varieties at sclereid formation, effects of ecological and growing factors

József Nyéki, PhD, DSc

Birth: 1944

Present post: full professor, department head
Pannon University of Agricultural Sciences
Georgikon Faculty of Agriculture
H-8361, Keszthely, Festetics u. 7, Hungary

Main fields of research:

- Floral biology and pollination of temperate zone fruit trees
- Planning of fruit orchards and associated planting of varietiees

Miklós Soltész, DSc

Birth: 1944

Present post: professor
University of Horticulture
and Food Industry, College Faculty
of Horticulture, Department of Fruit Growing
H-6000, Kecskemét, Erdei Ferenc tér 1–3,
Hungary

Main fields of research:

- Development of the basis of biology in integrated and variety-specific fruit growing
- Bloom phenology and floral biology of the apple

Zoltán Szabó, PhD

Birth: 1960

Present post: assistant professor
Pannon University of Agricultural Sciences
Georgikon Faculty of Agriculture
Department of Horticulture
H-8361 Keszthely, Festetics u. 7, Hungary

Main fields of research:

- Floral biology and pollination of fruit crops
- Evolution of stone fruit cultivars